Biocalorimetry

Foundations and
Contemporary Approaches

Biocalorimetry
Foundations and
Contemporary Approaches

Edited by
Margarida Bastos
Department of Chemistry & Biochemistry
Faculty of Sciences, University of Porto

CRC Press
Taylor & Francis Group
Boca Raton London New York

CRC Press is an imprint of the
Taylor & Francis Group, an **informa** business

CRC Press
Taylor & Francis Group
6000 Broken Sound Parkway NW, Suite 300
Boca Raton, FL 33487-2742

First issued in paperback 2019

ISBN-13: 978-1-4822-4665-0 (hbk)
ISBN-13: 978-0-367-87028-7 (pbk)

Library of Congress Cataloging-in-Publication Data

Names: Bastos, Margarida, 1959- editor.
Title: Biocalorimetry : foundations and contemporary approaches / edited by Margarida Bastos.
Description: Boca Raton, FL : CRC Press, Taylor & Francis Group, [2016] | "2015 | Includes bibliographical references and index.
Identifiers: LCCN 2015048064| ISBN 9781482246650 (alk. paper) | ISBN 1482246651 (alk. paper)
Subjects: LCSH: Calorimetry. | Biology--Technique.
Classification: LCC QH324.9.C3 B55 2016 | DDC 536/.6--dc23
LC record available at http://lccn.loc.gov/2015048064

Visit the Taylor & Francis Web site at
http://www.taylorandfrancis.com

and the CRC Press Web site at
http://www.crcpress.com

*To my children Rita and Tiago
and my grandson Tomás.*

Contents

Preface

Si j'avais une prière à formuler, ce serait moins "donnez-moi la force" que "donnez-moi le désir" de faire.

(If I had to say a prayer, it would be less "give me the strength" than "give me the desire" to do.)

FRANÇOIS JACOB
La statue intérieure

Biocalorimetry has presently become established as a most valuable research tool in the biological area, due to the appearance and commercialization of very-high-sensitivity instruments. The idea behind this book is to provide readers and users with biocalorimetry's fundamentals, history, and methodology background, brought to us by some of the most important developers in the field, representing an invaluable personal account of discovery, together with special and original applications of the technique, written by their creators or specialized users. Further, the book aims to have a very wide scope, as it deals with calorimetry use on membranes, nucleic acids, and proteins, addressing both thermodynamics and kinetics and its use as a tool in applied fields. Thus, the book builds a bridge between past, present, and future instrument developments and their use.

I am very lucky to have had the honor to work with all these exceptional scientists, who embraced this project and produced excellent chapters that I believe will be very useful to present and future users of calorimetry in the biological area. To all I express my deep gratitude; it has been an exhilarating experience to read each chapter and exchange comments throughout these months. To Sandro Keller, the first author to come onboard, further thanks are due for his support and suggestions on the inception of the book.

I am very grateful to Francesca McGowan, who invited me to put together the book, and to Stephanie Morkert, Sarfraz Kahn, Alex Edwards, and Amber Donley for their assistance throughout the process.

Finally and very importantly, I would like to thank my family for their unconditional support for this project: my daughter Rita for her continual help with picking up references from friends around the world whenever I needed them, and for checking the quality of the figures and working on some of them; my son Tiago for making me believe I could do it and for spending several nights working with me while I was writing the original proposal; my husband Luís for his calm and supporting presence in our lives; and my larger family and friends, for their companionship through life.

Margarida Bastos
Porto, Portugal

Contributors

Alicia Alonso
Departamento de Bioquímica
Unidad de Biofísica (Centro Mixto CSIC,
    UPV/EHU)
Universidad del País Vasco
Bilbao, Spain

Alexander Bachmann
Department of Urology
University Hospital Basel
Basel, Switzerland

Lina Baranauskienė
Department of Biothermodynamics and
    Drug Design
Institute of Biotechnology
Vilnius University
Vilnius, Lithuania

Margarida Bastos
Department of Chemistry and
    Biochemistry, CIQ-UP
University of Porto
Porto, Portugal

Alfred Blume
Martin-Luther-Universität Halle-Wittenberg
Institut für Chemie–Physikalische Chemie
Halle, Germany

Gernot Bonkat
Department of Urology
University Hospital Basel
and
Laboratory of Biomechanics and
    Biocalorimetry
University of Basel
Basel, Switzerland

Kim Borch
Novozymes A/S
Bagsværd, Denmark

Olivier Braissant
Department of Urology
University Hospital Basel
and
Laboratory of Biomechanics and
    Biocalorimetry (LOB2)
University of Basel
Basel, Switzerland

Patrick Connelly
Vertex Pharmaceuticals, Inc.
Boston, Massachusetts, U.S.A.

Philippe Dumas
Biophysique et Biologie Structurale
Architecture et Réactivité de
l'ARN
Institut de Biologie Moléculaire et
Cellulaire du CNRS
Université de Strasbourg
Strasbourg, France

Simon Ebbinghaus
Physical Chemistry II
Faculty of Chemistry and Biochemistry
Ruhr-University Bochum
Bochum, Germany

Simon Gaisford
School of Pharmacy
University College London
London, United Kingdom

Robert N. Goldberg
Biosystems and Biomaterials Division
National Institute of Standards and
Technology
Gaithersburg, Maryland, U.S.A.

Felix M. Goñi
Departamento de Bioquímica
Unidad de Biofísica (Centro Mixto CSIC,
UPV/EHU)
Universidad del País Vasco
Bilbao, Spain

Lee D. Hansen
Department of Chemistry and Biochemistry
Brigham Young University
Provo, Utah, U.S.A.

Christian Herrmann
Physical Chemistry I, Protein
Interactions
Faculty of Chemistry and Biochemistry
Ruhr-University Bochum
Bochum, Germany

Sandro Keller
Molecular Biophysics
University of Kaiserslautern
Kaiserslautern, Germany

Johannes Klingler
Molecular Biophysics
University of Kaiserslautern
Kaiserslautern, Germany

Jan P. Kraus
Department of Pediatrics
University of Colorado School of
Medicine
Aurora, Colorado, U.S.A.

Karl Lohner
Biophysics Division
Institute of Molecular Biosciences
University of Graz, NAWI Graz,
BioTechMed-Graz
Graz, Austria

Tomas Majtan
Department of Pediatrics
University of Colorado School of
Medicine
Aurora, Colorado, U.S.A.

Patrícia A. T. Martins
Department of Chemistry
Biological Chemistry Group
Coimbra Chemistry Center
University of Coimbra
Coimbra, Portugal

Daumantas Matulis
Department of Biothermodynamics and
Drug Design
Institute of Biotechnology
Vilnius University
Vilnius, Lithuania

Maria João Moreno
Department of Chemistry
Biological Chemistry Group
Coimbra Chemistry Center
University of Coimbra
Coimbra, Portugal

Vytautas Petrauskas
Department of Biothermodynamics and
    Drug Design
Institute of Biotechnology
Vilnius University
Vilnius, Lithuania

Angel L. Pey
Department of Physical Chemistry
Faculty of Sciences
University of Granada
Granada, Spain

Peter L. Privalov
Department of Biology
Johns Hopkins University
Baltimore, Maryland, U.S.A.

Michael Senske
Physical Chemistry II
Faculty of Chemistry and
    Biochemistry
Ruhr-University Bochum
Bochum, Germany

Jaak Suurkuusk
Soojus AB
Järfälla, Sweden

Malin Suurkuusk
TA Instruments
Sollentuna, Sweden

Adrian Velazquez-Campoy
Institute of Biocomputation and Physics
    of Complex Systems (BIFI), Joint-Unit
    IQFR-CSIC-BIFI, and Department of
    Biochemistry and Molecular and Cell
    Biology
University of Zaragoza
and
Fundacion ARAID, Government of
    Aragon
Zaragoza, Spain

Peter Vikegard (former name Peter Johansson)
TA Instruments
Sollentuna, Sweden

Ingemar Wadsö
Physical Chemistry
Chemical Center
Lund University
Lund, Sweden

Peter Westh
Research Unit for Functional Biomaterials,
    NSM
Roskilde University
Roskilde, Denmark

Asta Zubrienė
Department of Biothermodynamics and
    Drug Design
Institute of Biotechnology
Vilnius University
Vilnius, Lithuania

I

Introduction: Historical and Methodological Context

CHAPTER 1

Introduction

Margarida Bastos

CONTENTS

1.1 FROM EARLIER TO PRESENT DAYS OF BIOCALORIMETRY

Calorimetry is a long-established technique that through the years has provided invaluable data in many different areas. The onset of its use and the initial developments (at the macro level) are well documented (see e.g., Zielenkiewicz 2008). The evolution of calorimetry has been in the direction of increasingly sensitive instruments and the use of smaller amounts of substance, necessary attributes for many biological applications. Biological material, be it proteins, DNA, or lipids, is generally not available in large amounts having a high degree of purity, a necessary condition for the most careful measurements. Additionally, to avoid aggregation, dilute solutions must be used. In studies with living cellular systems, the rates of heat production are usually very low, and measurements are conducted over long periods of time. These concerns drove the quest, which started primarily in the 1960s, to develop high-sensitivity instruments that used small amounts of samples. The most remarkable aspect of this initial development phase was that it was driven by "problem solving" and took place in research laboratories, where instruments were constructed by the individual investigators to make possible the measurements that they wanted to perform. Examples of this are: (1) the vacuum, adiabatic heat-capacity calorimeter developed by Peter Privalov to measure the enthalpy denaturation of egg-white albumin (Privalov 1963); (2) the twin adiabatic calorimeter developed by Julian Sturtevant to measure thermally induced transitions in biopolymers (Danforth et al. 1967); (3) the various microcalorimeters developed in the 1970s by Ingemar Wadsö's laboratory to determine the thermodynamics of reactions involving proteins and enzymes or the thermodynamic properties of model compounds of biological relevance (Spink and Wadsö 1975, 1976; Suurkuusk and Wadsö 1972); (4) the differential scanning calorimeter (DSC) constructed by Phil Ross and Robert Goldberg (Ross and Goldberg 1974) for the measurement of energies of transition in solution using a simple and inexpensive design; (5) the design and construction of a DSC by Rodney Biltonen together with Jaak Suurkuusk and Donald Mountcastle (Biltonen et al. 1979), following Biltonen's previous attempts to measure the change in heat capacity of chymotrypsinogen in the native and denatured states by isoperibol calorimetry using the LKB 8700 (Biltonen et al. 1971); (6) the DSC developed by John Brandts and Michael Jackson (Jackson and Brandts 1970), which they used to measure the reversible denaturation of chymotrypsinogen. It was a time of rapid discovery and excitement, reported vividly in Frederic Richards' autobiographical investigation (Richards 1997), Julian Sturtevant's recollections (Sturtevant 1996), Peter Privalov's thoughts on microcalorimetry of biopolymers (Privalov 2007), and Ingemar Wadsö's reflections on calorimetry and calorimetrists through the years (Wadsö 1997). These examples show that it was an effort undertaken by scientists from different countries and continents. The cross-over of their knowledge and experience in papers, at scientific conferences, by short assignments/visits to each other's laboratories, or even by moving to different laboratories brought about the rapid development of new instruments.

Biocalorimetry is now well established, and excellent reviews are available in which the calorimetric principles and design are described in detail, a wide number of applications are provided, and the thermodynamics behind the processes that can be studied are explored: for example, Beezer (1980), Calvet and Prat (1963), Kemp (1999), Privalov (2012), Privalov and Dragan (2007), Wadsö (1985, 2002). Therefore, only a brief review of the first high-sensitivity instruments designed and constructed will be presented, along with some notes concerning the evolution of some of these calorimeters into commercially available instruments.

Following Calvet's introduction in 1948 of twinned calorimetric elements, essentially all high-sensitivity instruments have adopted a twin design, that is, what was measured was a differential property between the vessel with the solvent and the measuring vessel. Many of the early instruments date back to the 1960s. Among these were DSCs, which were built to study transitions that were induced by a continuous increase in temperature. A DSC working with solutions measures the excess heat per degree that is required to raise the temperature of a solution over that required for the reference buffer, thus yielding the heat capacity of the substance in solution as a function of temperature. The first high-sensitivity DSC was designed by Peter Privalov and collaborators (Privalov et al. 1964). This DSC was of the twin design and had a small sample volume. Groups in the United States also built twin DSC instruments in this period—Stanley Gill and colleagues (Gill and Beck 1965) and Julian Sturtevant and colleagues (Danforth et al. 1967)—followed by those built by John Brandts (Jackson and Brandts 1970), by Phil Ross and Robert Goldberg (Ross and Goldberg 1974), and by Rodney Biltonen and colleagues (Biltonen et al. 1979).

Another very important type of calorimeters were those used for measurements leading to the thermodynamic properties of biochemically important processes at constant temperature, such as protonation/deprotonation equilibria involving proteins, hydrolysis of peptide bonds and esters, and ligand binding involving proteins and enzymes. Most initial experiments were performed by measuring the enthalpies of mixing of the reactants in either batch or flow mode. Of particular note are the isothermal calorimeters of μW

sensitivity (batch and flow type) built by Ingemar Wadsö and coworkers (Nordmark et al. 1984; Spink and Wadsö 1976; Suurkuusk and Wadsö 1972, 1974; Wadsö 1970, 1976). This was an effort that paved the way for the multisample calorimeter designed by Suurkuusk and Wadsö (Suurkuusk and Wadsö 1982). During the same period, Stanley Gill built a flow-mixing microcalorimeter, and later worked together with Julian Sturtevant and Beckman Instruments on another flow-mixing instrument, which was used extensively by Sturtevant's group at Yale University (Sturtevant 1996). Meanwhile, it was felt necessary to move away from mixing experiments, as they required large amounts of sample and in some cases were not sensitive enough to measure the heats of protein association in very dilute solutions. This led to the measurement of enthalpies of reaction by stepwise (or continuous) addition of a solution of one of the reagents, contained in a syringe, into a solution of the other, which was in the calorimetric vessel where only minimal amounts of the substances were needed. Since the measurements were performed at constant temperature, this application became known as isothermal titration calorimetry (ITC) (Christensen et al. 1965), and several instruments appeared during this period. Some of the late versions of Wadsö's batch microcalorimeter allowed the performance of ITC experiments (Beezer et al. 1983; Suurkuusk and Wadsö 1982; Wadsö 1976). The group at Brigham Young University (BYU) designed and built their first isothermal titration calorimeter (continuous injection) in 1968 (Christensen et al. 1968) and an improved version in 1973 (Christensen et al. 1973). The Wadsö instruments and later versions of those from BYU employed vessels with a volume of 1–3 mL and performed at the μW/mJ sensitivity level, and were therefore not ideal for measurements with very dilute protein solutions. Stanley Gill designed and built the first high-sensitivity instrument aimed at such measurements: a differential, heat-compensation ITC without air in the reaction vessel (overflow type) and possessing a combined titration and stirring assembly (Spokane and Gill 1981). Three years later, Gill built a similar instrument with a much smaller sample volume of 200 μL (McKinnon et al. 1984). Finally, in 1989 Wiseman and collaborators built a new ITC sharing some similar principles but using a larger sample volume. This instrument used matched Hastelloy, lollipop-shaped 1.4 mL vessels (Wiseman et al. 1989).

Some of these instruments became commercially available: (1) the Russian Academy DASM instruments from Privalov's laboratory; (2) LKB calorimeters from the Lund laboratory, later to be manufactured by Thermometric AB, Sweden (the BAM/TAM multichannel isothermal calorimeters); (3) calorimeters from the BYU laboratory, further developed by Calorimetry Sciences Corporation (CSC) and by Tronac; and (4) the Wiseman, Brandts and collaborators instruments by MicroCal, USA (the OMEGA ITC). When Privalov and others from the Russian laboratory moved to the United States, their knowledge led to significant improvements in the instruments built by MicroCal and by CSC. All these instruments laid the ground for the DSC, ITC, and isothermal batch and flow calorimeters that are now commercially available.

This brought about a change in paradigm from users in research laboratories that specialized in calorimetry, where in some cases instruments were built for specific experimental needs, to users in a larger scientific community, where the technique is seen as a tool in biological studies. Thus, at present, most laboratories use commercially manufactured calorimeters made by large firms such as TA Instruments (owners of the former Thermometric AB and CSC) and by Malvern (which owns MicroCal). The availability of high-quality commercial calorimeters allows contemporary researchers to focus both on interesting research projects that use calorimetry to study biochemical processes and on clever experimental design, the chemistry involved, and careful data analysis. This is reflected in Parts 2–4 of this book, in which experts describe innovative approaches to the use of calorimetry in the biological field and novel data treatment and interpretation as applied to proteins, DNA, membranes and living matter, addressing both thermodynamics and kinetics. The following chapters in Part 1, on the other hand, describe part of the discovery process in microcalorimetry, written, in their own words, by some of the scientists who were at the beginning of the development of the calorimeters needed for biological research, and thus carrying their life experiences.

1.2 PRESENT INSTRUMENTS: PRINCIPLES AND APPLICATIONS

The calorimeters presently available and most widely used in biochemical/biological research are the VP-ITC, the VP-DSC, the iTC200 (and its automated version Auto-iTC200), the VP-Capillary DSC (Malvern, UK), the Nano ITC, the Nano DSC, and the TAM isothermal calorimeter (ITC, batch and flow) (TA Instruments, USA). The pressure perturbation calorimeter (PPC), available from both Malvern and TA, is less well known but has substantial capabilities. The use of these instruments is addressed in Parts 2–4. The choice of instrument depends critically on factors such as the type of measurements one wishes to make, the desired precision and accuracy, sample characteristics, and the amount of substances(s) available. Thus, by evaluating the information on the characteristics of the existing instruments and possibly performing test experiments, one can make a proper choice of instrument (Hansen and Russell 2006).

In brief, all of the aforementioned instruments are of twin design (measuring and reference vessels), and the measurements are differential (measuring the difference between the signals from the sample and the reference vessel). This serves to increase sensitivity. In terms of instrumental principles, the calorimeters use either *heat conduction* (or heat flux) or *power compensation*. Heat conduction instruments generally use thermopiles (Peltier elements) between the vessel and its surroundings (heat sink), and when the system is disturbed, the heat flows (in or out) till thermal equilibrium is again reached. The variable used to follow the reaction is thus the electric potential difference generated by the thermopiles (Bäckman et al. 1994; Calvet and Prat 1963). In power compensation instruments, a constant power is delivered to heaters in the measuring and reference vessels, which can be increased or decreased to maintain the difference in temperature between the vessels close to zero. In this case, the variable used to follow the experiment is the differential power supplied (Privalov 1980; Wiseman et al. 1989). Indeed, more involved compensation modes can also be used to match the temperature of the vessel and its surroundings. These modes depend on the detailed design and calorimetric principle chosen for each instrument. The Malvern instruments and the Nano ITC and Nano DSC from TA are of the power compensation type, whereas the TAM isothermal calorimeters from TA are heat-flow instruments. The TAM III instruments can also operate in power compensation mode to reduce instrument response time.

Another characteristic that varies among instruments is their flexibility—they are either modular (TAM instruments) or dedicated (all others). The dedicated instruments have permanent vessels, available at different volumes or shapes, as in ITC and DSC instruments (VP-ITC [1.4 mL] vs. ITC200 [200 μL], the Nano ITC Low Volume [190 μL] vs. Nano ITC Standard Volume [1.0 mL], Nano DSC with capillary or cylindrical vessel and VP-DSC [500 μL] and VP-Capillary DSC [130 μL]). The modular design uses removable vessels and allows a variety of vessels with different characteristics to be used within the same calorimeter. An increased flexibility brings about some decrease in sensitivity. Thus, one must consider the tradeoffs involved.

The commercial availability of a large number of high-sensitivity calorimetric instruments has led to considerable growth in biocalorimetry. The number of publications is large, and a proper review is beyond the scope of this chapter. Nevertheless, as a guide for new users, a number of applications with appropriate references (papers and reviews) will be provided. Some of the oldest fields of application involve protein/enzyme stability and ligand binding, the study of the effect of mutations on their stability, and DNA stability and interactions, using mainly DSC and ITC (Ababou and Ladbury 2007; Aguirre et al. 2014; Bruzzese and Connelly 1997; Connelly 1994; Cooper et al. 2000; Crespillo et al. 2014; Friis et al. 2014; Ladbury and Doyle 2004; Majtan et al. 2014; Pey et al. 2013; Privalov 2012; Vasilchuk et al. 2014; Velazquez Campoy and Freire 2005, 2006; Weingartner et al. 2012). These studies have been fundamental in providing thermodynamic background and information on biological systems, as well as pointing to new directions in health research. At present, these types of studies are in many cases related to disease research and to the search for new drugs (Ennifar et al. 2013; Garbett and Chaires 2008; Garbett et al. 2009; Schön et al. 2011; Yu et al. 2008). Indeed, their importance is continually expanding in the medical area and the pharmaceutical industry (Chaires 2008; Ladbury et al. 2010; Yadav et al. 2012). Pharmaceutical, food, and other industrial applications, as well as applied biology (involving cells and animals), have also been widely studied using isothermal calorimetry (heat flow or compensation type instruments) in its various options of batch, flow, and titration modes, in studies addressing both thermodynamics and kinetics. In such cases, the calorimeter is sometimes used as a "bio-activity" monitor. These studies show that calorimetry can be used as a tool to study drug stability and the compatibility of various formulations. This is in addition to its ability to follow a wide variety of biological processes, to characterize biological material, and to monitor biological activities *in vitro* (Alam et al. 2014; Alem et al. 2010; Almeida e Sousa et al. 2012; Bäckman et al. 1992; Beezer and Gaisford 2013; Braissant et al. 2010; Lamprecht 2009, 2013; Maskow et al. 2010; Nilsson et al. 2001). Another large field of research involves studies aimed at membrane characterization, mainly by DSC but also using ITC (Benesch and McElhaney 2014; Blicher et al. 2009; Blume 1988; Busto et al. 2014; Fidorra et al. 2009; Garidel and Blume 2000), as well as studies of the interactions of drugs, surfactants, peptides, and proteins with lipid membranes of different composition (Abrunhosa et al. 2005; Arouri et al. 2013; Bastos et al. 2008; Epand et al. 2007; Finger et al. 2014; Heerklotz and Seelig 2000, 2007; Heerklotz et al. 2009; Hickel et al. 2008; Hoernke et al. 2012; Ivanova et al. 2003; Keller et al. 2006; Klocek et al. 2009; Koller and Lohner 2014; Lewis et al. 2007; Moreno et al. 2010; Seeger et al. 2007; Seelig 2004; Vargas et al. 2013; Zweytick et al. 2014). These studies are important in the characterization of the basic components of cells, membranes, and their fundamental role in life (Mouritsen 2005).

1.3 DATA REPORTING, CALIBRATION, AND TEST PROCEDURES

Finally, some words on data reporting and on the regular use of calibration and test procedures, the latter being a key factor in the production of high-quality data. At present, a vast number of users of calorimetric techniques are not trained in calorimetry and are thus less aware of the importance of calibration and test procedures. It is thus important to implement the establishment of a culture of calibration and testing, and for that, it is important to understand the main causes of error and to have easy-to-use and well-established procedures for the various calorimeters and applications.

General recommendations on standards in biothermodynamics and data reporting were recently reviewed (Goldberg 2014) as part of the efforts made through the years to enhance standard scientific communication and to have consistent data incorporated into databases. Particular attention must be given to uncertainty assignment and to clearly reporting the way it was obtained. Repeatability is a measure of the imprecision of the measurement, reflecting random errors, and uncertainty is an estimate of the closeness of the reported value to the true value of a quantity, consisting of a combination of random and systematic errors, according to the terminology suggested in *GUM: Guide to the Expression of Uncertainty in Measurement* 2008 (JCGM/WG 1, 2008). Even though extensive repeat measurements are not always possible in bio-related measurements, an effort must be made to perform a reasonable number of experiments, as the natural variability of biochemical material already imposes a problem in obtaining high accuracy. Typical systematic errors to be considered in isothermal calorimetry are concentrations of reactants, vessel volume, baseline stability, heat calibration, gas bubbles in filled systems, pressure-volume (*PV*) work, adsorption of the reactants to the calorimeter, and heat flow values (Broecker et al. 2011; Keller et al. 2012; Tellinghuisen 2004; Tellinghuisen and Chodera 2011; Wadsö and Wadsö 2005). All calorimeters are calibrated for basic parameters when built and delivered, such as heat q, temperature T, vessel volume, and syringe delivery volume, but these must be regularly checked with appropriate test reactions and corrected for during use. Electrical calibration can also be used, and the results are almost always obtained with high accuracy. Nevertheless, due to construction constraints, the heat flow in many cases does not follow a pattern similar to the reaction under study, and this can result in significant systematic errors (Wadsö and Goldberg 2001; Wadsö and Wadsö 2005). This problem led to the use of chemical calibration procedures. Specific publications addressing these issues are available for ITC (Adão et al. 2012; Baranauskienė et al. 2009; Demarse et al. 2011; Hansen et al. 2011; Schwarz et al. 2008), DSC (Hinz and Schwarz 2001), and isothermal and solution microcalorimetry (Beezer et al. 2001; Hills et al. 2001;

Wadsö and Goldberg 2001). Altogether, they provide a detailed analysis of the factors to be considered in planning an experiment, possible random and systematic errors, and test reactions to check and correct for them, as well as recommendations for reporting data.

Some examples of the various techniques will be mentioned here as an initial guide. When using a DSC, calibration of the differential power and temperature should be performed regularly using the manufacturer's procedures, since the calibration constants may change with time. Nevertheless, it is also recommended that test solutions are used to further evaluate the operating performance of the DSC instrument, such as following the gel-to-liquid crystalline transition of phosphatidylcholines for temperature calibration and the denaturation of hen egg-white lysozyme, using 1–10 mg mL^{-1} solutions in 0.1 M (HCl + glycine) buffer at pH = 2.4 ± 0.1, to check enthalpy and overall performance (Hinz and Schwarz 2001).

Some test reactions suggested for ITC involve proteins or enzymes, and thus the results depend significantly on their purity, competent fraction, and accurate knowledge of concentration, pH, and ionic strength. Simple reactions involving chemicals that can be easily obtained in high purity are easier to use, such as propanol dilution (Adão et al. 2012; Wadsö and Goldberg 2001), the protonation of 2-amino-2-(hydroxymethyl)-1,3-propanediol (TRIS), or the reaction of Ba^{2+} with 1,4,7,10,13,26-hexaoxacyclooctadecane (18-Crown-6) (Briggner and Wadsö 1991; Demarse et al. 2011; Sgarlata et al. 2013). Further to the use of test reactions to ensure overall accuracy, when using ITC in ligand-binding experiments, two additional factors require attention: errors in fitting and in the concentration of the reactants. In these experiments one has to use models that are fitted to the measured data. The commercially available calorimeters are accompanied by software that can be used to calculate thermodynamic property values and estimates of possible errors. One should be aware that the quoted errors are those derived from the fitting procedure, and thus reflect only the goodness of fit of a model to a particular set of results. A more complete estimate of random error must be obtained from the repetition of the measurement. Also, the concentration of the reactant in the vessel is of key importance and can often be a major source of systematic error. The uncertainty in the equilibrium constant K and molar enthalpy change ΔH_m relies on an accurate knowledge of the titrant concentration. Clearly, the uncertainty in the concentration should be combined with the random errors in K and ΔH_m (Tellinghuisen and Chodera 2011). In cases where the stoichiometry can be assumed to be correct, the stoichiometry parameter N that is retrieved in current standard analysis algorithms can be used to correct for the concentration of the reactant in the vessel and thus absorb possible errors in vessel volume. Baseline correction is also an important source of possible error, and ways to optimize the treatment of the baseline have been recently suggested (Keller et al. 2012).

A test reaction for isothermal heat conduction microcalorimeters, particularly when used as thermal power meters, is the imidazole-catalyzed hydrolysis of triacetin, proposed by Wadsö (Briggner and Wadsö 1991; Chen and Wadsö 1982). Recently, an inter- and intralaboratory study provided validated values for the associated thermodynamic and kinetic parameters (Beezer et al. 2001), and another study showed how these values can be used to detect systematic errors in power calibration, baseline zero setting, or time-dependent errors and to correct for them (Hills et al. 2001).

Thus, the bottom line is that calibration of the instruments by use of appropriate test reactions is a key factor in the production of meaningful data and should become a regular practice in any laboratory that uses calorimetry, and care must be taken to correctly report the measured data so as to give a full uncertainty evaluation and a detailed description of the system and/or reaction(s) studied.

ACKNOWLEDGMENTS

I express my deepest thanks to all who were fundamental to my formation in biocalorimetry and instrument development, in particular to Ingemar Wadsö, my PhD supervisor, who introduced me to biocalorimetry and calorimeter design and its principles, and Gerd Olofsson, Rodney Biltonen, Jaak Suurkuusk, Donald Mountcastle, Robert N. Goldberg, George Mrevlishvili, and Peter Privalov, from whom I have learned so much through the years.

REFERENCES

Ababou, A., and Ladbury, J. E. (2007), Survey of the year 2005: Literature on applications of isothermal titration calorimetry, *J. Mol. Recognit.* 20, 4–14.

Abrunhosa, F., Faria, S., Gomes, P., Tomaz, I., Pessoa, J. C., Andreu, D., and Bastos, M. (2005), Interaction and lipid-induced conformation of two cecropin-melittin hybrid peptides depend on peptide and membrane composition, *J. Phys. Chem. B* 109, 17311–17319.

Adão, R., Bai, G., Loh, W., and Bastos, M. (2012), Chemical calibration of isothermal titration calorimeters: An evaluation of the dilution of propan-1-ol into water as a test reaction using different calorimeters, concentrations, and temperatures, *J. Chem. Thermodyn.* 52, 57–63.

Aguirre, Y., Cabrera, N., Aguirre, B., Perez-Montfort, R., Hernandez-Santoyo, A., Reyes-Vivas, H., Enriquez-Flores, S., et al. (2014), Different contribution of conserved amino acids to the global properties of triosephosphate isomerases, *Proteins* 82, 323–335.

Alam, S., Omar, M., and Gaisford, S. (2014), Use of heat of adsorption to quantify amorphous content in milled pharmaceutical powders, *Int. J. Pharm.* 459, 19–22.

Alem, N., Beezer, A. E., and Gaisford, S. (2010), Quantifying the rates of relaxation of binary mixtures of amorphous pharmaceuticals with isothermal calorimetry, *Int. J. Pharm.* 399, 12–18.

Almeida e Sousa, L., Beezer, A. E., Hansen, L. D., Clapham, D., Connor, J. A., and Gaisford, S. (2012), Calorimetric determination of rate constants and enthalpy changes for zero-order reactions, *J. Phys. Chem. B* 116, 6356–6360.

Arouri, A., Dathe, M., and Blume, A. (2013), The helical propensity of KLA amphipathic peptides enhances their binding to gel-state lipid membranes, *Biophys. Chem.* 180–181, 10–21.

Bäckman, P., Bastos, M., Hallen, D., Lönnbro, P., and Wadsö, I. (1994), Heat conduction calorimeters: Time constants, sensitivity and fast titration experiments, *J. Biochem. Biophys. Methods* 28, 85–100.

Bäckman, P., Kimura, T., Schön, A., and Wadsö, I. (1992), Effects of pH-variations on the kinetics of growth and energy metabolism in cultured T-lymphoma cells: A microcalorimetric study, *J. Cell Physiol.* 150, 99–103.

Baranauskienė, L., Petrikaitė, V., Matulienė, J., and Matulis, D. (2009), Titration calorimetry standards and the precision of isothermal titration calorimetry data, *Int. J. Mol. Sci.* 10, 2752–2762.

Bastos, M., Bai, G., Gomes, P., Andreu, D., Goormaghtigh, E., and Prieto, M. (2008), Energetics and partition of two cecropin-melittin hybrid peptides to model membranes of different composition, *Biophys. J.* 94, 2128–2141.

Beezer, A. E. (1980), *Biological Microcalorimetry*, Academic Press: London.

Beezer, A. E., and Gaisford, S. (2013), Isothermal calorimetry in the pharmaceutical sciences, *Eur. Pharm. Rev.* 18, 64–68.

Beezer, A. E., Hills, A. K., O'Neill, M. A. A., Morris, A. C., Kierstan, K. T. E., Deal, R. M., Waters, L. J., et al. (2001), The imidazole catalysed hydrolysis of triacetin: An inter- and intra-laboratory development of a test reaction for isothermal heat conduction microcalorimeters used for determination of both thermodynamic and kinetic parameters, *Thermochim. Acta* 380, 13–17.

Beezer, A. E., Hunter, W. H., and Storey, D. E. (1983), Enthalpies of solution of a series of m-alkoxy phenols in water, n-octanol and water-n-octanol mutually saturated: Derivation of the thermodynamic parameters for solute transfer between these solvents, *J. Pharm. Pharmacol.* 35, 350–357.

Benesch, M. G. K., and McElhaney, R. N. (2014), A comparative calorimetric study of the effects of cholesterol and the plant sterols campesterol and brassicasterol on the thermotropic phase behavior of dipalmitoylphosphatidylcholine bilayer membranes, *Biochem. Biophys. Acta* 1838, 1941–1949.

Biltonen, R. L., Schwartz, A. T., and Wadso, I. (1971), Calorimetric study of the chymotrypsinogen family of proteins, *Biochemistry* 10, 3417–3423.

Biltonen, R. L., Suurkuusk, J., and Mountcastle, D. B. (1979), Design and operation of a differential scanning calorimeter based on the heat leak principle. In Information Circular - United States, Bureau of Mines.

Blicher, A., Wodzinska, K., Fidorra, M., Winterhalter, M., and Heimburg, T. (2009), The temperature dependence of lipid membrane permeability, its quantized nature, and the influence of anesthetics, *Biophys. J.* 96, 4581–4591.

Blume, A. (1988), Applications of calorimetry to lipid model membranes. In C. Hidalgo (ed.), *Physical Properties of Biological Membranes and their Functional Implications*, 41–121, Plenum Press: New York.

Braissant, O., Wirz, D., Göpfert, B., and Daniels, A. U. (2010), Biomedical use of isothermal microcalorimeters, *Sensors (Basel, Switzerland)* 10, 9369–9383.

Briggner, L. E., and Wadsö, I. (1991), Test and calibration processes for microcalorimeters, with special reference to heat conduction instruments used with aqueous systems, *J. Biochem. Biophys. Methods* 22, 101–118.

Broecker, J., Vargas, C., and Keller, S. (2011), Revisiting the optimal c value for isothermal titration calorimetry, *Anal. Biochem.* 418, 307–309.

Bruzzese, F. J., and Connelly, P. R. (1997), Allosteric properties of inosine monophosphate dehydrogenase revealed through the thermodynamics of binding of inosine 5'-monophosphate and mycophenolic acid. Temperature dependent heat capacity of binding as a signature of ligand-coupled conformational equilibria, *Biochemistry* 36, 10428–10438.

Busto, J. V., García-Arribas, A. B., Sot, J., Torrecillas, A., Gómez-Fernández, J. C., Goñi, F. M., and Alonso, A. (2014), Lamellar gel (Lβ) phases of ternary lipid composition containing ceramide and cholesterol, *Biophys. J.* 106, 621–630.

Calvet, E., and Prat, H. (1963), *Recent Progress in Micro-Calorimetry*, Pergamon Press: New York.

Chaires, J. B. (2008), Calorimetry and thermodynamics in drug design, *Annu. Rev. Biophys.* 37, 135–151.

Chen, A., and Wadsö, I. (1982), A test and calibration process for microcalorimeters used as thermal power meters, *J. Biochem. Biophys. Methods* 6, 297–306.

Christensen, J. J., Gardner, J. W., Eatough, D. J., Izatt, R. M., Watts, P. J., and Hart, R. M. (1973), An isothermal titration microcalorimeter, *Rev. Sci. Instrum.* 44, 481–484.

Christensen, J. J., Izatt, R. M., and Hansen, L. D. (1965), New precision thermometric titration calorimeter, *Rev. Sci. Instrum.* 36, 779–783.

Christensen, J. J., Johnston, H. D., and Izatt, R. M. (1968), An isothermal titration calorimeter, *Rev. Sci. Instrum.* 39, 1356–1359.

Connelly, P. R. (1994), Acquisition and use of calorimetric data for prediction of the thermo-dynamics of ligand-binding and folding reactions of proteins, *Curr. Opin. Biotechnol.* 5, 381–388.

Cooper, A., Nutley, M. A., and Wadood, A. (2000), Differential scanning calorimetry. In S. E. Harding and B. Z. Chowdhry (eds.), *Protein–Ligand Interactions: Hydrodynamics and Calorimetry*, 287–318, Oxford University Press: Oxford, New York.

Crespillo, S., Casares, S., Mateo, P. L., and Conejero-Lara, F. (2014), Thermodynamic analysis of the binding of 2F5 (Fab and immunoglobulin G forms) to its gp41 epitope reveals a strong influ-ence of the immunoglobulin Fc region on affinity, *J. Biol. Chem.* 289, 594–599.

Danforth, R., Krakauer, H., and Sturtevant, J. M. (1967), Differential calorimetry of thermally induced processes in solution, *Rev. Sci. Instrum.* 38, 484–487.

Demarse, N. A., Quinn, C. F., Eggett, D. L., Russell, D. J., and Hansen, L. D. (2011), Calibration of nanowatt isothermal titration calorimeters with overflow reaction vessels, *Anal. Biochem.* 417, 247–255.

Ennifar, E., Aslam, M. W., Strasser, P., Hoffmann, G., Dumas, P., and van Delft, F. L. (2013), Structure-guided discovery of a novel aminoglycoside conjugate targeting HIV-1 RNA viral genome, *ACS Chem. Biol.* 8, 2509–2517.

Epand, R. F., Savage, P. B., and Epand, R. M. (2007), Bacterial lipid composition and the anti-microbial efficacy of cationic steroid compounds (ceragenins), *Biochem. Biophys. Acta* 1768, 2500–2509.

Fidorra, M., Heimburg, T., and Bagatolli, L. A. (2009), Direct visualization of the lateral struc-ture of porcine brain cerebrosides/POPC mixtures in presence and absence of cholesterol, *Biophys. J.* 97, 142–154.

Finger, S., Schwieger, C., Arouri, A., Kerth, A., and Blume, A. (2014), Interaction of linear poly-amines with negatively charged phospholipids: The effect of polyamine charge distance, *Biol. Chem.* 395, 769–778.

Friis, D. S., Johnsen, J. L., Kristiansen, E., Westh, P., and Ramløv, H. (2014), Low thermodynamic but high kinetic stability of an antifreeze protein from *Rhagium mordax*, *Protein Sci.* 23, 760–768.

Garbett, N. C., and Chaires, J. B. (2008), Binding: A polemic and rough guide, *Method. Cell Biol.* 84, 3–23.

Garbett, N. C., Mekmaysy, C. S., Helm, C. W., Jenson, A. B., and Chaires, J. B. (2009), Differential scanning calorimetry of blood plasma for clinical diagnosis and monitoring, *Exp. Mol. Pathol.* 86, 186–191.

Garidel, P., and Blume, A. (2000), Miscibility of phosphatidylethanolamine-phosphatidylglycerol mixtures as a function of pH and acyl chain length, *Eur. Biophys. J.* 28, 629–638.

Gill, S. J., and Beck, K. (1965), Differential heat capacity calorimeter for polymer transition studies, *Rev. Sci. Instrum.* 36, 274–276.

Goldberg, R. N. (2014), Standards in biothermodynamics, *Perspect. Sci.* 1, 7–14.

Hansen, L. D., and Russell, D. J. (2006), Which calorimeter is best? A guide for choosing the best calorimeter for a given task, *Thermochim. Acta* 450, 71–72.

Hansen, L. D., Fellingham, G. W., and Russell, D. J. (2011), Simultaneous determination of equi-librium constants and enthalpy changes by titration calorimetry: Methods, instruments, and uncertainties, *Anal. Biochem.* 409, 220–229.

Heerklotz, H., and Seelig, J. (2000), Titration calorimetry of surfactant-membrane partitioning and membrane solubilization, *Biochim. Biophys. Acta* 1508, 69–85.

Heerklotz, H., and Seelig, J. (2007), Leakage and lysis of lipid membranes induced by the lipopep-tide surfactin, *Eur. Biophys. J.* 36, 305–314.

Heerklotz, H., Tsamaloukas, A. D., and Keller, S. (2009), Monitoring detergent-mediated solu-bilization and reconstitution of lipid membranes by isothermal titration calorimetry, *Nat. Protoc.* 4, 686–697.

Hickel, A., Danner-Pongratz, S., Amenitsch, H., Degovics, G., Rappolt, M., Lohner, K., and Pabst, G. (2008), Influence of antimicrobial peptides on the formation of nonlamellar lipid mesophases, *Biochem. Biophys. Acta* 1778, 2325–2333.

Hills, A. K., Beezer, A. E., Mitchell, J. C., and Connor, J. A. (2001), Sources of error, and their correction, in the analysis of isothermal heat conduction microcalorimetric data: Applications of a newly developed test reaction, *Thermochim. Acta* 380, 19–26.

Hinz, H.-J., and Schwarz, F. P. (2001), Measurement and analysis of results obtained on biological substances with differential scanning calorimetry (IUPAC Technical Report), *Pure Appl. Chem.* 73, 745–759.

Hoernke, M., Schwieger, C., Kerth, A., and Blume, A. (2012), Binding of cationic pentapeptides with modified side chain lengths to negatively charged lipid membranes: Complex interplay of electrostatic and hydrophobic interactions, *Biochim. Biophys. Acta* 1818, 1663–1672.

JCGM/WG 1 (Joint Committee for Guides in Metrology). (2008), *GUM: Guide to the Expression of Uncertainty in Measurement 2008* (GUM 1995 with minor corrections).

Ivanova, V. P., Makarov, I. M., Schäffer, T. E., and Heimburg, T. (2003), Analyzing heat capacity profiles of peptide-containing membranes: Cluster formation of Gramicidin A, *Biophys. J.* 84, 2427–2439.

Jackson, W. M., and Brandts, J. F. (1970), Thermodynamics of protein denaturation. Calorimetric study of the reversible denaturation of chymotrypsinogen and conclusions regarding the accuracy of the two-state approximation, *Biochemistry* 9, 2294–2301.

Keller, S., Heerklotz, H., and Blume, A. (2006), Monitoring lipid membrane translocation of sodium dodecyl sulfate by isothermal titration calorimetry, *J. Am. Chem. Soc.* 128, 1279–1286.

Keller, S., Vargas, C., Zhao, H., Piszczek, G., Brautigam, C. A., and Schuck, P. (2012), High-precision isothermal titration calorimetry with automated peak-shape analysis, *Anal. Chem.* 84, 5066–5073.

Kemp, R. B. (1999), From macromolecules to man. In P. K. Gallagher (ed.), *Handbook of Thermal Analysis and Calorimetry*, vol. 4, Elsevier Science: Amsterdam, The Netherlands.

Klocek, G., Schulthess, T., Shai, Y., and Seelig, J. (2009), Thermodynamics of melittin binding to lipid bilayers. Aggregation and pore formation, *Biochemistry* 48, 2586–2596.

Koller, D., and Lohner, K. (2014), The role of spontaneous lipid curvature in the interaction of interfacially active peptides with membranes, *Biochem. Biophys. Acta* 1838, 2250–2259.

Ladbury, J. E., and Doyle, M. L. (2004), *Biocalorimetry 2. Applications of Calorimetry in the Biological Sciences*, John Wiley & Sons: England.

Ladbury, J. E., Klebe, G., and Freire, E. (2010), Adding calorimetric data to decision making in lead discovery: A hot tip, *Nat. Rev. Drug Discov.* 9, 23–27.

Lamprecht, I. (2009), The beauties of calorimetry, *J. Therm. Anal. Calorim.* 97, 7–10.

Lamprecht, I. (2013), Monitoring the heat production of small terrestrial animals by a twin calorimeter, *Eng. Life Sci.* 13, 510–519.

Lewis, R. N., Zweytick, D., Pabst, G., Lohner, K., and McElhaney, R. N. (2007), Calorimetric, X-ray diffraction, and spectroscopic studies of the thermotropic phase behavior and organization of tetramyristoyl cardiolipin membranes, *Biophys. J.* 92, 3166–3177.

Majtan, T., Pey, A. L., Fernandez, R., Fernandez, J. A., Martinez-Cruz, L. A., and Kraus, J. P. (2014), Domain organization, catalysis and regulation of eukaryotic cystathionine beta-synthases, *PLoS One* 9, e105290.

Maskow, T., Kemp, R. B., Buchholz, F., Schubert, T., Kiesel, B., and Harms, H. (2010), What heat is telling us about microbial conversions in nature and technology: From chip- to megacalorimetry, *Microb. Biotechnol.* 3, 269–284.

McKinnon, I. R., Fall, L., Parody-Morreale, A., and Gill, S. J. (1984), A twin titration microcalorimeter for the study of biochemical reactions, *Anal. Biochem.* 139, 134–139.

Moreno, M. J., Bastos, M., and Velazquez-Campoy, A. (2010), Partition of amphiphilic molecules to lipid bilayers by isothermal titration calorimetry, *Anal. Biochem.* 399, 44–47.

Mouritsen, O. G. (2005), *Life—as a Matter of Fat: The Emerging Science of Lipidomics*, Springer: Berlin.

Nilsson, A., Norbeck, J., Oelz, R., Blomberg, A., and Gustafsson, L. (2001), Fermentative capacity after cold storage of baker's yeast is dependent on the initial physiological state but not correlated to the levels of glycolytic enzymes, *Int. J. Food Microbiol.* 71, 111–124.

Nordmark, M. G., Laynez, J., Schön, A., Suurkuusk, J., and Wadsö, I. (1984), Design and testing of a new microcalorimetric vessel for use with living cellular systems and in titration experiments, *J. Biochem. Biophys. Methods* 10, 187–202.

Pey, A. L., Mesa-Torres, N., Chiarelli, L. R., and Valentini, G. (2013), Structural and energetic basis of protein kinetic destabilization in human phosphoglycerate kinase 1 deficiency, *Biochemistry* 52, 1160–1170.

Privalov, P. L. (1963), Investigation of the heat denaturation of egg white albumin, *Biophysica (USSR)* 8, 308–316.

Privalov, P. L. (1980), Scanning microcalorimeters for studying macromolecules, *Pure Appl. Chem.* 52, 479–497.

Privalov, P. L. (2007), Reflections on the origins of microcalorimetry of biopolymers, *Biophys. Chem.* 126, 13–15.

Privalov, P. L. (2012), *Microcalorimetry of Macromolecules*, John Wiley & Sons: Hoboken.

Privalov, P. L., and Dragan, A. I. (2007), Microcalorimetry of biological macromolecules, *Biophys. Chem.* 126, 16–24.

Privalov, P. L., Monaselidze, R. R., Mrevlishvili, G. M., and Magaldadze, V. A. (1964), Intramolecular heat of fusion of macromolecules, *J. Exp. Theor. Phys.* 47, 2073–2079.

Richards, F. M. (1997), Whatever happened to the fun? An autobiographical investigation, *Annu. Rev. Biophys. Biomol. Struct.* 26, 1–25.

Ross, P. D., and Goldberg, R. N. (1974), A scanning microcalorimeter for thermally induced transitions in solution, *Thermochim. Acta* 10, 143–151.

Schön, A., Madani, N., Smith, A. B., Lalonde, J. M., and Freire, E. (2011), Some binding-related drug properties are dependent on thermodynamic signature, *Chem. Biol. Drug. Des.* 77, 161–165.

Schwarz, F. P., Reinisch, T., Hinz, H.-J., and Surolia, A. (2008), Recommendations on measurement and analysis of results obtained on biological substances using isothermal titration calorimetry (IUPAC Technical Report), *Pure Appl. Chem.* 80, 2025–2040.

Seeger, H. M., Gudmundsson, M. L., and Heimburg, T. (2007), How anesthetics, neurotransmitters, and antibiotics influence the relaxation processes in lipid membranes, *J. Phys. Chem. B* 111, 13858–13866.

Seelig, J. (2004), Thermodynamics of lipid-peptide interactions, *Biochem. Biophys. Acta* 1666, 40–50.

Sgarlata, C., Zito, V., and Arena, G. (2013), Conditions for calibration of an isothermal titration calorimeter using chemical reactions, *Anal. Bioanal. Chem.* 405, 1085–1094.

Spink, C. H., and Wadsö, I. (1975), Thermochemistry of solutions of biochemical model compounds. 4. The partial molar heat capacities of some amino acids in aqueous solution, *J. Chem. Thermodyn.* 7, 561–572.

Spink, C., and Wadsö, I. (1976), Calorimetry as an analytical tool in biochemistry and biology, *Methods Biochem. Anal.* 23, 1–159.

Spokane, R. B., and Gill, S. J. (1981), Titration microcalorimeter using nanomolar quantities of reactants, *Rev. Sci. Instrum.* 52, 1728–1733.

Sturtevant, J. M. (1996), Calorimetric studies of biopolymers, *Protein Sci.* 5, 391–394.

Suurkuusk, J., and Wadsö, I. (1972), Thermochemistry of the avidin-biotin reaction, *Eur. J. Biochem.* 28, 438–441.

Suurkuusk, J., and Wadsö, I. (1974), Design and testing of an improved precise drop calorimeter for the measurement of the heat capacity of small samples, *J. Chem. Thermodyn.* 6, 667–679.

Suurkuusk, J., and Wadsö, I. (1982), Multichannel microcalorimetry system, *Chem. Scripta* 20, 155–163.

Tellinghuisen, J. (2004), Volume errors in isothermal titration calorimetry, *Anal. Biochem.* 333, 405–406.

Tellinghuisen, J., and Chodera, J. D. (2011), Systematic errors in isothermal titration calorimetry: Concentrations and baselines, *Anal. Biochem.* 414, 297–299.

Vargas, C., Klingler, J., and Keller, S. (2013), Membrane partitioning and translocation studied by isothermal titration calorimetry, *Method. Mol. Biol.* 1033, 253–271.

Vasilchuk, D., Pandharipande, P. P., Suladze, S., Sanchez-Ruiz, J. M., and Makhatadze, G. I. (2014), Molecular determinants of expansivity of native globular proteins: A pressure perturbation calorimetry study, *J. Phys. Chem. B* 118, 6117–6122.

Velazquez Campoy, A., and Freire, E. (2005), ITC in the post-genomic era …? Priceless, *Biophys. Chem.* 115, 115–124.

Velazquez Campoy, A., and Freire, E. (2006), Isothermal titration calorimetry to determine association constants for high-affinity ligands, *Nat. Protoc.* 1, 186–191.

Wadsö, I. (1970), Microcalorimeters, *Q. Rev. Biophys.* 3, 383–427.

Wadsö, I. (1976), A system of micro-calorimeters and its use in biochemistry and biology, *Biochem. Soc. Trans.* 4, 561–565.

Wadsö, I. (1985), Recent developments in micro solution calorimetry, *Thermochim. Acta* 88, 169–176.

Wadsö, I. (1997), Neither calorimeters nor calorimetrists are what they used to be, *Thermochim. Acta* 300, 1–5.

Wadsö, I. (2002), Isothermal microcalorimetry in applied biology, *Thermochim. Acta* 394, 305–311.

Wadsö, I., and Goldberg, R. N. (2001), Standards in isothermal microcalorimetry (IUPAC Technical Report), *Pure Appl. Chem.* 73, 1625–1639.

Wadsö, I., and Wadsö, L. (2005), Systematic errors in isothermal micro- and nanocalorimetry, *J. Therm. Anal. Calorim.* 82, 553–558.

Weingartner, H., Cabrele, C., and Herrmann, C. (2012), How ionic liquids can help to stabilize native proteins, *Phys. Chem. Chem. Phys.* 14, 415–426.

Wiseman, T., Williston, S., Brandts, J. F., and Lin, L.-N. (1989), Rapid measurement of binding constants and heats of binding using a new titration calorimeter, *Anal. Biochem.* 179, 131–137.

Yadav, S. P., Bergqvist, S., Doyle, M. L., Neubert, T. A., and Yamniuk, A. P. (2012), MIRG Survey 2011: Snapshot of rapidly evolving label-free technologies used for characterizing molecular interactions, *J. Biomol. Tech.* 23, 94–100.

Yu, H., Ren, J., Chaires, J. B., and Qu, X. (2008), Hydration of drug-DNA complexes: Greater water uptake for adriamycin compared to daunomycin, *J. Med. Chem.* 51, 5909–5911.

Zielenkiewicz, W. (2008), *Calorimetry*, 2nd edition, Institute of Physical Chemistry of the Polish Academy of Sciences: Poland.

Zweytick, D., Japelj, B., Mileykovskaya, E., Zorko, M., Dowhan, W., Blondelle, S. E., Riedl, S., Jerala, R., and Lohner, K. (2014), N-acylated peptides derived from human lactoferricin perturb organization of cardiolipin and phosphatidylethanolamine in cell membranes and induce defects in *Escherichia coli* cell division, *PLoS One* 9, e90228.

CHAPTER 2

From Classical Thermochemistry to Monitoring of Living Organisms

Ingemar Wadsö

CONTENTS

2.1 INTRODUCTION

This chapter reviews my development work in biocalorimetry. Following the Editor's suggestion, it will be a personal account of that process. I will also bring up some views and facts about problems in work on living systems, especially regarding systematic errors.

I was introduced to scientific research in the late 1950s in a classical thermochemistry laboratory. At that time, it was common that thermochemists were much engaged in method work, involving the design of new calorimeters, making adjustments to existing instruments, and refining working procedures. That picture is well illustrated by a passage in a chapter written by the three leading thermochemists in the early 1960s (Skinner et al. 1962):

> The design and construction of a suitable calorimeter is one of the first problems facing the experimental thermochemist planning to measure directly the heat of a chemical reaction. During the past 30 years over 300 papers on reaction calorimetry have been published, and more than 200 different reaction calorimeters have been described. This evident need for variety in calorimeter design reflects the very variegated nature of the chemical reactions that have been thermochemically studied.

Commercial instruments are now available, and that state of excessive construction of calorimeters is over. Earlier, calorimeters were almost exclusively used in thermodynamic measurements, but now some types of calorimeters are primarily designed for use as analytical instruments. Moreover, many calorimeters of different types are now used in biochemistry and in cell biology, which was rare in the early 1960s. My recollections will feature some contributions to those changes, but let me first give a brief historical background.

2.2 SOME HISTORICAL NOTES

In the late 1920s, Lennart Smith in the Chemistry Department at Lund University decided to start work in combustion calorimetry. In 1928, he visited College de France in Paris, to investigate the design of a calorimeter in the laboratory of Camille Matignon, a coworker and later the successor to Marcelin Berthelot, the father of bomb combustion calorimetry. After his return to Lund, Smith built a similar instrument and developed a new method, the quartz wool method, allowing accurate measurements of organic chlorine compounds. In 1949, his student Stig Sunner reported the design of the first rotating combustion calorimeter, by which accurate measurements of several "difficult" groups of compounds could be made, for example, organic sulfur compounds. A few years later, a group at the Bureau of Mines in Bartlesville, Oklahoma, independently developed a similar instrument. In 1952, Sunner spent some time in the Bartlesville laboratory and visited several other thermochemical groups in the United States. During his stay at the Bartlesville laboratory, he came into contact with the International Union of Pure and Applied Chemistry (IUPAC) and its Commission on Chemical Thermodynamics and with the Calorimetry Conference, at the time the most important forum for discussions and reports on methods in calorimetry (now often called Calcon). These contacts later turned out to be of great importance for the thermochemical activities in Lund.

Sunner soon made a name for himself in thermochemistry, but his working conditions in Lund were unsatisfactory till 1956, when he was awarded a research position jointly sponsored by the Swedish research councils Natural Research Council (NFR) and Technical Research Council (TFR) and financial support to set up the Thermochemistry Laboratory in the Chemistry Department. One year earlier, I had been accepted as a graduate student, with Sunner as my supervisor. At that time, he intended to start work in solution calorimetry to obtain some auxiliary data needed with some types of organic compounds for the calculation of the enthalpies of formation from combustion experiments. However, the laboratory had only combustion calorimeters, and he therefore suggested that I should take part in the construction of an instrument suitable for measurements of reactions in solution. To make a quick start, Sunner suggested that I should go to Manchester University to learn about solution calorimetric experiments from H. A. Skinner. In Manchester, I worked with a semiadiabatic calorimeter (sometimes called an isoperibol calorimeter), and it was decided that we should start building an instrument of that type in Lund. The Manchester visit turned out to be very important for my future work; in particular, I became aware of the risk of systematic errors associated with calorimetric work.

I had at that time some vague ideas that I wished to work on biochemical systems, but Sunner could not see any opening for such work within the resources of the laboratory. He therefore proposed that I should start working on the energetics of hydrolysis of some O-, S-, and N-acetyl compounds and regard them as models for biochemical reactions. Soon after my dissertation, I was able to form my own research group, continuing work on biochemical model substances and, later, on biochemical systems and living organisms.

2.3 A SEMIADIABATIC CALORIMETRIC SYSTEM

In an adiabatic system, there is no heat exchange with the surroundings. Therefore, the heat released in an experiment with an adiabatic calorimeter is equal to the measured temperature change multiplied by a calibration constant, which, in the ideal case, is equal to the heat capacity of the reaction vessel with its contents. In a semiadiabatic calorimeter, the heat exchange with the surroundings will be significant, and a correction term must be calculated. In the late 1950s, we started the construction of a series of these instruments for use in planned measurements of reactions of biochemical model compounds and of their dissolution processes. The first instrument we designed could be used in research, but we considered its equilibration time to be too long. That project was therefore followed by systematic development and testing of several instruments (Sunner and Wadsö 1959), some of which were used in my thesis work (Wadsö 1962).

Some slightly different versions of that instrument were constructed, for example, a titration calorimeter (Danielsson et al. 1964). A new type of vaporization calorimeter (Wadsö 1966) was also developed for this calorimetric system and was used in a long series of measurements, some of which were part of the biochemical model studies: see, for example, Starzewski et al. (1984).

Parts of the final versions of our semiadiabatic calorimeters were further developed into a commercial instrument by LKB Producer, Bromma, Sweden, where they were marketed under the name LKB Precision Calorimeter 8700 (Sunner and Wadsö 1966). This commercial instrument later came to be used in part of my work on biochemical model compounds. Typically, a large amount of substance, of the order of 1 mmol, was used in a measurement with these "macrocalorimeters." However, with the final design, the precision (repeatability) and accuracy in measurement of fast reactions were very high. For example, for the tris-test reaction, which we had developed at that time (Irving and Wadsö 1964), the reproducibility was 0.01%, and the accuracy was estimated to be about 0.02% (Hill et al. 1969; Sunner and Wadsö 1966).

Surprisingly, in the 1960s, very little attention was paid to the importance of interactions between water and the reactants in biochemical processes. For example, the highly negative ΔG values in the hydrolysis of ATP were taken as a sign of the breaking of an "energy-rich bond," without considering the role of water in the thermodynamic properties of such reactions. Edsall and Wyman's *Biophysical Chemistry* was of decisive importance for my future interest in the field (Edsall and Wyman 1958), in particular concerning the extremely high values for the heat capacities of hydrophobic compounds in contact with water. Therefore, for several years, values for ΔH_{sol} and ΔH_{vap} were measured for hydrophobic compounds by use of the semiadiabatic solution and vaporization calorimeters. From determinations of ΔH_{sol} at different temperatures, the changes in heat capacities ($\Delta C_{p,sol}$) were determined—time-consuming work requiring high accuracy in the measurements. An additivity scheme was derived for the partial molar heat capacities of nonionic compounds in infinitely dilute aqueous solution at 298.15 K (Nichols et al. 1976). That scheme has also been used in discussions of the results of ligand-binding reactions, in which hydrophobic groups are believed to be transferred from an aqueous to a nonaqueous environment (Bastos et al. 1990; Briggner and Wadsö 1990). Later, many values for ΔH_{sol} and $\Delta C_{p,sol}$ were determined for the dissolution of very slightly soluble compounds using a thermal activity monitor (TAM) (see Section 2.5) equipped with specially designed-microcalorimetric vessels (see Section 2.5.1.1).

2.4 ISOTHERMAL MICROCALORIMETRY

The term *isothermal microcalorimeter* is commonly used for calorimeters designed for experiments in the microwatt range and lower, under essentially isothermal conditions: see the IUPAC Technical Report "Standards in Isothermal Microcalorimetry" (Wadsö and Goldberg 2001). The name *nanocalorimeter* is sometimes used for instruments with a detection limit approaching the nanowatt region. However, that term is not used in this chapter, as the prefix "nano" today almost always represents "nanometer."

In 1959–1960, I spent a year working with J. M. Sturtevant at Yale University, where I took part in a study on enthalpy changes in enzymatic hydrolysis of peptides, using Sturtevant's home-made microcalorimeter (Rawitscher et al. 1961). A very important part of my stay in the United States was a 6 week tour, with a very old car, during which I visited all the important thermochemical laboratories I knew about. I also attended, for the first time, the Calorimetry Conference, a conference I would actively take part in for many years.

I thought my work in Lund on model compounds was interesting and meaningful. But I realized that I needed a more sensitive calorimetric technique to get into the field of "real biochemistry." Furthermore, I had become interested in studies of living cellular systems, which sometimes require measurements of low thermal powers over long periods of time (days), for which semiadiabatic calorimeters are not suitable. These ideas brought me around 1965 into the field of isothermal microcalorimetry, largely inspired by the work of Eduard Calvet in Marseille.

Calvet used isothermal microcalorimeters based on the heat flow (or heat conduction, heat flux) principle, which was developed by his predecessor Albert Tian (Calvet and Prat 1963). In these calorimeters, the heat flow between the reaction vessel and a surrounding heat sink (usually an aluminum block) is measured using a thermopile positioned between the vessel and the heat sink. In addition, Tian's calorimeter was equipped with a separate thermopile that could be used for Peltier effect, cooling or heating the reaction vessel (the power compensation principle; see Chapter 1 of this book). In the late 1940s, Calvet remodeled Tian's instrument to a differential ("twin") instrument, in which the difference between the thermopile potentials in the reaction unit and the reference unit is measured. The design of the Tian–Calvet calorimeter has greatly influenced the field of isothermal microcalorimetry and has been used in many investigations in chemistry, physics, and biology. The design of the heat flow microcalorimeter developed by Theodor Benzinger, Bethesda, MD, United States (Benzinger and Kitzinger 1963) was also important for my future development work.

2.4.1 A Rotating Batch Microcalorimeter

My first isothermal microcalorimeter was a rotating twin instrument positioned in a thermostated air bath (Wadsö 1968). The calorimeter had box-shaped gold vessels, each fitted with two open compartments holding solutions (or suspensions) of the two reagents. When the calorimeter was rotated one turn and back (repeated, if needed), the two solutions were brought together, and complete mixing was attained by the air passing through the liquid during rotation. This mixing principle is very efficient, even with suspensions of large particles.

This calorimeter was mainly intended for thermodynamic studies of different biochemical reactions. An article in *Scientific American* on high-resolution X-ray studies of lysozyme and its complexes with saccharide inhibitors (Phillips 1966) led me to a study of binding reactions between lysozyme and the saccharides NAG (*N*-acetyl-*D*-glucosamine), (NAG)$_2$, and (NAG)$_3$ using the new microcalorimeter (Bjurulf et al. 1970; Bjurulf and Wadsö 1972).

Solutions of different concentrations of the saccharides were mixed with lysozyme solution, and from analysis of the binding curves, it was possible to obtain values for the equilibrium constants and the enthalpy changes and thus also for corresponding entropy values. The determination of the binding curves was very time consuming, requiring 2–3 days, but the obtained thermodynamic data were good, even by today's standards.

2.4.2 Flow Microcalorimeters

The construction of the batch calorimeter was soon followed by similar instruments using different types of flow vessels. The first version (Monk and Wadsö 1968) was fitted with a mixing vessel and its reference unit. The mixing vessel, consisting of a gold tube formed into a flat spiral and imbedded between two thin copper plates, was mainly used in studies of fast chemical and biochemical reactions and in dilution experiments. Later, the instrument was equipped with two reaction vessels: one mixing vessel and one flow-through vessel (Wadsö 1980). In an experiment, one of them served as an active reaction vessel and the other was used as its reference, thus allowing different types of experiments to be performed using the same instrument. The flow-through vessels were primarily used in measurements of living organisms.

2.4.3 Insertion Vessels

Reaction vessels used with isothermal microcalorimeters can be permanently mounted in the calorimeters (called *dedicated* in Chapter 1) or can be detachable (here called *insertion vessels*) and are taken out from the instrument after each measurement and replaced after cleaning and recharging (called *modular* in Chapter 1). With insertion vessels, it is possible to use a variety of vessels that are specifically designed to allow certain types of experiments. A schematic picture of a microcalorimeter of this type (Wadsö 1974, 1980) is shown in Figure 2.1a. The simplest type of vessel used with this instrument is a cylindrical ampoule, volume 1–5 mL, which is closed with an O-ring seal. During introduction into the calorimeter, the ampoule is equilibrated in the copper constriction, B. A flow vessel, by which a gas or a liquid can be perfused through the sample compartment, P, is shown in Figure 2.1b. The equilibration unit, Q, contains a thin-walled steel tube formed into a spiral. During an experiment, the spiral is in thermal contact with the copper constriction.

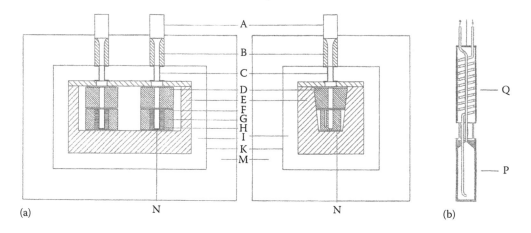

FIGURE 2.1 (a) Longitudinal and transverse sections of a twin isothermal microcalorimeter for use with insertion vessels. A, steel tube; B, copper constriction; C, steel tube; D, aluminum block; E, main heat sink; F, air space; G, aluminum block; H, thermopile; I, air space; K, steel container; M, thermostated water bath; N, holder for sample ampoule. (b) Perfusion vessel for gaseous or liquid flow. (From Wadsö, I., *Science Tools*, 21, 18–21, 1974.)

When gas, usually air, is perfused through the sample compartment, the spiral tube is kept wet (usually by a thin cotton thread wetted with medium) to avoid evaporation of water in the sample ampoule.

Both the batch and the flow instruments, as well as the ampoule insertion microcalorimeter, were further developed into commercial instruments by LKB Producter. In some cases, customers added valuable attachments to that system: for example, Beezer et al. (1982) equipped their rotating batch instrument with a motor-driven injection device, which greatly reduced the time for a titration experiment.

Slightly different versions of the instrument shown in Figure 2.1 (mainly using simple closed ampoules) were employed in a large number of measurements on living systems, for example, in most of our investigations related to the medical field (Section 2.5.1) and in all measurements on soil (Section 2.5.2). The baseline stability in such experiments was typically 1 μW for a 24 h period. Corresponding values for the perfusion vessel were about 2 μW at flow rates of 20 and 10 mL/h for water and air, respectively. The repeatability of the baseline value for the closed ampoule was typically ≤0.5 μW.

Calorimeters similar to that shown in Figure 2.1 were equipped with different peripherals: for example, various types of stirrers and injection devices. Two rather complex microcalorimeters were also developed using the same type of calorimeter as their basic unit: a drop heat capacity calorimeter and a vaporization-sorption calorimeter, which are briefly discussed in Sections 2.4.3.1 and 2.4.3.2.

2.4.3.1 Drop Heat Capacity Microcalorimeters

In measurements with a drop heat capacity calorimeter, the sample is dropped from a metal block, thermostated at a well-defined temperature, into a calorimeter, which is at a different, well-defined temperature. The heat quantity transferred between the two blocks is measured, and by using the temperature difference between the two blocks, the heat capacity of the sample ampoule with its contents can be calculated. We constructed two instruments of this type (Konicek et al. 1971; Suurkuusk and Wadsö 1974), which were used in several of our model compound projects. Values for heat capacities of pure compounds combined with data for $\Delta C_{p,sol}$ (at very low concentrations) lead to the important values for partial molar heat capacities.

The design of these instruments is discussed in some detail in Chapter 4, where initial applications to the study of biochemical model compounds are described. The second version, a double drop heat capacity calorimeter, has a precision of 0.01%. That instrument has been transferred to University of Porto, Portugal, where it has been equipped with modern electronics and used in the determination of partial molar heat capacities of alcohols in crowded media (Bai et al. 2014) and heat capacities of ionic liquids (Rocha et al. 2012).

2.4.3.2 Vaporization-Sorption Microcalorimeters

A new calorimetric method was developed for the determination of enthalpies of sorption of vapors (especially water vapor) on solid and liquid materials (Wadsö and Wadsö 1996, 1997). The calorimetric vessel consists of a vaporization chamber positioned above a sorption chamber, and the two units are combined by means of a diffusion tube. The vessel assembly fits into a double twin heat flow microcalorimeter, where the chambers are in thermal contact with separate thermopiles. In an experiment, water vapor formed in the vaporization chamber diffuses into the lower chamber, positioned in a calorimeter (Figure 2.1a), where it is absorbed by the sample. From the thermal power measured by the sorption calorimeter, the sorption isotherm can be calculated. Using information from both calorimeters, the enthalpy of sorption can be calculated as a function of the equilibrium vapor pressure. Measurements were made on the sorption of water vapor on cotton and on some technical materials.

2.5 MULTICHANNEL INSTRUMENTS

One long-term goal with most of our work on living cellular systems has been to develop methods for the assessment of their "activities" and to explore whether such techniques may be useful as an analytical tool in practical work (see Section 2.8). One problem with the (thermopile) heat flow instrument is its intrinsic low sample throughput, which has been a limiting factor, especially in practical analytical work. An attractive method to reduce this problem is to employ multichannel calorimeters. In such instruments, several calorimetric units ("channels") are joined into one instrument, where one (or several) of the channels is used as a reference. Alternatively, multichannel instruments are made up of several twin calorimeters that share one thermostated unit.

2.5.1 Four-Channel Microcalorimeter

Our experience from the work on biological systems discussed in Section 2.6 led up to the construction of a four-channel isothermal microcalorimeter (Suurkuusk and Wadsö 1982) (cf. Chapter 4 in this book). A thermostated unit consisting of a precisely thermostated water bath ($\pm 1 \times 10^{-4}$ K) holds four "channels," each consisting of a twin microcalorimeter (Figure 2.2). The instrument was designed as a modular system, and could be used with a wide range of insertion vessels with different sample containers, stirrers, injection (titration) devices, and other peripherals (Bäckman et al. 1994; Görman Nordmark et al. 1984; Suurkuusk and Wadsö 1982). The stirrers shown in Figure 2.2d and e are of a type that we call "turbine stirrers," which require a small space, yet move the liquid both vertically and horizontally (cf. other versions in Bäckman et al. 1994). Turbine stirrers are especially useful when sample compartments are crowded by, for example, different analytical sensors (Johansson and Wadsö 1999), and when it is desirable to keep particles in a uniform suspension (Schön and Wadsö 1986). Our four-channel instrument was primarily intended for characterization of biological systems, but its design also allowed accurate thermodynamic measurements; see for example, Bäckman et al. (1994). The instrument was further developed into a commercial instrument by LKB Producer and was marketed as Bio Activity Monitor (BAM). Later, the name was changed to Thermal Activity Monitor (TAM).

Our four-channel instrument turned out to be a milestone in our continued microcalorimetric development work. It is fair to state that it also has significantly influenced the advancement of commercial instruments and their use in biocalorimetry. One important

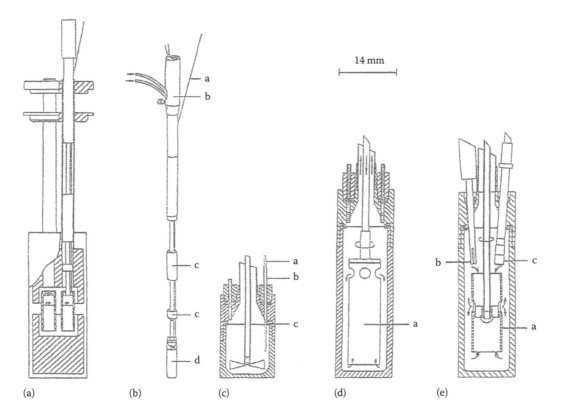

FIGURE 2.2 Examples of different insertion vessels. (a) Simplified picture of a twin microcalorimeter forming a "channel" in the four-channel instrument. Hashed areas indicate aluminum cylinders; devices bringing them into thermal contact are not shown. (b) Titration vessel (shown in measurement position in a): a, guide tube; b, stirring motor; c, aluminum cylinder (thermostating unit); d, sample compartment. (c) Sample compartment of 1 mL titration vessel: a, injection needle; b, guide tube; c, Teflon seal. (d) Sample compartment of 3 mL perfusion vessel: a, turbine stirrer. (e) Sample compartment of 3 mL vessel fitted with electrodes: a, turbine stirrer: b, pH electrode; c, oxygen electrode. (Adapted from Görman Nordmark et al., *J. Biochem. Biophys. Methods*, 10, 187–202, 1984 (a–c); Bäckman et al., *Pure Appl. Chem.* 66, 375–382, 1994 (d,e).)

element in that process was the foundation by Jaak Suurkuusk of Thermometric (Järfälla, Sweden), the instrument company that developed the TAM series of microcalorimeters, including a version equipped with 48 channels (TAM III) (cf. Chapter 4 in this book). A simpler instrument, TAM AIR, originally developed for the cement industry (Wadsö 2005), is now frequently used in measurements of living materials. It is equipped with 16 channels. The TAM series of instruments are now produced and marketed by TA Instruments (New Castle, DE, USA).

Isothermal multichannel microcalorimeters have also been reported from other laboratories, but among these, only Katsutada Takahashi's "multiplex calorimeters" (Yamaguchi et al. 1996) have been used in several important investigations on living systems; see, for example, Koga et al. (2003).

2.5.1.1 Vessels for Dissolution of Hydrophobic Compounds

Much of our work on biochemical model systems has been concerned with interactions between hydrophobic compounds and water. The abnormally large (apparent) heat capacities of hydrophobic groups in water are a useful criterion for the quantitative assessment of hydrophobic interaction in biopolymers, micelles, membranes, and, for example, drug–protein binding reactions. Several microcalorimeters for measurements of the enthalpy of dissolution of slightly soluble hydrophobic compounds were constructed. Special vessels were used for measurements of gases (Gill and Wadsö 1982; Hallén and Wadsö 1989), liquids (Gill et al. 1975; Hallén et al. 1989), wet solids (Nilsson and Wadsö 1986), and dry solids (Bastos et al. 2003). The designs by Gill and Wadsö (1982) and by Gill et al. (1975) were based on the twin microcalorimeter shown in Figure 2.2a, whereas the other instruments were used with the TAM.

As an example of this group of instruments, a schematic picture of the vessel for dissolution of slightly soluble liquids is shown in Figure 2.3, together with power–time curves for dissolution of benzene and diethyl disulfide. The derived values (298.15 K) were $\Delta H_{sol} = 2.21 \pm 0.02$ kJ/mol and $\Delta_{sol}C_p = 231 \pm 3$ J/K mol for benzene (Hallén et al. 1989) and $\Delta H_{sol} = 4.18 \pm 0.10$ kJ/mol and $\Delta_{sol}C_p = 301 \pm 15$ J/K mol for diethyl disulfide (Bastos et al. 1991). For benzene, the values are well documented—we even used benzene as a calibration substance for these kinds of instruments. The uncertainties for the values of diethyl disulfide are much larger. However, for this very slightly soluble compound (requiring 15 h experimental time at the level of a few microwatts), the result is satisfactory.

I think that the design of this group of instruments and the results obtained for several hydrocarbons and other hydrophobic compounds are important. For example, I believe it would be useful to have such data available in connection with thermodynamic studies of drug–biopolymer interactions; cf. our model work involving ligand-binding experiments with macrocyclic compounds (Bastos et al. 1990; Briggner and Wadsö 1990; Hallén et al. 1992; Stödeman and Wadsö 1995; Wadsö 1995a). However, it appears that there is currently little interest in this type of investigation.

2.5.1.2 Photomicrocalorimetric System for Studies of Plant Tissues

The photomicrocalorimetric instrument system shown in Figure 2.4 (Johansson and Wadsö 1997) was primarily designed for investigations of plant tissues. It was tested on leaf tissue from spinach during dark conditions and during photosynthesis. The system consists of three twin microcalorimeters (parts of a TAM): the main photocalorimeter A, photocalorimeter B, and calorimeter C, which serves as a CO_2 analyzer. Photocalorimeter B acts as a continuous monitor of the light power from the light source (a), which is conducted to calorimeter A.

The main calorimeter A is fitted with a removable gas perfusion vessel (Bäckman et al. 1995). The humidity of the gas (oxygen) is standardized by passing through equilibration vessels containing water (l) before reaching the 20 mL sample container (m). Light is conducted into the sample container (m) by means of a quartz rod. Gaseous or liquid reagents can be injected into (m) using a computer-controlled syringe (e). Photocalorimeter B is a standard TAM microcalorimeter fitted with a quartz rod. The third calorimeter, C, is also

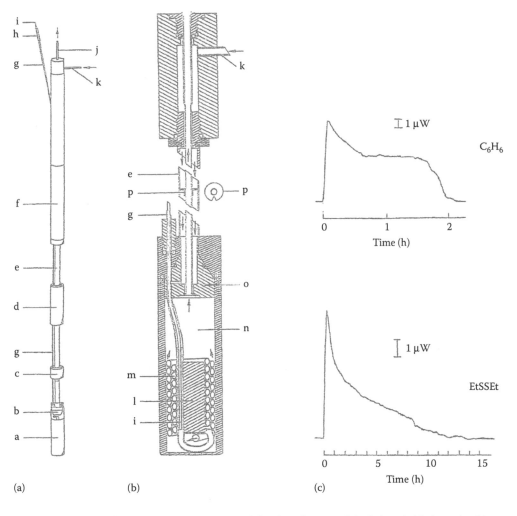

FIGURE 2.3 Microcalorimetric insertion vessel for dissolution of slightly soluble liquids. (a) insertion vessel: a, sample ampoule; b, lid; c, d, heat exchange units; e, steel tube; f, plastic tube; g, guide tube; h, silicone rubber tube; i, hypodermic needle; j, flow outlet tube; k, flow inlet tube. (b) sample compartment: e, steel tube; g, guide tube; i, hypodermic needle; k, flow inlet tube; l, silver core; m, double coiled silver (or gold) tube; n, hold up volume; o, aluminum cone; p, aluminum washer. (c) Calorimetric power-time curves for the dissolution of C_6H_6 and EtSSEt in water. (From Hallén, PhD Thesis, Lund University, 1989.)

a standard TAM microcalorimeter. It is connected to the gas outlet tube from calorimeter A, and is charged with an NaOH solution. The rate of production of CO_2 by the sample in calorimeter A will thus be continuously monitored by calorimeter C.

The photomicrocalorimeters A and B (Teixeira and Wadsö 1990) are shown in some detail in Figure 2.5. In initial experiments, light is supplied to both the photoreaction calorimeter (P in Figure 2.5) and its reference calorimeter (R) in conditions under which it is quantitatively transformed into heat in both calorimeters. The ratio between the measured thermal powers forms an instrument constant, L. During a photochemical experiment, light power to the reference vessel is again completely transformed into thermal power.

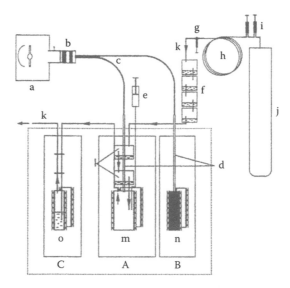

FIGURE 2.4 Schematic diagram of the photocalorimetric system and flow line. A: Main photo-calorimetric system. B: Photo reference calorimeter working as a continuous monitor of the light energy flow. C: Calorimeter used as a CO_2 analyzer. a, lamp house; b, cooled water filters; c, bifur-cated silica fiber bundles; d, quartz rods; e, motor-driven gastight injection syringe; f, prehumidi-fier cups; g, needle valve; h, silica capillary tube; i, pressure regulators; j, pressurized gas cylinder. (From Johansson, P., and Wadsö, I., *J. Biochem. Biophys. Methods*, 35, 103–114, 1997.)

By use of the constant *L*, the thermal power supplied to the photoreaction vessel can be calculated independently of fluctuations of the radiant power from the light source. For a more detailed description of the measurement principles and experimental procedures, see Johansson and Wadsö (1997) and Teixeira and Wadsö (1990).

2.5.1.3 Vessel Equipped with Three Analytical Sensors
The nonspecificity of isothermal microcalorimetric techniques can limit their use as mon-itors of living systems—it is often difficult to interpret calorimetric results for complex processes/systems. A combination of analytical tools can therefore be useful, leading to a calorimetric analytical technique that usually provides a general picture of an investigated process, which rarely is obtained by a specific analytical method. Examples are inserting oxygen and pH electrodes into a calorimetric vessel (Figure 2.2e) or placing them in series with a flow-through vessel: see, for example, (Wadsö 1980). Lee Hansen, in particular, has developed methods for combination of microcalorimetry and determination of CO_2 (*calo-respirometry*) (see Chapter 3 of this book). To support the analysis of results from experi-ments with small animals, Lamprecht and coworkers used an endoscope (Lamprecht and Becker 1988) and a microphone (Schulze-Motel and Lamprecht 1994) for the recording of movements and sound, respectively. A Tian–Calvet calorimeter fitted with 100 mL vessels was used in these experiments.

Figure 2.6a shows an outline of a microcalorimetric vessel equipped with three dif-ferent analytical sensors: a diode array detector, an oxygen electrode, and a pH electrode (Johansson and Wadsö 1999). The vessel is also equipped with an injection device and a

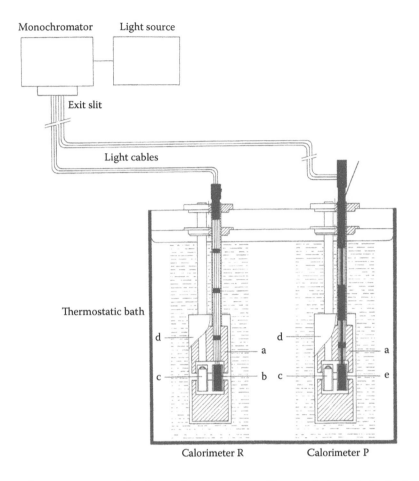

FIGURE 2.5 Schematic picture of a photocalorimeter assembly. a, calorimetric body; b, photoinert light absorption ampoule; d, steel can; e, photochemical reaction vessel. (From Teixeira, C., and Wadsö, I., *J. Chem. Thermodyn.*, 22, 703–713, 1990.)

turbine stirrer. Figure 2.6b shows recordings from a growth experiment with *Escherichia coli* using the vessel in Figure 2.6a and illustrates the high resolution of the thermal power–time curve compared with those obtained from the measurements with the electrodes and the photometer. Taken together, the four curves will support the interpretation of the experimental result. However, this kind of multifunction microcalorimetric vessel must be further developed before it will be useful in practical work.

2.5.2 Forty-Eight-Channel Microcalorimeter

The four-channel twin microcalorimeter, together with its different versions of reaction vessels, has proved to be a very useful instrument over a wide field of thermodynamic and analytical applications in chemistry, biochemistry, and biology. The sample throughput was increased by a factor of 4 compared with the different "ampoule calorimeters" used earlier. But it was still too low for most practical applications. However, the Thermometric instruments TAM III and TAM AIR, with 48 and 16 channels, respectively, have proved

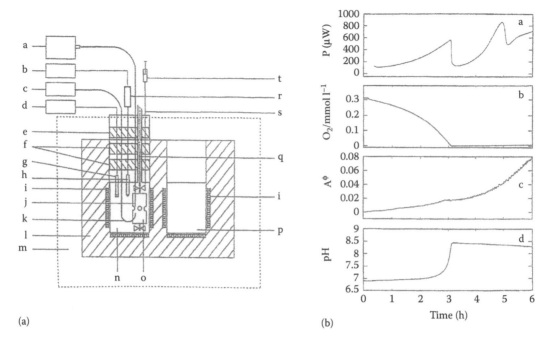

(a)   (b)

FIGURE 2.6   (a) Microcalorimetric vessel for continuous measurements of thermal power, pH, and oxygen with electrodes and a spectrophotometer: a, light source; b, pH meter; c, diode array detector; d, electronic unit for polarographic oxygen sensor; e and f, heat exchange bolts; g, oxygen electrode; h, pH electrode; i, thermopile; j and k, light guides; l, heat sink; m, thermostated water bath; n, sample compartment; o, turbine stirrer; p, reference vessel; q, steel tube; r, reference electrode; s, injection needle; t, syringe and syringe drive. (b) Records from a growth experiment with *Escherichia coli*. The sample compartment was filled with medium: a, thermal power–time curve; b, change of oxygen concentration versus time; c, pH-time curve; d, apparent optical density (turbidity) versus time. (Adapted from Johansson, P. and Wadsö, I., *Thermochim. Acta*, 342, 19–29, 1999.)

to be useful in practical applications involving the estimation of instabilities of technical products (Gaisford 2005). They are also used in work with living systems, although not yet in practical applications.

Following a discussion with my former PhD student Dan Hallén, in late 1999, I decided to construct a new multichannel microcalorimeter for use as a monitor of the thermal power from living organisms. Hallén, at that time employed by the pharmaceutical company Pharmacia & Upjohn in Stockholm, claimed there was a need in the pharmaceutical industry for new types of analytical methods for characterization of cell–drug interactions in developments of new drugs. Preferably, the methods should reflect some functional property related to the reactions. Furthermore, he put forward the idea of employing a well plate (microtiter plate) as a multisample calorimetric vessel. I thought this was a very interesting idea, and we agreed to start a joint project, whereby I would develop the instrument in Lund, and he would carry out tests of the instrument at Pharmacia & Upjohn in Stockholm and evaluate its properties in practical work, especially in the pharmaceutical industry. My own interest was mainly in the direction of medical and environmental applications.

Within a few months after our talk, it was possible for the development work in Lund to start, initially supported by a small grant from Pharmacia & Upjohn. I soon realized that a microtiter plate could not be used as a multisample vessel. However, a vessel holder, which was made in the format of a 48-well microplate and used with 1 mL sample ampoules, worked satisfactorily. The ampoules were made from glass or high-grade steel and were closed by means of steel lids fitted with O-ring seals. After about 3 years, a prototype with very satisfactory properties was ready.

A simplified picture of the longitudinal section of the instrument is shown in Figure 2.7. The main part of the calorimeter consists of two thermostated aluminum blocks, (j) and (h), which are separated by a plastic plate (i). The heat sink consisting of a third aluminum block (k) is suspended in the cavity of the main block (j). The 48 thermopile plates are attached to pillars (l), which are cut out from the heat sink. The block assembly is suspended in a Dewar vessel in horizontal position, which is closed by a thermostated aluminum lid (b) and a plastic lid (a). The Dewar vessel and the lids are enclosed by a layer of Styrofoam (not shown in the figure). The three thermostated units (j, h, and b) are thermostated separately, at nearly the same temperature. The heat sink (k) will attain the temperature of the surrounding block (j), which thus is the measurement temperature. The ampoule holder is introduced into the calorimeter through the channel (d). The ampoules and their holder are first equilibrated in each of blocks (h) and (j) and are then moved to the measurement zone (e), where they are in thermal contact with the thermopile plates (sample holder and ampoules are not shown in Figure 2.7). The thermopile plates are covered by a plastic film. The ampoules can move vertically in their holders, and the assembly can thus move smoothly over the thermopile plates. The calorimeter was primarily designed for use as a bioactivity monitor at 37°C and was therefore not fitted with a cooling unit. The 24 h baseline stability was about 0.2 μW. The design and properties of the instrument were reported

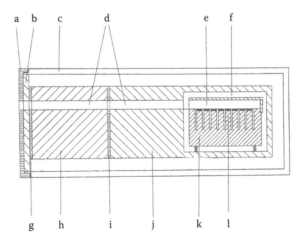

FIGURE 2.7 Section through the 48-channel isothermal microcalorimeter. a, plastic lid; b, thermostated aluminum lid; c, Dewar vessel; d, insertion channel for sample holder; e, measurement zone (samples and sample holder are not shown); f, air space; g, plastic plate; h, thermostated aluminum block; i, plastic plate; j, thermostated aluminum block with cavity in which the heat sink block, k, is suspended; l, squared pillar cut out from the heat sink block.

at the XIII ISBC conference in Würzburg, Germany (Wadsö and Hallén 2003; Hallén and Wadsö 2003).

The results of Hallén's test experiments, which were conducted at the pharmaceutical company Biovitrum, Stockholm, encouraged him to leave his employment and form a company, SymCel, for the development of our instrument into a commercial product. I supported his brave endeavor, and we agreed that he should make some final test experiments characterizing the instrument and further experiments with living cells as soon as a new instrument was ready—as a university scientist, I was under pressure to publish a scientific report of our work.

Except for the first year, I did not actively take part in the work at SymCel, but my close contacts with Hallén continued. The development project required a longer time than expected. Sadly, Hallén did not live to see the end of our project; he was struck by cancer and passed away in 2011. Our joint project had come to an end.

After Hallen's passing, SymCel was reconstructed. The development project, now managed by M. Jansson, continued, but till recently, my contact with the project was marginal. Using a preproduction model of the instrument, Olivier Braissant and coworkers in Basel, Switzerland, made a series of measurements on microorganisms, tumorous microtissues, and parasitic worms (Braissant et al. 2015). They demonstrated that the technique is very promising for practical use in, for example, the pharmaceutical industry and clinical laboratories (cf. Chapter 19 in this book). Jaak Suurkuusk (Soojus AB, Stockholm) performed supplementary test experiments, and a publication of the design and properties of the instrument is now in preparation—15 years after the start of the university project (Wadsö et al., manuscript to be submitted).

2.6 WORK ON LIVING SYSTEMS

Following the work with lysozymes (Bjurulf et al.1970; Bjurulf and Wadsö 1972), we conducted several more biochemical ligand-binding studies, but they were not part of a planned working scheme. The thermodynamic studies on model systems were more systematic and would continue for more than 20 years, with the focus on hydrophobic hydration. My interest in investigations on living materials was aroused by Paul Monk, a PhD student from Australia. Monk had been trained in microbiology at the Australian Wine Institute in Adelaide, and we agreed to test our flow microcalorimeter (Monk and Wadsö 1968) in measurements of microbial systems (Delin et al. 1969). I was intrigued by the results and the possibilities of further developing microcalorimetry as an analytical method in work with living organisms—I was well prepared for such thoughts by reading the book by Calvet and Prat (1963).

Here follow some comments on our extensive participation in projects oriented toward medicine and in the development of microcalorimetric methods for measurement of the microbial activity in soil.

2.6.1 Work in the Field of Medicine

In 1972, I gave a talk about microcalorimetry for a research group at Lund University Hospital, which resulted in a 20 year collaboration, especially with Mario Monti. Other groups at the University Hospital also took up studies in isothermal microcalorimetry. The general aim was to investigate differences between cellular systems obtained from healthy subjects and patients, thus exploring the potential use of the technique as an analytical method in the clinical laboratory. A large part of my contributions to these studies was concerned with investigations of different experimental factors that influence the measured thermal powers: calorimetric techniques, employed pH, temperature, cell concentrations, storage conditions, suspension media, and preparation methods. Work by the Monti group included, for example, investigations on different fractions of blood cells, cancer cells, fat cells, immune complexes, muscle fiber bundles, and biocompatibility problems (Monti 1999). A wealth of P-data was obtained, and in most cases, significant differences were found when results for patients and healthy subjects were compared. But values for different individuals vary widely, which limits the practical use of the technique as a diagnostic method. However, its use as a prognostic method appears to be more promising (cf. Section 2.10).

After about 90 research projects had been reported (I participated in only some of those studies), microcalorimetric work at the University Hospital ceased in the late 1990s, mainly because we realized that the sample throughput with our technique was too low for practical use in clinical laboratories. However, I trust that our developments of experimental procedures and the obtained thermal power data will be of lasting value. For a comprehensive review of microcalorimetric work directed toward the medical field and reported before the end of the 1990s, see Monti (1999). More recent work in this area has been conducted by Braissant (see Chapter 19 in this book).

2.6.2 Soil

My work on soil was, like that in the medical field, initiated during a seminar: this time at the Department of Microbiology, where the soil microbiologist Börje Norén became interested in our technique. As there had been no significant method development on microcalorimetric measurements on soil, we concentrated on this area. Microcalorimeters of the type shown in Figure 2.1a were used, with simple closed ampoules as reaction vessels. During the work, a new ampoule technique was developed (Wadsö 2009) by which measurements could be made on soil samples during long periods of time (months) without any significant changes in the concentrations of CO_2 or O_2 in the sample. Different types of homogenized samples of soil were studied, and the influence of several experimental factors on the calorimetric results was investigated to form a base for future work in this field. These factors included storage of the samples (time, temperature), changes of the water content, pH, sterilization procedures, depletion of O_2, accumulation of CO_2, content of organic matter, and effects caused by addition of glucose and of cellulose. Our work on soil, and work from other laboratories, is summarized in a review (Wadsö 2009), in which possible systematic errors in microcalorimetric work on soil—which can be very large—are also discussed. Since that time, important method development work has been done in this area, especially by Barros and Hansen and their coworkers (see Chapter 3 in this book).

2.7 SYSTEMATIC ERRORS IN MEASUREMENTS OF LIVING SYSTEMS

Practically all processes in physics, chemistry, and biology are accompanied by evolution or absorption of heat. This property can be most valuable, as it makes calorimetry a superb technique for the discovery and quantitative characterization, in terms of thermal power or heat, of unknown or unexpected processes. However, it also makes calorimetry very vulnerable to systematic errors and misinterpretations, which can easily ruin the results, in particular when measurements are conducted at a low level of thermal power (Wadsö 1980, 2009; Wadsö and Wadsö 2005). Since my early training in classical thermochemistry, I have been much engaged in matters of that kind, partly in connection with my work with IUPAC (Belaich et al. 1982; Wadsö 1986; Wadsö and Goldberg 2001).

Microcalorimeters are normally calibrated electrically. But in many vessels, for example the complex vessels shown in Figures 2.4 through 2.8, the path of the heat flow from an electrical heater may not accurately mimic the heat flow caused by the reaction. In such cases, calibrations are preferably performed using a suitable chemical reaction. For this purpose, we developed, for example, a method based on the hydrolysis of triacetin catalyzed by imidazole in acetate buffer (Chen and Wadsö 1982). The measurements were later refined and extended (Briggner and Wadsö 1991), and our triacetin test method was extensively used in the development of our 48-channel instrument (Wadsö et al. n.d.). Beezer and associates have conducted detailed thermodynamic and kinetic analyses of that reaction (see Chapter 1 of this book).

Water leakage is probably the most common cause of systematic as well as random errors in isothermal microcalorimetric work on biosystems. Water is always present in studies of cellular systems, and its enthalpy of vaporization is very high at 44 kJ/mol. This means that a rate of vaporization of 12 μg/h will cause a cooling power of 8 μW, such that in experiments with isothermal calorimeters on biosystems, it is common that virtually no leakage or vaporization of water can be accepted. Further, in measurements of living material, there are many problems (which are rarely serious in calorimetric measurements in chemistry): for example, oxygen depletion, sedimentation, adhesion, change of pH, and heat produced by the media (Wadsö 1980, 1993).

Figure 2.8 shows a striking example of a systematic error in experiments with fiber bundles from human skeletal muscle (Fagher et al. 1986). Two types of calorimetric vessels were used: a closed ampoule and the perfusion vessel shown in Figure 2.8a. In the perfusion experiment, the sample is positioned in a rotating cage that is open to the medium. The flow of medium is introduced into the sample container through the shaft of the stirrer, which consists of a stainless steel tube. The sample in the cage will thus be in continuous contact with fresh medium. In experiments with the closed ampoule, the sample will rest at its bottom. Presumably, the medium close to the sample will then rapidly be depleted of oxygen, and it is expected that the pH will be significantly reduced. Both factors will contribute to the low thermal power values shown in Figure 2.8b.

In the introduction to this chapter, I pointed out that at the time when I was introduced to scientific research, "it was common that thermochemists were much engaged in method work, involving the design of new calorimeters, making adjustments to existing instruments, or refining working procedures." Much of my calorimetric development work has been part of the transition process whereby home-made calorimeters were replaced by instruments manufactured and marketed commercially. This process has certainly led to improved instruments, simplified measurement procedures, and

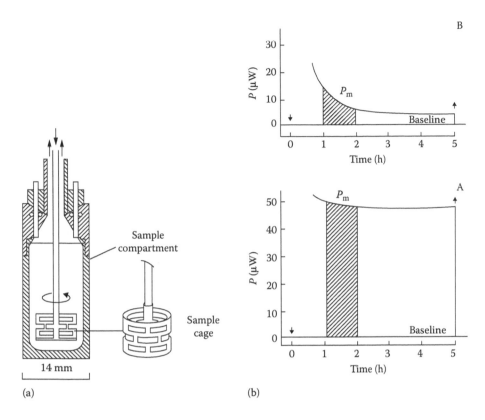

FIGURE 2.8 (a) Section through perfusion vessel with sample cage. (b) Thermal power–time curves for 100 mg samples of human skeletal muscle. Curve A is a recording of measurements using the perfusion vessel shown in Figure 2.8a. Perfusion rate was 20 mL/h, and stirring rate was 24 rev/min. Curve B shows results from an experiment with closed ampoules. (Adapted from Fagher, B., et al., *Clin. Sci. (Lond.)* 70, 63–72, 1986.)

increased productivity. However, I believe it has also led to a decreased awareness of possible systematic errors in calorimetric measurements, in particular by users with no earlier experience in calorimetry. I believe there is a need for more information on these problems, and I think that companies marketing microcalorimeters have a responsibility in this matter toward their customers. Do they take that responsibility? Are they competent to do so? I think we need more cooperation between the industry and scientists who have broad experience in this field.

2.8 WILL ISOTHERMAL MICROCALORIMETRY EVER BECOME A ROUTINE METHOD FOR ESTIMATION OF THE ACTIVITY OF LIVING ORGANISMS?

One long-term goal in my work has been, and is, to develop instruments that are useful for the estimation of the activity of living organisms and to explore their usefulness as analytical tools, especially in applied biology. Over the years, there have appeared many enthusiastic reports on the potential practical use of isothermal microcalorimetry as a general analytical tool in areas such as medicine, environmental work, and the pharmaceutical industry (Braissant et al. 2010; Wadsö 2002). Several very attractive properties of the technique have been listed, but also some factors that have limited the practical use of the technique (Wadsö 1993). Twenty years ago, I wrote a review article (Wadsö 1995b) with the title "Microcalorimetric Techniques for Characterization of Living Cellular Systems. Will There Be Any Important Practical Applications?" in which I pointed out that a wide field of applications, including the pharmaceutical and clinical areas, different medical techniques, agriculture, forestry, and ecology, have been explored, but it was concluded that the technique had not found any use in practical work. Now, 20 years later, still no practical applications in applied biology have been reported in which the technique is used routinely in the estimation of the activity of living organisms. However, it is clear that the use of multichannel instruments has finally brought us very close, especially in areas related to medicine (Braissant et al. 2010, 2015, and Braissant et al. in Chapter 19 of this book). Further, I believe that the emerging field of chip calorimeters (see especially work by the Lerchner group in Freiberg, Germany: Maskow et al. 2011; Wolf et al. 2015) will prove to be of practical importance.

ACKNOWLEDGMENTS

I have been fortunate to have had an exceptional person, Stig Sunner, as my mentor and friend. In addition to his achievements in thermochemistry and numerous engagements in international scientific commissions, his contributions to the advancements of chemical research in Lund were outstanding. He often mentioned that his visits to the thermochemistry laboratory in Bartlesville and to some major chemistry institutions in the United States had affected his views concerning the design and operation of scientific institutions. Those experiences certainly influenced his mind when in the early 1960s, he started lobbying for the formation of a Chemical Center at Lund University. After some years and a very heavy workload, his visions materialized. The Chemical Center soon became of tremendous importance for the advancement of chemical research at Lund University. For instance, for the Thermochemistry Laboratory, the creation of an excellent central mechanical workshop became of great importance.

Naturally, cooperation with students, guest-workers, and colleagues has been a necessary prerequisite for the calorimetric development work reported in this chapter. Here, I acknowledge my thanks to Per Bäckman, Margarida Bastos, Lars-Erik Briggner, Stanley Gill, Dan Hallén, Peter Johansson (now Vikegard), Takayoshi Kimura, Paul Monk, Sven-Ove Nilsson, Arne Schön, Jaak Suurkusk, and Lars Wadsö. I also acknowledge the vitally important role of different categories of technicians in my development projects. I especially thank Börje Pettersson and Jan-Erik Falk in the mechanical workshop, who suggested numerous improvements to my design ideas; Sven Hägg, who designed and built many electronic units that were not commercially available; and Eva Qvarnström, our treasured laboratory technician, who was for many years a key person in our team. Furthermore, thank you, Tony Beezer, Richard Kemp, Ingolf Lamprecht, Lee Hansen, and many other colleagues for inspiring ideas during our discussions and in your papers.

Soon after my dissertation, Sunner suggested that we should contact LKB Producer, an internationally well-known instrument producer in Stockholm, which might be interested in a joint development project aiming at the commercialization of our semiadiabatic solution calorimeter. We reached an agreement with LKB, but Sunner could not actively take part in the development program due to his vast engagement in different international and local projects. My cooperation with LKB, and later with Thermometric (founded by Jaak Suurkusk in 1980), lasted for about 40 years. Thermometric was then bought by TA Instruments. My very long collaboration with LKB and Thermometric was of great importance for my work in Lund. Both companies donated several instruments to the Thermochemical Laboratory. I learned much about instrument design. Our collaboration was very informal and totally different from commissioned projects; it was based on trust rather than on business agreements authored by lawyers. In fact, it was very similar to joint projects between university groups.

REFERENCES

Bäckman, P., Bastos, M., Briggner, L. E., Hägg, S., Hallen, D., Lönnbro, P., Nilsson, S. O., Olofsson, G., Schön, A., Suurkuusk, J., Teixeira, C., and Wadsö, I. (1994), A system of microcalorimeters, *Pure Appl. Chem.* 66, 375–382.

Bäckman, P., Bastos, M., Hallén, D., Lönnbro, P., and Wadsö, I. (1994), Heat conduction microcalorimeters: Sensitivity, time constants and fast titration experiments, *J. Biochem. Biophys. Methods* 28, 85–100.

Bäckman, P., Breidenbach, R. W., Johansson, P., and Wadsö, I. (1995), A gas perfusion microcalorimeter for studies of plant materials, *Thermochim. Acta* 251, 323–333.

Bai, G., Nunes, S. C. C., Rocha, M. A. A., Santos, L. M. N. B. F., Eusébio, M. E. S., Moreno, M. J., and Bastos, M. (2014), Enthalpies of solution, limiting solubilities and partial molar heat capacities of n-alcohols in water and in trehalose crowded media, *Pure Appl. Chem.* 86, 223–231.

Bastos, M., Bai, G., Qvarnström, E., and Wadsö, I. (2003), A new dissolution microcalorimeter: Calibration and test, *Thermochim. Acta* 405, 21–30.

Bastos, M., Briggner, L.-E., Shehatta, I., and Wadsö, I. (1990), The binding of α-, ω- alkanediols to α-cyclodextrin: A microcalorimetric study, *J. Chem. Thermodyn.* 22, 1181–1190.

Bastos, M., Kimura, T., and Wadso, I. (1991), Some thermodynamic properties of dialkylsulfides and dialkyldisulphides in aqueous-solution, *J. Chem. Thermodyn.* 23, 1069–1074.

Beezer, A. E., Hunter, W. H., Lipscombe, R. P., Newell, R. D., and Storey, D. E. (1982), Design and testing of a microtitration assembly for use with an LKB batch microcalorimeter, *Thermochim. Acta* 55, 345–349.

Belaich, J. P., Beezer, A. E., Prosen, E., and Wadsö, I. (1982), Calorimetric measurements on cellular systems: Recommendations for measurements and presentation of results, *Pure Appl. Chem.* 54, 671–679.

Benzinger, T. H., and Kitzinger, C. (1963), *Temperature—Its Measurement and Control in Science and Industry*, vol. 3, Reinhold: New York.

Bjurulf, C., Laynez, J., and Wadsö, I. (1970), Thermochemistry of lysozyme-inhibitor binding, *Eur. J. Biochem.* 14, 47–52.

Bjurulf, C., and Wadsö, I. (1972), Thermochemistry of lysozyme-inhibitor binding, *Eur. J. Biochem* 31, 95–102.

Braissant, O., Keiser, J., Meister, I., Bachmann, A., Wirz, D., Gopfert, B., Bonkat, G., and Wadso, I. (2015), Isothermal microcalorimetry accurately detects bacteria, tumorous microtissues, and parasitic worms in a label-free well-plate assay, *Biotechnol. J.* 10, 460–468.

Braissant, O., Wirz, D., Gopfert, B., and Daniels, A. U. (2010), Biomedical use of isothermal micro-calorimeters, *Sensors (Basel)* 10, 9369–9383.

Briggner, L.-E., and Wadsö, I. (1990), Some thermodynamic properties of crown ethers in aqueous solution, *J. Chem. Thermodyn.* 22, 143–148.

Briggner, L. E., and Wadsö, I. (1991), Test and calibration processes for microcalorimeters, with special reference to heat conduction instruments used with aqueous systems, *J. Biochem. Biophys. Methods* 22, 101–118.

Calvet, E., and Prat, H. (1963), *Recent Progresses in Micro-Calorimetry*, Pergamon Press: New York.

Chen, A., and Wadsö, I. (1982), A test and calibration process for microcalorimeters used as thermal power meters, *J. Biochem. Biophys. Methods* 6, 297–306.

Danielsson, I., Nelander, B., S., S., and Wadsö, I. (1964), An isothermal jacket titration calorimeter, *Acta Chem. Scand.* 18, 995–998.

Delin, S., Monk, P., and Wadsö, I. (1969), Flow microcalorimetry as an analytical tool in microbiology, *Science Tools* 16, 22–24.

Edsall, J. T., and Wyman, J. (1958), *Biophysical Chemistry*, vol. 1, Academic Press: New York.

Fagher, B., Monti, M., and Wadsö, I. (1986), A microcalorimetric study of heat production in resting skeletal muscle from human subjects, *Clin. Sci. (Lond.)* 70, 63–72.

Gaisford, S. (2005), Stability assessment of pharmaceuticals and biopharmaceuticals by isothermal calorimetry, *Curr. Pharm. Biotech.* 6, 181–191.

Gill, S. J., Nichols, N., and Wadsö, I. (1975), Calorimetric determination of heats of solution of slightly soluble liquids. Application to benzene in water, *J. Chem. Thermodyn.* 7, 175–183.

Gill, S. J., and Wadsö, I. (1982), Flow microcalorimetric techniques for solution of slightly soluble gases. Enthalpy of solution of oxygen in water, *J. Chem. Thermodyn.* 14, 905–919.

Görman Nordmark, M., Laynez, J., Schön, A., Suurkuusk, J., and Wadsö, I. (1984), Design and testing of a new microcalorimetric vessel for use with living cellular systems and in titration experiments, *J. Biochem. Biophys. Methods* 10, 187–202.

Hallén, D. (1989), Thermochemical studies of nonelectrolyte aqueous solutions. PhD Thesis, Lund University.

Hallén, D., Nilsson, S.-O., and Wadsö, I. (1989), A new flow microcalorimetric vessel for dissolution of small quantities of easily and slightly soluble liquids, *J. Chem. Thermodyn.* 21, 529–537.

Hallén, D., Schön, A., Shehatta., and Wadsö, I. (1992), Microcalorimetric titration of α-cyclodextrin with some straight-chain alkan-1-ols at 288.15, 298.15 and 308.15 K, *J. Chem. Soc. Faraday Trans.* 88, 2859–2863.

Hallén, D., and Wadsö, I. (1989), A new microcalorimetric vessel for dissolution of slightly soluble gases enthalpies of solution in water of carbon tetrafluoride and sulphur hexafluoride at 288.15, 298.15, and 308.15 K, *J. Chem. Thermodyn.* 21, 519–528.

Hallén, D., and Wadsö, I. (2003), A generic functional multisample calorimetric assay for biological systems, in D. Singer (Ed.), *The International Conference for Biological Calorimetry. (ISBC) XIII*, Abstract Book, p. 19, Veitshöchheim, Germany.

Hill, J. O., Öjelund, G., and Wadsö, I. (1969), Thermochemical results for "tris" as a test substance in solution calorimetry, *J. Chem. Thermodyn.* 1, 111–116.

Irving, R. J., and Wadsö, I. (1964), Use of tris(hydroxymethyl)aminomethane as a test substance in reaction calorimetry, *Acta Chem. Scand.* 18, 995–998.

Johansson, P., and Wadsö, I. (1997), A photo microcalorimeter for studies of plant tissues, *J. Biochem. Biophys. Methods* 35, 103–114.

Johansson, P., and Wadsö, I. (1999), An isothermal microcalorimetric titration/perfusion vessel equipped with electrodes and spectro-photometer, *Thermochim. Acta* 342, 19–29.

Koga, K., Osuga, Y., Yoshino, O., Hirota, Y., Ruimeng, X., Hirata, T., Takeda, S., Yano, T., Tsutsumi, O., and Taketani, Y. (2003), Elevated serum soluble vascular endothelial growth factor receptor 1 (sVEGFR-1) levels in women with preeclampsia, *J. Clin. Endocrinol. Metab.* 88, 2348–2351.

Konicek, J., Suurkuusk, J., and Wadso, I. (1971), A precise drop heat capacity calorimeter for small samples, *Chem. Scr.* 1, 217–220.

Lamprecht, I., and Becker, W. (1988), Combination of calorimetry and endoscopy for monitoring locomotor activities of small animals, *Thermochim. Acta* 130, 87–93.

Maskow, T., Schubert, T., Wolf, A., Buchholz, F., Regestein, L., Buechs, J., Mertens, F., Hauke, H., and Lerchner, J. (2011), Potentials and limitations of miniaturized calorimeters for bioprocess monitoring, *Appl. Microbiol. Biotechnol.* 92, 55–66.

Monk, P., and Wadsö, I. (1968), A flow micro reaction calorimeter, *Acta Chem. Scand.* 22, 1842–1852.

Monti, M. (1999), Calorimetric studies in medicine. In R. B. Kemp (Ed.), *Handbook of Thermal Analysis and Calorimetry*, pp. 657–710, Elsevier: Amsterdam.

Nichols, N. F., Sköld, R., Spink, C., Suurkuusk, J., and Wadsö, I. (1976), Additivity relations for the heat capacities of non-electrolytes in aqueous solution, *J. Chem. Thermodyn.* 8, 1081–1093.

Nilsson, S.-O., and Wadsö, I. (1986), A flow-microcalorimetric vessel for solution of slightly soluble solids, *J. Chem. Thermodyn.* 18, 1125–1133.

Phillips, D. C. (1966), The three-dimensional structure of an enzyme molecule, *Sci. Am.* 215, 78–90.

Rawitscher, M., Wadsö, I., and Sturtevant, J. M. (1961), Heats of hydrolysis of peptide bonds, *J. Am. Chem. Soc.* 83, 3180–3184.

Rocha, M. A. A., Bastos, M., Coutinho, J. A. P., and Santos, L. M. N. B. F. (2012), Heat capacities at 298.15 K of the extended [CnClim][Ntf2] ionic liquid series, *J. Chem. Thermodyn.* 53, 140–143.

Schön, A., and Wadsö, I. (1986), Microcalorimetric measurements of tissue cells attached to micro-carriers in stirred suspensions, *J. Biochem. Biophys. Methods* 13, 135–143.

Schulze-Motel, P., and Lamprecht, I. (1994), Correlation of sound generation and metabolic heat flux in the bumble bee *Bombus lapidarius*, *J. Exp. Biol.* 187, 315–318.

Skinner, H. A., Sturtevant, J., and Sunner, S. (1962), The design and operation of reaction calorimeters. In H. A. Skinner (Ed.), *Experimental Thermochemistry, Volume II*, pp. 157–219. Interscience: London.

Starzewski, P., Wadsö, I., and Zielenkiewicz, W. (1984), Enthalpies of vaporization of some *N*-alkylamides at 298.15 K, *J. Chem. Thermodyn.* 16, 331–334.

Stödeman, M., and Wadsö, I. (1995), Scope of microcalorimetry in the area of macrocyclic chemistry, *Pure Appl. Chem.* 67, 1059–1068.

Sunner, S., and Wadsö, I. (1959), On the design and efficiency of isothermal reaction calorimeters, *Acta Chem. Scand.* 13, 97–108.

Sunner, S., and Wadsö, I. (1966), A precision calorimetric system, *Science Tools* 13, 1–6.

Suurkuusk, J., and Wadsö, I. (1974), Design and testing of an improved precise drop calorimeter for the measurement of the heat capacity of small samples, *J. Chem. Thermodyn.* 6, 667–679.

Suurkuusk, J., and Wadsö, I. (1982), A multichannel microcalorimetry system, *Chem. Scr.* 20, 155–163.

Teixeira, C., and Wadsö, I. (1990), A microcalorimetric system for photochemical processes in solution, *J. Chem. Thermodyn.* 22, 703–713.

Wadsö, I. (1962), Enthalpy changes accompanying the hydrolysis of some *O*-, *S*-, and *N*-acetyl compounds, *Sv. Kem. Tidskr.* 74, 121–131.

Wadsö, I. (1966), A heat of vaporization calorimeter for work at 25°C and for small amounts of substances, *Acta Chem. Scand.* 20, 536–543.

Wadsö, I. (1968), Design and testing of a micro reaction calorimeter, *Acta Chem. Scand.* 22, 927–937.

Wadsö, I. (1974), A microcalorimeter for biological analysis, *Science Tools* 21, 18–21.

Wadsö, I. (1980), Some problems in calorimetric measurements on cellular systems. In A. E. Beezer (Ed.), *Biological Microcalorimetry*, Academic Press: London.

Wadsö, I. (1986), Recommendations for the presentation of thermodynamic and related data in biology, *Pure Appl. Chem.* 10, 1405–1410.

Wadsö, I. (1993), On the accuracy of results from microcalorimetric measurements on cellular systems, *Thermochim. Acta* 219, 1–15.

Wadsö, I. (1995a), Isothermal microcalorimetry for the characterization of interactions between drugs and biological materials, *Thermochim. Acta* 267, 45–59.

Wadsö, I. (1995b), Microcalorimetric techniques for characterization of living cellular systems. Will there be any important practical applications? *Thermochim. Acta* 269–270, 337–350.

Wadsö, I. (2002), Isothermal microcalorimetry in applied biology, *Thermochim. Acta* 394, 305–311.

Wadsö, I. (2009), Characterization of microbial activity in soil by use of isothermal microcalorimetry, *J. Therm. Anal. Cal.* 95, 843–850.

Wadsö, I., and Goldberg, R. N. (2001), Standards in isothermal microcalorimetry (IUPAC Technical Report), *Pure Appl. Chem.* 73, 1625–1639.

Wadsö, I., and Hallén, D. (2003), A new multichannel isothermal microcalorimeter for uses on living cellular systems, in D. Singer (Ed.), *The International Conference for Biological Calorimetry. (ISBC) XIII*, Abstract Book, p. 50, Veitshöchheim, Germany.

Wadsö, I., Hallén, D., Jansson, M., Suurkuusk, J., Wenzler, T., and Braissant, O., An isothermal multi-channel microcalorimeter, in a well-plate format, for monitoring the activity of living cells, tissues and small animals, *manuscript to be submitted*.

Wadsö, I., and Wadsö, L. (1996), A new method for the determination of vapour sorption isotherms using a twin double microcalorimeter, *Thermochim. Acta* 271, 179–187.

Wadsö, I., and Wadsö, L. (1997), A second generation twin double microcalorimeter for measurements of sorption isotherms, heats of sorption and sorption kinetics, *J. Therm. Anal. Calorim.* 49, 1045–1052.

Wadsö, I., and Wadsö, L. (2005), Systematic errors in isothermal micro- and nanocalorimetry, *J. Therm. Anal. Calorim.* 82, 553–558.

Wadsö, L. (2005), Applications of an eight-channel isothermal conduction calorimeter for cement hydration studies, *Cement International* 5, 94–101.

Wolf, A., Hartmann, T., Bertolini, M., Schemberg, J., Grodrian, A., Lemke, K., Förster, T., et al. (2015), Toward high-throughput chip calorimetry by use of segmented-flow technology, *Thermochim. Acta* 603, 172–183.

Yamaguchi, T., Wakizuka, T., and Takahashi, K. (1996), Application of calorimetry to investigate viability of crops seeds, *Netsu Sokuktei* 23, 2–4.

CHAPTER 3

A Journey in the Land of Biocalorimetry

A Quest for Knowledge

Lee D. Hansen

CONTENTS

3.1 INTRODUCTION

The journey described in this chapter is an intellectual quest full of adventure and exploration of new ideas during my 50+ years in the sometimes strange, new land of biocalorimetry. Journeying in new lands always requires goodly traveling companions who bring their skills, and I have enjoyed the companionship of many such people. Strange, new lands also always have dragons that must be slain before progress can be made and on this journey we encountered several along the way. Time will tell whether we have actually slain these dragons, or whether, like the phoenix, they will continue to reappear. The standard that went before us on this quest was accuracy. Measurements of heat are more prone to error than almost any other measurement in thermodynamics, requiring special care and attention to detail if accurate data are to be obtained. As a skeptic once told me, "You claim to be able to measure microwatts of heat, but you can't see it or feel it, and can only measure it when it is going from one place to another! I don't believe it." Achieving the necessary accuracy requires standards that mimic the measurements on unknowns. It became apparent during the early 1960s that solution calorimetry was sorely in need of such a standard; literature values for a key reaction, strong acid–strong base, disagreed. Through the auspices of the U.S. Calorimetry Conference, an accurate value was established in 1968 (Wagman et al. 1968) and became the first accepted standard reaction for solution calorimetry.

There are only three ways to measure heat: (1) by measuring a temperature change and multiplying by the heat capacity; (2) by measuring the temperature difference across a conductive path between the calorimetric vessel and a heat sink, which relies on Newton's law of cooling; and (3) by measuring the heat required to maintain a calorimetric vessel at the same temperature as a defined reference vessel (Hansen and Eatough 1983; Hansen et al. 1985). All three methods have found use in both isothermal and temperature-scanning calorimetry, and each method has its own advantages and disadvantages. Selecting the optimum calorimeter for a given measurement requires understanding of the issues and careful thought about vessel size, isothermal or temperature scanning, time constant, detection limit, necessary ancillary measurements, and sample physical properties (Hansen 1996; Hansen and Russell 2006).

Biocalorimetry has pursued two paths: one focuses on measurements of metabolic heat rates as a measure of the functions of organisms, while the other is focused on molecular biology and has two branches, one dealing with the interactions between biomolecules and the other dealing with the physical properties of biomolecules. Measuring the effects of temperature and pressure on the thermodynamics of these processes is key to understanding the chemistry, and calorimetry plays a central role in obtaining thermodynamics for these processes. The extensive work done on the thermodynamics of metal–ligand binding and of small molecules and ions in solution, which began early in the twentieth century, still provides many of the principles that underlie today's work on biomacromolecules (e.g., Hansen et al. 1966, 1971a, 1971b; Hansen and Hepler 1972; Lewis and Hansen 1973; Hansen and Lewis 1973; Lewis et al. 1976, 1984; Izatt et al. 1965, 1966, 1977).

3.2 CONTINUOUS-TITRATION, TEMPERATURE-CHANGE SOLUTION CALORIMETRY AND PROPERTIES OF SMALL MOLECULES

The first adventure on this journey was building a continuous-titration, temperature-change, isoperibol (constant surroundings) calorimeter as a graduate student at Brigham Young University (BYU) under the direction of James J. Christensen. (The calorimetric nomenclature used in Hansen [2001] is followed here.) The objective was to build a calorimeter to determine enthalpy changes for metal–ligand binding in solution. Obtaining enthalpy changes for systems with multiple equilibria requires measurements of heat as a function of composition of the system: a lengthy and tedious process with batch calorimetry, since one experiment only produces one data point at one composition. At the time, obtaining the enthalpy changes for three such systems was considered more than sufficient work for a PhD in chemistry.

The key to doing continuous titration in temperature-change calorimetry is the time constant for internal equilibration of the reaction vessel and contents. This time constant must be negligibly short compared with the titrant delivery rate if accurate results are to be obtained. This time constant includes mixing time, sensor response time, and thermal equilibration time of the calorimetric vessel. Mixing time is controlled by stirrer design and speed; a short sensor response time requires the sensor to be small and in good thermal contact with the solution; and thermal equilibration of the vessel requires minimal mass, good thermal conductivity, and a well-defined boundary between the system and its surroundings. A second requirement is a constant temperature environment for both the titrant and the reaction vessel. A simple calculation shows that the environment must be controlled to <0.001°C to minimize error in both the titrant temperature correction and the heat exchange correction on the measurement of temperature change in the vessel. A third requirement was accurate electrical calibration, since no agreed-on chemical standards yet existed. Several good chemical standards now exist (Hansen and Lewis 1971a; Demarse et al. 2011; Wadsö and Goldberg 2001), but there will always be an ongoing need for new chemical standards to meet the changing needs of new applications and new calorimetric instrumentation. The constant-temperature environment problem was solved by using two nested water baths that achieved a stability of ±0.0003°C in the inner bath. The current solution to this problem uses better temperature controllers and, where appropriate, metal blocks instead of water baths. Recognizing that constant-temperature baths are not equilibrium systems is important; whether liquid or metal, constant-temperature baths are steady-state systems. Both a heat source and a means of removing heat are necessary. Temperature gradients can only be avoided by close coupling of the heater and the cooler (Hansen and Hart 1983, 2004).

Designing and building the reaction vessel proved to be a particularly difficult dragon to slay. Commercially available Dewars have a poorly defined boundary, a long time constant, poor thermal conductivity of the construction materials, and much too large heat loss. The use of a metal for the inner vessel was ruled out because of issues with corrosion, potential for surface catalysis or absorption, expense, and difficulties of construction. A properly designed glass Dewar with a very thin inner wall appeared to be the only way forward; however, the glass blower told us, "Nobody could build that!" We were stymied till Jim happened to break the base off a 40 watt GE light bulb; the glass part was the right shape, size, and thickness. We took it to the glass blower to see whether he could help us silver it, put it into a metal can, and evacuate it. He studied it and said, "If GE can do it, I can." Nearly 6 months later, an angry, frustrated glass blower delivered us three blanks to silver. I silvered them and took them back to the glass blower to bake and evacuate. During the process, two were destroyed, leaving us with only one working Dewar. In later years, producing these Dewars became routine, but required a very skilled glass blower. These Dewars have an internal equilibration time of about 0.1 s and a heat exchange coefficient $k = 1 \times 10^{-3}\,min^{-1}$ (Christensen et al. 1965).

Over the next decade that it was in use, the original calorimeter (Christensen et al. 1965) produced a large body of data on solution reactions (e.g., Christensen et al. 1962, 1964, 1967a), particularly on acid–base reactions (e.g., Izatt et al. 1965, 1966; Hansen et al. 1966, 1971a, 1971b; Christensen et al. 1967b). An assembly-line process was established to determine pK and ΔH values in triplicate for three systems per day instead of three systems per graduate student lifetime. Reed M. Izatt was my mentor when the time came to interpret this large body of data in terms of the fundamental principles of solution chemistry. The interest in acid–base and metal–ligand reactions led to the publication of compilations of all the data on enthalpy changes that were then available for these systems (Christensen et al. 1976, Christensen and Izatt 1970). In 1966, Tronac began producing a commercial version of the prototype calorimeter, which continued to be available through Tronac and its successor companies, Hart Scientific and Calorimetry Sciences Corp. (CSC), till it was discontinued in 2007, when TA Instruments bought the company.

The large body of data on acid–base systems enabled us to establish a rule that the identity of a reactant can often be established from knowledge of the ΔH value and information on the composition of the reactants (e.g., Hansen and Lewis 1971b; Hansen and Eatough 1986; Matheson et al. 1989). This rule is a more general statement of Thornton's rule for oxidation of organic substances by O_2 (Thornton 1917) and has been used to determine the microstates of molecules (e.g., VanderJagt et al. 1972), the identities of air pollutants (Hansen et al. 1976, 1977; Eatough et al. 1977), and recently, the identities of alternative oxidants in soil (personal communication, Nieves Barros, University of Santiago, Spain).

3.3 CALORIMETRIC DETERMINATION OF EQUILIBRIUM CONSTANTS

The rapidity of data collection with continuous titration also allowed us to develop the method for calorimetric determination of equilibrium constants for reactions in solution. This method originated in the 1940s (Benzinger 1956; Sturtevant 1962) but only became practical with the advent of a calorimetric method for rapid collection of many data points over the entire composition range (Hansen et al. 1965, 1974; Christensen et al. 1966). By the early 1980s, data analysis was fully automated at BYU. Calorimetric data were collected as electronic files and sent directly to a mainframe computer, where the analyst could enter a model for the system in the form of balanced chemical reactions and process the data to obtain equilibrium constants and enthalpy changes for the system. Analyses of systems with multiple equilibria and competing reactions were routinely performed, and the limitations of the method were reasonably well delineated (Eatough et al. 1982b, 1985; Hansen et al. 1985). However, work on protein–ligand binding was constrained by the detection limit and the relatively large amounts of protein that were required, even in 3 mL Dewars (e.g., Hansen et al. 1975, 1980; Rehfeld et al. 1975; Eatough et al. 1982a; Sukow et al. 1980). The advent of heat conduction and power compensation calorimeters with 1 mL or smaller reaction vessels and nanowatt detection limits (Hansen et al. 1998b; Russell et al. 2006; Russell and Hansen 2006) overcame these instrumentation limits in the 1990s, but it is notable that current software for data analysis has not yet reached the ease of use and sophistication that was available in the early 1980s. Indeed, the method has been labeled with a new acronym (isothermal titration calorimetry [ITC]) and treated as though nothing had existed before 1990, and as a consequence, the old dragons of calibration issues (Demarse et al. 2011) and incorrect models have arisen anew (Hansen et al. 2011a).

3.4 DETERMINATION OF REACTION KINETICS BY HEAT CONDUCTION CALORIMETRY

Our first serious foray into kinetics came as a consequence of a problem with heart pacemakers. The electronics had improved to the point that the manufacturers were willing to guarantee the electrical circuitry for 10 years, but the batteries used to power pacemakers only had enough capacity for 2 years. As many as five new battery designs were competing for this application, but the only obvious way to test the capability was to power up some pacemakers and put them on the shelf for 10 years: much too long to wait to implement what was clearly a better technology. I have no idea where the idea originated, but someone thought that measuring the rate of open-circuit energy loss in a calorimeter could be used to qualify batteries. However, a simple calculation shows that a 1.0 V, 1.5 Ah battery contains 5,400 J of energy, and if the parasitic loss is 1 µW, the battery will lose 3,154 J or 58% over 10 years. If the parasitic loss is 0.1 µW, the battery will only lose 6% of the energy to parasitic losses and last close to 10 years. Thus, open-circuit heat rate measurements need to be made to ±0.1 µW to qualify pacemaker batteries for a 10 year lifetime.

The requirement to achieve a detection limit of 0.1 µW was a very small but very difficult dragon. Temperature-change calorimetry by simply attaching a temperature sensor to a battery and placing it in a well-insulated vessel such as a Dewar is not a viable method, because limiting heat exchange with the surroundings to this level requires a vacuum around the battery to avoid convection currents. Also, heat exchange through the necessary leads on the sensor could easily exceed 0.1 µW even with the best temperature control. The best available power-compensation calorimeter at the time was a Tronac system (Jensen et al. 1976), but the detection limit was 100 µW. Based on a report that Ingemar Wadsö had achieved a detection limit of 10 µW with a heat conduction calorimeter, Roger Hart visited the lab in Lund, Sweden to examine the calorimeter. On his return to Utah, Roger showed me how we could achieve 0.1 µW with Ingemar's design and upgraded electrical circuitry. Five days later, starting from a block of aluminum, we had a calorimeter operating with 0.1 µW baseline reproducibility and began making measurements on pacemaker batteries (Hansen and Hart 1978). But we discovered we had two more dragons to slay: how to get the batteries in and out and how to avoid heat effects from contaminants and other materials. We found that Scotch brand Magic Transparent tape produced <0.1 µW even when the calorimeter vessel was filled with it, and used it to attach a 60 denier cotton sewing thread to the battery so we could raise and lower it into the calorimeter. However, we soon learned that 36 cm of the thread conducted 1 µW/°C, and since the room temperature cycled approximately 1°C between day and night, we had to make a baseline correction depending on the time of day when we made the measurement. When Tronac later built a commercial version of the calorimeter, a detachable mechanism was included for placing and removing the batteries from the calorimeter.

The calorimeter we built for heart pacemaker batteries was a forerunner of a series of heat conduction calorimeters built first by Tronac and later by Hart Scientific and CSC. These calorimeters were used to qualify lithium–SOCl$_2$ batteries for space flight, where they powered the astronauts' space suits (the square block on the back of the neck) and the first Mars Rover (Hansen and Franks 1987). Searching for life on Mars has a direct connection to biocalorimetry!

Measurements of kinetics of decomposition by heat conduction calorimetry were critical in bringing the first drug used to block cholesterol synthesis into the marketplace. Bob Bergstrom invited me to visit Merck Pharmaceuticals in West Point, Pennsylvania, to discuss stability problems with a new drug. Drug shelf life is routinely determined by placing lots of the drug on a shelf in an oven, typically at 60°C, and periodically sampling for chemical analysis. In this accelerated test, the drug showed no chemical degradation, but it lost activity in biological tests. Further, some lots were stable, while some were very unstable, and lot degradation appeared to be a stochastic process, not a monotonic decline in properties. I returned to Utah with two lots, one stable and one not, but I did not know which was which. In the calorimeter at 25°C in air, one lot had a heat rate of a few microwatts, while the other had none, and in nitrogen, neither had a measurable heat rate. At 37°C in air, the difference became even greater. We assumed that the decomposition was due to oxidation by oxygen, but packaging the drug in an inert atmosphere stabilized some lots, but not others. We concluded that the source of the problem was in the manufacturing process and came up with a plan to use calorimetry to rapidly adjust the process to produce only stable material. The factory in Puerto Rico would send me a sample by express mail as soon as it was produced; I would put it in a calorimeter and 24 h later, report the heat rate to Merck. Merck would adjust the process and send me another sample. This procedure continued for several weeks, till finally, the samples all began to show no measurable heat rate. I was never privy to what caused the problem, but the chemistry was consistent with the presence of peroxide in the unstable lots, likely from organic solvents used in the manufacture. Further studies led to a now widely used general calorimetric method for preliminary predictions of the shelf life of drugs (Hansen et al. 1988, 1990; Hansen 1996). These methods are also widely used for determination of the stability of explosives and many other products (e.g., Fontana et al. 1993). To determine the stability of the polyvinyl chloride plastic in automobile bumpers, Tom Hofelich at Dow Chemical built a heat conduction calorimeter with a 1 m^3 chamber and a detection limit of 200 μW (Hansen 1996).

Tian and Calvet had very early (Calvet and Prat 1963) worked out many of the issues of using heat conduction calorimeters for kinetic measurements. Their calorimeters were the forerunners of many of the calorimeters used today for determination of heat rates in chemistry and biocalorimetry. Their designs continue to be used in calorimeters built by Setaram (Lyon, France). In the design of heat conduction calorimeters, the time constant for transfer of heat between the calorimeter and its surroundings is critical. This time constant is defined in terms of the heat transfer coefficient in Newton's law of cooling

$$\frac{-\mathrm{d}T}{\mathrm{d}t} = k\Delta T \tag{3.1}$$

where:
  k is the heat transfer coefficient between system and surroundings
  ΔT is the temperature difference between the calorimeter vessel and the reference block

The solution to Equation 3.1 is

$$\Delta T(t) = \Delta T_0 e^{-kt} = \Delta T_0 e^{-t/\tau} \tag{3.2}$$

where τ is the time constant and thus,

$$\tau = \frac{1}{k} \tag{3.3}$$

Both the sensitivity (output signal/watt input) and the detection limit (watts) of a heat conduction calorimeter can be improved by decreasing k, but only at the expense of increasing τ.

For analysis of kinetics, the Tian equation can be used to correct for the effect of calorimeter response time,

$$\frac{dQ}{dt_{corr}} = k\left[\frac{\Delta T + \tau d\Delta T}{dt + \ldots}\right] \quad \text{or} \quad \frac{dQ}{dt_{corr}} = \frac{dQ}{dt_{obs}} + \frac{\tau d^2Q}{dt^2 + \ldots} \tag{3.4}$$

where dQ/dt_{corr} is the corrected heat rate. However, note that τ depends on the rate of heat transport through the entire vessel and sample, not just through the sensor, and therefore is not independent of the sample. Accurate correction for the calorimeter response time requires simultaneous determination of the applicable value of τ and kinetic constants (Hansen et al. 2011b; Transtrum et al. 2015).

With the exception of the effect on τ, sample size has little effect on the absolute detection limit of heat conduction calorimeters expressed as the minimum detectable heat rate. However, detection limit can be defined in different ways: as the minimum detectable heat rate or as the minimum detectable heat rate relative to the mass or volume of sample. The size of the calorimetric vessel thus affects the relative detection limit; the larger the sample, the lower the relative detection limit (Hansen 1996).

A major problem in heat conduction calorimetry is the incorrect assumption that all the heat generated in the calorimetric vessel is conducted to the surroundings through the sensor. As a consequence of this assumption, it is also assumed that the location of the heat source does not affect the calibration constant (i.e., watts/volts). However, because a significant fraction of the heat generated in the reaction vessel is conducted away from the reaction vessel through electrical and other connections between the reaction vessel and the surroundings, the location of the heat source in relation to these other heat paths affects the output signal. Further, electrical calibration, which is typically done with a heater on the ampoule holder, which is in good thermal contact with the other heat paths, tends to give calibration constants that are too high. Chemical calibration with a reaction with a well-known heat effect that mimics measurements on unknowns must be used for accurate calibration. And finally, since nearly everything, biotic and abiotic, generates heat in the microwatt range, and it is impractical to achieve complete isolation from the surroundings, differential measurements between a sample vessel and a matched reference vessel are the norm for heat conduction calorimetry.

3.5 TEMPERATURE-SCANNING CALORIMETRY

Because of growing interest in DNA and protein structure in solution, a multicell, differential, temperature-scanning, heat conduction calorimeter (MC-DSC) was built in conjunction with Hart Scientific in 1982 with 1 mL removable ampoules made of Hastelloy (Hansen et al. 1998). These vessels are large compared with the differential scanning calorimeters (DSCs) used for material characterization (typically 10–100 μL), but a relatively large volume of solution was needed to obtain the necessary sensitivity. Thermal equilibration of such a large mass of material also dictated relatively slow scan rates: 2°C/min maximum compared with 2°C/s in low-volume DSCs. The dragons to be slain proved to be materials for gaskets and cements that did not have phase transitions between −40°C and 150°C. After much searching, a Viton rubber was found that was suitable for a gasket to seal the ampoules. Cementing the ampoule holders to the Seebeck sensors was more difficult: either the cements had phase transitions even after high-temperature curing or the thermal expansion coefficient was so different from that of the sensor that the ampoule holder broke loose from the sensor after only a few temperature cycles. We finally found an iron oxide–filled silicone rubber cement used to seal head gaskets on engines that worked and could be obtained from auto parts stores.

The MC-DSC was used in studies of biopolymers for a time (e.g., Bell et al. 1995; Huang et al. 1997; Borges et al. 2003; Vest et al. 2004), but after a few years, was displaced by power compensation calorimeters that required much less material because of smaller volumes and better sensitivity. However, the MC-DSC proved to be close to an optimum calorimeter for metabolic studies on plant tissues and small animals. With both isothermal and temperature-scanning capability in the microwatt range, it is perhaps the most useful general-purpose calorimeter yet designed and has been used for many applications (e.g., Criddle et al. 1993; Fontana et al. 1993; Pyne et al. 1996; Hansen et al. 1996a, 1996b, 1996c; Larsen et al. 1997; Hansen 2000; Hein et al. 2003).

3.6 METABOLIC CALORIMETRY

My full-time entry into metabolic calorimetry began with a call from Richard Criddle at UC Davis, California. He wanted to know whether we could use calorimetry to characterize the growth properties of cell cultures of plants. Obtaining cell cultures of plants is quick and easy, and thousands of mutants can be generated in days by exposing the cultures to mutagens. The problem from there is two-fold: first, it is very difficult, time consuming, and more art than science to regenerate a plant from tissue culture; second, determining the properties of the mutant plant, that is, response to temperature, water, salt, nutrients, and so on, by traditional methods requires many years. By the end of the call, I had made a list of more than 20 reasons why calorimetric measurements would never work for this purpose, and only two reasons why we should try it anyway; I had an MC-DSC available, and Richard had access to suspended cell cultures of carrot and tomato. We decided to try it despite the doubts, and the 20+ dragons on my list proved to be mirages. The real dragons appeared later, when our results began to challenge decades-old and widely accepted paradigms in biology.

We spent 25+ years working to find answers to questions in plant physiology and ecology, but never, during that entire time, did we receive any significant funding. The dragon that always blocked funding was a tautology: photosynthesis fixes carbon, and carbon equals growth; therefore, improving photosynthesis will increase growth. And as a corollary dragon, respiration burns fixed carbon back to CO_2, so decreasing respiration will increase growth. Photosynthesis good! Respiration bad! So, the money was all bet on increasing photosynthesis rates and decreasing respiration rates. However, photosynthesis only produces sugar, not plant structural tissue. The actual process of growth is

$$Photosynthesis : CO_2 + H_2O + h\nu \rightarrow sugar \rightarrow starch$$

$$Respiration : starch \rightarrow sugar \text{ or } C(H_2O) \qquad (3.5)$$

and

$$C(H_2O) + xO_2 + (N, P, K, \text{ etc.}) \rightarrow \varepsilon C \text{structural biomass} + (1-\varepsilon)CO_2 \qquad (3.6)$$

where ε is substrate carbon conversion efficiency, making it obvious that respiration rate, not photosynthesis rate, determines growth rate. Plants really do respire and grow! Even in the dark!

The MC-DSC was a near-ideal calorimeter for measuring respiratory heat rates on small samples of plant tissue in both scanning and isothermal operation. From measurements of heat rates alone, we learned much about the effects of temperature, salt, and other environmental variables on plant respiration. We found that cultivars and accessions of plant species could be rapidly selected for growth rates and tolerance to environmental stresses, and that respiratory properties of autochthonous plants are correlated with environmental variables (Criddle et al. 1989, 1991a, 1994, 1999; Hansen et al. 1992, 1994b; Breidenbach et al. 1990; Rank et al. 1991; Smith et al. 1992, 2000a; Hemming et al. 2000; Lytle et al. 2000; Qiao et al. 2005; Mukhanov et al. 2012).

Chilling-sensitive species, such as tomato, were once believed to undergo a phase transition in cell membranes that damaged the tissues at temperatures below 15°C (59°F). To test this hypothesis, we measured respiratory heat rate in temperature scans and looked for a simultaneous phase transition and change in metabolic heat rate at 15°C. Although the abrupt decrease in respiration rate was readily apparent, there was no accompanying phase transition. (Hansen and Criddle 1990). We had slain a dragon! The cause of the chilling response is still unknown; so-called chilling-tolerant cultivars of tomato simply have a slower loss of respiratory capacity below 15°C than less tolerant cultivars (Breidenbach et al. 1990; Rank et al. 1991; Hansen et al. 1994b). During these studies, we also discovered a way to preserve fresh cauliflower (Hansen and Criddle 1989) and fresh-cut pineapple fruit. If kept below about 18°C in a sealed container, cauliflower emits a gas-phase inhibitor that completely stops respiration when it reaches a critical concentration. We were never able to identify the inhibitor, but the information was sufficient to develop a method for shipping cauliflower from California to Japan by boat with no refrigeration. Fresh-cut pineapple develops off colors and flavors because of growth of endophytes from spores incorporated in the fruit when the flower closes. We found we could prevent this growth by dipping the cut fruit in 0.001% sodium benzoate solution.

Reminder: The original purpose of this journey was to find a rapid way to determine plant tissue growth rate as a function of environmental variables. But, getting to that destination was blocked by a particularly bad dragon: the accepted respiration model for plant growth adopted by many big names in plant physiology. Many observations of respiration rates taken on plant tissue of varying age show that mass-specific respiration rate increases linearly with mass-specific growth rate and with a nonzero intercept at zero growth rate. The explanation, borrowed from microbiology, partitioned respiration between growth and maintenance. The maintenance respiration rate was assumed to be given by the intercept, and the slope equated to a growth coefficient, that is, the amount of respiration required to grow a unit of mass. Proponents of this model missed the hidden variable in the correlation: plant tissue grows by expanding cells, mostly with inert cell-wall material, not by cell division as microbes do. Cell division only occurs in meristem tissue. Thus, what was being measured in the plant growth–respiration correlation was simply the dilution rate of the active part of the tissue (Hansen et al. 2002). Further, defining and measuring maintenance respiration in plant tissues has proved to be impossible, and division of respiration between growth and maintenance is not a useful model. We did not succeed in slaying this dragon till 2004 (Matheson et al. 2004).

Respiration can be divided between oxidative phosphorylation, in which fixed carbon is oxidized to CO_2 by O_2, and anabolic reactions, producing structural biomass or growth.

$$\text{Oxidative phosphorylation}: C(H_2O) + O_2 + ADP + Pi \rightarrow CO_2 + H_2O + ATP \quad (3.7)$$

$$\text{Anabolic reactions}: C(H_2O) + (N, P, K, \text{ etc.}) \rightarrow yC_{\text{structural biomass}} + zCO_2 \quad (3.8)$$

These reactions constitute a model or map that tells us where to go, but by what conveyance and by what route? Looking at Equations 3.7 and 3.8, the easiest things to measure are the O_2 uptake rate and the CO_2 production rate, but that leaves y and z as unknowns unless there is a way to distinguish the CO_2 from reaction Equation 3.7 from the CO_2 from reaction Equation 3.8. Thornton's rule to the rescue! Equation 3.7 has an enthalpy change of -455 kJ/mole CO_2, and Equation 3.8 has an enthalpy change of approximately 0 kJ/mole CO_2. Thus, Equations 3.7 and 3.8 suggest that if we measure the respiratory heat rate and the CO_2 production rate, we can calculate the growth rate from two simple measurements.

From Equation 3.6, the ratio $\varepsilon/(1-\varepsilon)$ is the variable stoichiometric ratio of carbon going into structural biomass to carbon lost as CO_2:

$$\frac{\varepsilon}{(1-\varepsilon)} = \left(\frac{dC_{\text{structural biomass}}}{dt}\right) \bigg/ \left(\frac{dCO_2}{dt}\right) \quad (3.9)$$

and the respiratory heat rate is

$$dq/dt = (470)(dO_2/dt) \qquad (3.10)$$

where Thornton's constant is fine-tuned to the value 470 kJ/mole CO_2 for oxidation of sugars in solution. But there is a big problem: three unknowns (ε, $d(C_{\text{structural biomass}})/dt$, and $dO_2)/dt$), only two equations, and no useful way to relate $C_{\text{structural biomass}}$ to the measured variables, even if we measure all three rates dCO_2/dt, dO_2/dt, and dq/dt. Hess's law to the rescue! Find some way to divide Equation 3.6 such that unknowns and measured variables are related in a useful way. After exploring many blind alleys, we stumbled into this one:

$$(1-\varepsilon)C(H_2O)+(1-\varepsilon)O_2 \rightarrow (1-\varepsilon)CO_2 \quad \Delta H_{CO2} = \frac{-470\text{ kJ}}{\text{mole }CO_2} \qquad (3.11)$$

$$\varepsilon C(H_2O) \rightarrow \varepsilon C_{\text{structural biomass}} + \varepsilon yO_2 \quad \Delta H_B = \left(\frac{-\gamma_B}{4}\right)\left(\frac{470\text{ kJ}}{\text{mole }C_{\text{structural biomass}}}\right) \qquad (3.12)$$

where γ_B is the oxidation state of carbon in $C_{\text{structural biomass}}$.

The equation for heat rate from Equation 3.6 now becomes

$$\frac{dq}{dt} = 470\left(\frac{dCO_2}{dt}\right)+\left(\frac{-\gamma_B}{4}\right)\left(\frac{-470\text{ kJ}}{\text{mole }C_{\text{structural biomass}}}\right)\left(\frac{dC_{\text{structural biomass}}}{dt}\right) \qquad (3.13)$$

and solving for the growth rate,

$$\frac{dC_{\text{structural biomass}}}{dt} = \frac{\left[470\left(dCO_2/dt - dq/dt\right)\right]}{\left[\left(-\gamma_B/4\right)\left(470\text{ kJ/mole }C_{\text{structural biomass}}\right)\right]} \qquad (3.14)$$

So, measuring respiratory heat and CO_2 rate allows calculation of the growth rate if the oxidation state of the structural biomass is known or can be estimated. As a bonus, ε can be calculated from Equation 3.9, or from combining Equations 3.9 and 3.14:

$$\frac{\varepsilon}{(1-\varepsilon)} = \frac{\left[470-\dfrac{dq}{dt}\bigg/\dfrac{dCO_2}{dt}\right]}{\left[\left(\dfrac{-\gamma_B}{4}\right)\left(\dfrac{470\text{ kJ}}{\text{mole }C_{\text{structural biomass}}}\right)\right]} \qquad (3.15)$$

Since γ_B can be estimated with sufficient accuracy from knowledge of the composition of the structural biomass, Equations 3.14 and 3.15 represent a viable way to determine both plant growth rate and substrate carbon conversion efficiency from relatively rapid and simple measurements of respiratory heat rate and CO_2 production rate. Only one assumption was made in this derivation: the oxidation state of the substrate is zero, that is, carbohydrate, which is generally true for plants. The first paper describing this model went through 65+ major revisions and three journals before it was accepted for publication in 1994 (Hansen et al. 1994a, 2004; Ellingson et al. 2003).

The MC-DSC was nearly ideal for measurements of heat and CO_2 rates in plant tissues. Measure the heat rate with about 100 mg of tissue, add about 50 µL of 0.4 M NaOH solution to the ampoule to measure the CO_2 production rate, remove the vial of NaOH, remeasure the heat rate to verify that the tissue has not changed, change the temperature, and repeat. With three measuring vessels and the ability to change temperature rapidly, a tissue could be measured in triplicate in less than 90 min, and a complete and well-defined temperature response curve could be obtained in about 2 days (Criddle et al. 1990; Hansen et al. 2005). The concentration of NaOH was chosen to match the osmotic pressure of plant tissues and thus avoid transfer of water between the tissue and the NaOH solution, which would produce an extraneous heat effect. The enthalpy change for reaction of 0.4 M NaOH with $CO_2(g)$ is −108.5 kJ/mole CO_2, which limits the precision of the measurement of the CO_2 production rate, and it would be desirable to have a reaction with a larger enthalpy change. $Ba(OH)_2$ and $Ca(OH)_2$ have larger enthalpy changes for reaction with CO_2, but quickly form a film of insoluble $BaCO_3$ or $CaCO_3$ over the solution that prevents further reaction. To date, we have not been able to find a better reagent than NaOH.

While waiting to get the model published, we continued to develop methods for obtaining the respiratory triad of heat rate, CO_2 rate, and O_2 rate (Criddle et al. 1991b; Fontana et al. 1995), an effort that has continued ever since (Criddle and Hansen 1999; Hansen 2000; Wadsö 2015; Wadsö and Hansen 2015; Itoga and Hansen 2009; Neven et al. 2014). Hansen et al. (2004) gives the equations for the relations among these variables. These methods and models enabled further exploration of the growth, respiration, and stress landscape for plants: work that continues today (Smith et al. 2000a, 2001, 2002, 2006; Criddle et al. 1991a, 1996, 1997, 2000; Hansen et al. 1995, 1996a, 1997, 1998, 2005, 2007, 2008, 2009a, 2009b; Anekonda et al. 1994, 1994a, 1996; Hemming et al. 1999; Jones et al. 2000; Lytle et al. 2000; Harris et al. 2001; Stradling et al. 2001, 2002; Thygerson et al. 2002; McCarlie et al. 2003; Macfarlane et al. 2002, 2005, 2009; Ellingson et al. 2003; Matheson et al. 2004; Yu et al. 2008; Summers et al. 2009; Nogales et al. 2013).

Beyond the photosynthesis dragon that ate all the funding, we found four other dragons blocking the way to acceptance of the calorimetric method for determining growth rates and substrate carbon conversion efficiencies: wound healing, CO_2 suppression of respiration rate, representing growth of a whole plant with a small piece of tissue, and communicating with biologists. We showed that wound healing was not an issue for most tissues by demonstrating that making many cuts to a tissue sample did not change the respiration rate, and in those tissues that did show a response to wounding, calorimetry could be used to study the wound response (Smith et al. 2000b; Lytle et al. 2000). Because the calorimetric measurements have to be made in sealed ampoules to avoid evaporation of water, CO_2 builds up to as much as 10% during the measurement of respiratory heat rate, and because measurements of CO_2 made with a commonly used LI-COR Biosciences (Lincoln, NE) infrared instrument apparently showed decreasing rates of CO_2 production with time in a sealed chamber, our claim that respiration rate did not decrease during our calorimetric measurements was disputed. This dragon died when others discovered that the measurement chamber in the LI-COR instrument leaked. We showed that small samples of growing tissue could be used to predict the growth of large trees if the morphology of the tree was taken into account. One dragon still lives and has prevented widespread adoption of the calorimetric methods for studying plants: communication with biologists. Biologists use the state of reduction of carbon instead of chemical oxidation states, and as a consequence, do not understand stoichiometry when other elements are involved; biologists typically use mass units instead of moles—"What is a watt?" was a common question in reviews; and biologists' understanding of heats of reactions and enthalpy changes for reactions is poor to nonexistent. Thornton's rule is a foreign concept to the large majority of biologists. A severe backlash came from a paper outlining the misconceptions about plant respiration and growth we published in 1998 (Hansen et al. 1998). I took comfort in a book, *The Structure of Scientific Revolutions* by Thomas S. Kuhn, and encouragement by the International Society for Biological Calorimetry (ISBC) became a refuge during this period.

Because of the slow acceptance of the methods and models by plant biologists, we tried branching out into other opportunities in biology. One brief excursion into animals led to a discovery that has yet to be explained: increasing total gas pressure or O_2 partial pressure on mussel gill tissue, mouse brain tissue, and humans approximately doubles respiration rates in a sigmoidal response (Cheney et al. 1996). Calorimetric data showed the effect in mussel gill tissue and mouse brain tissue, and the effect on humans is commonly noted by scuba divers. Insects were obvious poikilothermic animals for study, and we found more ready acceptance by entomologists than we had by plant biologists. These models and methods have been found useful for determining the effects of environment on competition among species (Acar et al. 2001, 2004, 2005), for understanding the processes of egg and larval development (Acar et al. 2004; Joyal et al. 2005), and for developing methods for disinfestation of produce (Downes et al. 2003; Neven and Hansen 2010; Neven et al. 2014). And more recently, working with Nieves Barros in Spain, the same methods and models are being applied to understanding soil carbon dynamics (Barros et al. 2008, 2010, 2011).

A sabbatical year in Jean-Pierre Grolier's lab at Université Blaise Pascal in Clermont-Ferrand, France gave me the opportunity to spend time thinking about where to go next with the plant research. A wonderful year, good food, good friends, and because of my limited, bad French, little time spent socializing, just quiet time in a place with a very diverse ecosystem of plants, insects and small animals. The only movie we went to during the entire year was *Microcosmos* (*A Day in the Life of a Meadow*). Richard Criddle came to stay for a couple of months, and we decided to see where the knowledge gained from the calorimetric studies would take us in understanding and modeling global-scale ecology. Two significant papers eventually came from those conversations. The first was "Thermodynamic Law for Adaptation of Plants to Environmental Temperatures" (Criddle et al. 2005), which demonstrated that the ratio of mean temperature to the temperature range during the growing season was a critical variable in understanding the global distribution of plant species. The second was "Biological Calorimetry and the Thermodynamics of the Origination and Evolution of Life" (Hansen et al. 2009a), which was based on calorimetric data showing that the entropy change was zero, the enthalpy change was zero to slightly exothermic, and hence, the Gibbs energy change for conversion of a random collection of molecules into a living structure was zero to slightly negative. The first challenged many dragons in plant ecology, and the second challenged two very old dragons: common interpretations of Schrödinger's book *What is Life?* and the idea that Shannon's information entropy can be equated with Boltzmann's thermodynamic entropy. In response, I continue to develop a more complete law of adaptation of poikilotherms. A mathematical proof that congruency between the performance curve (i.e., a plot of the rate of an activity versus temperature) and the temperature distribution is an optimum state has been achieved, and a model equation for systems controlled by a single enzyme has been developed. But we still lack a biochemical model for indirectly coupled systems such as respiration. A definitive definition of the difference between information and entropy was developed by Jaynes (1957), but I leave it to the reader with more knowledge of statistics and statistical thermodynamics than I possess to pursue that issue.

In the meantime, a standard for determination of equilibrium constants by current small volume, power compensation, isothermal titration calorimeters has been developed (Demarse et al. 2011), and application of nano-DSCs to disease diagnosis continues (Chagovetz et al. 2013). Although I officially retired several years ago, slaying dragons has been challenging and so much fun that my plan is to continue as long as possible. My journey into the land of biocalorimetry has not yet ended!

ACKNOWLEDGMENTS

In exchange for teaching beginning college chemistry and technical writing, Brigham Young University generously supplied a laboratory; support for electronics, secretarial help, and machining; freedom to choose whatever research area I was interested in; and frequent leaves to travel the world. The stalwarts who accompanied me on this particular journey were Richard S. Criddle, a biophysicist at the University of California in Davis, and Bruce N. Smith, a plant physiologist at BYU. Richard retired from UCD in 2000, but continued working with me till Third World eye glass clinics became a nearly full-time occupation. After Bruce retired in 2004, he worked nearly full time as a volunteer for the National Alliance for the Mentally Ill till he died of Alzheimer's in 2014. T. A. Golovko invited Bruce and me to a workshop on plant respiration in Syktyvkar, Russia, where we were introduced to the beauties of the taiga forests of the area (Golovko et al. 1996). Craig MacFarland and Mark Adams were my guides to the fascinating flora of southern Australia. Weili Yu showed me the wonderful diversity of plants in the steppes and mountains of northwest China. Nieves Barros introduced me to the complexities and dynamics of soil carbon biology. And lately, Birgit Arnhold-Schmitt has begun to push the biocalorimetric methods into selection of crop plants for sustainable yields (Nogales et al. 2013; Arnholdt-Schmitt et al. 2015). Thanks to Ed Lewis and Rusty Russell, who are students, colleagues, and friends, and Tony Beezer, Richard Kemp, Ingolf Lamprecht, Tom Hofelich, and Ingemar Wadsö for their company and discussions at many conferences. Also, this work could not have been completed without the hard work of the many students who spent innumerable hours in the laboratory collecting and analyzing data and asking hard questions that made me think in new ways.

REFERENCES

Acar, E.B., Mill, D.D., Smith, B.N., Hansen, L.D., and Booth, G.M. (2004), Calorespirometric determination of the effects of temperature on metabolism of *Harmonia axyridis* (Col:Coccinellidae) from second instars to adults, *Environ. Entomol.* 33, 832–838.

Acar, E.B., Mill, D.D., Smith, B.N., Hansen, L.D., and Booth, G.M. (2005), Comparison of respiration in adult *Harmonia axyridis* Pallas and *Hippodamia convergens* Guerrin-Manaville (Coleoptera:Coccinelidae), *Environ. Entomol.* 34, 241–245.

Acar, E.B., Smith, B.N., Hansen, L.D., and Booth, G.M. (2001), Use of calorespirometry to determine effects of temperature on metabolic efficiency of an insect, *Environ. Entomol.* 30, 811–816.

Anekonda, T.S., Criddle, R.S., Bacca, M., and Hansen, L.D. (1994a), Contrasting adaptation of two *Eucalyptus* subgenera is related to differences in respiratory metabolism, *Funct. Ecol.* 13, 675–682.

Anekonda, T.S., Criddle, R.S., Libby, W.J., Breidenbach, R.W., and Hansen, L.D. (1994b), Respiration rates predict differences in growth of coast redwood, *Plant Cell Environ.* 17, 197–203.

Anekonda, T.S., Hansen, L.D., Bacca, M., and Criddle, R.S. (1996), Selection for biomass production based on respiration parameters in eucalypts: Effects of origin and growth climates on growth rates, *Can. J. For. Res.* 26, 1556–1568.

Arnholdt-Schmitt, B., Hansen, L.D., and Nogales, A. (2015), Calorespirometry, oxygen isotope analysis and functional-marker-assisted selection ('CalOxy-FMAS') for genotype screening: A novel concept and tool kit for predicting stable plant growth performance and functional marker identification. *Brief. Funct. Genomics*, 1–6.

Barros, N., Feijoo, S., and Hansen, L.D. (2011), Calorimetric determination of metabolic heat, CO_2 rates and the calorespirometric ratio of soil basal metabolism, *Geoderma* 160, 542–547.

Barros, N., Feijoo, S., Salgado, J., Ramajo, B., Garcia, J.R., and Hansen, L.D. (2008), The dry limit of microbial life in the Atacama Desert revealed by calorimetric approaches, *Eng. Life. Sci.* 8, 447–486.

Barros, N., Salgado, J., Rodriguez-Añón, J.A., Proupín, J., Villanueva, M., and Hansen, L.D. (2010), Calorimetric approach to metabolic carbon conversion efficiency in soils, *J. Therm. Anal. Calorim.* 99, 771–777.

Bell, J.D., Baker, M.L., Bent, E.D., Ashton, R.W., Hemming, D.J.B., and Hansen, L.D. (1995), Effects of temperature and glycerides on the enhancement of *Agkistrodon piscivorus* phospholipase A_2 activity by lysolecithin and palmitic acid, *Biochemistry* 34, 11551–11560.

Benzinger, T.H. (1956), Equations to obtain, for equilibrium reactions, free-energy, heat, and entropy changes from two calorimetric measurements, *Proc. Natl. Acad. Sci. USA* 42, 109–113.

Borges, J.C., Fischer, H., Craivichi, A.F., Hansen, L.D., and Ramos, C.H.I. (2003), Free human mitochondrial GrpE is a symmetric dimer in solution, *J. Biol. Chem.* 278, 35337–35344.

Breidenbach, R.W., Rank, D.R., Fontana, A.J., Hansen, L.D., and Criddle, R.S. (1990), Calorimetric determination of tissue responses to thermal extremes as a function of time and temperature, *Thermochim. Acta* 172, 179–186.

Calvet, E. and Prat, H. (1963), *Recent Progress in Microcalorimetry*, Pergamon: NY.

Chagovetz, A.A., Quinn, C., Demarse, N., Hansen, L.D., Chagovetz, A.M., and Jensen, R.L. (2013), Differential scanning calorimetry of gliomas: A new tool in brain cancer diagnostics? *Neurosurgery* 73, 289–295.

Cheney, M.A., Hansen, L.D., Breidenbach, R.W., Wilhelmsen, E., and Criddle, R.S. (1996), Pressure effects on metabolism in tissues from mice (*Mus muscalis*) and freshwater mussel (*Elliptio complanata*), *Comp. Biochem. Physiol.* 114B, 69–76.

Christensen, J.J., Hansen, L.D., and Izatt, R.M. (1976), *Handbook of Proton Ionization Heats and Related Thermodynamic Quantities*, John Wiley: New York.

Christensen, J.J., Hansen, L.D., Izatt, R.M., and Partridge, J.A. (1967a), Application of high precision thermometric titration calorimetry to several chemical systems, in *Microcalorimetrie et Thermogenese*, Pub. 156, Centre National de la Recherche Scientifique, Paris, France, 207–221.

Christensen, J.J., and Izatt, R.M. (1970), *Handbook of Metal Ligand Heats and Related Thermodynamic Quantities*, Marcel Dekker: New York.

Christensen, J.J., Izatt, R.M., and Hansen, L.D. (1962), A thermometric titration procedure for determination of consecutive heats of complex ion formation, in *Proceedings of the Seventh International Conference of Coordination Chemistry*, Stockholm, Paper 7 Fl, 344–346.

Christensen, J.J., Izatt, R.M., and Hansen, L.D. (1965), New precision thermometric titration calorimeter, *Rev. Sci. Instrum.* 36, 779–783.

Christensen, J.J., Izatt, R.M., and Hansen, L.D. (1967b), Thermodynamics of proton ionization in dilute aqueous solution. VII. $\Delta H°$ and $\Delta S°$ values for proton ionization from carboxylic acids at 25°C, *J. Am. Chem. Soc.* 89, 213–222.

Christensen, J.J., Izatt, R.M., Hansen, L.D., and Hale, J.D. (1964), Thermodynamics of metal halide coordination. II. $\Delta H°$ and $\Delta S°$ values for stepwise formation of HgX_2 (X = Cl, Br, I) in aqueous solution at 8, 25, and 40°C, *Inorg. Chem.* 3, 130–133.

Christensen, J.J., Izatt, R.M., Hansen, L.D., and Partridge, J.A. (1966), Entropy titration: A calorimetric method for the determination of $\Delta G°$, $\Delta H°$, and $\Delta S°$ from a single thermometric titration, *J. Phys. Chem.* 70, 2003–2010.

Criddle, R.S., Anekonda, T.S., Sachs, R.M., Breidenbach, R.W., and Hansen, L.D. (1996), Selection for biomass production based on respiration parameters in eucalypts: Acclimation of growth and respiration to changing growth temperature, *Can. J. For. Res.* 26, 1569–1576.

Criddle, R.S., Anekonda, T.S., Tong, S., Church, J.N., Ledig, F.T., and Hansen, L.D. (2000), Effects of climate on growth traits of river red gum are determined by respiration parameters, *Aust. J. Plant Physiol.* 27, 435–443.

Criddle, R.S., Breidenbach, R.W., Fontana, A.J., and Hansen, L.D. (1993), Reaction rates as a function of pressure, temperature and concentration by heat conduction differential scanning calorimetry, *Thermochim. Acta* 216, 147–155.

Criddle, R.S., Breidenbach, R.W., and Hansen, L.D. (1991a), Plant calorimetry: How to quantitatively compare apples and oranges, *Thermochim. Acta* 193, 67–90.

Criddle, R.S., Breidenbach, R.W., Rank, D.R., Hopkin, M.S., and Hansen, L.D. (1990), Simultaneous calorimetric and respirometric measurements on plant tissues, *Thermochim. Acta* 172, 213–221.

Criddle, R.S., Fontana, A.J., Rank, D.R., Paige, D., Hansen, L.D., and Breidenbach, R.W. (1991b), Simultaneous measurement of metabolic heat rate, CO_2 production, and O_2 consumption by microcalorimetry, *Anal. Biochem.* 194, 413–417.

Criddle, R.S., and Hansen, L.D. (1999), Calorimetric methods for analysis of plant metabolism, In R.B. Kemp (Ed.), *Handbook of Thermal Analysis and Calorimetry: From Macromolecules to Man*, vol. 4, 711–763, Elsevier Science: Amsterdam.

Criddle, R.S., Hansen, L.D., Breidenbach, R.W., Ward, M.R., and Huffaker, R.C. (1989), Effects of NaCl on metabolic heat evolution rates by barley roots, *Plant Physiol.* 90, 53–58.

Criddle, R.S., Hansen, L.D., Smith, B.N., Macfarlane, C., Church, J.N., Thygerson, T., Jovanovic, T., and Booth, T. (2005), Thermodynamic law for adaptation of plants to environmental temperatures, *Pure Appl. Chem.* 77, 1425–1444.

Criddle, R.S., Hopkin, M.S., McArthur, E.D., and Hansen, L.D. (1994), Plant distribution and the temperature coefficient of metabolism, *Plant Cell Environ.* 17, 233–243.

Criddle, R.S., Smith, B.N., and Hansen, L.D. (1997), A respiration based description of plant growth rate responses to temperature, *Planta* 201, 441–445.

Demarse, N.A., Quinn, D.F., Eggett, D.L., Russell, D.J., and Hansen, L.D. (2011), Calibration of nanowatt isothermal titration calorimeters with overflow reaction vessels, *Anal. Biochem.* 417, 247–255.

Downes, C.J., Carpenter, A., Hansen, L.D., and Lill, R.E. (2003), Microcalorimetric and mass spectrometric methods for determining the effects of controlled atmospheres on insect metabolism, *Thermochim. Acta* 397, 19–29.

Eatough, D.J., Hansen, L.D., Izatt, R.M., and Mangelson, N.F. (1977), Determination of acidic and basic species in particulates by thermometric titration calorimetry, in *Methods and Standards for Environmental Measurement*. Proceedings of the Eighth IMR Symposium, 20–24 September 1976, Gaithersburg, MD, Special Publication 464, National Bureau of Standards, pp. 643–650.

Eatough, D.J., Izatt, R.M., and Christensen, J.J. (1982b), Titration and flow calorimetry: Instrumentation and data calculation. In N.D. Jesperson (Ed.), *Biochemical and Clinical Applications of Thermometric and Thermal Analysis, Part B*, vol. XII *Thermal Analysis*, in G. Svehla (Ed.), *Wilson and Wilson's Comprehensive Analytical Chemistry*, 3–38, Elsevier: Amsterdam.

Eatough, D.J., Lewis, E.A., and Hansen, L.D. (1985), Solution calorimetry: Determination of ΔH_R and K_{eq} values. In K. Grime (Ed.), *Analytical Solution Calorimetry*, 137–161, John Wiley: New York.

Eatough, D.J., Rehfeld, S.J., Izatt, R.M., and Christensen, J.J. (1982a), Titration and flow calorimetry: Application to proteins and lipids. In N.D. Jesperson (Ed.), *Biochemical and Clinical Applications of Thermometric and Thermal Analysis, Part B*, vol. XII *Thermal Analysis*, in G. Svehla (Ed.), *Wilson and Wilson's Comprehensive Analytical Chemistry*, 112–135, Elsevier: Amsterdam.

Ellingson, D., Olson, A., Matheson, S., Criddle, R.S., Smith, B.N., and Hansen, L.D. (2003), Determination of the enthalpy change for anabolism by four methods, *Thermochim. Acta* 400, 79–85.

Fontana, A.J., Hilt, K.L., Paige, D., Hansen, L.D., and Criddle, R.S. (1995), Calorespirometric analysis of plant tissue metabolism using calorimetry and pressure measurement, *Thermochim. Acta* 258, 1–15.

Fontana, A.J., Howard, L., Criddle, R.S., Hansen, L.D., and Wilhelmsen, E. (1993), Kinetics of deterioration of pineapple concentrate, *J. Food Sci.* 58, 1411–1417.

Golovko, T.A., Semikhatova, O.A., Ivanova, T.I., et al. (1996), Report of work done in Syktyvkar, Komi Republic of Russia, Workshop: Respiration in Plants: Physiological and Environmental Aspects, *Russian J. Plant Physiol.* 43(2), 275–277.

Hansen, C.W., Hansen, L.D., Nicholson, A.D., Chilton, M.C., Thomas, N., Clark, J., and Hansen, J.C. (2011b), Correction for instrument time constant and baseline in determination of chemical kinetics, *Int. J. Chem. Kinetics* 43, 53–61.

Hansen, L.D. (1996), Instrument selection for calorimetric drug stability studies, *Pharm. Technol.* 20, 64–74.

Hansen, L.D. (2000), Calorimetric measurement of the kinetics of slow reactions, *Ind. Eng. Chem. Res.* 39, 3541–3549.

Hansen, L.D. (2001), Toward a standard nomenclature for calorimetry, *Thermochim. Acta* 371, 19–22.

Hansen, L.D., Afzal, M., Breidenbach, R.W., and Criddle, R.S. (1994b), High and low temperature limits to growth of tomato cells, *Planta* 195, 1–9.

Hansen, L.D., Breidenbach, R.W., Smith, B.N., Hansen, J.R., and Criddle, R.S. (1998a), Misconceptions about the relation between plant growth and respiration, *Bot. Acta* 111, 255–260.

Hansen, L.D., Christensen, J.J., and Izatt, R.M. (1965), Entropy titration: A calorimetric method for the determination of $\Delta G°(K)$, $\Delta H°$ and $\Delta S°$, *J. Chem. Soc. Chem. Commun.* 3, 36–37.

Hansen, L.D., Crawford, J.W., Keiser, D.R., and Wood, R.W. (1996b), Calorimetric method for rapid determination of critical water vapor pressure and kinetics of water sorption on hygroscopic compounds, *Int. J. Pharm.* 135, 31–42.

Hansen, L.D., and Criddle, R.S. (1989), Batch-injection attachment for the Hart DSC, *Thermochim. Acta* 154, 81–88.

Hansen, L.D., and Criddle, R.S. (1990), Determination of phase changes and metabolic rates in plant tissues as a function of temperature by heat conduction DSC, *Thermochim. Acta* 160, 173–192.

Hansen, L.D., Criddle, R.S., and Battley, E.H. (2009a), Biological calorimetry and the thermody-namics of the origination and evolution of life, *Pure Appl. Chem.* 81, 1843–1855.

Hansen, L.D., Criddle, R.S., and Smith, B.N. (2005), Calorespirometry in plant biology. In H. Lambers and M. Ribas-Carbo (Eds.), *Plant Respiration: From Cell to Ecosystem*, 17–30, Springer: Dordrecht.

Hansen, L.D., Criddle, R.S., Smith, B.N., and MacFarlane, C. (2002), Growth-maintenance compo-nent models are an inaccurate representation of plant respiration, *Crop Sci.* 42, 659.

Hansen, L.D., and Eatough, D.J. (1983), Comparison of the detection limits of microcalorimeters, *Thermochim. Acta* 70, 257–268.

Hansen, L.D., and Eatough, D.J. (1986), Determination of ammonia, amides, and the degree of hydrolysis of partially hydrolyzed polyacrylamide by thermometric titration and direct injec-tion enthalpimetry, *Thermochim. Acta* 111, 57–65.

Hansen, L.D., Eatough, D.J., and Lewis, E.A. (1990), Shelf-life prediction from induction period calorimetric measurements on materials undergoing autocatalytic decomposition, *Can. J. Chem.* 68, 2111–2114.

Hansen, L.D., Eatough, D.J., Mangelson, N.F., and Izatt, R.M. (1977), Determination of reducing agents and sulfate in airborne particulates by thermometric titration calorimetry, in *Methods and Standards for Environmental Measurement*. Proceedings of the Eighth IMR Symposium, 20–24 September 1976, Gaithersburg, MD, Special Publication 464, National Bureau of Standards, pp. 637–641.

Hansen, L.D., Farnsworth, L.K., Itoga, N.K., Nicholson, A., Summers, H.L., Whitsitt, M.C., and McArthur, E.D. (2008), Two subspecies and a hybrid of big sagebrush: Comparison of respira-tion and growth characteristics, *J. Arid Environ.* 72, 643–651.

Hansen, L.D., Fellingham, G.W., and Russell, D.J. (2011a), Simultaneous determination of equilib-rium constants and enthalpy changes by titration calorimetry: Methods, instruments, and uncertainties, *Anal. Biochem.* 409, 220–229.

Hansen, L.D., and Franks, H. (1987), Kinetics and thermodynamics of chemical reactions in Li/ $SOCl_2$ cells, *J. Electrochem. Soc.* 134(1), 1–7.

Hansen, L.D., and Hart, R.M. (1978), The characterization of internal power losses in pacemaker batteries by calorimetry, *J. Electrochem. Soc.* 125, 842–845.

Hansen, L.D., and Hart, R.M. (1983), Constant temperature baths. In P.J. Elving (Ed.), *Treatise on Analytical Chemistry*, 12, 2nd edition, 135–164, John Wiley: New York.

Hansen, L.D., and Hart, R.M. (2004), The art of calorimetry, *Thermochim. Acta*, 417, 257–273.

Hansen, L.D., and Hepler, L.G. (1972), Field and resonance components of substituent effects in relation to the Hammett equation, *Can. J. Chem.* 50, 1030–1035.

Hansen, L.D., Hopkin, M.S., and Criddle, R. S. (1997), Plant calorimetry: A window to plant physi-ology and ecology, *Thermochim. Acta* 300, 183–197.

Hansen, L.D., Hopkin, M.S., Rank, D.R., Anekonda, T.S., Breidenbach, R.W., and Criddle, R.S. (1994a), The relation between plant growth and respiration: A thermodynamic model, *Planta* 194, 77–85.

Hansen, L.D., Hopkin, M.S., Taylor, D.K., Anekonda, T.S., Rank, D.R., Breidenbach, R.W., and Criddle, R.S. (1995), Plant calorimetry. Part 2. Modeling the differences between apples and oranges, *Thermochim. Acta* 250, 215–232.

Hansen, L.D., Izatt, R.M., and Christensen, J.J. (1974), Application of thermometric titrimetry to analytical chemistry. In J. Jordan (Ed.), *New Developments in Titrimetry*, 1–89, Marcel Dekker: New York.

Hansen, L.D., Jensen, T.E., and Eatough, D.J. (1980), Applications of continuous titration iso-peribol and isothermal calorimetry to biological problems. In A.V. Beezer (Ed.), *Biological Microcalorimetry*, 453–476, Academic: New York.

Hansen, L.D., Jensen, T.E., Mayne, S., Eatough, D.J., Izatt, R.M., and Christensen, J.J. (1975), Heat-loss corrections for small, isoperibol calorimeter reaction vessels, *J. Chem. Thermodyn.* 7, 919–926.

Hansen, L.D., and Lewis, E.A. (1971a), The evaluation of aqueous TRIS solutions as a standard for titration calorimetry, *J. Chem. Thermodyn.* 3, 35–41.

Hansen, L.D., and Lewis, E.A. (1971b), Analysis of binary mixtures by thermometric titration calorimetry, *Anal. Chem.* 73, 1393–1397.

Hansen, L.D., and Lewis, E.A. (1973), Electrostatic effects on the proton ionization reactions of 2-, 3-, and 4-piperidinemonocarboxylic acids, *J. Phys. Chem.* 77, 286–289.

Hansen, L.D., Lewis, E.A., Christensen, J.J., Izatt, R.M., and Wrathall, D.P. (1971a), Electrostatic and resonance-energy effects in proton ionization from pyridinecarboxylic and anilinesulfonic acids, *J. Am. Chem. Soc.* 93, 1099–1101.

Hansen, L.D., Lewis, E.A., and Eatough, D.J. (1985), Instrumentation and data reduction. In K. Grime (Ed.), *Analytical Solution Calorimetry*, 57–95, John Wiley: New York.

Hansen, L.D., Lewis, E.A., Eatough, D.J., Bergstrom, R.G., and DeGraft-Johnson, D. (1988), Kinetics of drug decomposition by heat conduction calorimetry, *Pharm. Res.* 6, 20–27.

Hansen, L.D., Macfarlane, C., McKinnon, N., Smith, B.N., and Criddle, R.S. (2004), Use of calorespirometric ratios, heat per CO_2 and heat per O_2, to quantify metabolic paths and energetic of growing cells, *Thermochim. Acta* 422, 55–61.

Hansen, L.D., Partridge, J.A., Izatt, R.M., and Christensen, J.J. (1966), Thermodynamics of proton dissociation in dilute aqueous solution. IV. pK, $\Delta H°$, $\Delta S°$ values for proton ionization from $[1,12\text{-}B_{12}H_{10}(COOH)_2]^{2-}$, *Inorg. Chem.* 5, 569–573.

Hansen, L.D., Pyne, M.T., and Wood, R.W. (1996c), Water vapor sorption by cephalosporins and penicillins, *Int. J. Pharm.* 137, 1–9.

Hansen, L.D., and Russell, D.J. (2006), Which calorimeter is best? A guide for choosing the best calorimeter for a given task, *Thermochim. Acta* 450, 71–72.

Hansen, L.D., Russell, D.J., and Choma, C.T. (2007), From biochemistry to physiology: The calorimetry connection, *Cell Biochem. Biophys.* 49, 125–140.

Hansen, L.D., Russell, D.J., and Lewis, E.A. (1998b), Calorimetry Sciences Corporation and related calorimeters. In M. E. Brown (Ed.), *Handbook of Thermal Analysis and Calorimetry: Principles and Practice*, 595–608, Elsevier: Amsterdam.

Hansen, L.D., Smith, B.N., and Criddle, R.S. (1998c), Calorimetry of plant metabolism: A means to rapidly increase agriculture biomass production, *Pure Appl. Chem.* 70, 687–694.

Hansen, L.D., Taylor, D.K., Smith, B.N., and Criddle, R.S. (1996a), The relation between plant growth and respiration: Applications to ecology and crop cultivar selection, *Russian J. Plant Physiol.* 43, 691–697.

Hansen, L.D., Thomas, N.R., and Arnholdt-Schmitt, B. (2009b), Temperature responses of substrate carbon conversion efficiencies and growth rates of plant tissues, *Physiol. Plant.* 137, 446–458.

Hansen, L.D., Whiting, L., Eatough, D.J., Jensen, T.E., and Izatt, R.M. (1976), Determination of sulfur(IV) and sulfate in aerosols by thermometric methods, *Anal. Chem.* 48, 634–638.

Hansen, L.D., Woodward, R.A., Breidenbach, R.W., and Criddle, R.S. (1992), Dark metabolic heat rates and integrated growth rates of coast redwood clones are correlated, *Thermochim. Acta* 211, 190–193.

Harris, L.C., Gul, B., Khan, M.A., Hansen, L.D., and Smith, B.N. (2001), Seasonal changes in respiration of halophytes in salt playas in the Great Basin, U.S.A, *Wetl. Ecol. Manag.* 9, 463–468.

Hein, D.K., Ploeger, B.J., Hartup, J.K., Wagstaff, R.S., Palmer, T.M., and Hansen, L.D. (2003), In-office vital tooth bleaching: What do lights add? *Compendium* 24, 340–352.

Hemming, D.J.B., Meyer, S.E., Smith, B.N., and Hansen, L.D. (1999), Respiration characteristics differ among cheatgrass (*Bromus tectorum* L.) populations, *Great Basin Nat.* 59(4), 355–360.

Hemming, D.J.B., Monaco, T.A., Hansen, L.D., and Smith, B.N. (2000), Respiration as measured by scanning calorimetry reflects the temperature dependence of different soybean cultivars, *Thermochim. Acta* 349, 131–134.

Huang, W., Vernon, L.P., Hansen, L.D., and Bell, J.D. (1997), Interactions of thionin from *Pyrularia pubera* with dipalmitoylphosphatidylglycerol large unilamellar vesicles, *Biochemistry* 36, 3860–3866.

Itoga, N.K., and Hansen, L.D. (2009), Gas-phase optrode measurements of oxygen in calorimetric vessels, *Thermochim. Acta* 490, 78–81.

Izatt, R.M., Hansen, L.D., Eatough, D.J., Bradshaw, J.S., and Christensen, J.J. (1977), Cation selectivities shown by cyclic polyethers and their thia derivatives. In B. Pullman and N. Goldblum (Eds.), *Metal-Ligand Interactions in Organic Chemistry and Biochemistry, Part I*, 337–361, D. Reidel: Dordrecht, Holland.

Izatt, R.M., Hansen, L.D., Rytting, J.H., and Christensen, J.J. (1965), Proton ionization from adenosine, *J. Am. Chem. Soc.* 87, 2760–2761.

Izatt, R.M., Rytting, J.H., Hansen, L.D., and Christensen, J.J. (1966), Thermodynamics of proton dissociation in dilute aqueous solution. V. An entropy titration study of adenosine, pentoses, hexoses, and related compounds, *J. Am. Chem. Soc.* 88, 2641–2645.

Jaynes, E.T. (1957), Information theory and statistical mechanics, *Phys. Rev.* 106, 620–630.

Jensen, T.E., Hansen, L.D., Eatough, D.J., Sagers, R.D., Izatt, R.M., and Christensen, J.J. (1976), Influence of cytotoxic agents on thermogenesis in *Streptococcus faecalis*, *Thermochim. Acta* 17, 65–71.

Jones, A.R., Lytle, C.M., Stone, R.L., Hansen, L.D., and Smith, B.N. (2000), Methylcyclopentadienyl manganese tricarbonyl (MMT), plant uptake and effects on metabolism, *Thermochim. Acta* 349, 141–146.

Joyal, J.J., Hansen, L.D., Coons, D.R., Booth, G.M., Smith, B.N., and Mill, D.D. (2005), Calorespirometric determination of the effects of temperature, humidity, low O_2 and high CO_2 on the development of *Musca domestica* pupae, *J. Therm. Anal. Calorim.* 82, 703–709.

Larsen, M.J., Hemming, D.J.B., Bergstrom, R.G., Wood, R.W., and Hansen, L.D. (1997), Water-catalyzed crystallization of amorphous acadesine, *Int. J. Pharm.* 154, 103–107.

Lewis, E.A., Barkley, T.J., Reams, R.R., and Hansen, L.D. (1984), Thermodynamics of proton ionization from poly(vinylammonium) salts, *Macromolecules* 17, 2874–2881.

Lewis, E.A., and Hansen, L.D. (1973), Thermodynamics of binding of guest molecules to α- and β-cyclodextrins, *J. Chem. Soc. Perkin Trans.* II, 2081–2085.

Lewis, E.A., Hansen, L.D., Baca, E.J., and Temer, D.J. (1976), Effects of alkyl chain length on the thermodynamics of proton ionization from arsonic and arsinic acids, *J. Chem. Soc. Perkin Trans.* II, 125–128.

Lytle, C.M., Smith, B.N., Hopkin, M.S., Hansen, L.D., and Criddle, R.S. (2000), Oxygen-dependence of metabolic heat production in the appendix tissue of the voodoo lily (*Sauromatum guttatum* Schott), *Thermochim. Acta* 349, 135–140.

Macfarlane, C., Adams, M.A., and Hansen, L.D. (2002), Application of an enthalpy balance model of the relation between growth and respiration to temperature acclimation of *Eucalyptus globulus* seedlings, *Proc. Biol. Sci.* 269, 1499–1507.

Macfarlane, C., Hansen, L.D., Edwards, J., White, D.A., and Adams, M.A. (2005), Growth efficiency increases as relative growth rate increases in shoots and roots of *Eucalyptus globulus* deprived of nitrogen or treated with salt, *Tree Physiol.* 25, 571–582.

Macfarlane, C., Hansen, L.D., Florez-Sarasa, I., and Ribas-Carbo, M. (2009), Plant mitochondria electron partitioning is independent of short-term temperature changes, *Plant Cell Environ.* 32, 585–591.

Matheson, J.D., Hansen, L.D., Eatough, D.J., and Lewis, E.A. (1989), Calorimetric titration analysis of a coal-derived liquid: Determination of phenolic, pyridinic, and carboxylic compounds, *Thermochim. Acta* 154, 145–160.

Matheson, S., Ellingson, D.J., McCarlie, V.W., Smith, B.N., Criddle, R.S., Rodier, L., and Hansen, L.D. (2004), Determination of growth and maintenance coefficients by calorespirometry, *Funct. Plant Biol.* 31, 929–939.

McCarlie, V.W., Hansen, L.D., Smith, B.N., Monsen, S.B., and Ellingson, D.J. (2003), Anabolic rates measured by calorespirometry for eleven subpopulations of *Bromus tectorum* match temperature profiles of local microclimates, *Russian J. Plant Phys.* 50, 205–214.

Mukhanov, V.S., Hansen, L.D., and Kemp, R.B. (2012), Nanocalorimetry of respiration in microorganisms in natural waters, *Thermochim. Acta* 531, 66–69.

Neven, L.G., and Hansen, L.D. (2010), Effects of temperature and controlled atmospheres on codling moth metabolism, *Ann. Entomol. Soc. Am.* 130, 418–423.

Neven, L.G., Lehrman, N., and Hansen, L.D. (2014), Effects of temperature and modified atmospheres on diapausing 5th instar codling moth metabolism, *J. Therm. Biol.* 42, 9–14.

Nogales, A., Muñoz-Sanhueza, L., Hansen, L.D., and Arnholdt-Schmitt, B. (2013), Calorespirometry as a tool for studying temperature response in carrot (*Daucus carota* L.), *Eng. Life Sci.* 13, 541–548.

Pyne, M.T., Strathdee, G., and Hansen, L.D. (1996), Water vapor sorption by potash fertilizers studied by a kinetic method, *Thermochim. Acta* 273, 277–285.

Qiao, Y., Wang, R., Bai, Y., and Hansen, L.D. (2005), Characterizing critical phases of germination in winterfat and malting barley with isothermal calorimetry, *Seed Sci. Res.* 15, 229–238.

Rank, D.R., Breidenbach, R.W., Fontana, A.J., Hansen, L.D., and Criddle, R.S. (1991), Time-temperature responses of tomato cells during high- and low-temperature inactivation, *Planta* 185, 576–582.

Rehfeld, S.J., Eatough, D.J., and Hansen, L.D. (1975), The interaction of albumin and concanavalin A with normal and sickle human erythrocytes, *Biochem. Biophys. Res. Commun.* 66, 586–591.

Russell, D.J., and Hansen, L.D. (2006), Calorimeters for biotechnology, *Thermochim. Acta* 445, 151–159.

Russell, D.J., Thomas, D., and Hansen, L.D. (2006), Batch calorimetry with solids, liquids and gases in less than 1 mL total volume, *Thermochim. Acta* 446, 161–167.

Smith, B.N., Criddle, R.S., and Hansen, L.D. (2000a), Plant growth, respiration and environmental stress, *J. Plant Biol.* 27, 89–97.

Smith, B.N., Hansen, L.D., Breidenbach, R.W., Criddle, R.S., Rank, D.R., Fontana, A.J., and Paige, D. (2000b), Metabolic heat rate and respiratory substrate changes in aging potato slices, *Thermochim. Acta* 349, 121–124.

Smith, B.N., Harris, L.C., Keller, E.A., Gul, B., Khan, M.A., and Hansen, L.D. (2006), Calorespirometric metabolism and growth in response to seasonal changes of temperature and salt. In M.A. Khan and D.J. Weber (Eds.), *Ecophysiology of High Salinity Tolerant Plants*, 115–125, Springer: Netherlands.

Smith, B.N., Harris, L.C., McCarlie, V.W., Stradling, D.L., Thygerson, T., Walker, J.L., Criddle, J., R.S., and Hansen, L.D. (2001), Time, plant growth, respiration, and temperature. In M. Pessarakli (Ed.), *Handbook of Plant and Crop Physiology*, 1–12, Marcel Dekker: New York.

Smith, B.N., Lytle, C.M., and Hansen, L.D. (1992), Isotopic fractionation, respiration, and growth in seedlings of a cold-desert shrub, *Suppl. Bull. Ecol. Soc. Am.* 73(2), 347.

Smith, B.N., Monaco, T.A., Jones, C., Holmes, R.A., Hansen, L.D., McArthur, E.D., and Freeman, D.C. (2002), Stress-induced metabolic differences between populations and subspecies of *Artemisia tridentata* (sagebrush) from a single hillside, *Thermochim. Acta* 394, 205–210.

Stradling, D.A., Thygerson, T., Smith, B.N., Hansen, L.D., Criddle, R.S., and Pendleton, R.L. (2001), Calorimetric studies of cryptogamic crust metabolism in response to temperature, water vapor, and liquid water, in *Shrubland Ecosystem Genetics and Biodiversity. Proceedings*, USDA Forest Service, RMRS-P-21, 280–282.

Stradling, D.A., Thygerson, T., Walker, J.A., Smith, B.N., Hansen, L.D., Criddle, R.S., and Pendleton, R.L. (2002), Cryptogamic crust metabolism in response to temperature, water vapor, and liquid water, *Thermochim. Acta* 394, 219–225.

Sturtevant, J. (1962), The calorimetric evaluation of equilibrium constants in heats of biochemical reactions. In H.A. Skinner (Ed.), *Experimental Thermochemistry*, vol. 2, 439–440, Interscience: New York.

Sukow, W.W., Sandberg, H.E., Lewis, E.A., Eatough, D.J., and Hansen, L.D. (1980), Binding of the triton X series of nonionic surfactants to bovine serum albumin, *Biochemistry* 19, 912–917.

Summers, H.A., Smith, B.N., and Hansen, L.D. (2009), Comparison of respiratory and growth characteristics of two co-occurring shrubs from a cold desert, *Coleogyne ramosissima* (Blackbrush) and *Atriplex confertifolia* (Shadscale), *J. Arid Environ.* 73, 1–6.

Thornton, W.M. (1917), The relation of oxygen to the heat of combustion of organic compounds, *Phil. Mag.* 33, 196–203.

Thygerson, T., Harris, J.M., Smith, B.N., Hansen, L.D., Pendleton, R.L., and Booth, D.T. (2002), Metabolic response to temperature for six populations of winterfat (*Eurotia lanata*), *Thermochim. Acta* 394, 211–217.

Transtrum, M.K., Hansen, L.D., and Quinn, C. (2015), Enzyme kinetics determined by single-injection isothermal titration calorimetry, *Methods* 76, 194–200.

VanderJagt, D.L., Hansen, L.D., Lewis, E.A., and Han, L.-P.B. (1972), Calorimetric determination of the micro ionization constants of glutathione, *J. Chem. Thermodyn.* 4, 621–636.

Vest, R.S., Gonzales, L.J., Permann, S.A., Spencer, E., Hansen, L.D., Judd, A.M., and Bell, J.D. (2004), Divalent cations increase lipid order in erythrocytes and susceptibility to secretory phospholipase A_2, *Biophys. J.* 86, 2251–2260.

Wadsö, L. (2015), A method for time-resolved calorespirometry of terrestrial samples, *Methods* 76, 20–26.

Wadsö, I., and Goldberg, R.N. (2001), Standards in isothermal microcalorimetry, *Pure Appl. Chem.* 73, 1625–1639.

Wadsö, L., and Hansen, L.D. (2015), Calorespirometry of terrestrial organisms and ecosystems, *Methods* 76, 11–19.

Wagman, D.D., Evans, W.H., Parker, V.B., Bailey, S.M., and Schumm, R.H. (1968), Selected values of chemical thermodynamic properties. National Bureau of Standards, Technical Note 270-3, US Government Printing Office, Washington, DC, 1968. (Reference to the BYU work was inadvertently left out of this publication; Hale, J.D., Izatt, R.M., and Christensen, J.J. 1963, *Proc. Chem. Soc. (London)* 240 and *J. Phys. Chem.* 67, 2605–2608.)

Yu, W., Hansen, L.D., Fan, W., Zhao, W., and McArthur, E.D. (2008), Adaptation of growth and respiration of three varieties of caragana to environmental temperature, *Asian J. Plant Sci.* 7, 67–72.

CHAPTER 4

An Odyssey in Calorimetric Development

From Past to Future

Jaak Suurkuusk, Malin Suurkuusk, and Peter Vikegard

CONTENTS

4.1 INTRODUCTION

For many years, it has been necessary or desirable to measure heat associated with changes in materials and organisms to provide answers to fundamental questions about matter and the processes it undergoes. Even 60 years or so before Joule's classical experiment proving the equivalence of mechanical energy and heat, the concept of heat of fusion had been established, which made it possible for Lavoisier and Laplace to construct their ice fusion calorimeter in the last few decades of the eighteenth century. One driving force for this development was the wish to understand the nature of the biological process of respiration just after the old phlogiston theory had been made obsolete. In a series of classical experiments, they proved that respiration is a combustion process: "*La respiration est donc une combustion, à la vérité fort lente,*" as they stated in their famous paper from 1780 (Lavoisier and Laplace 1780).

With the Industrial Revolution, a desire to understand the properties and efficiency of heat engines emerged, which resulted in the formulation of several natural laws by scientists and engineers such as Carnot, Thomson, and Clausius, to mention a few: discoveries that subsequently led to the birth of thermodynamics as a scientific discipline. The main driving force for calorimetric development in those days of the nineteenth century came from the need to know the energy contents of various materials, in particular those that could be used as fuels.

Modern calorimetric technologies employed as fine analytical laboratory tools began to be developed in the first half of the twentieth century with the instruments by Tian and later Calvet (Calvet and Prat 1963). What many consider as conventional or *pan-type* DSC instruments were developed in the early 1960s and modern highly sensitive isothermal calorimeters in the latter half of the 1960s and the early 1970s (Wadsö 1968). An intriguing scientific quest was (and still is) to understand the behavior and thermodynamics of biological macromolecules in aqueous environments. This has triggered a demand for very sensitive and precise heat measurements. For this purpose, a differential drop-C_p calorimeter working as a "single-point" heat capacity calorimeter based on the heat flow principle was developed for measurement of the partial heat capacity of water associated with various model compounds as well as whole protein systems (Suurkuusk and Wadsö 1974). A decade earlier, highly sensitive DSC instruments (Privalov 1963) began to be developed for studies of biological macromolecules in solution; see Privalov and Potehkin (1986) for a review of scanning calorimetry.

In the early 1980s, one of the first calorimetric systems with multisample capacity emerged from academic work (Suurkuusk and Wadsö 1982) and was commercialized under the name Thermal Activity Monitor (TAM), initially by LKB Instruments, later by Thermometric AB, and currently by TA Instruments. Although it was initially intended to be applied for measurements of biological systems, it soon became a multipurpose instrument with a wide range of applications in various disciplines. The technique continued to develop with respect to sensitivity and multisample capacity, and in the first years of the twenty-first century the third-generation TAM was introduced with detection limits in the microwatt to nanowatt range and a measuring capacity of up to 48 samples (Suurkuusk et al. 2015). Alongside the development of the TAM system, various sample handling systems were developed that enable control of the sample environment *in situ*, for example, injection and mixing of liquids, isothermal titration calorimetry (ITC), and changes in gas composition including vapor activity (Bakri 1993; Briggner 1993). A recent trend has been to combine calorimetric measurements with sensors such as fiber-optic spectrophotometers, electrodes, or pressure sensors into hyphenated units to gain more specific information from the measured processes (Johansson and Wadsö 1999).

This chapter will review calorimetric development—the driving forces behind the development, the technical challenges, and where we stand today—and give a hint on what we expect to see in the future.

4.2 THE IMPORTANCE OF HEAT MEASUREMENTS: A SHORT OVERVIEW

Let us briefly reflect on how we make use of the heat we measure with our calorimeters and what types of information we gain from it. The measurement of evolution or consumption of heat associated with changes in materials has found a wide range of uses ever since the very early studies of living species by Lavoisier and Laplace (1780). Even though their studies had the practical purpose of investigating the equivalency between respiration and combustion of organic compounds, they were measuring what we may regard today as catalytic processes far from equilibrium. This falls into the realm of irreversible thermodynamics, a field that started developing a couple of centuries later; see, for example, Prigogine (1961). Since then, calorimetry has found wide practical as well as academic use for studies of life processes, regarding complex biological species as well as microorganisms; see Braissant et al. (2010) and Wadsö (2002) for reviews on the subject.

Heat evolution or consumption under non-steady-state conditions always indicates a change of the specimen: that is, with calorimetry, we measure changes occurring in materials. Although heat evolution is not a necessity for a change to be spontaneous, the overwhelming majority of processes result in heat production or consumption. A very simple process is a change of temperature of a sample without any change of its composition. This forms the basis for calorimetric determinations of the absolute heat capacity (or specific heat) of a material. Naturally, the specific heat of a material is important for several engineering applications, but very sensitive and precise heat capacity measurements have also been employed to explain the very subtle forces that are responsible for the structure of biological macromolecules. Hydrophobicity is an important phenomenon in this respect, and the role of water in conjunction with hydrophobic surfaces has been widely studied through precise heat capacity determinations (see Section 4.4).

Heat production/consumption of a material at isothermal and isobaric conditions in a closed system is due to a change of the composition of a specimen, including changes of phase or molecular specie(s). Depending on the measuring conditions, the heat generated or consumed by a process is equivalent to changes in the enthalpy or internal energy, which are thermodynamic state functions, thus forming the basis of calorimetry as a thermodynamic tool. With the development of highly sensitive calorimeters exhibiting the possibility of mixing or titrating liquid components *in situ*, it is possible to determine Gibbs energy change from stoichiometric complexation phenomena through model calculations. In this sense, the measured heat is used as a parameter that quantifies the extent of the complexation reaction for each increment of the reagent through a series of discrete titrations. The measured heat is related to the change of enthalpy of the process (cf. ITC).

In addition to thermodynamics, heat evolution/consumption rates have often been used as a measure of process rates. This applies especially to calorimeters that measure heat exchange rates directly, that is, instruments working according to the heat flow or power compensation principle. Since rates are measured directly with such calorimeters, they are generally referred to as rate-sensitive techniques. The most direct way of using heat rates is in applications for which the heat in itself is the quantity of interest. This applies to assessments of safe storage of materials that undergo spontaneous exothermic decomposition, which may result in accumulation of heat with rising temperatures as a consequence. Accumulation of heat may in turn result in runaway reactions with ignition in the case of explosives or propellants, or critical pressure buildup due to gas evolution in closed containers. To make safe storage assessments, the heat production rate is a critical parameter when constructing energy balances for storage dimensions.

An indirect way of using heat rates is to assess the rates and mechanisms of chemical reactions by comparing the calorimetric data with the appropriate kinetic model. For the heat rate to be used in this way, it must be in approximate parity with the rate of change of reactants, intermediates, and products. A main advantage of heat rate measurements, at least when monitored by most modern calorimetric instruments, is that they provide real-time data, resulting in strong discrimination between mechanistic differences as well as precise determination of kinetic parameters. The example in Figure 4.1 is adapted from Angberg et al. (1988) and describes the pseudo-first-order reaction of the hydrolysis of aspirin in basic conditions.

In some instances, heat measurement has been found to be very useful as a quantitative analytical measurable. This applies in particular to quantifying different phases in solid materials, such as amorphous regions in predominantly crystalline solids; see Briggner et al. (1994), Hogan and Buckton (2000), and Sebhatu et al. (1994) for details on different approaches.

In summary, further to relating heat exchange to the heat capacity of materials, the heat generated by a sample that undergoes a change can be expressed either as a rate in energy units per time unit or as the integrated quantity. This has been successfully used to gain

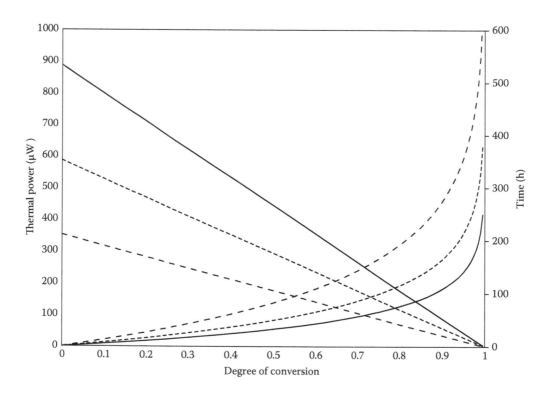

FIGURE 4.1   Kinetic illustration of the hydrolysis of aspirin measured by isothermal microcalorimetry at three different temperatures: 298 K (- - -), 303 K (----), and 308 K (——). The straight lines refer to the heat rate axis.

thermodynamic and kinetic information about processes as well as quantitative analytical information on materials, illustrated by Equation 4.1:

$$P(\xi) = \Delta H \cdot k \cdot f(\xi) \tag{4.1}$$

where:

P is the rate of heat production or consumption at isothermal conditions
ΔH is the change of enthalpy, representing the thermodynamic state function
k is a rate constant
$f(\xi)$ represents a time-dependent function that describes how the rate of a process varies with the extent of the process, ξ

Since the amount of reactants depleted or products formed in a process depends on the extent of the reaction, the heat exchanged in the process can be used as a quantitative analytical measurable:

$$\Delta n \propto \Delta \xi = \int_{t_0}^{t} P(\xi) dt / \Delta H \tag{4.2}$$

In Equation 4.2, Δn represents the change of the amount of specie or phase to be analyzed.

4.3 HOW DO WE MEASURE HEAT? A SHORT OVERVIEW

The term *calorimetry* has a wide definition and essentially includes all technologies used to make assessments of heat evolution or consumption caused by a process. Calorimetry can be conducted either isothermally or during dynamic temperature conditions, as in differential scanning calorimetry (DSC). The fundament of heat measurement is based on temperature measurements and can be divided in three main groups: (1) the heat flow principle, (2) the adiabatic principle, and (3) the power compensation (or power-feedback) principle. Within each of these groups, instruments can be designed in a multitude of different ways only limited by the creativity of the designer.

4.3.1 Heat Flow Principle

The heat flow principle is based on the measurement of the temperature difference created between a thermally active sample and a temperature-controlled environment. Differential temperature is typically measured by thermoelectric modules. A temperature difference across the thermoelectric module will generate a voltage, which in turn is proportional to the heat flow generated by the temperature gradient. For processes faster than the time constant of the instrument, corrections for temporary heat accumulation/depletion need to be done to obtain true process rates (Calvet and Prat 1963; Randzio and Suurkuusk 1980). Calorimeters based on the heat flow principle are normally designed as twin or differential instruments, in which a sample side is mirrored by a reference side containing an inert material to match the thermal properties of the sample, and what is measured is the difference between the two sides. This means that the influence of a temperature disturbance in the environment is more or less cancelled.

4.3.2 Adiabatic Measuring Principle

The adiabatic principle is based on temperature shift measurement. In an adiabatic calorimeter, no heat is allowed to leave or enter the system, which is achieved either by insulation or by using an adiabatic shield. As no heat is exchanged with the surroundings, the temperature will increase or decrease as a result of a reaction, and this temperature shift is measured. In practice, many calorimeters are not completely thermally insulated; hence, some heat leakage is accepted and corrected for in semiadiabatic calorimeters. An adiabatic or semiadiabatic calorimeter needs to be calibrated before each experiment, as the system's heat capacity needs to be known.

4.3.3 Principles for Power Compensation

The principle of the power compensation calorimeter is to keep the temperature of the system, or the temperature difference between sample and reference, constant. This is achieved by the use of one or more heaters, adjusted to compensate for the heat evolved or absorbed by a reaction that occurs in the calorimeter. The benefit of the power compensation or feedback principle is the fast response of the calorimeter to the heat evolved/absorbed by a reaction.

4.4 HEAT CAPACITY INVESTIGATIONS OF BIOCHEMICAL MODEL COMPOUNDS AND BIOLOGICAL STRUCTURES AND THE DEVELOPMENT OF A HIGHLY SENSITIVE SINGLE-POINT HEAT CAPACITY CALORIMETER

The quest to understand the thermodynamics of biological macromolecules in proximity to their natural state has been a main driving force in the development of highly sensitive calorimetric instrumentation over the last decades. Many of the early fundamental studies in this respect focused on the role of water in biological systems and its responsibility for the stabilization of native biological structures.

Frank and Evans (1945) postulated that the solvation of nonpolar molecules in water is accompanied by structuring of water resulting in increased hydrogen bonding in the vicinity of the solvated species, a phenomenon that results in a large decrease in entropy. The abnormally large apparent heat capacity of solvated nonpolar surfaces due to structuring of water is also a feature associated with the solvation process, as observed by Edsall (1935). Although calorimetric solvation studies of nonpolar molecules show favorable enthalpy changes, in some cases, this is not sufficient to counterbalance the decrease in entropy above a critical concentration. Hence, to avoid this decrease in entropy, these systems have a spontaneous tendency to form aggregates to minimize the area exposed to water. The less surface area is exposed, the fewer water molecules have to suffer the loss of conformational entropy, and thus the lower the energetic penalty.

One approach to understanding thermodynamic results on biological macromolecules has been to study smaller compounds in aqueous solution as models for more complex systems. In this respect, a systematic search for relationships between component groups in individual molecules and the same groups covalently bound into larger and more complex structures was undertaken.

From the point of view of calorimetry, the partial molar enthalpy and the partial molar heat capacity have been interesting to study. The partial molar heat capacity at infinite dilution, $C_{p,2}^{\infty}$, has been of particular interest due to its sensitivity to interactions between hydrophobic groups and water. This quantity can be obtained indirectly by measuring the heat of solution of the compound into water, $\Delta_{sol}H$, at two or more different temperatures, thus obtaining the heat capacity change, $\Delta C_{p,2}$, for the dissolution process. This is then combined with the heat capacity of the pure compound, C_p^*, to obtain the partial molar heat capacity of the solute, $C_{p,2}^{\infty}$:

$$C_{p,2}^{\infty} = \Delta C_{p,2} + C_p^* \tag{4.3}$$

A more straightforward approach to obtain partial molar heat capacities of solutions would be to measure the heat capacity directly. For slow dissolution processes and for proteins normally obtained in small quantities, the direct approach has clear advantages. Protein solutions can be considered ideal or exhibit only limited intermolecular interactions at concentrations below 0.3%. For such dilute solutions, the apparent heat capacity of the protein solute falls in the range of 0.1% of the total. In addition, studies of the model compounds generally involve dilute solutions on the order of 0.05 M. For direct heat capacity measurement of such systems, a calorimeter with a precision on the order of 0.01% or better is required.

A solution to this problem was the improvement of a previously described instrument, a single-point heat capacity calorimeter or "drop-C_p" operating according to the heat flow principle (Konicek et al. 1971). The technical principle of the improved calorimeter is the same as that of its predecessor, except that the earlier version was operated in single mode, while the new version was operated in twin or differential mode, resulting in more precise determinations (Suurkuusk and Wadsö 1974). A stainless steel vessel containing the sample is initially equilibrated in a furnace at temperature T_1, situated above the calorimeter. The temperature difference between the furnace and the calorimeter is typically 10 K or less. After a period of thermal equilibration, the vessel is dropped into the receiving calorimeter, which has been set to temperature T_2. The amount of heat, q, thus exchanged between the sample vessel and the temperature-controlled environment of the calorimeter is measured by a set of thermoelectric modules. The calorimeter was calibrated with pure water and verified with reference materials of known specific heat, such as heptane or sapphire.

$$q = C_p \cdot (T_2 - T_1) \tag{4.4}$$

The obtained heat capacity, C_p, is the mean heat capacity for the temperature interval T_1 to T_2. Systematic measurements with samples of known heat capacities showed that the accuracy of the calorimeter is better than 0.1%. Although it is labor consuming compared with temperature-scanning calorimeters, the single-point approach with the drop-C_p calorimeter measures heat capacity at true equilibrium conditions, thus eliminating kinetic effects due to the dynamic temperature changes of the calorimeter.

The studies of heat capacity changes of model compounds and their additivity relations described in Section 4.4 was extended to whole proteins by the use of this new, improved drop-C_p calorimeter (Suurkuusk 1974). The proteins studied were lysozyme, chymotrypsinogen, and ovalbumin in the solid state with different water contents as well as in the dilute liquid state. The measured systems were considered to be constituted by two components: the dry protein and water. Hence, the total measured heat capacity, C_p, was the sum of the two components' partial specific heats multiplied by their respective mass:

$$C_p = C_{p,s,\text{protein}} \cdot m_{\text{protein}} + C_{p,s,\text{water}} \cdot m_{\text{water}} \qquad (4.5)$$

By varying the water content, a relationship between the measured heat capacity and the fractional water content was established at low water fractions (below 0.35), and the partial specific heats of the protein component at zero water content could be obtained. Only small differences between the three proteins were observed, with a mean value of 1.22 J K^{-1}g^{-1} and variation within 2%.

The experimentally obtained values of the dry protein states were compared with the sum of the heat capacities of the component amino acids corrected for the formation of the peptide bonds. This showed that it is possible to calculate the specific heat of proteins from their amino acid sequence with a precision of a few percent. Due to the fact that the specific heats of different amino acids vary only slightly, most proteins are expected to have a specific heat near 1.2 J K^{-1}g^{-1} in the dry state.

In a series of measurements, the partial specific heats of lysozyme, chymotrypsinogen, and ovalbumin at infinite dilution were determined (Suurkuusk 1974). Protein solutions in the dilute region down to 0.6 weight % were measured, and an essentially linear relation was observed between the partial specific heats and the protein concentrations in the measured region. The partial specific heat for the proteins could therefore be extrapolated with a good precision to obtain the quantity at infinite dilution. The results from these measurements are summarized in Table 4.1.

The interesting part of these results is the large and, within error limits, identical change in specific heat of transferring the protein from its solid state to the solvated state. The obtained average $\Delta C^{\circ}_{p,\text{solv}}$ was 0.305 J K^{-1}g^{-1} with estimated error limits of $\pm$0.015 J K^{-1}g^{-1}.

TABLE 4.1　Partial Specific Heats of Water and Protein at Infinite Dilution for Lysozyme, Chymotrypsinogen, and Ovalbumin at 298 K

	$C^{\circ}_{p,\text{water}}$/J K^{-1}g^{-1}	$C^{\circ}_{p,\text{protein}}$/J K^{-1}g^{-1}
Lysozyme	4.1792$\pm$0.0004	1.494$\pm$0.007
Chymotrypsinogen	4.1808$\pm$0.0008	1.529$\pm$0.015
Ovalbumin	4.1778$\pm$0.0004	1.534$\pm$0.014

Source: Suurkuusk, J., *Acta Chem. Scand.* B 28, 409–417, 1974.

The experimentally derived partial specific heat of each protein was analyzed by grouping the macromolecule into a number of constituent parts using simple additivity relations to sum up the total partial specific heat according to Equation 4.6:

$$C^{\circ}_{p,\text{protein}} = C_{p,\text{backbone}} + C^{\circ}_{p,\text{polar}} + \Delta C_{p,\text{ionization}} + (C^{\circ}_{p,\text{non-polar}} + \alpha \cdot \Delta C_{p,\text{solv non-polar}}) \quad (4.6)$$

The contribution from the backbone of the protein was assumed to be unaffected by solvation, while that of the polar groups was estimated from their respective specific heat at infinite dilution as obtained from solvated model compounds compiled in Nichols et al. (1976). The term $\Delta C_{p,\text{ionization}}$ describes the change in heat capacity as polar groups undergo ionization, a process that is generally accompanied by a decrease in heat capacity. This contribution was estimated from the pI values of the respective protein.

The sum $(C^{\circ}_{p,\text{non-polar}} + \alpha \cdot \Delta C_{p,\text{solv non-polar}})$ describes the contribution from the nonpolar groups, where the term $C^{\circ}_{p,\text{non-polar}}$ is equivalent to the heat capacity of the nonpolar fraction of the protein in its pure solid state. The term $\alpha \cdot \Delta C_{p,\text{solv non-polar}}$ quantifies the contribution of the solvated nonpolar groups according to the additivity relations obtained from the model compounds, as described in the first part of this section. The factor α is a parameter that describes the fraction of the nonpolar groups that is solvated in the native protein. This parameter is unknown and was calculated from the difference between the sum of the different contributions and the experimentally obtained specific heat. The values thus obtained were 0.27 for lysozyme and 0.20 for chymotrypsinogen, values of α showing good agreement with estimates by Tanford (1970).

The drop-C_p calorimeter was later employed in a study of the heat capacity of the different levels of bound and free water associated with biological plant tissue (Hedlund and Johansson 2000). Recently, the instrument was reassembled and modernized at Universidade do Porto, where it was tested with a number of test substances (Santos et al. 2011). In the same publication, the heat capacity of a number of polyphenyls was investigated, and it was shown that high-precision heat capacity measurements with this instrument could lead to distinguishing between *ortho-*, *para-*, and *meta*-phenyl isomers.

A highly sensitive temperature-scanning calorimeter (DSC) to measure processes at near-equilibrium conditions was required in a work aimed at understanding the stability of small single-lamellar dipalmitoylphosphatidylcholine (Suurkuusk et al. 1976). It was shown that the thermal characteristics of the gel-to-liquid crystalline transitions in small unilamellar liposomes differed markedly from the Bangham-type multilamellar liposomes regarding the transitions and time-dependent stability. Furthermore, the kinetics of fusion into multilamellar structures of the unilamellar vesicles below the transition temperature was investigated and resolved with a proposed thermodynamic and kinetic model. The instrument, which is briefly described in the same article, is a differential heat flow calorimeter, the design of which was initiated by Ross and Goldberg (1974). The temperature range for the instrument is 273–358 K, with scanning rates from 3 to 50 K h^{-1}. The precision in terms of baseline noise was shown to be better than ± 100 μJ K^{-1}.

In a later publication, the same calorimeter was employed for studies of the thermotropic behavior of sphingomyelins using a scanning rate of 15 K h^{-1} (Barenholz et al. 1976). In this work, the complex phase transitions exhibited by natural sphingomyelins were revealed and compared with those of synthetic sphingomyelins. An important aspect of the described thermodynamic investigations using the DSC instrument was the ability to scan the temperature very slowly so as to keep the measured systems close to equilibrium, that is, in quasi-equilibrium conditions. This creates a high demand on the calorimeter temperature control system and the detection limit of the measured heat. That the system was indeed close to equilibrium in the case of the sphingomyelin studies was confirmed by measurements with a drop heat capacity calorimeter similar to the design of Konicek et al. (1971).

With the development of the new multichannel calorimeter system described in Section 4.5 a new temperature control system was developed that enabled temperature control on the order of 10 μK, making it possible to scan with temperature resolution that is yet to be matched.

4.5 EVOLUTION OF A MULTICHANNEL CALORIMETER SYSTEM

The main initial driving force for the development of a multichannel calorimeter system, initially named the Bio Activity Monitor (BAM) (Suurkuusk and Wadsö 1982), was a problem related to septicemia, or blood poisoning, as a clinical condition in humans. By measuring the heat production rate of blood samples, questions such as the presence of bacteria, bacterial strain, and antibiotic susceptibility could be addressed (Kemp 1999). The idea of using calorimetry to measure the life processes of microorganisms dates back to the nineteenth century with the macrocalorimetric studies of fermentation by Dubrunfaut (1856), who was the first researcher to estimate the heat of alcoholic fermentation. The idea of using calorimetry to measure the metabolism and growth of microorganisms gained new interest in some academic laboratories in the early 1970s, resulting from an attempt to introduce a novel method for identification of bacterial infection in the bloodstream (Monk and Wadsö 1975). The advantages that seemed attractive with calorimetry were the speed of analysis and the possibility of using the calorimetric heat flow-time profiles as a fingerprint to identify specific bacterial strains and for quick assessment of antibiotic susceptibility (Beezer et al. 1980; Hoffner et al. 1990; Mårdh et al. 1976). However, at that time, calorimeters were instruments requiring relatively large sample volumes and were more or less dedicated to calorimetrists, specialists in the field of performing accurate and precise heat measurements. A challenge from the technical point of view was to develop a calorimeter system that could be managed by a laboratory technician as well as by an expert in the field. In addition, an instrument that is sensitive enough to handle small sample amounts and has the ability to run several samples in parallel is very important from the point of view of real-time comparative measurements as well as increased measuring capacity.

The idea for the multichannel calorimetric system was based on a modular system: a highly regulated thermostat that could accommodate multiple heat flow calorimeters that could be operated in parallel, that is, it should be possible to initiate experiments independently with minimal cross-talk between channels. The sample should be contained in removable reaction vessels that could be replaced and sterilized. These requirements resulted in a four-channel calorimetric system consisting up to four independent heat flow calorimeters immersed in a highly regulated thermostat; see Suurkuusk and Wadsö (1982) for a detailed technical description. A calorimetric system based on the original prototype was soon commercialized by the company LKB Instruments under the name BAM. It was soon realized that the potential of the technique was not limited to living systems, and its name was changed to the more generic Thermal Activity Monitor (TAM).

Besides a range of biological applications, TAM started to be used for thermodynamic and kinetic investigations of physical or chemical processes in academic work. The technique also spread to industry and applied fields such as pharmacy and applied chemistry. Examples of practical applications in industry included chemical or physical stability problems related to product shelf life or safe storage of bulk chemicals.

Soon after the introduction of TAM to the scientific and industrial community, various sample handling systems, from simple closed reaction vessels to more complex devices, started to emerge. Examples include open vessels to which liquid reactants could be added and flow-through vessels for perfusion of gases or liquids. With such devices, it became possible to initiate processes *in situ*. Although technologies for setting up such experiments had existed previously, for example, with the LKB 2107 (Wadsö 1968), the new development greatly extended the possibilities. Nordmark and colleagues (Görman Nordmark et al. 1984) introduced a reaction vessel with an automated system to generate a series of liquid injections, a technique widely used for interaction studies of dilution, dissolution, and complexation phenomena in the liquid state or to initiate a chemical or biological response on injection of a reactant. Another development, mainly driven by problems in the pharmaceutical field, came from a need to study solids under the influence of water or solvent vapors. This required an open reaction vessel, containing the sample specimen, through which a carrier gas with a controlled vapor pressure is passed. A few different designs were suggested to this end (Bakri 1993; Briggner 1993; Suurkuusk et al. 2015). The design that became commercialized consists of a flow system in which dry carrier gas is separated into two flow lines, one of which becomes humidified before entering the reaction vessel. The humidification system consists of two 3 mL chambers placed in series on a shaft that extends from the lid of the reaction vessel (see Figure 4.2). The two chambers are precharged with the liquid that is used to wet the gas. The dry gas enters the wet flow line and passes through the humidification system, where it takes up vapor and is saturated close to 100% before entering the reaction vessel, where it is mixed with dry gas. The position of the humidifier chambers is within the boundaries of the calorimetric thermostat, thus creating a vapor pressure at a temperature that is within 10^{-4} K of that of the sample. By controlling the relative mass flow rates of the dry and wet flow lines, stepwise changes or linear ramps of vapor activity can be programmed, from 0 to 100%. This enables various sorption characteristics of samples to be studied, such as critical water activity points of solids, absorption/adsorption, and secondary events initiated by sorption, such as chemical reactions, hydrate formation, and crystallization of amorphous phases.

Approximately two decades after the initial TAM was introduced to the market, the technology has found widespread use across several scientific and technical areas. In pharmaceutical academia and industry, methods have been developed, which in some instances have

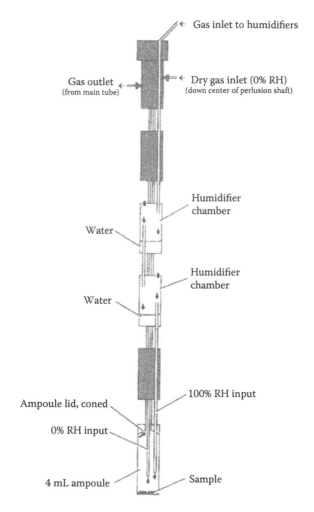

FIGURE 4.2   Schematic picture of the RH perfusion ampoule (TA Instruments, United States).

been transferred to quality control laboratories (USP 2012). The technology has also found important use in defense laboratories for stability, service life, and safety assessments of propellants, pyrotechnics, and explosives. Development of the methodology in this field resulted in a NATO standard, STANAG 4582, a stability test procedure for nitrocellulose-based propellants. Increased interest in this field triggered an extensive makeover of the TAM regarding measuring capacity, sample throughput, detection limit, and modernization of the electronics and software. The result was the third-generation TAM, or TAM III, launched little more than a decade ago. Apart from some important differences, this generation has many of the basic features of the original TAM regarding the modular design, with the thermostat as the basic module and calorimeters and sample handling systems as independent modules. The temperature control system of the thermostat has been improved with the introduction of a new temperature sensing system that makes it possible to control the temperature down to about 1 μK, a two-fold improvement from the original instrument (Suurkuusk and Thisell 2006). Other changes were improvements to the thermostat, including a double-sided wall for increased isolation from surrounding disturbances, and a new inbuilt Peltier cooling device to replace the external circulator required in the original system. Another important development was the introduction of the axial calorimeter design (Suurkuusk and Svensson 2004), which significantly reduced the size of the calorimeter, enabling up to 48 independent units to fit into a single TAM thermostat. In the axial design, the reference side is mirrored vertically below the sample side as compared with the horizontally sidewise placement in the original design. In addition to the axial placement of the reference side, the size of the metallic block that surrounds the measuring assembly and functions as the primary heat sink was greatly reduced without much loss in calorimeter performance due to the improved temperature control system. For details of the TAM III system, see Suurkuusk et al. (2016).

The measuring capacity of the third-generation TAM resulted in a renewed interest in infection assessments and antibiotic susceptibility tests. Due to the increased sample throughput, analysis in clinical environments is becoming a more realistic alternative to conventional techniques. (Braissant et al. 2010).

In addition to operating as an isothermal calorimeter system like its predecessor, the third-generation TAM can be operated in temperature-scanning mode limited to scanning rates below 2 K h^{-1}. The low scanning rate enables many reversible changes in materials to be initiated in near-equilibrium conditions. Furthermore, the resulting temperature resolution greatly enhances the possibility of resolving events that are very close in temperature as compared with conventional DSC techniques. n-Hexatriacontane has been proposed as a test substance to quantify the temperature resolution of DSC instruments (Marti et al. 2004). This compound is well suited for that purpose due to a number of solid–solid transitions and melting to the liquid state appearing within a temperature interval of

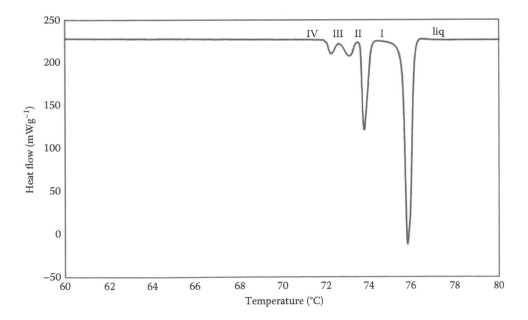

FIGURE 4.3 The result of a temperature scan measurement with n-hexatriacontane $(CH_3(CH_2)_{34}CH_3)$ at a scanning rate of 2 K h^{-1}.

only 3.3 K. The example in Figure 4.3 shows four transitions in the displayed temperature interval: three reversible solid-state transitions between 345 and 346.6 K and subsequent melting to the liquid state at 348.3 K. The reversibility of the transitions was verified by a number of consecutive heating/cooling cycles (Suurkuusk et al. 2016). The example shows the high temperature resolution of the slow temperature-scanning feature of TAM III. Previous investigations have revealed two different solid-state forms in the approximate temperature interval 345–349 K (Marti et al. 2004; Wang et al. 2006).

4.6 CONCLUDING REMARKS

Calorimetric technology and applications have developed extensively over the last few decades from often home-made, specialized devices that needed to be operated by calorimeter specialists to fine analytical tools for the general laboratory. Several factors have driven this development, but one of the strongest has been the need to increase our understanding of the driving forces behind biological recognition processes requiring very sensitive heat measurements (in the nanowatt or microwatt range).

It seems that we have reached a point in calorimeter development at which detection limits are limited by today's electronics and thermocouple sensor technology. With the improvement of temperature control systems with control levels toward 1 μK, the detection limit of heat flow calorimeters is no longer limited by thermal fluctuations of the surroundings but, rather, by electronics, together with the performance of the thermoelectric devices used as heat flow detectors.

Now that calorimetry is an established and valuable technique in many disciplines, with sample throughput significantly increased by the availability of multichannel calorimeter systems, allowing samples to be analyzed simultaneously and independently, the future of calorimetry depends less on instrumentation improvement than on the development of automation and interpretation of data from complex systems.

For routine high-throughput calorimeter tests, as in the pharmaceutical and biomedical fields, automation of the measuring setup is the next step to further increase the usability of calorimetry. Automation is already available for DSC and dedicated ITC instruments, and we predict that automation will also be needed for isothermal measurements of complex samples.

Further, calorimetry today is no longer a technique used only for studying pure systems and measuring thermodynamic parameters that describe reaction and model processes. It is mainly used as a process monitor in complex systems. As almost all processes (chemical, physical, or biological) are accompanied by heat exchange, calorimetry can be and is used as a reaction detector as well as for continuous monitoring of a reacting system. With increased knowledge of the processes going on in a complex system, the possibility of interpreting calorimetric data not only in phenomenological terms but also quantitatively will be the future challenge of calorimetry.

REFERENCES

Angberg, M., Nystrom, C., and Castensson, S. (1988), Evaluation of heat-conduction microcalorimetry in pharmaceutical stability studies. I. Precision and accuracy for static experiments in glass vials, *Acta Pharm. Suec.* 25, 307–320.

Bakri, A. (1993), Design, testing and pharmaceutical applications of a gas pressure controller device for solid-gas microcalorimetric titration, *TA Instruments Application Note* M136. TA Instruments, DE, U.S.A.

Barenholz, Y., Suurkuusk, J., Mountcastle, D., Thompson, T. E., and Biltonen, R. L. (1976), A calorimetric study of the thermotropic behavior of aqueous dispersions of natural and synthetic sphingomyelins, *Biochemistry* 15, 2441.

Beezer, A. E., Miles, R. J., Shaw, E. J., and Vickerstaff, L. (1980), Antibiotic bioassay by flow microcalorimetry, *Experientia* 36, 1051–1052.

Braissant, O., Wirz, D., Gopfert, B., and Daniels, A. U. (2010), Use of isothermal microcalorimetry to monitor microbial activities, *FEMS Microbiol. Lett.* 303, 1–8.

Briggner, L. E. (1993), Microcalorimetric characterisation of physical changes in solid state drugs, *TA Instruments Application Note* M149. TA Instruments, DE, U.S.A..

Briggner, L. E., Buckton, G., Bystrom, K., and Darcy, P. (1994), The use of isothermal microcalorimetry in the study of changes in crystallinity induced during the processing of powders, *Int. J. Pharm.* 105, 125–135.

Calvet, E., and Prat, H. (1963), *Recent Progress in Microcalorimetry*, Pergamon: New York.

Dubrunfaut, C. R. (1856), Note sur la chaleur et le travail mécanique produit par la fermentation vineuse. *Hebd. Séances Acad. Sci.* 42, 945–948.

Edsall, J. T. (1935), Apparent molal heat capacities of amino acids and other organic compounds, *J. Am. Chem. Soc.* 57, 1506–1507.

Frank, H. S., and Evans, M. W. (1945), Free volume and entropy in condensed systems III. Entropy in binary liquid mixtures; partial molal entropy in dilute solutions; structure and thermodynamics in aqueous electrolytes, *J. Chem. Phys.* 13, 507–532.

Görman Nordmark, M., Laynez, J., Schon, A., Suurkuusk, J., and Wadso, I. (1984), Design and testing of a new microcalorimetric vessel for use with living cellular systems and in titration experiments, *J. Biochem. Biophys. Methods* 10, 187–202.

Hedlund, H., and Johansson, P. (2000), Heat capacity of birch determined by calorimetry: Implications for the state of water in plants, *Thermochim. Acta* 349, 79–88.

Hoffner, S. E., Svenson, S. B., and Beezer, A. E. (1990), Microcalorimetric studies of the initial interaction between antimycobacterial drugs and *Mycobacterium avium*, *J. Antimicrob. Chemother.* 25, 353–359.

Hogan, S. E., and Buckton, G. (2000), The quantification of small degrees of disorder in lactose using solution calorimetry, *Int. J. Pharm.* 207, 57–64.

Johansson, P., and Wadsö, I. (1999), Towards more specific information from isothermal microcalorimetric measurements on living systems, *J. Therm. Anal. Calorim.* 57, 275.

Kemp, R. B. (1999), *Handbook of Thermal Analysis and Calorimetry: From Macromolecules to Man*, vol. 4, pp. 1–1032, Elsevier: Amsterdam.

Konicek, J., Suurkuusk, J., and Wadsö, I. (1971), A precise drop heat capacity calorimeter for small samples, *Chem. Scripta* 1, 217.

Lavoisier, A. L., and Laplace, P. S. (1780), Mémoire sur la Chaleur, *Mém. Acad. Roy. Sci.*, 355–408.

Mårdh, P., Andersson, K., Ripa, T., and Wadsö, I. (1976), Microcalorimetry as a tool for evaluation of antibacterial effects of doxycycline and tetracycline, *Scand. J. Infect. Dis. Suppl.* 9, 12–16.

Marti, E., Kaisersberger, E., and Emmerich, W. D. (2004), New aspects of thermal analysis, Part I. Resolution of DSC and means for its optimization, *J. Therm. Anal. Calorim.* 77, 905.

Monk, P., and Wadsö, I. (1975), The use of microcalorimetry for bacterial classification, *J. Appl. Microbiol.* 38, 71–74.

Nichols, N., Sköld, R., Spink, C., Suurkuusk, J., and Wadsö, I. (1976), Additivity relations for the heat capacities of non-electrolytes in aqueous solution, *J. Chem. Thermodyn.* 8, 1081–1093.

Prigogine, I. (1961), *Thermodynamics of Irreversible Processes*, 2nd edition, Interscience: New York.

Privalov, P. L. (1963), Investigation of the heat denaturation of egg white albumin, *Biophysica (USSR)* 8, 308–316.

Privalov, P. L., and Potehkin, S. A. (1986), Scanning microcalorimetry in studying temperature-induced changes in proteins, *Meth. Enzymol.* 131, 4–51.

Randzio, S. L., and Suurkuusk, J. (1980), Interpretation of calorimetric thermograms and their dynamic corrections, in *Biological Microcalorimetry*, 311–341, Academic: London.

Ross, P. D., and Goldberg, R. N. (1974), A scanning microcalorimeter for thermally induced transitions in solution, *Thermochim. Acta* 10, 143–151.

Santos, L. M. N. B. F., Rocha, M. A. A., Rodrigues, A. S. M. C., Štejfa, V., Fulem, M., and Bastos, M. (2011), Reassembling and testing of a high-precision heat capacity drop calorimeter. Heat capacity of some polyphenyls at T = 298.15 K, *J. Chem. Thermodyn.* 43, 1818–1823.

Sebhatu, T., Angberg, M., and Ahlneck, C. (1994), Assessment of the degree of disorder in crystalline solids by isothermal microcalorimetry, *Int. J. Pharm.* 104, 135–144.

Suurkuusk, J. (1974), Specific heat measurements on lysozyme, chymotrypsinogen, and ovalbumin in aqueous solution and in solid state, *Acta Chem. Scand. B* 28, 409–417.

Suurkuusk, J., Lentz, B. R., Barenholz, Y., Biltonen, R. L., and Thompson, T. E. (1976), A calorimetric and fluorescent probe study of the gel-liquid crystalline phase transition in small single-lamellar dipalmitoylphosphatidylcholine vesicles, *Biochemistry* 15, 1393–1401.

Suurkuusk, J., Suurkuusk, M., and Vikegard, P. (2016), A multichannel microcalorimetric system: The third generation Thermal Activity Monitor (TAM III), submitted to *J. Therm. Anal. Calorim.*

Suurkuusk, J., and Svensson, L.-G. (2004), Patent SE0102505-5, Kalorimeter, Sweden.

Suurkuusk, J., and Thisell, N. (2006), Patent US6994467, Absolute temperature measuring apparatus and method.

Suurkuusk, J., and Wadsö, I. (1974), Design and testing of an improved precise drop calorimeter for the measurement of the heat capacity of small samples, *J. Chem. Thermodyn.* 6, 667–679.

Suurkuusk, J., and Wadsö, I. (1982), A multichannel microcalorimetric system, *Chem. Scripta* 20, 155–163.

Tanford, C. (1970), Protein denaturation: Part C. Theoretical models for the mechanism of denaturation, *Adv. Protein Chem.* 24, 1–95.

USP (2012), Characterization of crystalline solids by microcalorimetry and solution calorimetry. Edited by Official December 1 US Pharmacopeia document, Stage 6 Harmonisation.

Wadsö, I. (1968), Design and testing of a micro reaction calorimeter, *Acta Chem. Scand.* 22, 927–937.

Wadsö, I. (2002), Isothermal microcalorimetry in applied biology, *Thermochim. Acta* 394, 305–311.

Wang, S., Tozaki, K.-I., Hayashi, H., Inaba, H., and Yamamoto, H. (2006), Observation of multiple phase transitions in some even *n*-alkanes using a high resolution and super-sensitive DSC, *Thermochim. Acta* 448, 73–81.

CHAPTER 5

Enzyme-Catalyzed Reactions[*]

Robert N. Goldberg

CONTENTS

5.1 INTRODUCTION

The first value of an equilibrium constant for an enzyme-catalyzed reaction was reported by Quastel and Woolf (1926) for the reaction {L-aspartate(aq) = fumarate(aq) + ammonia(aq)} catalyzed by aspartate ammonia-lyase. The first calorimetric measurement for an enzyme-catalyzed reaction was done by Meyerhof and Lohmann (1935). They used the enzymes fructose-biphosphate aldolase and triose-phosphate isomerase to catalyze the reaction {D-fructose 1,6-bisphosphate(aq) = 2 glycerone phosphate(aq)}. Approximately 1200 studies that report thermodynamic results for enzyme-catalyzed reactions (Goldberg et al. 2007) have been published through 2007. Interest in this subject is due to the fundamental importance of these reactions to biochemistry and, in particular, to metabolism. Additional interest arises from the large-scale use of enzymes in biomanufacturing. This chapter reviews the principles that underlie the thermodynamics of both chemical and overall biochemical reactions, and the generic types of measurements and calculations pertinent to these reactions. The status of the literature on the thermodynamics of enzyme-catalyzed reactions is summarized along with what the author considers to be the most critical research needs in this area. The discussion of thermodynamic principles

---

[*] This is an official contribution of the National Institute of Standards and Technology (NIST) and is not subject to copyright in the United States. Any mention of commercial products is not intended to imply recommendation or endorsement by NIST, nor is it intended to imply that the products identified are necessarily the best available for the purpose.

presented herein applies to all reactions and not just enzyme-catalyzed reactions. In general, the enzyme serves the important role of lowering the energy of activation of a reaction, thus permitting a reaction to take place more rapidly than it would in the absence of the enzyme.

5.2 THERMODYNAMIC PRINCIPLES

5.2.1 Chemical Reactions Involve Specific Species

The importance of specifying the system under study is central to all of thermodynamics. In this regard, the distinction between chemical reactions and overall biochemical reactions is a fundamental matter (Alberty et al. 1994, 2011). Chemical reactions involve specific chemical species (these species may be of biochemical importance) and are the subject of our initial discussion. Later, we will discuss the thermodynamics of overall biochemical reactions and substances where one or more of the reactants is a mixture of species. In its general form, a chemical reaction can be written as an equation:

$$\Sigma v_i R_i = 0,\qquad(5.1)$$

where:

R_i are the chemical reactants, which are distinct chemical species
v_i are the respective stoichiometric numbers

Note that the R_i must be defined in terms of the specific atoms and charges present and their respective numbers. The summation in Equation 5.1 runs over all the N_r reactants, and this convention also pertains to the summations and products in the equations that follow below. If v_i is positive, the reactant lies on the right side of the equation and is a product of the reaction. If v_i is negative, the reactant lies on the left side of the equation. Clearly, when writing a chemical reaction, one must balance both atoms and charges. If necessary, one must also specify the isotopes and, perhaps, provide other fine chemical details. The thermodynamics of chemical reactions are standard fare in most elementary chemistry textbooks, and one has the well-known thermodynamic relations

$$\Delta_r G_m^\circ = -RT\ln K,\qquad(5.2)$$

$$K = \Pi a_1^{v_i},\qquad(5.3)$$

$$\Delta_r G_m^\circ = \Delta H_m^\circ - T\Delta_r S_m^\circ,\qquad(5.4)$$

$$\Delta_r H_m^\circ = RT^2 \left(\frac{\partial \ln K}{\partial T} \right)_p, \tag{5.5}$$

$$\Delta_r C_{p,m}^\circ = \left(\frac{\partial \Delta_r H_m^\circ}{\partial T} \right)_p. \tag{5.6}$$

where:
  $\Delta_r G_m^\circ$ is the standard molar Gibbs energy change
  R is the molar gas constant
  T is the thermodynamic temperature
  K is the thermodynamic equilibrium constant
  a_i is the activity of reactant i

The quantity $\Delta_r H_m^\circ$ is the standard molar enthalpy change, $\Delta_r S_m^\circ$ is the standard molar entropy change, p is pressure, and $\Delta_r C_{p,m}^\circ$ is the standard molar heat capacity change at constant pressure. The subscript r denotes a reaction that *must be specified*, and the subscript m denotes a molar quantity. When it is clear that molar quantities are being used, the subscript *m* is often omitted. Values of the standard molar formation properties can be used to calculate values of $\Delta_r G_m^\circ$, $\Delta_r H_m^\circ$, $\Delta_r S_m^\circ$, $\Delta_r C_{p,m}^\circ$:

$$\Delta_r G_m^\circ = \Sigma v_i \Delta_f G_{m,i}^\circ, \tag{5.7}$$

$$\Delta_r H_m^\circ = \Sigma v_i \Delta_f H_{m,i}^\circ, \tag{5.8}$$

$$\Delta_r S_m^\circ = \Sigma v_i S_{m,i}^\circ, \tag{5.9}$$

$$\Delta_r C_{p,m}^\circ = \Sigma v_i C_{p,m,i}^\circ. \tag{5.10}$$

The quantities $\Delta_f G_{m,i}^\circ$, $\Delta_f H_{m,i}^\circ$, $S_{m,i}^\circ$, and $C_{p,m,i}^\circ$ are, respectively, the standard molar Gibbs energy of formation, the standard molar enthalpy of formation, the standard molar entropy, and the standard molar heat capacity of the chemical reactant R_i.

Note that the thermodynamic equilibrium constant is defined in terms of activities. This requires a choice of standard state that must be specified. The molality-based standard state that is generally used in chemical thermodynamics (Wagman et al. 1982) is now described. The standard state for a solute i is the hypothetical ideal (activity coefficient $\gamma_i = 1$) solute at unit molality ($m = 1$ mol·kg⁻¹). The standard state for the solvent is the pure solvent, that is, the activity a of the pure solvent → 1 as the molality of all solutes → 0. The standard state for a gas is the hypothetical ideal gas at $p = 0.1$ MPa. The standard states for a pure solid and for a pure liquid are, respectively, the pure solid and pure liquid at $p = 0.1$ MPa. The concentration-based standard state is like the molality-based standard state except that the standard state for a solute is the hypothetical ideal solute at unit concentration ($c = 1$ mol dm⁻³). Clearly, the choice of standard state is arbitrary, and one may need to choose a different standard state as circumstances dictate. However, given the large body of thermochemical data that uses the aforementioned molality standard state, it would seem advantageous to adhere to it. Finally, many biochemical reactions can be carried out in non-aqueous media. If water is a reactant in such reactions, it is necessary to measure the molality or concentration of the water at equilibrium and include its actual molality or concentration in the calculation of the equilibrium constant for that reaction (Tewari et al. 1995).

The thermodynamic equilibrium constant can also be written in terms of molalities m_i and activity coefficients γ_i:

$$K = \Pi m_i^{\nu_i} \cdot \Pi \gamma_i^{\nu_i} = K_m \cdot \Pi \gamma_i^{\nu_i}. \tag{5.11}$$

Thus, the molality basis equilibrium constant K_m is obtained by measuring the molalities of the reactants at equilibrium. In the ideal case, one measures the molalities of all of the reactants. However, in most situations, it will suffice to measure the molality of one of the reactants and use it to calculate the extent of reaction ξ, which is defined as

$$\xi = \frac{(n_i - n_{i,0})}{\nu_i}, \tag{5.12}$$

where:

n_i is the amount of R_i

$n_{i,0}$ is the amount of R_i at the initiation of reaction

Thus, by knowing ξ *at equilibrium*, one can calculate the values of n_i for all of the reactants. By considering the reaction involving 1 kg of solvent, molalities m_i replace the n_i's in Equation 5.12. Thus, ξ, the m_i's, and K_m are determined. The calculation of the thermodynamic equilibrium constant K, done by using Equation 5.11, requires a knowledge or estimation of the activity coefficients or an extrapolation of values of K_m to the standard state. For solutions that involve electrolytes, this extrapolation is generally done by extrapolation to the ionic strength $I = 0$.

Extended Debye–Hückel theory provides a convenient model for charged biochemical species. In this model, the activity coefficient of a solute species γ_i is given by

$$\ln\gamma_i = \frac{-A_m z_i^2 I^{1/2}}{1 + BI^{1/2}}, \qquad (5.13)$$

where:

A_m is the Debye–Hückel constant (see Clarke and Glew [1980] for values of Debye–Hückel constants as a function of temperature)

z_i is the signed charge of species i

B is a constant that is often referred to as the "ion-size" parameter

We have generally used the value $B = 1.6 \text{ kg}^{1/2} \text{mol}^{-1/2}$ based on the values of this parameter obtained in fitting data on a series of electrolytes of charge types 1–1, 1–2, and 2–1 (Goldberg and Tewari 1991). The Debye–Hückel adjustment for the enthalpy is obtained by starting with the excess Gibbs energy,

$$G_i^{ex} = G_i - G_i^{id} = G_i - G_i^\circ = RT\ln\gamma_i = \frac{-A_m RT z_i^2 I^{1/2}}{1 + BI^{1/2}}. \qquad (5.14)$$

Here, G_i^{id} is the ideal partial molar Gibbs energy of reactant R_i. For the hypothetical ideal 1 mol kg^{-1} standard state, $G_i^{id} = G_i^\circ$. Then, by assuming that B does not depend on temperature, the excess entropy S_i^{ex} is obtained by taking the temperature derivative of G_i^{ex}. Thus,

$$S_i^{ex} = \left(\frac{\partial G_i^{ex}}{\partial T}\right)_p = \frac{A_m RT z_i^2 I^{1/2}}{1 + BI^{1/2}} + \frac{\left(\partial A_m / \partial T\right)_p RT z_i^2 I^{1/2}}{1 + BI^{1/2}}. \qquad (5.15)$$

The combination of Equations 5.14 and 5.15 gives

$$H_i^{ex} = G_i^{ex} + TS_i^{ex} = \frac{\left(\partial A_m / \partial T\right)_p RT z_i^2 I^{1/2}}{1 + BI^{1/2}}. \qquad (5.16)$$

Thus, the molar enthalpy change for a reaction at an ionic strength I is given by

$$\Delta_r H_m (I) = \Delta_r H_m^\circ + \Sigma \nu_i H_i^{ex}. \qquad (5.17)$$

By using the Gibbs–Duhem relation together with Equations 5.14 and 5.16, one obtains the following equations (Goldberg and Tewari 1991) for, respectively, the excess Gibbs energy of the solvent G_w^{ex}, the activity of the solvent a_w, and the excess enthalpy of the solvent H_w^{ex}:

$$G_w^{ex} = \frac{2A_m RT\sigma}{m_1^* B^3}, \tag{5.18}$$

$$\ln a_w = \left(\frac{G_w^{ex}}{RT}\right) - \frac{\Sigma_j m_j}{m_1^*}, \tag{5.19}$$

$$H_w^{ex} = \frac{-2RT^2\sigma(\partial A_m / \partial T)_p}{m_1^* B^3}. \tag{5.20}$$

Here, m_1^* is the amount of water in a kilogram of water (55.508 mol) and

$$\sigma = \left(1 + BI^{1/2}\right) - 2\ln\left(1 + BI^{1/2}\right) - \left(1 + BI^{1/2}\right)^{-1}. \tag{5.21}$$

Equations 5.18 through 5.21 are needed when water is a reactant in the reaction under study.

Note that while molalities were used in the above discussion, concentrations and mole fractions can also be used. In such cases, one uses either the concentration basis equilibrium constant K_c or the mole fraction basis equilibrium constant K_x:

$$K_c = \Pi c_1^{v_i}, \tag{5.22}$$

$$K_x = \Pi x_1^{v_i}. \tag{5.23}$$

Here, c_i and x_i are, respectively, the concentration and mole fraction of R_i. Also, if one or more reactants are in the gas phase, one has an equilibrium constant that has a pressure basis. The specification of the standard state (which includes the pressure units) for reactants in the gas phase is of particular importance. We note that there are two advantages to the use of molalities. Specifically, molality, unlike concentration, does not depend on temperature. Also, solutions can generally be prepared with better precision and accuracy by using gravimetric, as opposed to volumetric, methods.

Several additional fine, but important, points are now made. First, for very exact work, one must recognize that the activity coefficients needed to calculate the thermodynamic equilibrium constant from measured values of K_m, K_c, and K_x differ from each other. These activity coefficients are designated by the International Union of Pure and Applied Chemistry (IUPAC) (Cohen et al. 2008), respectively, as γ_m, γ_c, and γ_x, where the subscripts denote the basis for the γ's. Relations between these activity coefficients are given by Robinson and Stokes (1955). Second, for symmetric reactions, K_m and K_c are dimensionless, and K_x is always dimensionless. However, if a reaction is not symmetric, the unit molality $m° = 1$ mol kg^{-1} or unit concentration $c° = 1$ mol dm^{-3} can be introduced to make K_m and K_c, respectively, dimensionless. The literature contains a number of reports where concentrations expressed in micromoles have been used to calculate values of K_c. In such cases, a standard state has been selected that is substantially different from the conventional standard state based on $c° = 1$ mol·dm^{-3} or $m° = 1$ mol·kg^{-1}. This does not cause a problem for symmetric reactions. However, for reactions that are not symmetric, the calculated values of K_c and $\Delta_r G_m^°$ that have used micromoles will differ by at least a factor of 1000 from the value of K_c and by at least 17.1 kJ mol^{-1} from the value of $\Delta_r G_m^°$ that is obtained by using the conventional standard state. Thus, it is essential that the standard state be specified when reporting the value of an equilibrium constant. Third, the relations between the concentration and molality of a solute i are

$$c_i = \frac{m_i \rho}{1 + m_i M_i},$$
(5.24)

$$m_i = \frac{c_i}{\rho - c_i M_i},$$
(5.25)

where M_i is the molar mass of solute i and ρ is the density of the solution.

Note that for dilute solutions ($m_i < 0.25$ mol kg^{-1}) and for solutes having M_i less than 0.3 kg mol^{-1}, the difference between m_i and c_i will generally be less than 0.10 m_i. This difference should not be neglected for accurate work. Finally, single-ion activities and single-ion activity coefficients have been used in this discussion. While the existence of these properties has not been resolved (Goldberg and Frank 1972; Zarubin 2011), they are a necessary part of the model used to describe chemical and overall biochemical reactions. The important point is that the model, which includes the use of single-ion activities, leads to properties such as equilibrium constants and enthalpy and heat capacity changes that can be measured.

If the quantities $\Delta_r G_m^°$, $\Delta_r H_m^°$, and $\Delta_r C_{p,m}^°$ are known at a reference temperature θ, one can calculate the equilibrium constant as a function of temperature by using the following equation (Clarke and Glew 1966):

$$R \ln K_T = -\left(\frac{\Delta_r G_{m,\theta}^°}{\theta}\right) + \Delta_r H_{m,\theta}^° \left\{\left(\frac{1}{\theta}\right) - \left(\frac{1}{T}\right)\right\} + \Delta_r C_{p,m}^° \left\{\left(\frac{\theta}{T}\right) - 1 + \ln\left(\frac{T}{\theta}\right)\right\}$$
(5.26)

Here, $\Delta_r G_{m,\theta}^°$ and $\Delta_r H_{m,\theta}^°$ are, respectively, the values of $\Delta_r G_m^°$ and $\Delta_r H_m^°$ at the temperature θ. The quantity $\Delta_r C_{p,m}^°$ is assumed to be constant over the temperature range of interest. The reference temperature θ is often taken to be 298.15 K. While the above equation works well over a moderate range of temperature, derivatives of $\Delta_r C_{p,m}^°$ with respect to temperature are needed if one is dealing with a wide temperature range

(Clarke and Glew 1966). Measured values of K, $\Delta_r H_m^\circ$, and $\Delta_r C_{p,m}^\circ$ can be used together with Equation 5.26 to calculate values of $\Delta_r G_{m,\theta}^\circ$, $\Delta_r H_{m,\theta}^\circ$, and $\Delta_r C_{p,m}^\circ$ by using weighted least-squares regression methods. Note that the rigorous application of Equation 5.26 for the calculation of standard molar enthalpy changes requires the use of the thermodynamic equilibrium constant K. Thus, errors can arise if measured values of K_m are used in Equation 5.26 (or in the van't Hoff equation) to calculate $\Delta_r H_{m,\theta}^\circ$ without any consideration of the activity coefficients that are needed in Equation 5.11 to calculate $\Delta_r G_m^\circ$. Use of the van't Hoff equation assumes that $\Delta_r C_{p,m}^\circ$ is equal to zero. Also, if one uses values of K_c and if the reaction is not symmetric, one must apply a correction for the temperature dependence of the density of the solution.

It is extremely important to recognize that the definition of the equilibrium constant pertains to a reaction that is at equilibrium. The reaction quotient Q, when written in the form of ratios of molalities or concentrations or mole fractions, has the same form as the equilibrium constant written on a molality, concentration, or mole fraction basis. And, since it is reaction quotients that are measured, it is necessary to establish that the reaction is at equilibrium so that Q can be identified with K. In the case of enzyme-catalyzed reactions, a failure to achieve equilibrium can be due to loss of enzyme activity caused by a change in the enzyme, for example, product inhibition.

5.2.2 Overall Biochemical Reactions

Overall biochemical reactions are characterized by the fact that one or more of the reactants is a mixture of species that are related via chemical reactions. Very often, the species differ in the number of hydrogen or metal atoms that are bound to a central ligand. A frequently used example of this (see the most recent International Union of Biochemistry and Molecular Biology [IUBMB] recommendations [Alberty et al. 2011]) is adenosine 5′-triphosphate (ATP), where the total molality of ATP is given by

$$m\left(\text{ATP}\right)_{\text{Total}} = m(\text{ATP}^{4-}) + m(\text{HATP}^{3-}) + m(\text{H}_2\text{ATP}^{2-}) + m(\text{MgATP}^{2-})$$

$$+ m(\text{MgHATP}^-) + m\left(\text{Mg}_2\text{ATP}\right). \tag{5.27}$$

Calcium ions could also be added to the above equation, and one may also need to include Na^+ and K^+ ions at high concentrations of these latter two cations. The ATP species shown on the right-hand side of Equation 5.27 are referred to as pseudoisomers (Alberty 2003), as they differ only in the number of H^+ and Mg^{2+} ions attached to a central ATP^{4-} species. Equations similar to the above can be written for many reactants, for example, phosphate, carbonate, adenosine 5′-diphosphate (ADP), glutamate, ammonia, asparagine, and glutamine. The hydrolysis reaction of ATP, written as an overall biochemical reaction, is

$$\text{ATP}\left(\text{aq}\right) + \text{H}_2\text{O}\left(\text{l}\right) = \text{ADP}\left(\text{aq}\right) + \text{phosphate}\left(\text{aq}\right). \tag{5.28}$$

Note that ADP and phosphate also refer to the total amounts of the species of which they are comprised:

$$m(\text{ADP})_{\text{Total}} = m(\text{ADP}^{3-}) + m(\text{HADP}^{2-}) + m(\text{H}_2\text{ADP}^-) + m(\text{MgADP}^-) + m(\text{MgHADP}),$$

(5.29)

$$m(\text{phosphate})_{\text{Total}} = m(\text{PO}_4^{3-}) + m(\text{HPO}_4^{2-}) + m(\text{H}_2\text{PO}_4^-) + m(\text{H}_3\text{PO}_4)$$

$$+ m(\text{MgPO}_4^-) + m(\text{MgHPO}_4).$$

(5.30)

Since most experimental methods measure total amounts of reactants and do not distinguish between the species, the apparent equilibrium constant K' arises naturally:

$$K' = \frac{m(\text{ADP})_{\text{Total}} \cdot m(\text{phosphate})_{\text{Total}}}{m(\text{ATP})_{\text{Total}} \cdot m^\circ}.$$

(5.31)

A subscript m could be added to the K' in Equation 5.31 to denote that the equilibrium constant is based on molality. Since $\text{H}^+(\text{aq})$ and $\text{Mg}^{2+}(\text{aq})$ bind to $\text{ATP}^{4-}(\text{aq})$, $\text{ADP}^{3-}(\text{aq})$, and HPO_4^{2-} (aq) (see Equations 5.27, 5.29, and 5.30), the thermodynamic properties for overall biochemical reactions, such as that in Equation 5.28, depend on pH and pMg, in addition to their dependence on temperature, pressure, and ionic strength. Here, pMg is defined in an analogous way to pH. One can choose a chemical reaction such as

$$\text{ATP}^{4-}(\text{aq}) + \text{H}_2\text{O}(\text{l}) = \text{ADP}^{3-}(\text{aq}) + \text{HPO}_4^{2-}(\text{aq}) + \text{H}^+(\text{aq}),$$

(5.32)

for the system of chemical reactions that is serving as the model for the overall biochemical reaction (Equation 5.28). Note that the choice of the chemical reference reaction is arbitrary and several other chemical reactions could have been chosen in place of Equation 5.32. The chemical reference reaction is especially useful in that once it and all of the $\text{H}^+(\text{aq})$ and $\text{Mg}^{2+}(\text{aq})$ binding constants and standard molar enthalpies of reaction are known, one can calculate the thermodynamic properties for the overall biochemical reaction as a function of T, pH, pMg, and I (see below). Also, since the chemical reference reaction involves specific chemical species, it can be used in thermodynamic network calculations to connect with other chemical species that have known values of standard formation properties.

A very general way to treat the thermodynamics of overall biochemical reactions is to assemble the following array of data: all of the individual chemical reactions that take place in the system—this includes reactions that involve the solvent and the buffer; the initial molalities of all substances in the system; the thermodynamic equilibrium constants, the standard molar enthalpy changes, and the standard molar heat capacity changes for all of the specified individual chemical reactions; and the temperature, pressure, and pH. If the above information is known, one can write for each chemical reaction in the system a single equation that involves the thermodynamic equilibrium constant and the activities of the pertinent species. This results in a system of non-linear equations. Next, the initial molalities and the extents of reaction ξ_j for each reaction j can be introduced into this system of equations. These equations can be solved numerically, for example, by using the Newton–Raphson method, to calculate the molalities of all of the species in solution at equilibrium. Total molalities of biochemical reactants such as ATP, ADP, and phosphate, and then K', can be calculated. Note that in performing such calculations, it is first necessary to adjust the thermodynamic equilibrium constants and standard molar enthalpy changes for the individual chemical reactions from the reference temperature to the desired temperature if it differs from the reference temperature, which is usually taken to be 298.15 K (see Equation 5.26). Then, it is necessary to either have the actual activity coefficients on hand or use a model for the activity coefficients (see Equation 5.13). Consequently, one must make an initial guess for the ionic strength I. This initial guess allows one to solve for the molalities of all of the species and then calculate an improved guess for I. This second guess for I can be improved by iterative refinement until the value of I converges to within a specified tolerance. After this has been done, one has values of the equilibrium constants and molar enthalpy changes for all the chemical reactions that pertain to the specified temperature and calculated ionic strength. The molalities of all of the chemical species can then be calculated, as well as n_r', the amount of overall biochemical reaction. Note that $m(Mg^{2+})$ and pMg are also calculated in this iterative procedure.

Thus, K' can be calculated at desired values of T, I, pH, and pX, where X is Mg, Ca, and so forth. This calculation also yields values of the extents of reaction ξ_j for the individual reactions, where j designates the jth reaction in the system of chemical reactions. Then, one can calculate the molar enthalpy change for the overall biochemical reaction, that is, what one would measure calorimetrically:

$$\Delta_r H(\text{cal}) = \frac{\Sigma \xi_j \Delta_r H_{m,j}^\circ}{n_r'}. \tag{5.33}$$

The summation is made over all of the j reactions that occur, including any reactions of the buffer with others species in the system, such as $H^+(aq)$ and $Mg^{2+}(aq)$. Since pH is generally measured for the final equilibrium state, one can use this pH value to calculate $m\{H^+(aq)\}$ by using an activity coefficient model (see Equation 5.13). In the absence of a reliable electrode for $Mg^{2+}(aq)$, $m\{Mg^{2+}(aq)\}$ and pMg must be calculated by solving the system of equations. Thus, the initial and final pH and pMg can be calculated for both the initial state and the final equilibrium state. Then, the changes in binding for H^+ and Mg^{2+}, and $\Delta_r N_H$ and $\Delta_r N_{Mg}$, respectively, can be calculated.

An alternative approach is to calculate values of K' as a function of T, pH, and pMg and then use the following relations (Alberty 2003):

$$\Delta_r H_m^{\prime\circ} = RT^2 \left(\frac{\partial \ln K'}{\partial T} \right)_{p,I,\text{pH,pMg}}, \tag{5.34}$$

$$\Delta_r N_H = -\left(\frac{\partial \log_{10} K'}{\partial \text{pH}} \right)_{T,p,I,\text{pMg}}, \tag{5.35}$$

$$\Delta_r N_{Mg} = -\left(\frac{\partial \log_{10} K'}{\partial \text{pMg}} \right)_{T,p,I,\text{pH}}. \tag{5.36}$$

Additionally, $\Delta_r H_m$(cal) can be calculated by using (Alberty and Goldberg 1993)

$$\Delta_r H_m \left(\text{cal} \right) = \Delta_r H_m^{\prime\circ} + \Delta_r N_H \cdot \Delta_r H_m^{\circ} \left(\text{H} \cdot \text{Buffer} \right) + \Delta_r N_{Mg} \cdot \Delta_r H_m^{\circ} \left(\text{Mg} \cdot \text{Buffer} \right). \tag{5.37}$$

Here, $\Delta_r H_m^{\circ} \left(\text{H} \cdot \text{Buffer} \right)$ and $\Delta_r H_m^{\circ} \left(\text{Mg} \cdot \text{Buffer} \right)$ are the standard molar enthalpy changes for the reactions

$$\text{H} \cdot \text{Buffer} \left(\text{aq} \right) = \text{H}^+ \left(\text{aq} \right) + \text{Buffer}^- \left(\text{aq} \right), \tag{5.38}$$

$$\text{Mg} \cdot \text{Buffer} \left(\text{aq} \right) = \text{Mg}^{2+} \left(\text{aq} \right) + \text{Buffer}^{2-} \left(\text{aq} \right). \tag{5.39}$$

The prime (') was used above to distinguish the apparent equilibrium constant K' from the equilibrium constants used for chemical reactions, as well as other thermodynamic properties for overall biochemical reactions. It is also used (Alberty et al. 2011) to denote standard transformed thermodynamic properties such as $\Delta_r H_m^{\prime\circ}$ (see Equation 5.34) and $\Delta_r G_m^{\prime\circ}$ where

$$\Delta_r G_m^{\prime\circ} = -RT \ln K'. \tag{5.40}$$

Equations 5.34 through 5.37 are based on the Legendre transform (Alberty 1992, 2003),

$$G' = G - n_H \cdot \mu_{H^+}, \tag{5.41}$$

where:

G' is the transformed Gibbs energy

n_H is the total amount of hydrogen in the system

μ_{H^+} is the chemical potential of H^+ at the specified T, p, I, and pH

Note that this Legendre transform makes pH a constraint on a reaction in the same formal way that the specification of T and p serves in the derivation of many thermodynamic relations. Thus, based on the Legendre transform shown in Equation 5.41, Alberty (1969b, 2003) derived several Maxwell-type relations that are important to biothermodynamics.

Akers and Goldberg (2001) have developed a *Mathematica* package that allows for many of the calculations discussed in this section. Thus, one can calculate how K', $\Delta_r H_m'^{\,\circ}$, $\Delta_r N_H$, and $\Delta_r N_{Mg}$ vary with T, pH, pMg, and I. Clearly, one must specify the reactions and provide the essential thermodynamic property data for the chemical reference reaction and related ionic reactions. If desired, one can constrain pH, pMg, and I in the calculations. Many of the calculations in the *Mathematica* package are relevant to the treatment of experimental results. Specifically, one can calculate values of the thermodynamic equilibrium constant K and $\Delta_r H_m^\circ$ for a specified chemical reference reaction from measured values of K' and $\Delta_r H_m$(cal). The package also allows the user to plot thermodynamic properties as a function of T, pH, pMg, and I. Finally, note that reactions of *any type* can be introduced into the equilibrium equations that model an overall biochemical reaction. Thus, one can also treat overall biochemical reactions where one or more reactants is a mixture of different anomers, for example, α-and β-D-glucose. The inclusion of anomeric equilibria is an extended use of the apparent equilibrium constant beyond that discussed in IUBMB and IUPAC recommendations (Wadsö et al. 1976; Alberty et al. 1994, 2011).

The approach to biochemical thermodynamic properties that is obtained by writing all of the pertinent equilibria and then solving these equations is extremely general. The description of the reacting system uses a conservation or formula matrix and a stoichiometric number matrix (Klein 1971; Cohen et al. 2008, p. 53). This method has been in use for gas-phase reactions for many years, and it is of particular importance to complex combustion problems (Klein 1971). Note that this approach does not yield analytic expressions for thermodynamic properties, as in earlier approaches to solving this generic biothermodynamic problem (Alberty 1968, 1969a; Goldberg and Tewari 1991). However, analytic solutions are not always possible for all classes of chemical equilibria. The non-analytic approach of solving a system of non-linear equilibrium equations is extremely powerful and allows for the solution of the most complex problems involving chemical equilibria and for the calculation of associated thermodynamic properties, such as apparent equilibrium constants, molar enthalpies of reaction, and changes in binding. Finally, it is important to note that the thermodynamic principles discussed above are not limited to enzyme-catalyzed reactions. *These principles apply to all types of reactions.*

5.3 EXPERIMENTAL PRINCIPLES

5.3.1 Equilibrium Measurements

We now turn to a discussion of the experimental principles that underlie the measurement of the primary thermodynamic properties of interest, namely, the apparent equilibrium constant K' and the calorimetrically determined molar enthalpy of reaction $\Delta_r H_m(cal)$. The measurement of K' requires an analytical method or methods that allow for the measurement of the molalities (or concentrations) of the reactants. However, it is often not practical to measure the molalities of all of the reactants. In fact, in most of the literature, the molality of a single reactant is measured, and this measurement is often done by using a single analytical method. The molalities of the remaining reactants can then be calculated by using the initially known molalities of the reactants together with the extent of reaction variable ξ' for the overall biochemical reaction. Clearly, if the determination of the molality of any of the reactants is based on ξ', the presence of a side reaction that involves any of the reactants will be a source of systematic error. Barring side reactions, any method that yields the molality of one or more of the reactants will serve for the determination of the molalities of the other reactants and, in turn, for the calculation of K'. Indeed, numerous methods have been used in the literature to determine values of K'. The most commonly used methods are spectrophotometry, chromatography, and chemical and enzymatic methods of analysis. Nuclear magnetic resonance (NMR), radioactive methods, and mass spectrometry have also played a role in equilibrium measurements. In cases where the value of K' is not too far from unity, calorimetry can also be used to measure ξ', which, in turn, leads to a value for K'. The general rule for equilibrium measurements is to use *any analytical method that works*. Clearly, when reporting a value of K', it is critical that the reaction, purities of reagents, and conditions of measurement be reported. These conditions include the temperature T, the pressure p, the pH, the molalities of all substances in the solution, and the calculated ionic strength. Note that it is rare that K' varies much over the narrow range of pressure encountered at $p \approx 0.1$ MPa.

When performing equilibrium measurements for any reaction, especially for enzyme-catalyzed reactions, one must take care to establish that equilibrium has, in fact, been reached. This can be done by approaching equilibrium from both directions of reaction. Thus, if the measured values of the apparent reaction quotients Q' obtained from both directions of reaction are equal to each other within their respective uncertainties, one has good evidence that equilibrium has been reached. Then, the average of the measured values of Q' can be identified with K'. However, since some enzymes are subject to product inhibition or may lose activity, it may not be possible to establish that equilibrium has been reached by performing experiments in which equilibrium is approached by starting from both directions of reaction. In such cases, one can prepare synthetic mixtures of the reactants (and products) such that the initial values of Q' are not too distant from an *estimated* value of K'. One then adds the enzyme to the reaction mixtures and measures either the molality of one of reactants or a signal that is a measure of a change in the molality of one of the reactants. If one does this for several synthetic mixtures having different initial values of Q', one can plot the initial values of Q' as a function of either the change in signal or the change in molality (note that the change in signal could also be plotted as a function of the Q' initial values). Since the starting solutions were prepared synthetically, the initial Q' values are well known. Now, if by some chance the value of Q' for an initial synthetic mixture was equal to K', there would be no change in signal or molality, and one then knows that the value of K' is equal to that initial Q' value. In general, for a series of measurements that involve several synthetic mixtures, one can plot the well-known initial Q' values as a function of the change in signal or molality and draw a fitted curve through the measured values. If the fitted curve crosses the axis where the change in molality or signal equals zero, the value of Q' where this occurs is equal to K'. If the fitted curve does not cross the axis where the change in molality or signal equals zero, one must repeat the experiment using a new set of synthetic mixtures. This method of synthetic mixtures was used by Weischet and Kirschner (1976) and Kishore et al. (1998) to measure K' for the reaction {indole(aq) + D-glyceraldehyde 3-phosphate(aq) = 1-(indol-3-yl)glycerol 3-phosphate(aq)}. In this particular case, the necessity for using this approach arose due to the instability of the reactant D-glyceraldehyde 3-phosphate.

Values of K' can also be obtained from rate measurements on enzyme-catalyzed reactions and the use of the Haldane equations, which are expressions for K' in terms of kinetic parameters in the steady-state rate equations (Cornish-Bowden 2004; Alberty et al. 2011). Note that the kinetic parameters are dependent on the enzyme. However, K' is independent of the enzyme.

5.3.2 Calorimetric Measurements

The general aim of a calorimetric measurement performed on an enzyme-catalyzed reaction is the measurement of the molar enthalpy change $\Delta_r H_m(\text{cal})$ for the overall biochemical reaction. In such measurements, one must prepare a *substrate solution* that contains the substrates and cofactors that are necessary for the reaction to proceed. A separate *enzyme solution* that contains the necessary enzymes is also prepared. Both solutions are generally prepared in a buffered solution having a suitable pH. It is obvious that the buffer solutions used for the substrate and enzyme solutions should be as identical as possible. In this regard, it is important to use either a lyophilized enzyme that contains a minimal amount of buffer salts or an enzyme that has been dialyzed against the buffer being used for the substrate solution. The substrate and enzyme solutions can then be loaded into separate compartments in the calorimeter, allowed to equilibrate, and then mixed. Note that the enzyme serves to trigger the reaction just as a spark or combustion of a piece of thread does in a combustion calorimeter. Upon mixing, the enzyme-catalyzed reaction will proceed, and one uses the calorimetric instrumentation to measure $\Delta_r H$. Following reaction, the resultant solution is removed from the calorimeter and the molalities of the substrates in this solution can be measured. It is good practice to perform measurements of "blank" enthalpies by mixing the substrate solution with the buffer and the enzyme solution with the buffer. These blank enthalpies of mixing can then be subtracted from the measured value of $\Delta_r H$.

Clearly, there are two aspects associated with such a measurement, namely, the measurement of the enthalpy change $\Delta_r H$ (units of J) measured with the calorimeter and that is associated with the reaction of interest and the measurement of the amount of reaction n'_r (units of mol). The quantity $\Delta_r H_m(\text{cal})$, with units of J mol^{-1}, is the ratio of these two quantities. The quantity n'_r can be measured using either gravimetric or volumetric methods. In both cases, it is also necessary to measure the fraction of substrate that has reacted. Thus, when a gravimetric method is used,

$$n'_r = m \cdot w_{H_2O} \cdot x \cdot m_{sol}. \tag{5.42}$$

Here, m is the molality of the substrate in the substrate solution, w_{H_2O} is the mass fraction of water in the substrate solution and prior to any mixing with the enzyme solution, x is the fraction of the substrate that has undergone reaction, and m_{sol} is the mass of substrate solution placed in the calorimeter. If a volumetric method is used,

$$n'_r = c \cdot V \cdot x, \tag{5.43}$$

where:
  c is the concentration of the substrate in the substrate solution
  V is the volume of substrate solution placed in the calorimeter

Just as in the case of equilibrium measurements, it is critical that the conditions of measurement be reported for calorimetric measurements.

From the above discussion, one can see that there are several possible sources of systematic error that may arise in the equilibrium measurements. Specifically, errors may arise due to impurities in the samples used, measurements of molalities (or concentrations), failure to achieve equilibrium, and the possibility of side reactions that interfere with the accurate measurement of K'. Errors due to weighing, pH measurement, and assigned reaction temperature can usually be made negligible by using a reasonable degree of care. For calorimetric measurements, one must also be concerned about heat measurement error, including corrections for blank enthalpies of mixing. Indeed, when $\Delta_r H$ is small, the blank enthalpies of mixing may be a large source of error.

For many enzyme-catalyzed reactions, a very practical consideration is the cost or availability of the enzymes and substrates needed to study a reaction. Fortunately, equilibrium measurements can generally be carried out by using solutions having a volume of less than 1 cm³. The early calorimetric measurements involving enzyme-catalyzed reactions used macrocalorimeters that required substantial amounts of substrate and enzyme. However, the development of microcalorimeters in the 1960s and 1970s (Benzinger and Kitzinger 1963; Wadsö 1968; Evans 1969; Prosen et al. 1974) that were suitable for mixing relatively small amounts of solution made possible the measurement of values of $\Delta_r H_m$(cal) for many enzyme-catalyzed reactions. In such measurements, the enzyme and substrate solutions are loaded into two separate compartments of the reaction vessel. The two compartments are separated from each other by a partition but share a common vapor space. Following a suitable equilibration time in the heat sink of the microcalorimeter, the two solutions are mixed together by rotating the entire microcalorimeter heat sink.

The early microcalorimeters developed by Tian and Calvet (1963) required the tedious construction of multijunction thermocouples. However, the commercial availability of inexpensive solid-state Peltier cooling modules made it possible to avoid the very difficult construction problems associated with the construction of multijunction thermocouples. During these same years, solid-state electronic amplifiers became available. These instruments were much easier to use and more reliable than traditional galvanometers. And in the 1990s, digital multimeters with nanovolt sensitivity became available, and these instruments eliminated the need for separate amplifiers. Finally, when desktop computers became available, one could use them for the data acquisition and control associated with microcalorimetry experiments (Steckler et al. 1986a, 1986b). The ability to remove the reaction vessel from a mixing microcalorimeter is of substantial utility. Specifically, the empty reaction vessel can be weighed, loaded with solution, and weighed again so that one has a gravimetric determination of the masses of the enzyme and substrate solutions in the vessel. Additionally, a thorough cleaning of the removable vessel is substantially easier than the cleaning of a vessel that is left permanently in place in a microcalorimeter. Most mixing microcalorimeters are calibrated by using a resistance heater located in the calorimeter and in close proximity to the reaction vessel. Thus, it is good practice to use a test reaction that has a well-known molar enthalpy of reaction in order to check the accuracy of these calorimeters. In this regard, the use of (acid + base) reactions and the enzyme-catalyzed reaction of sucrose to form (**D**-glucose + **D**-fructose) has been recommended (Wadsö and Goldberg 2001) as test reactions for isothermal mixing microcalorimeters.

5.4 SCIENTIFIC LITERATURE ON THE THERMODYNAMICS OF ENZYME-CATALYZED REACTIONS

The scientific literature on the thermodynamics of enzyme-catalyzed reactions has been summarized in a series of reviews—see Goldberg et al. (2004, 2007) and the several references that are cited therein. The property values given in these reviews are also available at the website http://xpdb.nist.gov/enzyme_thermodynamics/ (accessed January 7, 2016). This substantial collection of thermodynamic property data embraces ≈1200 scientific publications. The reported thermodynamic properties are interrelated due to the fact that the Gibbs energy, enthalpy, and entropy are state functions. Thus, there is an opportunity and need to tie these property data together by using thermodynamic network calculations. The NBS Tables of Chemical Thermodynamic Properties (Wagman et al. 1982) are based on such calculations—albeit most of the property values were obtained by solving the equations sequentially. This subject was discussed in a recent publication (Goldberg 2010, p. 214):

> This large set of [biochemical] property values can be used to establish a thermodynamic network, that is, a system of linear equations that can be solved for the desired standard formation properties. Such an undertaking requires extensive literature work, a substantial amount of analysis and computation on the results of the individual studies, and a careful fitting together of the property values by means of a judicious weighting of the property values. It can be viewed as a very large "jig-saw puzzle" of information. But the proper construction of such a network would serve to bring together a large body of related property values and would be of immense practical value to the scientific community.

Note that such an effort is not a trivial endeavor, as one also needs to bring in, as needed, property values for standard molar enthalpies of solution, standard molar enthalpies of combustion, standard molar entropies, saturation molalities, and equilibrium constants and standard molar enthalpies for proton and metal–ion binding reactions for the substances in the network. By doing this, one can create a "reaction catalog" (Goldberg and Tewari 1989; Goldberg 2010) that consists of all pertinent thermodynamic property values for the substances and reactions that are contained in the existing Thermodynamics of Enzyme-Catalyzed Reactions Database (Goldberg et al. 2004, 2007). This reaction catalog is a system of linear equations that can be solved by least squares to yield values of standard formation properties for the species in the reactions (Goldberg and Tewari 1989; Ruscic et al. 2004).

A successful effort would include several major benefits to the scientific community:

- The ability to calculate thermodynamic property values for substances and reactions for which direct measurements are not available.

- The ability to identify discrepancies in existing property values and to identify which new measurements would be most valuable.

- The capability of having a database that can be added to and corrected at any time and which can be used to provide the most up-to-date values of thermodynamic property values on demand.

- The basis for the development of accurate methods and parameters for the estimation of thermodynamic property values for a wide variety of biochemical substances and reactions and where direct measurements have not been performed. For example, these estimation methods could be based on correlations with structural features in the substances of interest and use the Benson group contribution approach (Benson and Buss 1958; Benson 1968; Benson et al. 1969; Domalski and Hearing 1993; Domalski 1998).

- The capability to calculate thermodynamic property values K', $\Delta_r H_m'^{\circ}$, $\Delta_r H_m(\text{cal})$, $\Delta_r N(\text{H}^+)$, $\Delta_r N(\text{Mg}^{2+})$, and so on, as a function of T, pH, pMg, and I simply by specifying the reaction and the desired conditions.

In summary, the above capabilities would be of immense value to bioprocess and metabolic engineering, to metabolic control analysis, and to many aspects of quantitative biochemistry. Much of the proposed database rests on the equilibrium and calorimetric measurements performed by many investigators over the past 90 years.

GLOSSARY

a: activity (dimensionless)
c: concentration (mol dm^{-3})
A_m: Debye–Hückel constant, molality basis (kg$^{1/2}$ mol$^{-1/2}$)
B: parameter in the extended Debye–Hückel equation (kg$^{1/2}$ mol$^{-1/2}$)
C_p: heat capacity at constant pressure (J K^{-1})
G: Gibbs energy (J)
H: enthalpy (J)

I: ionic strength (mol kg^{-1})

K: thermodynamic equilibrium constant (dimensionless)

K_m: equilibrium constant, molality basis (mol kg^{-1})$^{\Sigma v_i}$

K_c: equilibrium constant, concentration basis (mol m^{-3})$^{\Sigma v_i}$

K_x: equilibrium constant, mole fraction basis (dimensionless)

K′: apparent equilibrium constant*

m: mass (kg)

m: molality (mol kg^{-1})

M: molar mass (kg mol^{-1})

n: amount of substance (mol)

N_r: number of reactants in a chemical reaction (dimensionless)

N_H: number of hydrogen atoms (dimensionless)

N_{Mg}: number of magnesium atoms (dimensionless)

p: pressure (Pa)

pH: $-\lg\{a(H^+)\} = -\lg\{m(H^+)\gamma(H^+)/m°\}$ (dimensionless)[†]

pMg: $-\lg\{a(Mg^{2+})\} = -\lg\{m(Mg^{2+})\gamma(Mg^{2+})/m°\}$ (dimensionless)[‡]

pX: $-\lg\{a(X)\} = -\lg\{m(X)\gamma(X)/m°\}$ (dimensionless)[§]

Q: reaction quotient (dimensionless)[¶]

R: denotes a reactant R

R: molar gas constant (8.314 4621 J mol K^{-1})

S: entropy (J K^{-1})

T: thermodynamic temperature (K)

V: volume (m^3)

w: mass fraction (dimensionless)

x: mole fraction (dimensionless)

z: signed charge (dimensionless)

Subscripts

c: denotes a concentration basis

f: denotes a formation reaction

i: denotes a specific reactant

j: denotes a reaction

---

[*] In general, one must specify the basis for the apparent equilibrium constant, that is, molality, concentration, or mole fraction. Thus, one can write K'_m, K'_c, or K'_x in analogy to what is done for the equilibrium constant for a chemical reaction. The dimensions of K'_m are (mol kg^{-1})$^{\Sigma v_i}$, and the dimensions of K'_c are (mol m^{-3})$^{\Sigma v_i}$, where the v_i are the stoichiometric numbers for the reactants in the reaction. All values of the thermodynamic equilibrium constant and all values of K'_x are dimensionless. Also, all values of K'_m and K'_c are dimensionless for symmetric reactions. The author believes that a degree of neatness is achieved by making all equilibrium constants dimensionless by the appropriate use of either the standard molality ($m° = 1$ mol kg^{-1}) or the standard concentration ($c° = 1$ mol dm^{-3}). In all cases, it is critical to define how one calculates the value of an equilibrium constant and to specify the standard state.

[†] This definition is a notional one in that single-ion activities have not yet been measured.

[‡] See previous footnote.

[§] See previous footnote.

[¶] We consider the reaction quotient Q to be dimensionless by appropriate use of the standard molality $m°$ or the standard concentration $c°$.

m: denotes a molality basis

m: denotes a molar quantity[*]

r: denotes a reaction

x: denotes a mole fraction basis

w: denotes the solvent

θ: denotes a reference temperature

Superscripts

ex: denotes an excess quantity

id: denotes an ideal quantity

°: denotes a standard quantity

′: denotes a transformed quantity that pertains to an overall biochemical reaction

*: denotes the pure solvent

Greek

γ: activity coefficient (dimensionless)

Δ: denotes a change

ξ: extent of reaction (mol)

θ: reference temperature (K)

μ: chemical potential (J mol^{-1})

υ: stoichiometric number (dimensionless)

ρ: density (kg m^{-3})

σ: parameter defined in Equation 5.20 (dimensionless)

REFERENCES

Akers, D. L. and Goldberg, R. N. 2001. BioEqCalc: A package for performing equilibrium calculations on biochemical reactions. *Mathematica J.* 8:86–113. The *Mathematica* package can be downloaded from http://xpdb.nist.gov/enzyme_thermodynamics/ (accessed January 7, 2016).

Alberty, R. A. 1968. Effect of pH and metal ion concentration on the equilibrium hydrolysis of adenosine triphosphate to adenosine diphosphate. *J. Biol. Chem.* 243:1337–1343.

Alberty, R. A. 1969a. Standard Gibbs free energy, enthalpy, and entropy changes as a function of pH and pMg for several reactions involving adenosine phosphates. *J. Biol. Chem.* 244:3290–3302.

Alberty, R. A. 1969b. Maxwell relations for thermodynamic quantities of biochemical reactions. *J. Am. Chem. Soc.* 91:3899–3903.

Alberty, R. A. 1992. Equilibrium calculations on systems of biochemical reactions at specified pH and pMg. *Biophys. Chem.* 42:117–131.

Alberty, R. A. 2003. *Thermodynamics of Biochemical Reactions*. Wiley-Interscience: Hoboken, NJ.

Alberty, R. A., Cornish-Bowden, A., Gibson, Q. H., Goldberg, R. N., Hammes, G., Jencks, W., Tipton, K. F., Veech, R., Westerhoff, H. V., and Webb, E. C. 1994. Recommendations for nomenclature and tables in biochemical thermodynamics. *Pure Appl. Chem.* 66:1641–1666. This manuscript was also published in *Eur. J. Biochem.* 240:1–14 (1966).

Alberty, R. A., Cornish-Bowden, A., Goldberg, R. N., Hammes, G. G., Tipton, K., and Westerhoff, H. V. 2011. Recommendations for terminology and databases for biochemical thermodynamics. *Biophys. Chem.* 155:89–103.

---

[*] The subscript *m* is often omitted in cases where it is clear that the quantity of interest is a molar quantity.

Alberty, R. A. and Goldberg, R. N. 1993. Calorimetric determination of the standard transformed enthalpy of a biochemical reaction at specified pH and pMg. *Biophys. Chem.* 47:213–223. Also see errata published in *Biophys. Chem.* 47:213–223.

Benson, S. W. 1968. *Thermochemical Kinetics.* John Wiley: New York.

Benson, S. W. and Buss, J. H. 1958. Additivity rules for the estimation of molecular properties. Thermodynamic properties. *J. Chem. Phys.* 29:546–572.

Benson, S. W., Cruickshank, F. R., Golden, D. M., Haugen, G. R., O'Neal, H. E., Rodgers, A. S., Shaw, R., and Walsh, R. 1969. Additivity rules for the estimation of thermochemical properties. *Chem. Rev.* 69:279–324.

Benzinger, T. H. and Kitzinger, C. 1963. Microcalorimetry, new methods and objectives. In J. D. Hardy (ed.), *Temperature—Its Measurement and Control in Science and Industry*, vol. 3, part 3, 43–60. Reinhold: New York.

Calvet, E. 1963. *Recent Progress in Microcalorimetry.* Macmillan: New York.

Clarke, E. C. W. and Glew, D. N. 1966. Evaluation of thermodynamic functions from equilibrium constants. *Trans. Faraday Soc.* 62:539–547.

Clarke, E. C. W. and Glew, D. N. 1980. Evaluation of Debye-Hückel limiting slopes for water between 0 and 150°C. *J. Chem. Soc. Faraday Trans. I* 76:1911–1916.

Cohen, E. R., Cvitaš, T., Frey, J. G., et al. 2008. *Quantities, Units and Symbols in Physical Chemistry*, 3rd ed. IUPAC & RSC Publishing: Cambridge, UK.

Cornish-Bowden, A. 2004. *Fundamentals of Enzyme Kinetics.* Portland Press: London.

Domalski, E. S. 1998. Estimation of enthalpies of formation of organic compounds at infinite dilution in water at 298.15 K. In K. K. Irikura and D. J. Frurip (eds.), *Computational Thermochemistry—Prediction and Estimation of Molecular Thermodynamics.* ACS Symposium Series No. 677. American Chemical Society: Washington, DC.

Domalski, E. S. and Hearing, E. D. 1993. Estimation of the thermodynamic properties of C-H-N-O-S-halogen compounds at 298.15 K. *J. Phys. Chem. Ref. Data* 22:805–1159.

Evans, W. J. 1969. The conduction-type microcalorimeter. In H. D. Brown (ed.), *Biochemical Microcalorimetry*, 257–273. Academic: New York.

Goldberg, R. N. 2010. Thermodynamic network calculations applied to biochemical substances and reactions. In M. G. Hicks and C. Kettner (eds.), *Proceedings of the 4th International Beilstein Workshop on Experimental Standard Conditions of Enzyme Characterizations*, 213–230. Logos Verlag Berlin: Berlin.

Goldberg, R. N. and Frank, H. S. 1972. Liquid junction potentials and single-ion activities by computer simulation. I. The concentration cell with transference. *J. Phys. Chem.* 76:1758–1762.

Goldberg, R. N. and Tewari, Y. B. 1989. Thermodynamic and transport properties of carbohydrates and their monophosphates: The pentoses and hexoses. *J. Phys. Chem. Ref. Data* 18:809–880.

Goldberg, R. N. and Tewari, Y. B. 1991. Thermodynamics of the disproportionation of adenosine 5'-diphosphate to adenosine 5'-triphosphate and adenosine 5'-monophosphate. I. Equilibrium model. *Biophys. Chem.* 40:241–261.

Goldberg, R. N., Tewari, Y. B., and Bhat, T. N. 2004. Thermodynamics of enzyme-catalyzed reactions—A database for quantitative biochemistry. *Bioinformatics* 16:2874–2877.

Goldberg, R. N., Tewari, Y. B., and Bhat, T. N. 2007. Thermodynamics of enzyme-catalyzed reactions: Part 7—2007 update. *J. Phys. Chem. Ref. Data.* 36:1347–1397.

Kishore, N., Tewari, Y. B., Akers, D. L., Goldberg, R. N., and Miles, E. W. 1998. A thermodynamic investigation of reactions catalyzed by tryptophan synthase. *Biophys. Chem.* 73:265–280.

Klein, M. 1971. Practical treatment of coupled gas equilibrium. In H. Eyring, W. Jost, and D. Henderson (eds.), *Physical Chemistry—An Advanced Treatise*, vol. 1, 489–544. Academic: New York.

Meyerhof, O. and Lohmann, K. 1935. Über die enzymatische Gleichgewichtreaktion zwischen Hexosediphosphate und Dioxyacetonphophorsäure. IV. *Biochem. Z.* 275:430–432.

Prosen, E. J., Goldberg, R. N., Staples, B. R., Boyd, R. N., and Armstrong, G. T. 1974. Microcalorimetry applied to biochemical processes. In H. Kambe and P. D. Garn (eds.), *Thermal Analysis: Comparative Studies on Materials*, 253–289. Kodansha Ltd. and John Wiley: Tokyo and New York.

Quastel, J. H. and Woolf, B. 1926. LXXII. The equilibrium between L-aspartic acid, fumaric acid and ammonia in presence of resting bacteria. *Biochem. J.* 20:545–555.

Robinson, R. A. and Stokes, R. H. 1955. *Electrolyte Solutions*. Academic: New York.

Ruscic, B., Pinzon, R. E., Morton, M. L., von Laszewski, G., Bittner, S., Nijsure, S. G., Amin, K. A., Minkoff, M., and Wagner, A. F. 2004. Introduction to active thermochemical tables: Several "key" enthalpies of formation revisited. *J. Phys. Chem. A* 108:9979–9997.

Steckler, D. K., Goldberg, R. N., Tewari, Y. B., and Buckley, T. J. 1986a. High precision microcalorimetry: Apparatus, procedures, and biochemical applications. *J. Res. Natl. Bur. Stand.* 91:113–121.

Steckler, D. K., Goldberg, R. N., Tewari, Y. B., and Buckley, T. J. 1986b. Computer software for the acquisition and treatment of calorimetric data. National Bureau of Standards Technical Note 1224. U.S. Government Printing Office: Washington, DC.

Tewari, Y. B., Schantz, M. M., Pandey, P. C., Rekharsky, M. V., and Goldberg, R. N. 1995. Thermodynamics of the hydrolysis of N-acetyl-L-phenylalanine ethyl ester in water and in organic solvents. *J. Phys. Chem.* 99:1594–1601.

Wadsö, I. 1968. Design and testing of a micro reaction calorimeter. *Acta Chem. Scand.* 22:927–937.

Wadsö, I. and Goldberg, R. N. 2001. Standards in isothermal microcalorimetry. *Pure Appl. Chem.* 73:1625–1639.

Wadsö, I., Gutfreund, H., Privalov, P., Edsall, J. T., Jencks, W. P., Armstrong, G. T., and Biltonen, R. L. 1976. Recommendations for measurement and presentation of biochemical equilibrium data. *J. Biol. Chem.* 251:6879–6885. This manuscript was also published in *Q. Rev. Biophys.* 2:439–456.

Wagman, D. D., Evans, W. H., Parker, V. B., Schumm, R. H., Halow, I., Bailey, S. M., Churney, K. L., and Nuttall, R. L. 1982. The NBS tables of chemical thermodynamic properties. Selected values for inorganic and C_1 and C_2 organic substances in SI units. *J. Phys. Chem. Ref. Data* 11(2).

Weischet, W. O. and Kirschner, K. 1976. The mechanism of the synthesis of indoleglycerol phosphate catalyzed by tryptophan synthase from *Escherichia coli*. Steady-state kinetic studies. *Eur. J. Biochem.* 65:365–373.

Zarubin, D. P. 2011. The nature of single-ion activity coefficients calculated from potentiometric measurements on cells with liquid junctions. *J. Chem. Thermodyn.* 43:1135–1152.

II

Membrane Characterization and Partition to Membranes

Temperature-Induced and Isothermal Phase Transitions of Pure and Mixed Lipid Bilayer Membranes Studied by DSC and ITC

Alfred Blume

CONTENTS

6.1 BIOLOGICAL AND MODEL MEMBRANES AND THEIR LIPID CONSTITUENTS

Differential scanning calorimetry (DSC) was used early on for the determination of thermotropic phase transitions in lyotropic lipid phases. The interest in lipid behavior stemmed from the fact that the basic structure of the biological membrane is a bimolecular leaflet of lipids, in which proteins are either incorporated in or peripherally bound to the surface of the lipid bilayers. After several models had been proposed for the structure of biological membranes, the Singer and Nicolson fluid-mosaic model presented in 1972 was finally widely accepted as a structural model for a biological membrane (see Figure 6.1) (Singer and Nicolson 1972; Nicolson 2014). Today, this model is still used, but with modifications taking into account inhomogeneous distributions of lipids and proteins in the plane of the membrane and the presence of dynamic lipid/protein clusters having different composition (Vaz and Almeida 1993; Vaz 1994; Vereb et al. 2003; Goni 2014; Nicolson 2014).

Biological membranes are composed of a multitude of different lipids, phospholipids being one of the major lipid classes. The chemical structures of some typical phospholipids with saturated fatty acyl chains are shown in Figure 6.2. In biological membranes, phospholipids have a high proportion of unsaturated fatty acids, and the two chains also differ in their length and degree of unsaturation.

When lipids are brought into contact with water, the spontaneous formation of lyotropic phases in the form of myelin figures is observed (Lehmann 1904). It was recognized very early that changes in concentration and/or temperature can lead to different phases, a process called *lyotropic and thermotropic mesomorphism* (Chapman 1968). DSC is well suited to the observation of the thermotropic transitions between the different phases. However, in the beginning, it was quite difficult to obtain reliable DSC data because of the limited availability of calorimetric instruments suited to this purpose. In the late 1960s, Dennis Chapman was one of the pioneers in applying calorimetric

FIGURE 6.1 Fluid-mosaic model of a biological membrane as suggested by Singer and Nicolson. (From Singer, S. J., and Nicolson, G. L., *Science* 175, 720–731, 1972; Nicolson, G. L., *Biochim. Biophys. Acta* 1838, 1451–1466, 2014; modified from Singer, S. J., and Nicolson, G. L., *Science* 175, 720–731, 1972; Blume, A., and Tuchtenhagen, J., *Biochemistry* 31, 4636–4642, 1992.)

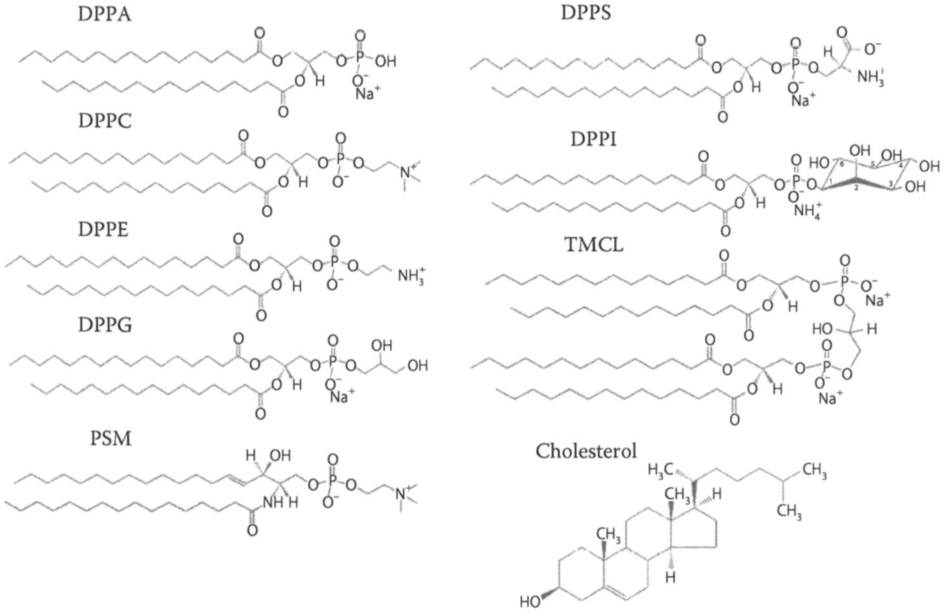

FIGURE 6.2 Chemical structures of various phospholipids. DPPA: 1,2-dipalmitoyl-*sn*-glycero-3-phosphoric acid; DPPC: 1,2-dipalmitoyl-*sn*-glycero-3-phosphocholine; DPPE: 1,2-dipalmitoyl-*sn*-glycero-3-phosphoethanolamine; DPPG: 1,2-dipalmitoyl-*sn*-glycero-3-phosphoglycerol; DPPS: 1,2-dipalmitoyl-*sn*-glycero-3-phosphoserine; DPPI: 1,2-dipalmitoyl-*sn*-glycero-3-phosphoinositol; TMCL: 1′,3′-*bis*[1,2-dimyristoyl-*sn*-glycero-3-phospho]-*sn*-glycerol; PSM: *N*-(hexadecanoyl)-sphing-4-enine-1-phosphocholine; cholesterol: Cholest-5-en-3β-ol.

methods to the study of these phase transitions in lipid bilayers, applying differential thermal analysis (DTA) as well as DSC, using one of the commercial instruments available at that time, the Perkin-Elmer DSC 1B. The thermotropic and lyotropic properties of 1,2-dipalmitoyl-*sn*-glycero-3-phosphocholine (DPPC), as a dry compound and suspended in water, were extensively characterized as a function of water content, and the first DPPC/water phase diagram was constructed (Chapman et al. 1967; Chapman 1968, 1975; Ladbrooke and Chapman 1969; Lee 1977a).

For investigation of the lyotropic behavior of lipids in water, the dry lipid is usually suspended in water. Either the lyotropic phase forms spontaneously at room temperature or the suspension has to be sonicated or vortexed at a temperature above its transition temperature into the liquid-crystalline phase. Phospholipids then form lyotropic lamellar phases in the form of either multilamellar or smaller oligolamellar or unilamellar vesicles. Changes of temperature can then lead to temperature-induced transitions between different types of lamellar phases or to other lyotropic phases, such as inverted hexagonal or bicontinuous cubic phases. Figure 6.3 shows as an example the phase diagram for the typical phospholipid DPPC with the stability regions of the different lyotropic phases (Small 1986; Caffrey and Cheng 1995).

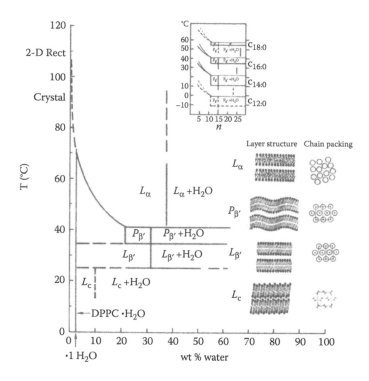

FIGURE 6.3   Phase diagram for DPPC as a function of water content showing the different lamellar phases. Insert: phase diagrams obtained for PCs with other saturated acyl chains. (From Small, D. M., *Handbook of Lipid Research*, vol. 4, Plenum Press: New York, 1986.)

6.2 DIFFERENTIAL SCANNING CALORIMETRY

6.2.1 Lyotropic and Thermotropic Properties of Lipid–Water Mixtures

In DSC experiments, investigations of the thermotropic behavior of lipids are usually carried out in excess water, so that the experiment corresponds to a vertical cut through the phase diagram shown in Figure 6.3. In this chapter, we will describe only transitions between different lamellar phases and will not discuss the formation of other phases, such as various cubic or inverted hexagonal phases, because the lamellar phase is the most relevant for understanding the structure of membranes. Figure 6.4 shows the DSC curve observed for DPPC in excess water with the corresponding structures of the lamellar phases. Each peak represents a transition into a different lamellar phase. The major endothermic peak is due to the so-called main transition from the $P_{\beta'}$- to the L_α-phase. From DSC curves of lipid samples and the corresponding water–water base lines, the apparent molar heat capacities of the phospholipids in their different phases can also be calculated, provided the apparent molar volume of the lipids is known. These heat capacity values contain information on the amount of hydration of the polar as well as the nonpolar moieties of the lipids (Blume 1983). DSC data for the thermotropic behavior of phosphatidylcholines (PCs) with different fatty acyl chains can be found in a comprehensive review (Koynova and Caffrey 1998).

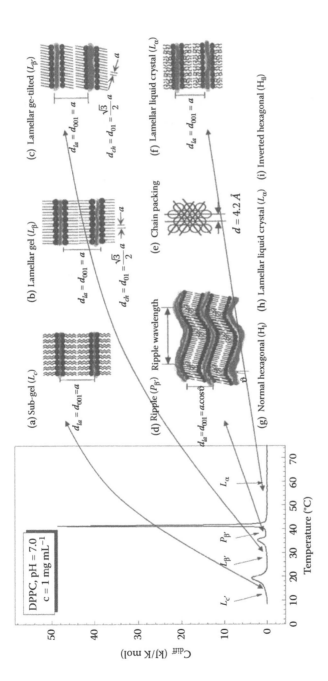

FIGURE 6.4 Differential scanning calorimetry curve of multilamellar DPPC vesicles in excess water and the structures of the different lamellar phases. (Scheme for phases with water layers in light grey adapted from Caffrey, M., and Cheng, A., *Curr. Opin. Struct. Biol.* 5, 548–555, 1995.)

The lyotropic and thermotropic properties of phospholipids depend not only on the length and saturation of the acyl chains and their connection to the glycerol backbone, but also on the chemical structure of the headgroups (Cevc and Marsh 1987; Cevc 1993). Figure 6.5 shows DSC scans of four different types of phospholipids (see Figure 6.2) with either C_{14} or C_{16} acyl chains (Blume 1983, 1991; Blume and Garidel 1999). It is evident that the temperature of the main endothermic peak due to the transition from an ordered lamellar gel phase (L_β-, $L_{\beta'}$- or $P_{\beta'}$-phase) into a disordered liquid-crystalline lamellar phase (L_α-phase) depends on the chain length of the saturated acyl chains and on the chemical structure of the headgroup. Phosphatidylethanolamines (PEs) and phosphatidic acids (PAs) have significantly higher transition temperatures than the corresponding PCs and phosphatidylglycerols (PGs). This is due to the fact that the former convert directly from a lamellar L_β-phase (see Figure 6.4) with untilted chains to the liquid-crystalline L_α-phase, whereas the transition for the latter two lipids occurs from a $P_{\beta'}$-phase (see Figure 6.4), the so-called ripple phase with surface undulations of the lamellae, to the L_α-phase. The dependence of the main transition temperature on the length of the saturated acyl chains for phospholipids carrying identical chains is depicted in Figure 6.5. For data on the influence of chain length, headgroup structure, and unsaturation on the thermotropic properties of lyotropic lipid phases, the reader is referred to a recent review by Marsh (2010).

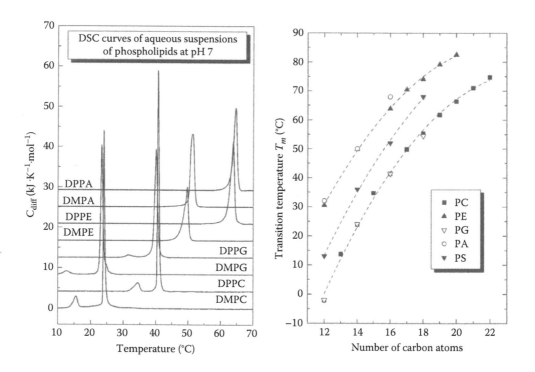

FIGURE 6.5 (a) DSC curves for different phospholipids with myristoyl and palmitoyl chains. (b) Temperature of the main gel to liquid-crystalline phase transition as a function of chain length. (Adapted from Blume, A., *Thermochim. Acta* 193, 299–347, 1991; Blume, A., and Garidel, P., *Handbook of Thermal Analysis and Calorimetry*, 109–173, Elsevier: Amsterdam, 1999.)

6.2.2 Thermotropic Behavior of Biological Membranes and Model Membranes of Lipid Mixtures

As many different lipids occur in biological membranes, the systematic investigation of the thermotropic behavior of lipid mixtures also started relatively early (Ladbrooke and Chapman 1969; Blume and Ackermann 1974; Chapman 1975; Mabrey and Sturtevant 1976; Lee 1977b, 1978). Steim investigated membranes, extracted lipids of *Mycoplasma laidlawii* grown on media supplemented with different fatty acids, and found that the temperature of the thermal transitions observed in these membranes was identical to the transition temperature found for the extracted lipids suspended in water, proving that the lipids are in a similar state in the cell membranes as in the lyotropic lamellar phase in water (Steim et al. 1969). Similar results were obtained in experiments with *Escherichia coli* cells performed by Jackson and Sturtevant (1977). Figure 6.6 shows as an example DSC traces of natural membranes and their lipid extracts from *Salmonella minnesota*. The thermal transitions are very broad and differ between the whole cell envelope and the outer membrane of the gram-negative bacterium *S. minnesota* grown at 37°C, while the lipopolysaccharide extract of the outer membrane shows the same transition range as the complete outer membrane (Brandenburg and Blume 1987). The inner membrane has a lower transition temperature, well below the growth temperature. The finding that the growth temperature is usually above or at the upper end of the transition range has been seen with many cells and shows that the membrane has to be mainly in its liquid-crystalline state for the cell to be viable. Lowering the growth temperature usually causes thermoadaptation of the cells, leading to a shift of the membrane phase transition to a lower temperature as more unsaturated and/ or shorter chains are incorporated into the cell membranes (Hazel 1995).

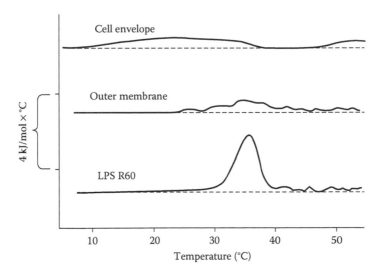

FIGURE 6.6 DSC curves of natural membranes and lipopolysaccharide preparations (approximately 10^{-3} M) from *S. minnesota* from mutant strain R60. (From Brandenburg, K., and Blume, A., *Thermochim. Acta* 119, 127–142, 1987.)

6.2.2.1 Pseudo-Binary Lipid Mixtures

As lipid extracts of biological membranes are usually complex mixtures of lipids differing in chemical structure of the headgroups as well as the chains, the question arose quite early of whether immiscibility could be observed in the plane of the lamellae, and which factors determine the miscibility of lipids in bilayer membranes. Systems with two lipids are called pseudo-binary mixtures, because the third component, water, is usually in excess, so that only the mixing of the two lipids is considered without taking into account possible phase changes due to a reduction of water content. As observed for three-dimensional systems, it was to be expected that miscibility in a liquid, that is, the liquid-crystalline lamellar phase, would be better than in an ordered lamellar phase, because packing two different molecules in the same ordered lamellar "lattice" requires the two compounds to have similar shapes. Therefore, immiscibility in lamellar gel phases consisting of two different lipids is quite common. For instance, it was found that in the gel phase, homogeneous mixing of two lipids having identical headgroups but different chain lengths is possible as long as this difference does not exceed four CH_2-groups. For larger differences, phase separation into two different lamellar gel phases is observed. When one of the lipid components has unsaturated chains with, for instance, one double bond, the *cis*-double bond induces a bend in the acyl chain and a drastic lowering of the main phase transition temperature (Cevc 1993; Marsh 1999). Mixing with lipids with saturated chains in the ordered gel phase is thus prevented. Due to the fluid nature of the bilayers in the L_α-phase caused by the *trans*-gauche isomerization of the acyl chains, immiscibility in the L_α-phase is observed only in special cases (see Section 6.2.2.1). Figure 6.7 shows schematically different types of pseudo-binary phase diagrams that can be expected and are in fact observed for the mixing of two lipids, A and B, where both components mix homogeneously in the liquid-crystalline (l.c.) phase, but differently in the gel phase. The regions where two phases coexist are indicated.

The mixing properties also depend on the chemical structure of the headgroups. Not only the size but also the net charge of the headgroup is important, as is its ability to form intermolecular hydrogen bonds in the plane of the bilayer headgroup region. In our group, we have intensively studied the behavior of various phospholipid mixtures with saturated chains to elucidate the influence of structure, size, and charge of the headgroups on the mixing behavior. As an example, Figure 6.8a shows DSC scans of a variety of lipid mixtures of 1,2-dimyristoyl-*sn*-glycero-3-phosphoglycerol (DMPG) with 1,2-dimyristoyl-*sn*-glycero-3-phosphoethanolamine (DMPE) at two different pH values. As the chains have the same length, only the headgroup properties have a decisive influence on the mixing behavior (Garidel and Blume 2000b).

From the DSC curves, phase diagrams can be constructed by taking the onset and end temperatures of the observed endothermic peaks and plotting them as a function of mole fraction, as shown in Figure 6.8b. Problems arise when broad endothermic peaks are observed. Then these temperatures are not so easily detectable, so that various other methods have been proposed to construct phase diagrams from experimental DSC curves (Lee 1978; Von Dreele 1978; Brumbaugh et al. 1990). In our laboratory, we have extensively used the method of simulation of the DSC curves, taking into account the broadening of the phase transition by limited cooperativity (Blume 1988, 1991; Johann et al. 1996;

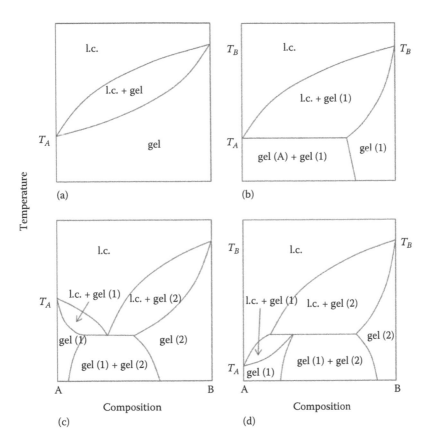

FIGURE 6.7 Simplified schemes of possible types of phase diagrams observed for pseudo-binary phospholipid mixtures with components A and B having transition temperatures T_A and T_B, respectively. (a) Diagram with complete miscibility in both phases. (b) Monotectic phase diagram with miscibility gap in the gel phase. (c) Eutectic phase diagram with eutectic phase transition temperature minimum and miscibility gap in the gel phase. (d) Peritectic phase diagram with miscibility gap in the gel phase.

Garidel et al. 1997a, 1997b, 2005, 2011; Garidel and Blume 1998, 1999, 2000b). This method makes the determination of the onset and end temperatures less arbitrary. The phase diagrams determined by this method can then be simulated using thermodynamic mixing models, which yield nonideality parameters for both the gel and the liquid-crystalline phase. In Figure 6.8a, the simulated C_p curves are shown as dashed lines, and Figure 6.8b shows the phase diagrams based on these simulations and also the phase boundaries determined in the classical way (Garidel and Blume 2000a).

The simulation of phase diagrams yields information on the type of nonideal mixing in both phases. Whether ideal or nonideal mixing occurs depends on the difference in interaction parameters between like and unlike molecules. This is usually described by a parameter $\omega_{AB} = g_{AB} - 1/2(g_{AA} + g_{BB})$, where g_{AA}, g_{BB}, and g_{AB} are the Gibbs free energies for the interactions between molecules A, molecules B, and molecules A and B, respectively (Hill 1985). These Gibbs free energies of interaction are also related to the nonideality

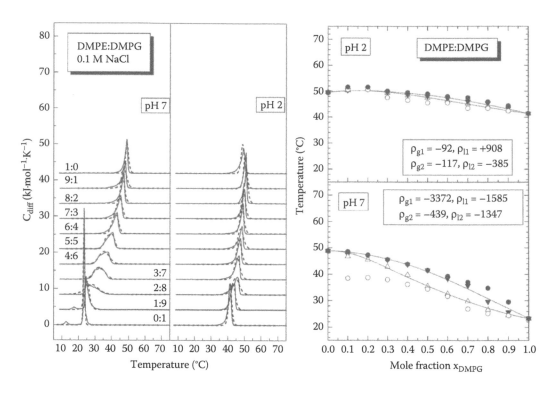

FIGURE 6.8   (a) DSC curves of mixtures of DMPE with DMPG in 0.1 M NaCl at two different pH values and various molar ratios. Solid line: experimental C_p curves; dotted line: simulated C_p curves. At pH = 2, the PG headgroup is almost completely protonated. (b) Pseudo-binary phase diagrams for the DMPE:DMPG system at pH 2 and 7. Triangles are T(−) and T(+) values obtained from the simulation of the C_p curve; circles are Texp(−) and Texp(+) values obtained by the usual empirical procedure. The solid lines are the coexistence lines obtained by a nonlinear least square fit of the T(−) and T(+) values using the four-parameter nonideal, nonsymmetric mixing model, yielding the nonideality parameters as indicated. (Adapted from Garidel, P., and Blume A., *Eur. Biophys. J.* 28, 629–638, 2000.)

parameters, ρ, derived from thermodynamic models assuming a certain type of lattice for the arrangement of the molecules in the plane of the bilayers, that is, $\omega_{AB} = z\,\rho$, with z being the coordination number, that is, the number of nearest neighbors (Hill 1985). For lipid lamellar phases, z is usually taken to be six for the L_α as well as for the gel phase. In the case of positive ρ, a tendency for demixing is present, leading to the formation of clusters of like molecules. If ρ have negative values, the preferential formation of pairs of unlike molecules is favored, sometimes also called *complex formation*. In our simulations, we used the model of regular solution theory or Bragg–Williams approximation, where nonideality is caused solely by an excess enthalpy of mixing, and the excess mixing entropy is zero. Nonideal and nonsymmetric mixing for both phases was assumed, meaning that the nonideality parameters depend on the composition, that is, $\rho = \rho_1 + \rho_2(2x-1)$, with x being the mole fraction of component A (Johann et al. 1996). The value ρ_2 describes the deviation from the case of symmetric mixing. For $x = 0.5$, $\rho = \rho_1$, that is, the nonideality parameter ρ for the simple symmetric mixing model is obtained. For DMPG/DMPE mixtures at pH 7 (see Figure 6.8), all ρ values are negative, indicating that preferential formation of unlike pairs is present, and that this pair formation is more favored with increasing DMPG content (Garidel and Blume 2000b). When DMPG becomes protonated and thus uncharged at pH 2, the mixing behavior changes drastically, particularly for the liquid-crystalline phase, where now a positive ρ_1 value is found, indicating a tendency for slight demixing. In other pseudo-binary lipid mixtures, even the formation of miscibility gaps in the liquid-crystalline phase can sometimes be observed (Garidel et al. 1997a, 2011). This observation is of particular relevance for the behavior of biological membranes, which are mostly in the liquid-crystalline phase. It is thus possible to induce the formation of clusters or domains with different composition in the plane of the membrane and thus also a heterogeneous distribution of membrane proteins by lipid demixing.

For pseudo-binary systems containing other components, such as fatty acids, monoacylglycerols, or detergents, the mixing behavior is usually much more complicated, as also inverted hexagonal, cubic, or micellar phases can be formed. The determination of the phase diagrams is then only possible using additional methods such as spectroscopic or scattering techniques (Carion-Taravella et al. 2002; Majhi and Blume 2002).

6.2.2.2 Pseudo-Ternary Lipid Mixtures

When more than two lipids are mixed, the presentation of the results as a simple binary phase diagram as shown in Figure 6.7 is no longer possible, as an additional concentration axis has to added. The phase diagram is now represented by a three-dimensional phase prism, and the phase boundaries are now surfaces in the phase prism. A horizontal cut through the prism gives an equilateral triangle for the phase diagram at a constant temperature, with the pure components being located at the vertices.

We studied recently the behavior of the ternary system DMPE/DMPG/1,2,-1′,2′-tetramyristoylcardiolipin (TMCl), a system relevant to the lipid composition of inner membranes of Gram-negative bacteria such as *E. coli* or *Bacillus subtilis* (Finger et al. 2013). The determination of the phase diagram was quite time consuming, as three pure compounds and approximately 33 binary and 27 ternary mixtures had to be investigated by DSC. Figure 6.9 shows the phase boundary surfaces of this pseudo-ternary system with the two-phase coexistence region of the L_α and gel phases located between the two surfaces. A particularity of this pseudo-ternary system is that the lower surface is almost horizontal for samples rich in DMPG, indicating limited miscibility in the gel phase for samples with 80–90 mol% DMPG, the rest being DMPE and/or TMCL. This is remarkable, as no differences in chain length exist.

As stated in Section 6.2.2.1, phase diagrams for more complicated systems cannot be determined by using only DSC. Other methods, such as x-ray scattering and spectroscopic methods, are needed. A compilation of more than 2000 phase diagrams of phospholipid mixtures, including also other components such as fatty acids, cationic lipids, detergents, or acylglycerols, has been published by Koynova and Caffrey (2002).

6.2.2.3 Lipid–Cholesterol Mixtures

A special, widely studied case is the thermotropic behavior of phospholipid–cholesterol mixtures. Cholesterol is an abundant typical component of many eukaryotic membranes,

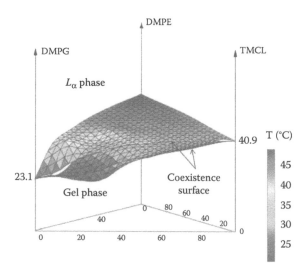

FIGURE 6.9 Ternary phase diagram obtained by DSC for mixtures of DMPG, DMPE, and TMCL. (From Finger, S. et al., *Eur. Biophys. J.* 42, S115, 2013.)

whereas it is missing in prokaryotic membranes. As cholesterol has a completely different shape, with its four-membered ring system containing only a small hydrophilic OH-group and, on the opposite end, a short branched aliphatic chain (see Figure 6.2), it can be expected that mixing with lipids will lead to quite different behavior compared with mixtures of normal phospholipids. It was found quite early on that in mixed bilayers of cholesterol with PCs or PEs having saturated chains, the main phase transition is gradually broadened and finally eliminated when the cholesterol content reaches a value between 40 and 50 mol% (Ladbrooke et al. 1968; Oldfield and Chapman 1972; Blume 1980). Figure 6.10 shows as a typical example DSC curves for the 1,2-distearoyl-*sn*-glycero-3-phosphocholine (DSPC)/ cholesterol system (Huang et al. 1993). As can be seen from the DSC scans, the peak for the main transition is gradually broadened with increasing cholesterol content, till it almost vanishes at 50 mol% cholesterol. For a sample with 40 mol% cholesterol, however, a very broad endothermic transition covering a range of more than 50°C is still visible. In addition to DSC, spectroscopic methods were used to detect changes in motional and conformational behavior of the lipid molecules in these mixtures, and it was found that cholesterol can induce an intermediate state of the lipids, in which the chains are almost in an all-*trans* conformation, that is, almost as ordered as in the gel state, whereas the rotational motions of the lipids are still as fast as in the L_α-phase. Likewise, the lateral diffusion coefficient is only slightly lower than those found for lipids in the L_α-phase (Oldfield and Chapman 1972). This particular state or phase was later called the liquid-ordered phase (L_0-phase).

For binary mixtures of PCs with cholesterol, phase diagrams or temperature-composition diagrams have been proposed that show possible regions of phase coexistence of L_α and L_0-phase (Pink and Chapman 1979; Ipsen et al. 1987; Thewalt and Bloom 1992). However, it turned out to be difficult to prove this L_α–L_0-phase coexistence by spectroscopic methods, as most likely the domains of different composition and order are small, and the exchange of molecules between the different domains is fast. For domains of the size of 30 nm, nuclear magnetic resonance (NMR) methods see only averaged signals, and the resolution of confocal fluorescence microscopy, and even more advanced methods such as stimulated emission depletion (STED), is still not high enough to detect these very small clusters. A critical assessment of the presence of the L_α–L_0-phase coexistence in these pseudo-binary mixtures has been performed recently (Goni et al. 2008; Almeida 2009; Feigenson 2009; Marsh 2009; Quinn 2010, 2012; Hirst et al. 2011). The phase diagrams published for these pseudo-binary lipid–cholesterol systems still fail to explain the broad endothermic transition present at high cholesterol content, where the existence of only a single L_0-phase is proposed (Blume 1980; Huang et al. 1993). This transition has a well-defined maximum at a certain temperature (see Figure 6.10) and a transition enthalpy of 29 kJ mol⁻¹ for bilayers composed of DSPC/40 mol% cholesterol. It is likely that the conformational disorder of the acyl chains is increasing over this wide temperature range with a very low cooperativity, though the plot of the chain order versus temperature determined by ²H-NMR shows a more or less linear and not a sigmoidal decrease (Huang et al. 1993).

The situation is different in pseudo-ternary lipid mixtures containing cholesterol, such as those containing a saturated phospholipid with a high transition temperature (PC or sphingomyelin [SM]) and a phospholipid with unsaturated chains, for instance, DOPC. In these cases, phase diagrams at constant temperature with the delineation of the phase boundaries were determined by using other methods. For instance, a wealth of information

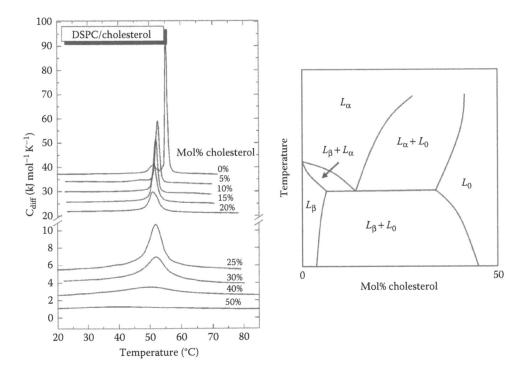

FIGURE 6.10 (a) DSC curves of DSPC–cholesterol mixtures with increasing cholesterol concentration. (Adapted from Huang, T. H. et al., *Biochemistry* 32, 13277–13287, 1993.) (b) Generic phase diagram suggested for PC–cholesterol mixtures with the different phases as indicated. (From Ipsen, J. H. et al., *Biochim. Biophys. Acta* 905, 162–172, 1987.)

has been obtained by confocal fluorescence microscopy (CFM) of giant unilamellar vesicles (GUVs) with fluorescent probes partitioning into domains of different order. Extensive phase separation into large domains of L_0- and L_d-phases could be visualized by CFM (Dietrich et al. 2001; Veatch and Keller 2005; Bagatolli 2006; Veatch et al. 2007). In these systems composed of PCs with unsaturated chains, SM, and cholesterol, an association of cholesterol with SM, the lipid having the long saturated chains, is observed. SM/cholesterol then forms the L_0-phase domain, because an all-*trans* conformation of the lipid chains of SM in the vicinity of the cholesterol can be induced, whereas this is not possible in the lipid with unsaturated chains. For more detailed information, the reader is referred to recent reviews on these systems and the various phase diagrams obtained for ternary lipid systems containing cholesterol (de Almeida et al. 2003; Goni et al. 2008; Almeida 2009; Feigenson 2009; Marsh 2009; Quinn 2010, 2013).

It should be mentioned again that the determination of phase diagrams of these pseudo-ternary systems using DSC alone is not possible, as the DSC endotherms are very broad and a determination of phase boundaries becomes virtually impossible. It is therefore absolutely necessary to use other methods in combination, such as CFM on GUVs and various spectroscopic techniques together with x-ray scattering experiments (Almeida et al. 1993; Nicolini et al. 2006; Pokorny et al. 2006). For the mixing properties of physiologically important ceramides with phospholipids, the reader is referred to Chapter 7 of this volume (Alonso and Goñi 2015).

6.2.3 Binding of Inorganic and Organic Cations to Membranes of Negatively Charged Lipids

The binding of divalent inorganic or multivalent organic ions to charged lipid bilayers can drastically change the phase behavior of single-component vesicles, but also the mixing behavior in lipid membranes containing, for instance, negatively charged lipids such as PG, phosphatidylserines (PS), or PA. The binding of divalent inorganic cations with their high charge density to the negatively charged phosphate group leads to complex formation, with a concomitant reduction in hydration of the headgroup region and a shift of the L_β–L_α transition to higher temperature. The binding effects depend on the size of the divalent ion, as they coordinate with the phosphate groups of the phospholipids and form inner-sphere or outer-sphere complexes (Garidel 1997; Garidel and Blume 1999, 2005). In mixtures with zwitterionic phospholipids, the binding of divalent cations usually induces phase separation, that is, the formation of a complex of the charged lipids separating from the uncharged ones (Garidel 1997; Garidel and Blume 2000a). The binding constant of singly charged cations such as Na^+ or K^+ to negatively charged phospholipids is much lower. In this case, up-shifts of the lipid phase transition temperature are observed only at very high concentrations of 1 M or more of these cations (Tatulian 1993).

Multivalent organic cations, such as cationic polyamines, oligopeptides, and polypeptides, also bind to negatively charged lipid bilayers. For organic cations, the driving force for binding is not solely the electrostatic attraction to the negatively charged surface; hydrophobic effects can also play a role. Electrostatic binding alone usually increases the phase transition temperature due to dehydration of the headgroup. The incorporation of hydrophobic moieties of polyamines or oligopeptides into the headgroup region of the lipid bilayers can perturb chain packing, causing a broadening of the phase transition and a concomitant shift to lower temperature. These two competing effects can be disentangled by systematically reducing the length of the hydrophobic segment of, for instance, the amino acid lysine in pentalysine (Hoernke et al. 2012). Similar differences can be seen when the effects of polylysines on the phase transition are compared with those of polyarginines (Schwieger and Blume 2007, 2009). The cationic charge is more delocalized in the guanidinium side chain of arginine, so that it can more easily insert into the headgroup region of the bilayers compared with the lysine side chain. Therefore, a shift of the phase transition to lower temperature is observed for polyarginine binding to PGs.

Recently, we studied the effect of binding of various positively charged polyamines on the thermal behavior of negatively charged PG bilayers (Finger et al. 2014). For diamines, the extent of the shift of the phase transition depends on the distance between the two charges (see Figure 6.11). When the shift of the phase transition is plotted as a function of diamine concentration, the apparent binding constant can also be determined. Two compensating effects are operating. Firstly, the charge distance in ethylenediamine and propylenediamine fits best to the charge distance between the DMPG headgroups, so that the electrostatic attraction is optimal for these two diamines. An increase in charge distance leads to an increase in hydrophobicity and a less efficient charge interaction. The incorporation of the CH_2 groups into the headgroup region of the bilayers can then perturb chain packing, thus decreasing the phase transition temperature (Finger et al. 2014).

The binding of polycations can also induce phase separation in lipid mixtures consisting of negatively charged and zwitterionic lipids, for instance, binding of polylysines to PC–PG mixed bilayers (Franzin and Macdonald 2001; Schwieger and Blume 2007). These

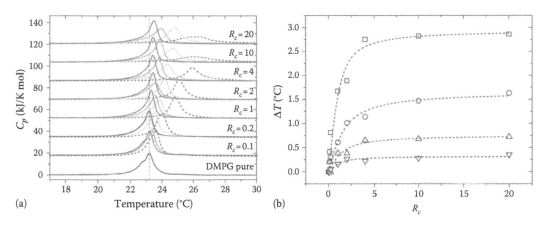

FIGURE 6.11 (a) DSC curves of pure DMPG (solid black) and DMPG with different polyamines at different charge ratios (R_c): ethylenediamine (dashed line), propylenediamine (light gray), putrescine (mid-gray), and cadaverine (dark gray). (b) ΔT_m as a function of charge ratio R_c of divalent polyamines to DMPG: ethylenediamine (squares), propylenediamine (circles), putrescine (upward triangles), and cadaverine (downward triangles). Dotted lines: simulation of experimental data using a binding model assuming a 1:1 binding of divalent amines to two DMPG molecules. (Adapted from Finger, S. et al., *Biol. Chem. Hoppe-Seyler* 395, 769–778, 2014.)

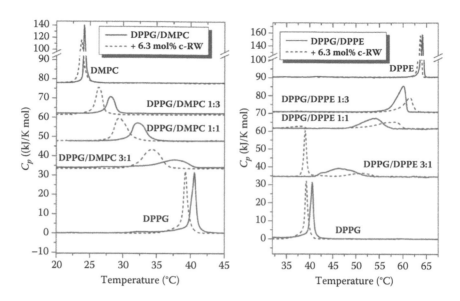

FIGURE 6.12 DSC curves of DPPG–DMPC and DPPG–DPPE mixtures without (solid lines) and after addition of cyclo-RRWWRF (c-RW) (dashed lines). (Adapted from Arouri, A. et al., *Biochim. Biophys. Acta* 1788, 650–659, 2009.)

demixing effects after binding of polycations are, however, not easily observable by DSC alone, so that normally spectroscopic methods are used in addition to obtain detailed information.

Whether binding of cationic oligopeptides can lead to phase separation in mixed bilayers depends not only on the properties of the oligopeptides but also on the chemical structure of the two lipids in the bilayer. For instance, in mixtures of negatively charged DPPG with DMPC, the binding of the cyclic antimicrobial hexapeptide cyclo-RRWWRF (c-RW) leads only to a shift of the transition peak, whereas phase separation is induced in mixtures of DPPG with DPPE, as two transition peaks appear after peptide binding (see Figure 6.12) (Arouri et al. 2009, 2010; Finger et al. 2015). This induction of phase separation may possibly contribute to the antimicrobial action of c-RW, as the inner membranes of Gram-negative bacteria contain PE and not PC. The examples presented here show that a complex interplay between different types of interactions between lipids in the mixed bilayer and the ligand has to be considered. Electrostatic binding and hydrophobic interactions with the acyl chains are competing with hydrogen bonding and charge interactions between different lipids in the bilayer and also of lipids with the oligopeptides.

6.3 ISOTHERMAL TITRATION CALORIMETRY (ITC)

ITC is now a widespread method to study the heats of binding of ligands to macromolecules and to lipid model and biological membranes, and the incorporation and partitioning of hydrophobic or amphiphilic molecules into membranes (Blume 1988, 1991; Blume et al. 1993; Blume and Garidel 1999). In many cases, the binding of molecules or ions to lipid bilayer vesicles at constant temperature can trigger phase changes. For instance, a change in pH can induce a phase transition into a different lamellar phase, because the phospholipid headgroup is protonated or deprotonated with a concomitant change of the transition temperature (see Figure 6.8) (Blume and Tuchtenhagen 1992; Johann et al. 1996; Garidel et al. 1997a; Garidel and Blume 2000b). Another example is the electrostatic binding of divalent ions leading to shifts of the main phase transition temperature or even to a transition into another lyotropic phase (Garidel and Blume 1999, 2000a). A third example is the detergent-induced solubilization or micellization of lipid bilayer vesicles, that is, a phase change from lyotropic lamellar phases to mixed micellar aggregates (Heerklotz et al. 1996, 1997; Keller et al. 1997; Keller 2001; Hildebrand et al. 2002, 2004; Majhi and Blume 2002; Heerklotz and Blume 2012).

6.3.1 Heats of Dissociation, Protonation, and pH-Induced Phase Transitions

The first example for the application of ITC is the pH titration of phospholipid vesicles in systems where the transition temperature depends on the charge of the headgroup, for instance, in PGs or PAs. Figure 6.13 shows the heats of reaction as a function of added NaOH for titration of a 1,2-dimyristoyl-*sn*-glycero-3-phosphoric acid (DMPA) vesicle suspension at different temperatures (Blume and Tuchtenhagen 1992). Addition of base leads to dissociation of the second proton of PAs. At high and low temperature, the heat of reaction is due only to the heat of dissociation and the heat of neutralization, as the released proton recombines with OH⁻ ions to form H_2O. When the titration experiment is

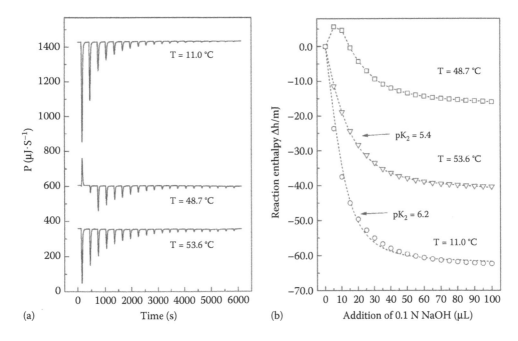

FIGURE 6.13 (a) ITC titration curves obtained by titrating a 0.1 M NaOH solution stepwise (5 µL) into a 1 mM DMPA (pH 7) vesicle suspension. (b) Total integrated heat of reaction (Δh) calculated after subtraction of the heat of dilution of NaOH. The dashed curves for T = 11°C and T = 53.6°C are calculated curves using the Gouy–Chapman theory, yielding the pK_2 values as indicated. (Adapted from Blume, A., and Tuchtenhagen, J., *Biochemistry* 31, 4636–4642, 1992.)

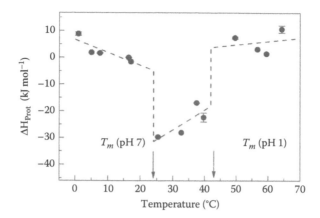

FIGURE 6.14   Total reaction enthalpy for the protonation of DMPG vesicles as a function of temperature. In the temperature range between the two arrows, the reaction enthalpy includes the transition enthalpy of DMPG from the liquid-crystalline to the gel phase induced by protonation of DMPG. (From Tuchtenhagen, J., Kalorimetrische und FT-IR-spektroskopische Untersuchungen an Phospholipidmodellmembranen. PhD, Department of Chemistry, University of Kaiserslautern, 1994.)

performed at a temperature between the phase transition temperatures observed at pH 7 and 12, respectively, then titration with base leads additionally to a phase transition into the L_α-phase. The shape of the titration curve observed at a temperature of 48.7°C is therefore a superposition of the endothermic heat of transition with the exothermic heats arising from the deprotonation and the formation of water. The curves obtained at low and high temperature can be fitted to determine the pK_2 of the phosphoester group of DMPA, using an appropriate dissociation model including electrostatic corrections based on the Gouy–Chapman theory (Figure 6.13) (Blume and Tuchtenhagen 1992).

The lipid DMPG is negatively charged at pH 7 and becomes completely protonated at a very low pH of 1 (Tuchtenhagen 1994). In this case, a pH titration with HCl is difficult to perform, as the protonation of the headgroup leads to a precipitation of the vesicles. Therefore, only the reversed experiment is feasible: the vesicle suspension is injected into a solution of 0.1 N HCl, and thus the total heat of protonation is observed for each injection. The heat of protonation was found to be endothermic when DMPG was in the gel or the L_α-phase. At intermediate temperatures, strongly exothermic reaction heats were observed due to the triggering of a phase transition from the L_α to the gel phase. Figure 6.14 shows the total reaction heat for the protonation of DMPG as a function of temperature (Tuchtenhagen 1994).

6.3.2 Heats of Ion Binding and Ion Binding–Induced Phase Transitions

The binding of divalent cations to negatively charged lipid vesicles can drastically change their phase transition temperature and the structure of the lamellar phase (Garidel and Blume 1999, 2000a; Garidel et al. 2000a). For instance, complexes of negatively charged DMPG with Mg^{2+} or Ca^{2+} form very stable lamellar phases, melting into the L_α-phase only at very high temperature (Garidel et al. 2000a). A transition from the uncomplexed L_α-phase of DMPG to the lamellar gel phase can thus be induced by addition of divalent cations. For DMPA-, the binding of Mg^{2+} leads to a dissociation of the second proton, so that a 1:1 complex between doubly charged $DMPA^{2-}$ with Mg^{2+} is observed if the suspension is buffered at pH 7. This "complex" shows a very high transition temperature of more than 95°C. However, in an unbuffered suspension, the binding of Mg^{2+} leads to a lowering of the pH due to the dissociation of the second proton, so that mostly a 2:1 complex of $DMPA^{1-}$ with Mg^{2+} is formed. This complex has a transition temperature of ca. 55°C (Garidel 1997).

Figure 6.15 a shows the DSC curves obtained for pure DMPA and pH 4 and 7, together with the transition curves of the DMPA 2:1 complex with Mg^{2+}. A normal titration with Mg^{2+} as shown for the pH titration in Figure 6.13 is not possible, as addition of Mg^{2+} ions immediately leads to the formation and precipitation of the complex. Therefore, only the total heat of binding can be measured by injecting a vesicle suspension into an Mg^{2+}- containing solution. The total reaction heat for Mg^{2+} binding to DMPA in water at pH 7 is shown in Figure 6.15b as a function of temperature. Endothermic reaction heats are observed for Mg^{2+} binding to gel or L_α-phase DMPA vesicles. In the temperature range between the transition of pure DMPA and the $(DMPA-)_2$:Mg^{2+} complex, exothermic heats of reaction are seen due to the induction of the phase transition of $DMPA^-$ from the L_α- to the gel phase of the complex. The temperatures where the vertical jumps in the reaction

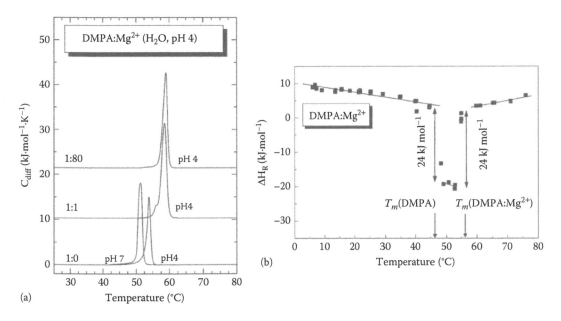

FIGURE 6.15 (a) DSC curves for DMPA without and with bound Mg^{2+} in a stoichiometric 2:1 ratio. (b) Total reaction enthalpy for the binding of Mg^{2+} to DMPA as a function of temperature. In the temperature range between the two arrows, the reaction enthalpy includes the transition enthalpy of DMPA from the liquid-crystalline to the gel phase induced by Mg^{2+} binding. (From Garidel, P., The negatively charged phospholipids phosphatidic acid and phosphatidylglycerol. PhD, Department of Chemistry, University of Kaiserslautern, 1997.)

enthalpy are observed coincide with the DSC transition temperatures, and the height of the jumps with the transition enthalpies of pure DMPA- and the $(DMPA-)_2:Mg^{2+}$ complex, respectively (Garidel 1997).

6.3.3 Binding of Cationic Oligo- and Polypeptides

While the binding of inorganic cations to negatively charged lipids is usually an endothermic reaction (Garidel 1997; Garidel and Blume 1999), this can be different for the binding of multivalent organic cations such as polyamines or oligo- and polypeptides. The binding of peptides will be covered in Chapter 8. Therefore, I will only briefly present two examples: the binding of an amphipathic cationic antimicrobial peptide and the binding of homopolypeptides such as poly-L-lysine (PLL) and poly-L-arginine (PLA).

The cationic oligopeptide KLA1 (KLAL KLAL KAW KAAL KLA-NH$_2$) is amphipathic with six positive charges and has antimicrobial activity (Dathe et al. 2002; Erbe et al. 2009; Arouri et al. 2010, 2013). The binding constant depends on the surface charge density of the bilayer; that is, when DPPG is diluted by the zwitterionic DPPC or DPPE, the apparent binding constant decreases. Likewise, screening the charges of the lipids by addition of salt reduces the apparent binding affinity. These experiments show that for KLA1, the binding is mainly driven by electrostatic attractions between the cationic peptide and the negatively charged vesicle surface. The binding stoichiometry of six also supports this notion. The ITC experiments show that the binding enthalpy is endothermic but relatively low for binding to gel-phase vesicles and changes only slightly with temperature. At temperatures above the phase transition temperature, the binding enthalpy is almost zero, despite the fact that binding still occurs.

The DSC curves of lipid–peptide mixtures show that the phase transition is somewhat broadened and its temperature lowered, indicating a perturbation in chain packing caused by the insertion of hydrophobic side chains of the peptide into the bilayer. Therefore, the observed vanishing binding enthalpy above T_m is obviously due to a compensation of endothermic effects due to electrostatic binding by exothermic effects arising from the insertion of hydrophobic residues into the lipid headgroup region (Arouri et al. 2013). As DPPG has a relatively high phase transition temperature of 41°C, a study of the temperature dependence of the binding enthalpy in the L_α-phase above T_m was difficult to perform. This was easier for a lower-melting lipid such as 1-palmitoyl-2-oleoyl-sn-glycero-3-phosphoglycerol (POPG). Figure 6.16 shows the heats of reaction at different temperatures (Arouri 2009). With increasing temperature, the binding enthalpy becomes less endothermic and finally negative at high temperature. This temperature dependence of the binding enthalpy clearly shows that hydrophobic effects contribute to the binding, because the negative ΔC_p is characteristic for the hydrophobic effect, that is, the removal of hydrophobic moieties from water after binding.

The binding enthalpy of cationic PLA to negatively charged bilayers is also endothermic at temperatures below the lipid phase transition of DPPG, but becomes clearly exothermic when DPPG is in the fluid phase. Figure 6.17 shows as an example two titration curves of PLA to lipid vesicles below and above T_m of DPPG (Schwieger and Blume 2009). Here, the two effects, endothermic electrostatic binding and exothermic effects due to shielding

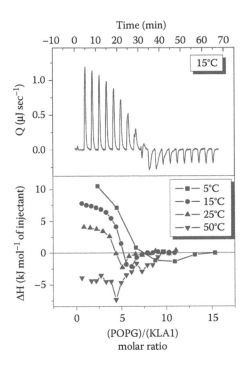

FIGURE 6.16 ITC titration curves for a titration of POPG vesicles into a 5–10 μM solution of KLA1 at different temperatures as indicated. (From Arouri, A., Interaction of antimicrobial peptides with model lipid membranes. PhD, Institute of Chemistry, Martin-Luther University Halle-Wittenberg, 2009.)

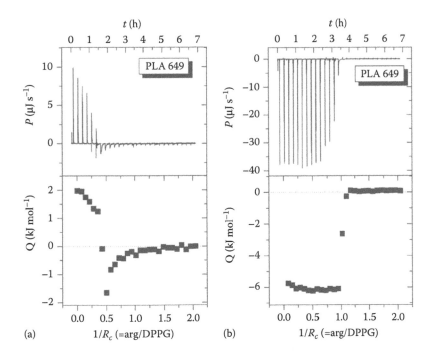

FIGURE 6.17 ITC titration curves of DPPG vesicles with poly-L-arginine (PLA 649) at 30°C (a) and 50°C (b). (Adapted from Schwieger, C., and Blume, A., *Biomacromolecules* 10, 2152–2161, 2009.)

of hydrophobic groups from water, do not compensate, as the exothermic contribution is much larger. For binding of PLA to DPPG vesicles, not only the sign of the binding enthalpy changes, but also the stoichiometry of binding.

For binding to gel-phase vesicles, the stoichiometry ratio arg/DPPG is 0.5, indicating that only binding to the outside of the vesicles occurs; the gel-phase vesicles seem to be impermeable to PLA. This is different for binding to DPPG vesicles in the fluid phase. Here, the stoichiometry is clearly 1; that is, PLA can now permeate the vesicle bilayer, and binding occurs also to the inside of the vesicles. The exothermic heat of binding indicates that hydrophobic interactions between the arginine side chains and the lipid alkyl chains are present, leading to the observed higher permeability of the bilayers to PLA (Schwieger and Blume 2009). This complex interplay of electrostatic interactions with hydrophobic effects has also been observed for much simpler cationic pentapeptides in which the side chain has been reduced in length, starting from penta-L-lysine (Hoernke et al. 2012).

6.3.4 Micellization of Detergents, Detergent Partitioning, and Detergent-Induced Bilayer–Micelle Transitions

ITC is particularly suited to study the micellization or demicellization of detergents and their partitioning into lipid vesicles, leading with increased detergent concentration to so-called *solubilization*, that is, the transformation of mixed vesicles into mixed micelles. Several review articles have appeared on this subject (Heerklotz and Seelig 2000b; Heerklotz et al. 2003; Heerklotz 2004, 2008; Heerklotz and Blume 2012).

The discovery of the usefulness of ITC for this process resulted from investigations of the micellization of detergents at different temperatures (Blume et al. 1993; Paula et al. 1995; Garidel et al. 2000b; Majhi and Blume 2001; Hildebrand et al. 2003b; Beyer et al. 2006). ITC, as a probe-free method, was found to be ideally suited for determination of the critical micelle concentration (cmc) and its temperature dependence. A further advantage is that not only the cmc, but also the micellization enthalpy and, at least for some cases, the aggregation number can be determined. Micellization is accompanied by enthalpic effects whose sign depends on temperature, being endothermic below and exothermic above the temperature of the cmc minimum. From the temperature dependence of the micellization enthalpy, the extent of hydrophobic surfaces still in contact with water in a micelle could be determined, as ΔC_p is correlated with the amount of hydrophobic surface being removed from water during the micellization process (Blume et al. 1993; Paula et al. 1995).

The detergent-induced transformation of bilayer vesicles into mixed micelles was previously studied mainly by optical techniques, for instance by light scattering, as the transformation of vesicles with a diameter of 100 nm or more into much smaller micelles leads to a drastic decrease in scattering intensity. On the basis of these investigations, a partial generic "phase diagram" for highly diluted samples was suggested, as shown in Figure 6.18 (Lichtenberg et al. 1983; Lichtenberg 1985).

The numbered arrows in Figure 6.18 designate specific types of experiments that can be performed by ITC. Arrow 1 describes experiments for the determination of the cmc

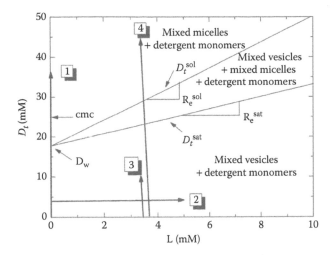

FIGURE 6.18 Schematic phase diagram for a phospholipid/detergent/water mixture in excess water. The regions where vesicles and micelles exist and the coexistence region are indicated. The arrows designate calorimetric titration experiments described in the text. D_t: total detergent concentration; L: lipid concentration. (From Lichtenberg, D. et al., *Biochim. Biophys. Acta* 737, 285–304, 1983; Lichtenberg, D., *Biochim. Biophys. Acta* 821, 470–478, 1985.)

of detergents. Arrows 2 and 3 describe ITC titration experiments for the partitioning of detergents into the lipid bilayers, while Arrow 4 crosses the phase boundaries and thus describes the ITC experiment for the solubilization of vesicles into micelles.

For the partitioning of the detergent into lipid bilayers, a similar temperature dependence is observed as for the aggregation of the detergent into micelles (Keller et al. 1997; Wenk et al. 1997). A correlation between the cmc and the partition coefficient was found, showing that in both cases, the detergent is removed from water and transferred into more hydrophobic surroundings, which is similar in micelles and in fluid bilayers (Heerklotz and Seelig 2000a; Tan et al. 2002). Figure 6.19 shows ITC curves for the experiment described by Arrow 2 in Figure 6.18 for partitioning of the detergent octylglucoside (OG) into DMPC bilayers at two different temperatures. The enthalpy of partitioning is endothermic at low and exothermic at high temperature. From the shape of the curves, the partition coefficient, P, and the transfer enthalpy, ΔH^T, for the detergent from water to the bilayer can be calculated using a thermodynamic partitioning model (Keller et al. 1997; Wenk et al. 1997; Heerklotz and Seelig 2000b; Heerklotz and Blume 2012). This simple analysis is only possible when the flip-flop of the detergent, here OG, is fast on the time scale of the ITC experiment. However, in many cases, flip-flop is much slower, and initially only outside binding is observed, for instance, when charged detergents are studied. In these cases, ITC can also be applied to study the kinetics of detergent incorporation and flip-flop (Keller et al. 2006a; Martins and Moreno 2015).

The experiment according to Arrow 4 in Figure 6.18 leads to a transformation of the lipid vesicles to mixed micelles. The characteristic concentrations D_t^{sat}, for the saturation of the lipid bilayer membranes with detergent, where the first mixed micelles

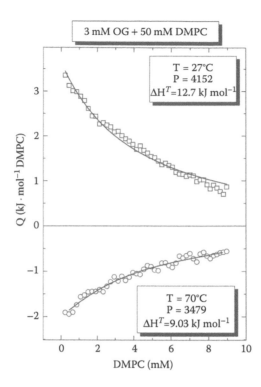

FIGURE 6.19 Heats of reaction from a partitioning experiment at two different temperatures according to arrow 2 in Figure 6.18, where a DMPC vesicle suspension was titrated into an OG solution. The curves can be fitted, yielding the enthalpies of detergent incorporation ΔH^T and the partition coefficient P. (Adapted from Keller, M. et al., *Biochim. Biophys. Acta* 1326, 178–192, 1997.)

appear, and D_t^{sol}, where all vesicles have disappeared, can be easily determined from the ITC titration curves, and numerous lipid/detergent systems have been investigated in recent years (Heerklotz et al. 1995, 1996, 1997, 2003; Keller et al. 1997, 2006b; Keller 2001; Hildebrand et al. 2002, 2003a, 2004; Majhi and Blume 2002; Tan et al. 2002; Garidel et al. 2007). Figure 6.20 shows as an example the observed heats of detergent incorporation and solubilization of soybean PC with OG at different temperatures (Keller 2001).

The dashed line in Figure 6.20 connects all concentration values, D_t^{sat}, and shows that D_t^{sat} is temperature dependent in a similar way to the cmc of OG, having a minimum at a temperature of ca. 45°C (Paula et al. 1995; Majhi and Blume 2001). The same applies for the concentration values, D_t^{sol} (connection line not shown in Figure 6.19), and the ratios $R_t^{sat} = D_t^{sat}/L$ and $R_t^{sol} = D_t^{sol}/L$, shown in Figure 6.20b. This is to be expected, because partitioning of the detergent and the transformation from vesicles to micelles are all processes driven mainly by the hydrophobic effect.

The resulting phase diagram for the solubilization of soybean PC by OG is shown in Figure 6.21 (Keller 2001). The slopes of the phase boundaries are steeper than in the phase

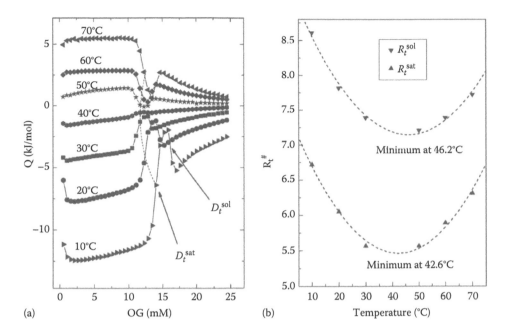

FIGURE 6.20 (a) Heats of reaction of an ITC titration experiment in which a vesicle suspension of soybean PC was titrated with OG at different temperatures. The arrows point to the inflection points that designate the phase boundaries, that is, the values D_t^{sat} and D_t^{sol} indicated in Figure 6.18. (b) $R_t^{sat} = D_t^{sat}/L$ and $R_t^{sol} = D_t^{sol}/L$ values as a function of temperature. The temperature dependence shows a clear minimum at a temperature where the cmc of OG has its minimum. (From Paula, S. et al., *J. Phys. Chem.* 99, 11742–11751, 1995; Keller, M., Thermodynamik der Demizellisierung und Solubilisierung von Alkylglucosiden mit ausgewählten Phospholipiden sowie das rheologische Verhalten der Solubilisierung am Beispiel Octylglucosid/Dimyristoylphosphatidylglyzerol. PhD, Department of Chemistry, University of Kaiserslautern, 2001; Majhi, P. R., and Blume, A., *Langmuir* 17, 3844–3851, 2001.)

diagram of the related system DMPG/OG (Keller et al. 1997). This means that soybean PC vesicles having unsaturated chains can tolerate more OG before becoming unstable and transforming into mixed micelles compared with DMPC vesicles with completely saturated chains.

As shown in Figure 6.20, the saturation concentrations depend on temperature, and therefore also the phase diagrams are shifted with changing temperature (Keller et al. 1997). This leads to the possibility of inducing micelle–vesicle transition by increasing or decreasing temperature, depending on whether one starts from temperatures below or above the temperature of the saturation concentration. This has been shown in detail for the DMPC/sodium n-dodecyl sulfate (SDS) and DMPC/n-dodecyl-tetramethylammonium bromide (DTAB) systems, in which temperature-induced micelle–vesicle transitions could be followed by DSC and by light-scattering methods (Majhi and Blume 2002).

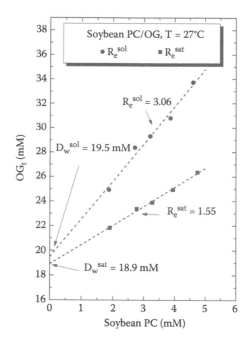

FIGURE 6.21 Phase diagram obtained from ITC experiments for the soybean PC/OG system. (From Keller, M., Thermodynamik der Demizellisierung und Solubilisierung von Alkylglucosiden mit ausgewählten Phospholipiden sowie das rheologische Verhalten der Solubilisierung am Beispiel Octylglucosid/Dimyristoylphosphatidylglyzerol. PhD, Department of Chemistry, University of Kaiserslautern, 2001.)

6.4 CONCLUSIONS

DSC and ITC are now standard techniques for investigations of the behavior of model and biological membranes. Due to the increased sensitivity of DSC instruments and their small cell volumes, systematic studies of the effects of peptides and proteins incorporated into membranes or peripherally bound to the membrane surface have become feasible. This applies also to ITC, which is now one of the standard methods to determine binding constants of peptides and proteins to membranes and for studies of the incorporation of amphiphiles into membranes.

The interpretation of ITC titration curves, however, is still not straightforward, as the observed reaction heats are composed of heat effects originating from multiple processes. The analysis of the data relies on thermodynamic binding models, making the outcome ambiguous, as different binding models can result in similar titration curves. Various algorithms for models assuming multiple binding sites with different binding constants and different binding enthalpies, or models with independent or sequential binding, having cooperativity or anti-cooperativity, have been proposed (Martinez et al. 2013; Velazquez-Campoy 2015). All these models lead to complicated expressions for the observed heat of reaction, so that systematic studies with variation of concentrations of reactants, reversal of the titration procedure, and change in temperature are needed to be sure that data interpretation is correct. For the special case of experiments with lipids, the heterogeneity of the system has to be taken into account, as lipids are assembled in bilayer membranes forming closed vesicles. When the ligand cannot permeate the lipid bilayer, titration of ligands to lipid vesicles leads only to outside binding. However, the binding of the ligand can, at a certain threshold, lead to permeabilization or even rupture of the vesicles, so that during the course of the titration, inside binding can also take place (Schwieger and Blume 2009). These sequential effects complicate the analysis but also enable the determination of the concentration threshold for vesicle permeabilization. An example of this possibility is the partitioning and permeation of the detergent SDS through lipid vesicle membranes (Keller et al. 2006a).

ITC experiments in which ligand binding induces vesicle aggregation and/or fusion are notoriously difficult to perform. In many cases, the aggregated vesicles sediment and are no longer completely accessible for reaction with the ligand injected into the cell, despite the fact that the suspension is stirred. In some cases, the addition of the vesicle suspension to the solution of the ligand in the cell can partially circumvent the problem, but in many cases, only the total heat of reaction can be measured and not a complete titration curve.

Despite these difficulties, DSC and ITC instruments have become technically mature enough over recent years that both methods are now standard techniques and can be used for model membrane systems with increased complexity, as other chapters in this book show (see Chapters 7–10).

ACKNOWLEDGMENTS

This work was supported by grants from the Deutsche Forschungsgemeinschaft (DFG), the state of Saxonia-Anhalt, and the Phospholipid Research Center, Heidelberg. I express my gratitude to all members of my group for their contributions.

REFERENCES

Almeida, P. F. (2009), Thermodynamics of lipid interactions in complex bilayers, *Biochim. Biophys. Acta* 1788, 72–85.

Almeida, P. F., Vaz, W. L., and Thompson, T. E. (1993), Percolation and diffusion in three-component lipid bilayers: Effect of cholesterol on an equimolar mixture of two phosphatidylcholines, *Biophys. J.* 64, 399–412.

Alonso, A., and Goñi, F. M. (2015), The thermotropic properties of ceramides in aqueous or lipidic environments. In M. Bastos (Ed.), *Biocalorimetry*, Taylor and Francis: Boca Raton, FL.

Arouri, A. (2009), Interaction of antimicrobial peptides with model lipid membranes. PhD, Institute of Chemistry, Martin-Luther University Halle-Wittenberg.

Arouri, A., Dathe, M., and Blume, A. (2009), Peptide induced demixing in PG/PE lipid mixtures: A mechanism for the specificity of antimicrobial peptides towards bacterial membranes? *Biochim. Biophys. Acta* 1788, 650–659.

Arouri, A., Dathe, M., and Blume, A. (2013), The helical propensity of KLA amphipathic peptides enhances their binding to gel-state lipid membranes, *Biophys. Chem.* 180–181, 10–21.

Arouri, A., Kiessling, V., Tamm, L., Dathe, M., and Blume, A. (2010), Morphological changes induced by the action of antimicrobial peptides on supported lipid bilayers, *J. Phys. Chem. B* 115, 158–167.

Bagatolli, L. A. (2006), To see or not to see: Lateral organization of biological membranes and fluorescence microscopy, *Biochim. Biophys. Acta* 1758, 1541–1556.

Beyer, K., Leine, D., and Blume, A. (2006), The demicellization of alkyltrimethylammonium bromides in 0.1 M sodium chloride solution studied by isothermal titration calorimetry, *Colloids Surf. B. Biointerfaces* 49, 31–39.

Blume, A. (1980), Thermotropic behavior of phosphatidylethanolamine-cholesterol and phosphatidylethanolamine-phosphatidylcholine-cholesterol mixtures, *Biochemistry* 19, 4908–4913.

Blume, A. (1983), Apparent molar heat capacities of phospholipids in aqueous dispersion. Effects of chain length and head group structure, *Biochemistry* 22, 5436–5442.

Blume, A. (1988), Applications of calorimetry to lipid model membranes. In C. Hidalgo, *Physical Properties of Biological Membranes and Their Functional Implications*, Plenum: New York.

Blume, A. (1991), Biological calorimetry: Membranes, *Thermochim. Acta* 193, 299–347.

Blume, A., and Ackermann, T. (1974), A calorimetric study of the lipid phase transitions in aqueous dispersions of phosphorylcholine-phosphorylethanolamine mixtures, *FEBS Lett.* 43, 71–74.

Blume, A., and Garidel, P. (1999), Lipid model membranes and biomembranes. In R. B. Kemp (Ed.), *Handbook of Thermal Analysis and Calorimetry*, 109–173, Elsevier: Amsterdam.

Blume, A., and Tuchtenhagen, J. (1992), Thermodynamics of ion binding to phosphatidic acid bilayers. Titration calorimetry of the heat of dissociation of DMPA, *Biochemistry* 31, 4636–4642.

Blume, A., Tuchtenhagen, J., and Paula, S. (1993), Application of titration calorimetry to study binding of ions, detergents, and polypeptides to lipid bilayers, *Prog. Colloid Polym. Sci.* 93, 118–122.

Brandenburg, K., and Blume, A. (1987), Investigations into the thermotropic phase behaviour of natural membranes extracted from gram-negative bacteria and artificial membrane systems made from lipopolysaccharides and free lipid A, *Thermochim. Acta* 119, 127–142.

Brumbaugh, E. E., Johnson, M. L., and Huang, C.-h. (1990), Non-linear least squares analysis of phase diagrams for non-ideal binary mixtures of phospholipids, *Chem. Phys. Lipids* 52, 69–78.

Caffrey, M., and Cheng, A. (1995), Kinetics of lipid phase changes, *Curr. Opin. Struct. Biol.* 5, 548–555.

Carion-Taravella, B., Lesieur, S., Chopineau, J., Lesieur, P., and Ollivon, M. (2002), Phase behavior of mixed aqueous dispersions of dipalmitoylphosphatidylcholine and dodecyl glycosides: A differential scanning calorimetry and x-ray diffraction investigation, *Langmuir* 18, 325–335.

Cevc, G. (1993), *Phospholipids Handbook*, 1st edition, Marcel Dekker: New York.

Cevc, G., and Marsh, D. (1987), *Phospholipid Bilayers: Physical Principles and Models*, Wiley: New York.

Chapman, D. (1968), *Biological Membranes*, Academic: London.

Chapman, D. (1975), Phase transitions and fluidity characteristics of lipids and cell membranes, *Q. Rev. Biophys.* 8, 185–235.

Chapman, D., Williams, R. M., and Ladbrooke, B. D. (1967), Physical studies of phospholipids. VI. Thermotropic and lyotropic mesomorphism of some 1,2-diacyl-phosphatidylcholines (lecithins), *Chem. Phys. Lipids* 1, 445–475.

Dathe, M., Meyer, J., Beyermann, M., Maul, B., Hoischen, C., and Bienert, M. (2002), General aspects of peptide selectivity towards lipid bilayers and cell membranes studied by variation of the structural parameters of amphipathic helical model peptides, *Biochim. Biophys. Acta* 1558, 171–186.

de Almeida, R. F. M., Fedorov, A., and Prieto, M. (2003), Sphingomyelin/phosphatidylcholine/cholesterol phase diagram: Boundaries and composition of lipid rafts, *Biophys. J.* 85, 2406–2416.

Dietrich, C., Bagatolli, L. A., Volovyk, Z. N., Thompson, N. L., Levi, M., Jacobson, K., and Gratton, E. (2001), Lipid rafts reconstituted in model membranes, *Biophys. J.* 80, 1417–1428.

Erbe, A., Kerth, A., Dathe, M., and Blume, A. (2009), Interactions of KLA amphipathic model peptides with lipid monolayers, *ChemBioChem* 10, 2884–2892.

Feigenson, G. W. 2009. Phase diagrams and lipid domains in multicomponent lipid bilayer mixtures, *Biochim. Biophys. Acta* 1788:47–52.

Finger, S., Kerth, A., Dathe, M., and Blume, A. (2015), The efficacy of trivalent cyclic hexapeptides to induce lipid clustering in PG/PE membranes correlates with their antimicrobial activity, *Biochim. Biophys. Acta* 1848, 2998–3006.

Finger, S., Kerth, A., Meister, A., Schwieger, C., Dathe, M., and Blume, A. (2013), The effect of cyclic antimicrobial hexapeptides on model bacteria membranes, *Eur. Biophys. J.* 42, S115.

Finger, S., Schwieger, C., Arouri, A., Kerth, A., and Blume, A. (2014), Interaction of linear polyamines with negatively charged phospholipids: The effect of polyamine charge distance, *Biol. Chem. Hoppe-Seyler* 395, 769–778.

Franzin, C. M., and Macdonald, P. M. (2001), Polylysine-induced 2H NMR-observable domains in phosphatidylserine/phosphatidylcholine lipid bilayers, *Biophys. J.* 81, 3346–3362.

Garidel, P. (1997), The negatively charged phospholipids phosphatidic acid and phosphatidylglycerol. PhD thesis, Department of Chemistry, University of Kaiserslautern.

Garidel, P., and Blume, A. (1998), Miscibility of phospholipids with identical headgroups and acyl chain lengths differing by two methylene units: Effects of headgroup structure and headgroup charge, *Biochim. Biophys. Acta* 1371, 83–95.

Garidel, P., and Blume, A. (1999), Interaction of alkaline earth cations with the negatively charged phospholipid 1,2-dimyristoyl-*sn*-glycero-3-phosphoglycerol: A differential scanning and isothermal titration calorimetric study, *Langmuir* 15, 5526–5534.

Garidel, P., and Blume A. (2000a), Calcium induced nonideal mixing in liquid-crystalline phosphatidylcholine–phosphatidic acid bilayer membranes, *Langmuir* 16, 1662–1667.

Garidel, P., and Blume A. (2000b), Miscibility of phosphatidylethanolamine-phosphatidylglycerol mixtures as a function of pH and acyl chain length, *Eur. Biophys. J.* 28, 629–638.

Garidel, P., and Blume, A. (2005), 1,2-Dimyristoyl-*sn*-glycero-3-phosphoglycerol (DMPG) monolayers: Influence of temperature, pH, ionic strength and binding of alkaline earth cations, *Chem. Phys. Lipids* 138, 50–59.

Garidel, P., Forster, G., Richter, W., Kunst, B. H., Rapp, G., and Blume A. (2000a), 1,2-Dimyristoyl-*sn*-glycero-3-phosphoglycerol (DMPG) divalent cation complexes: An x-ray scattering and freeze-fracture electron microscopy study, *Phys. Chem. Chem. Phys.* 2, 4537–4544.

Garidel, P., Hildebrand, A., Knauf, K., and Blume, A. (2007), Membranolytic activity of bile salts: Influence of biological membrane properties and composition, *Molecules* 12, 2292–2326.

Garidel, P., Hildebrand, A., Neubert, R., and Blume, A. (2000b), Thermodynamic characterization of bile salt aggregation as a function of temperature and ionic strength using isothermal titration calorimetry, *Langmuir* 16, 5267–5275.

Garidel, P., Johann, C., and Blume, A. (1997a), Nonideal mixing and phase separation in phosphatidylcholine-phosphatidic acid mixtures as a function of acyl chain length and pH, *Biophys. J.* 72, 2196–2210.

Garidel, P., Johann, C., and Blume, A. (2005), The calculation of heat capacity curves and phase diagrams based on regular solution theory, *J. Therm. Anal. Calorim.* 82, 447–455.

Garidel, P., Johann, C., and Blume, A. (2011), Non-ideal mixing and fluid–fluid immiscibility in phosphatidic acid–phosphatidylethanolamine mixed bilayers, *Eur. Biophys. J.* 40, 891–905.

Garidel, P., Johann, C., Mennicke, L., and Blume, A. (1997b), The mixing behavior of pseudobinary phosphatidylcholine-phosphatidylglycerol mixtures as a function of pH and chain length, *Eur. Biophys. J.* 26, 447–459.

Goni, F. M. (2014), The basic structure and dynamics of cell membranes: An update of the Singer-Nicolson model, *Biochim. Biophys. Acta* 1838, 1467–1476.

Goni, F. M., Alonso, A., Bagatolli, L. A., Brown, R. E., Marsh, D., Prieto, M., and Thewalt, J. L. (2008), Phase diagrams of lipid mixtures relevant to the study of membrane rafts, *Biochim. Biophys. Acta* 1781, 665–684.

Hazel, J. R. (1995), Thermal adaptation in biological membranes: Is homeoviscous adaptation the explanation? *Annu. Rev. Physiol.* 57, 19–42.

Heerklotz, H. (2004), The microcalorimetry of lipid membranes, *J. Phys.: Condens. Matter* 16, R441–R467.

Heerklotz, H. (2008), Interactions of surfactants with lipid membranes, *Q. Rev. Biophys.* 41, 205–264.

Heerklotz, H., Binder, H., Lantzsch, G., Klose, G., and Blume, A. (1997), Lipid/detergent interaction thermodynamics as a function of molecular shape, *J. Phys. Chem. B* 101, 639–645.

Heerklotz, H., and Blume, A. (2012), Detergent interactions with lipid bilayers. In E. G. Egelman (Ed.), *Comprehensive Biophysics*, vol. 5, Academic: Oxford.

Heerklotz, H., Lantzsch, G., Binder, H., Klose, G., and Blume, A. (1995), Application of isothermal titration calorimetry for detecting lipid membrane solubilization, *Chem. Phys. Lett.* 235, 517–520.

Heerklotz, H., Lantzsch, G., Binder, H., Klose, G., and Blume, A. (1996), Thermodynamic characterization of dilute aqueous lipid/detergent mixtures of POPC and $C_{12}EO_8$ by means of isothermal titration calorimetry, *J. Phys. Chem.* 100, 6764–6774.

Heerklotz, H., and Seelig, J. (2000a), Correlation of membrane/water partition coefficients of detergents with the critical micelle concentration, *Biophys. J.* 78, 2435–2440.

Heerklotz, H., and Seelig, J. (2000b), Titration calorimetry of surfactant–membrane partitioning and membrane solubilization, *Biochim. Biophys. Acta* 1508, 69–85.

Heerklotz, H., Szadkowska, H., Anderson, T., and Seelig, J. (2003), The sensitivity of lipid domains to small perturbations demonstrated by the effect of triton, *J. Mol. Biol.* 329, 793–799.

Hildebrand, A., Beyer, K., Neubert, R., Garidel, P., and Blume, A. (2003a), Temperature dependence of the interaction of cholate and deoxycholate with fluid model membranes and their solubilization into mixed micelles, *Colloids Surf. B. Biointerfaces* 32, 335–351.

Hildebrand, A., Beyer, K., Neubert, R., Garidel, P., and Blume, A. (2004), Solubilization of negatively charged DPPC/DPPG liposomes by bile salts, *J. Colloid Interface Sci.* 279, 559–571.

Hildebrand, A., Garidel, P., Neubert, R., and Blume, A. (2003b), Thermodynamics of demicellization of mixed micelles composed of sodium oleate and bile salts, *Langmuir* 20, 320–328.

Hildebrand, A., Neubert, R., Garidel, P., and Blume, A. (2002), Bile salt induced solubilization of synthetic phosphatidylcholine vesicles studied by isothermal titration calorimetry, *Langmuir* 18, 2836–2847.

Hill, T. L. (1985), *An Introduction to Statistical Thermodynamics*, Dover: New York.

Hirst, L. S., Uppamoochikkal, P., and Lor, C. (2011), Phase separation and critical phenomena in biomimetic ternary lipid mixtures, *Liq. Cryst.* 38, 1735–1747.

Hoernke, M., Schwieger, C., Kerth, A., and Blume, A. (2012), Binding of cationic pentapeptides with modified side chain lengths to negatively charged lipid membranes: Complex interplay of electrostatic and hydrophobic interactions, *Biochim. Biophys. Acta* 1818, 1663–1672.

Huang, T. H., Lee, C. W. B., Das Gupta, S. K., Blume, A., and Griffin, R. G. (1993), A carbon-13 and deuterium nuclear magnetic resonance study of phosphatidylcholine/cholesterol interactions: Characterization of liquid-gel phases, *Biochemistry* 32, 13277–13287.

Ipsen, J. H., Karlström, G., Mouritsen, O. G., Wennerström, H., and Zuckermann, M. J. (1987), Phase equilibria in the phosphatidylcholine-cholesterol system, *Biochim. Biophys. Acta* 905, 162–172.

Jackson, M. B., and Sturtevant, J. M. (1977), Studies of the lipid phase transitions of *Escherichia coli* by high sensitivity differential scanning calorimetry, *J. Biol. Chem.* 252, 4749–4751.

Johann, C., Garidel, P., Mennicke, L., and Blume, A. (1996), New approaches to the simulation of heat-capacity curves and phase diagrams of pseudobinary phospholipid mixtures, *Biophys. J.* 71, 3215–3228.

Keller, M. (2001), Thermodynamik der Demizellisierung und Solubilisierung von Alkylglucosiden mit ausgewählten Phospholipiden sowie das rheologische Verhalten der Solubilisierung am Beispiel Octylglucosid/Dimyristoylphosphatidylglyzerol. PhD thesis, Department of Chemistry, University of Kaiserslautern.

Keller, M., Kerth, A., and Blume, A. (1997), Thermodynamics of interaction of octyl glucoside with phosphatidylcholine vesicles: Partitioning and solubilization as studied by high sensitivity titration calorimetry, *Biochim. Biophys. Acta* 1326, 178–192.

Keller, S., Heerklotz, H., and Blume, A. (2006a), Monitoring lipid membrane translocation of sodium dodecyl sulfate by isothermal titration calorimetry, *J. Am. Chem. Soc.* 128, 1279–1286.

Keller, S., Heerklotz, H., Jahnke, N., and Blume, A. (2006b), Thermodynamics of lipid membrane solubilization by sodium dodecyl sulfate, *Biophys. J.* 90, 4509–4521.

Klingler, J., and Keller, S. (2015), Peptide–membrane interactions studied by ITC. In M. Bastos (Ed.), *Biocalorimetry*, Taylor and Francis: Boca Raton, FL.

Koynova, R., and Caffrey, M. (1998), Phases and phase transitions of the phosphatidylcholines, *Biochim. Biophys. Acta* 1376, 91–145.

Koynova, R., and Caffrey, M. (2002), An index of lipid phase diagrams, *Chem. Phys. Lipids* 115, 107–219.

Ladbrooke, B. D., and Chapman, D. (1969), Thermal analysis of lipids, proteins, and biological membranes. A review and summary of some recent studies, *Chem. Phys. Lipids* 3, 304–367.

Ladbrooke, B. D., Williams, R. M., and Chapman, D. (1968), Studies on lecithin-cholesterol-water interactions by differential scanning calorimetry and x-ray diffraction, *Biochim. Biophys. Acta* 150, 333–340.

Lee, A. G. (1977a), Lipid phase transitions and phase diagrams I. Lipid phase transitions, *Biochim. Biophys. Acta* 472, 237–281.

Lee, A. G. (1977b), Lipid phase transitions and phase diagrams II. Mixtures involving lipids, *Biochim. Biophys. Acta* 472, 285–344.

Lee, A. G. (1978), Calculation of phase diagrams for non-ideal mixtures of lipids, and a possible non-random distribution of lipids in lipid mixtures in the liquid crystalline phase, *Biochim. Biophys. Acta* 507, 433–444.

Lehmann, O. (1904), *Flüssige Kristalle: Sowie Plastizität von Kristallen im allgemeinen, molekulare Umlagerungen und Aggregatzustandsänderungen*, Verlag von Wilhelm Engelmann: Leipzig, Germany.

Lichtenberg, D. (1985), Characterization of the solubilization of lipid bilayers by surfactants, *Biochim. Biophys. Acta* 821, 470–478.

Lichtenberg, D., Robson, R. J., and Dennis, E. A. (1983), Solubilization of phospholipids by detergents. Structural and kinetic aspects, *Biochim. Biophys. Acta* 737, 285–304.

Mabrey, S., and Sturtevant, J. (1976), Investigation of phase transitions of lipids and lipid mixtures by sensitivity differential scanning calorimetry, *Proc. Natl. Acad. Sci. USA* 73, 3862–3866.

Majhi, P. R., and Blume, A. (2001), Thermodynamic characterization of temperature-induced micellization and demicellization of detergents studied by differential scanning calorimetry, *Langmuir* 17, 3844–3851.

Majhi, P. R., and Blume, A. (2002), Temperature-induced micelle-vesicle transitions in DMPC–SDS and DMPC–DTAB mixtures studied by calorimetry and dynamic light scattering, *J. Phys. Chem. B* 106, 10753–10763.

Marsh, D. (1999), Thermodynamic analysis of chain-melting transition temperatures for mono-unsaturated phospholipid membranes: Dependence on cis-monoenoic double bond position, *Biophys. J.* 77, 953–963.

Marsh, D. (2009), Cholesterol-induced fluid membrane domains: A compendium of lipid-raft ternary phase diagrams, *Biochim. Biophys. Acta* 1788, 2114–2123.

Marsh, D. (2010), Structural and thermodynamic determinants of chain-melting transition temperatures for phospholipid and glycolipids membranes, *Biochim. Biophys. Acta* 1798, 40–51.

Martinez, J. C., Murciano-Calles, J., Cobos, E. S., Iglesias-Bexiga, M., Luque, I., and Ruiz-Sanz, J. (2013), Isothermal titration calorimetry: Thermodynamic analysis of the binding thermograms of molecular recognition events by using equilibrium models. In A. A. Elkordy (Ed.), *Applications of Calorimetry in a Wide Context: Differential Scanning Calorimetry, Isothermal Titration Calorimetry and Microcalorimetry*, InTech: Rijeka, Croatia.

Martins, P. A. T., and Moreno, M. J. (2015), Kinetics of the interaction of amphiphiles with lipid bilayers using ITC. In M. Bastos (Ed.), *Biocalorimetry*, Taylor and Francis: Boca Raton, FL.

Nicolini, C., Kraineva, J., Khurana, M., Periasamy, N., Funari, S. S., and Winter, R. (2006), Temperature and pressure effects on structural and conformational properties of POPC/SM/cholesterol model raft mixtures—A FT-IR, SAXS, DSC, PPC and Laurdan fluorescence spectroscopy study, *Biochim. Biophys. Acta* 1758, 248–258.

Nicolson, G. L. (2014), The fluid-mosaic model of membrane structure: Still relevant to understanding the structure, function and dynamics of biological membranes after more than 40 years, *Biochim. Biophys. Acta* 1838, 1451–1466.

Oldfield, E., and Chapman, D. (1972), Dynamics of lipids in membranes: Heterogeneity and the role of cholesterol, *FEBS Lett.* 23, 285–297.

Paula, S., Sues, W., Tuchtenhagen, J., and Blume, A. (1995), Thermodynamics of micelle formation as a function of temperature: A high sensitivity titration calorimetry study, *J. Phys. Chem.* 99, 11742–11751.

Pink, D. A., and Chapman, D. (1979), Protein-lipid interactions in bilayer membranes: A lattice model, *Proc. Natl. Acad. Sci. USA* 76, 1542–1546.

Pokorny, A., Yandek, L. E., Elegbede, A. I., Hinderliter, A., and Almeida, P. F. (2006), Temperature and composition dependence of the interaction of delta-lysin with ternary mixtures of sphingomyelin/cholesterol/POPC, *Biophys. J.* 91, 2184–2197.

Quinn, P. J. (2010), A lipid matrix model of membrane raft structure, *Prog. Lipid Res.* 49, 390–406.

Quinn, P. J. (2012), Lipid-lipid interactions in bilayer membranes: Married couples and casual liaisons, *Prog. Lipid Res.* 51, 179–198.

Quinn, P. J. (2013), Structure of sphingomyelin bilayers and complexes with cholesterol forming membrane rafts, *Langmuir* 29, 9447–9456.

Schwieger, C., and Blume, A. (2007), Interaction of poly(L-lysines) with negatively charged membranes: An FT-IR and DSC study, *Eur. Biophys. J.* 36, 437–450.

Schwieger, C., and Blume, A. (2009), Interaction of poly(l-arginine) with negatively charged DPPG membranes: Calorimetric and monolayer studies, *Biomacromolecules* 10, 2152–2161.

Singer, S. J., and Nicolson, G. L. (1972), The fluid mosaic model of the structure of cell membranes, *Science* 175, 720–731.

Small, D. M. (1986), The physical chemistry of lipids. In D. M. Small (Ed.), *Handbook of Lipid Research*, vol. 4, Plenum: New York.

Steim, J. M., Tourtelotte, M. E., Reinert, J. C., McElhaney, R. N., and Rader, R. L. (1969), Calorimetric evidence for the liquid-crystalline state of lipids in a biomembrane, *Proc. Natl. Acad. Sci. USA* 63, 104–109.

Tan, A., Ziegler, A., Steinbauer, B., and Seelig, J. (2002), Thermodynamics of sodium dodecyl sulfate partitioning into lipid membranes, *Biophys. J.* 83, 1547–1556.

Tatulian, S. A. (1993), Ionization and ion binding. In G. Cevc (Ed.), *Phospholipids Handbook*, Marcel Dekker: New York.

Thewalt, J. L., and Bloom, M. (1992), Phosphatidylcholine: Cholesterol phase diagrams, *Biophys. J.* 63, 1176–1181.

Tuchtenhagen, J. (1994), Kalorimetrische und FT-IR-spektroskopische Untersuchungen an Phospholipidmodellmembranen. PhD thesis, Department of Chemistry, University of Kaiserslautern.

Vaz, W. L. C. (1994), Diffusion and reaction in phase-separated membranes, *Biophys. Chem.* 50, 139–145.

Vaz, W. L. C., and Almeida, P. F. F. (1993), Phase topology and percolation in multi-phase lipid bilayers: Is the biological membrane a domain mosaic? *Curr. Opin. Struct. Biol.* 3, 482–488.

Veatch, S. L., and Keller, S. L. (2005), Seeing spots: Complex phase behavior in simple membranes, *Biochim. Biophys. Acta* 1746, 72–85.

Veatch, S. L., Soubias, O., Keller, S. L., and Gawrisch, K. (2007), Critical fluctuations in domain-forming lipid mixtures, *Proc. Natl. Acad. Sci. USA* 104, 17650–17655.

Velazquez-Campoy, A. (2015), Allostery and cooperative interactions in proteins assessed by isothermal titration calorimetry. In M. Bastos (Ed.), *Biocalorimetry*, Taylor and Francis: Boca Raton, FL.

Vereb, G., Szollosi, J., Matko, J., Nagy, P., Farkas, T., Vigh, L., Matyus, L., Waldmann, T. A., and Damjanovich, S., (2003), Dynamic, yet structured: The cell membrane three decades after the Singer-Nicolson model, *Proc. Natl. Acad. Sci. USA* 100, 8053–8058.

Von Dreele, P. H. (1978), Estimation of lateral species separation from phase transitions in nonideal two-dimensional lipid mixtures, *Biochemistry* 17, 3939–3943.

Wenk, M. R., Alt, T., Seelig, A., and Seelig, J. (1997), Octyl-beta-D-glucopyranoside partitioning into lipid bilayers: Thermodynamics of binding and structural changes of the bilayer, *Biophys. J.* 72, 1719–1731.

CHAPTER 7

Thermotropic Properties of Ceramides in Aqueous or Lipidic Environments

Alicia Alonso and Felix M. Goñi

CONTENTS

7.1 CERAMIDES: AN INTRODUCTION

Ceramides are N-acylsphingosines. In turn, sphingosine is 2S,3R,4E-2-amino-4-octadecene-1,3-diol. Some ceramides are known to contain sphingosine analogs, for example, the saturated form, known as sphinganine, or the 20 C-atom eicosasphingosine, or the 4-hydroxylated and saturated phytosphingosine. The fatty acyl chain is linked through an amide bond to the 2-amino group of sphingosine and can contain from 2 to 24 C atoms, although palmitic (C16:0), stearic (C18:0), and nervonic (C24:1) fatty acids are the most common in mammals.

Ceramides have been known for decades as intermediates in sphingolipid metabolism and as minor membrane components (Stoffel 1971). The most commonly found ceramides, those with a fatty acyl chain of 16 carbon atoms or longer, are among the least polar, most hydrophobic lipids in membranes. Indeed, their hydrophobicity explains their abundance in the stratum corneum, the barrier that prevents water evaporation through the skin (Wertz and Downing 1989). Their solubility in water is negligible. Thus, free ceramides cannot exist in solution in biological fluids or in cytosol. In addition, long-chain ceramides (e.g., with a fatty acid C12 or longer) belong to the category of *nonswelling amphiphiles*

(Small 1970), implying that they cannot even give rise to micelles or other aggregates in aqueous suspension, unlike, for example, common phospholipids, glycosphingolipids, and surfactants. Short-chain ceramides (e.g., N-acetyl or C2 ceramide) appear to "swell" in water, giving rise to homogeneous dispersions (Hannun et al. 1987; Simon and Gear 1998) that are treated, in practice, as solutions.

Sphingolipids have long been known as cell components. They were given their name by J. L. W. Thudichum in 1884 because of their enigmatic nature. Recently, with the discovery of the sphingolipid signaling pathway (Kolesnick 1987; Okazaki et al. 1989), interest in these lipids was renewed. At the time of the discovery of sphingolipid signaling, very little was known about the properties of these lipids in membranes (for a review, see Maggio 1994). Our own studies on ceramides started in 1996 with the description of their ability to permeabilize lipid bilayers (Ruiz-Arguello et al. 1996).

Because of length restrictions, some relevant aspects of ceramide calorimetry could not be included in this chapter; nor could the relevant principles of differential scanning calorimetry (DSC), which can be found in the first chapters of the book and also in Goñi and Alonso (2006b), a discussion of ceramides present in the skin stratum corneum (Bouwstra and Ponec 2006), or calorimetric studies of ceramide produced by sphingomyelinase hydrolysis of sphingomyelin, briefly discussed in a previous review (Goñi and Alonso 2009).

7.2 CALORIMETRIC STUDIES OF PURE CERAMIDES

The thermotropic properties and phase behavior of anhydrous and hydrated pure ceramides were first examined in detail by Shipley and coworkers. A synthetic N-palmitoylsphingosine (C16 ceramide) was studied at various degrees of hydration using DCS and x-ray diffraction techniques (Shah et al. 1995a). The fully hydrated ceramide exhibits a complex polymorphic behavior (Figure 7.1). At room temperature, a well-ordered metastable bilayer phase exists. On increasing the temperature, an exothermic transition occurs at 64.2°C to form a stable bilayer phase with crystalline chain packing. The transition produces an increased chain tilting with respect to the bilayer normal and/or partial dehydration. Further heating converts the crystalline phase into a disordered, melted-chain phase, of an undefined structure, through an endothermic transition at 90°C ($\Delta H = 57.7$ kJ/mol). In contrast, the anhydrous C16 ceramide shows a single, broad endothermic transition at 95.4°C ($\Delta H = 43.5$ kJ/mol). The complex behavior of the hydrated ceramide is also found in more polar sphingolipids, for example sphingomyelin or cerebroside, and suggests hydrogen bonding of the hydroxyl groups of sphingosine to the amide group of adjacent molecules. However the presence of an α-hydroxy group in the fatty acyl residues induces a simpler phase behavior, and the fully hydrated hydroxy ceramides exhibit a single, reversible endothermic transition between a bilayer gel and a hexagonal fluid phase (Shah et al. 1995b). Similar results have been independently obtained by Han and coworkers (Han et al. 1995).

Years later, Prieto and coworkers (Pinto et al. 2008) examined the thermotropic properties of N-nervonoyl (C24:1) ceramide. The main structural difference between this and N-palmitoyl ceramide is the asymmetric chain length of N-nervonoyl: the fatty acyl chain is far longer than the sphingosine chain. This kind of chain-length asymmetry, with one

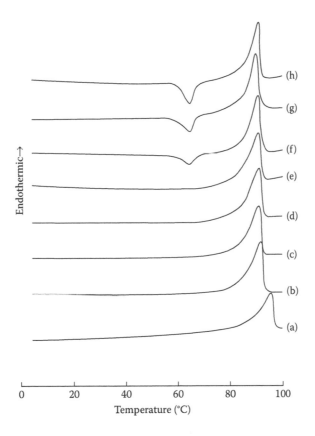

FIGURE 7.1 DSC heating curves; heating rate, 5°C/min. (a) 0 wt.% water; (b) 3.8 wt.% water; (c) 8.0 wt.% water; (d) 9.3 wt.% water; (e) 12.0 wt.% water; (f) 23.1 wt.% water; (g) 34.0 wt.% water; and (h) 74.3 wt.% water. This research was originally published in the *Journal of Lipid Research*. Shah, J., Atienza, J. M., Duclos, R. I. J., Rawlings, A. V., Dong, Z., and Shipley, G. G. 1995. Structural and thermotropic properties of synthetic C16:0 (palmitoyl) ceramide: Effect of hydration. *J. Lipid Res.* 36:1936–44. © the American Society for Biochemistry and Molecular Biology.

chain much longer than C16, often leads to intermonolayer interdigitation. DSC of C24:1 ceramide revealed a minor transition at 20°C from a mixed interdigitated gel phase to a partially interdigitated gel phase, and a broad main transition to the fluid phase at approximately 52°C. The same ceramide was studied more recently by Jiménez-Rojo and coworkers, with different results (Jiménez-Rojo et al. 2014a). The latter authors found the main gel–fluid transition at ~74°C. The difference was probably due to sample preparation: Pinto et al. included several freeze–thaw cycles, while Jiménez-Rojo et al. used the procedure of forcing the water/ceramide mixture through a narrow tubing at high *T* to facilitate dispersion. Many authors (Maggio et al. 1985; Shah et al. 1995a; Westerlund et al. 2010) have pointed out the relevance of the sample thermal history in ceramide studies, particularly for the long-chain ones.

In fact, Jiménez-Rojo and coworkers performed a systematic study of the thermotropic properties of aqueous dispersions of ceramides with N-acyl chains varying from C6:0 to C24:1 using DSC (Jiménez-Rojo et al. 2014a). In general, ceramide showed a main narrow endotherm, sometimes accompanied by small, wider endotherms and even small exotherms (Figure 7.2). T_m of the main endotherm increased with chain length of the saturated species, up to C16:0, and ΔH also increased in a similar way. With the longer-chain ceramides, C18:0 and C24:0, smaller exotherms were found, which represented in both cases ~12%–13% of the total heat absorbed. Small exotherms were also observed in several samples, at temperatures always below that of the main transition (Table 7.1). These

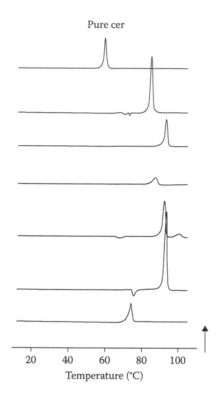

FIGURE 7.2 DSC thermograms of pure ceramides of different N-acyl lengths. Chain lengths are, from top to bottom, 6:0, 12:0, 16:0, natural egg ceramide, 18:0, 24:0, and 24:1. Representative results (third scans) of two very similar measurements. (Modified from Jiménez-Rojo, N. et al., *Biochim. Biophys. Acta* 1838, 456–464, 2014.)

TABLE 7.1 Thermodynamic Parameters of Thermotropic Transitions of Ceramides

N-acyl chain	T_m (°C)			ΔH (kJ/mol)			WHH[a] (°C)		
6	59.6	—	—	7.1	—	—	1.0	—	—
12	70.3	**85.2**	—	−0.29	15.61	—	1.1	—	—
16	93.1	—	—	9.37	—	—	1.4	—	—
Egg	87.5	—	—	3.81	—	—	3.0	—	—
18	67.5	**92.2**	99.9	−0.46	9.87	1.30	1.1	3.3	—
24	75.3	**93.2**	—	−1.75	21.00	—	—	—	—
24:1	63.1	**73.7**	81.7	0.67	6.11	0.13	4.1	1.6	0.9

Note: T_m of the main transitions is indicated in bold. Data obtained from thermograms as shown in Figure 7.2. Average values of three closely similar experiments.

[a] Width at half height.

exotherms were present even when the most painstaking efforts were made to achieve a homogeneous hydration and did not show major changes when the sample was stored for up to 48 h at room temperature or at 4°C before the measurements. These exotherms have been attributed to metastable crystalline phases that change into more stable gel phases before melting (Shah et al. 1995a; Westerlund et al. 2010). Our T_m values for the main endotherms (Table 7.1, figures in bold) are in good agreement with those published by Shah and coworkers (Shah et al. 1995a) and by Westerlund and coworkers (Westerlund et al. 2010).

As mentioned in Section 7.1, ceramides and dihydroceramides are N-acyl derivatives of sphingosine and sphinganine, respectively, which are the major sphingoid-base backbones of mammals. Recent studies have found that mammals, like certain other organisms, also produce 1-deoxy-(dihydro)ceramides (1-deoxyDHCers) that contain sphingoid bases lacking the 1-hydroxyl or 1-hydroxymethyl groups. The amounts of these compounds can be substantial; indeed, Jiménez-Rojo and coworkers have found comparable levels of 1-deoxyDHCers and ceramides in RAW 264.7 cells maintained in culture (Jiménez-Rojo et al. 2014b). These lipids might play important roles in normal cell regulation and in the pathology of diseases in which they are elevated, such as hereditary sensory autonomic neuropathies and diabetes. Jiménez-Rojo and coworkers used several approaches, including surface-pressure measurements, DSC, and confocal microscopy, to study the behavior of 1-deoxyDHCers of different N-acyl-chain lengths and their interaction with sphingomyelin (SM) (Jiménez-Rojo et al. 2014b). The gel–fluid transition temperatures of the pure compounds increased in the order 1-deoxyceramide < ceramide ≈ 1-deoxyD-HCer < 1-(deoxymethyl)DHCer (Figure 7.3). Of these, 1-(deoxymethyl)DHCers were the most hydrophobic, not even capable of forming monolayers at the air–water interface. These properties suggest that 1-deoxyDHCer can influence the properties of cellular membranes in ways that might affect biological function/malfunction.

7.3 MIXTURES WITH PHOSPHOLIPIDS

In mixtures with phospholipids, ceramides have two main effects: they increase the molecular order of phospholipids, and they give rise to lateral phase separation and domain formation (Goñi and Alonso 2006a).

7.3.1 Ceramide/Glycerophospholipid Mixtures

Separation into ceramide-rich domains was first observed by Huang and coworkers, who examined the structure of bilayers composed of fully deuterated dipalmitoylphosphatidylcholine (DPPC) and bovine brain ceramide using ^{2}H-nuclear magnetic resonance (NMR) spectroscopy (Huang et al. 1996). These authors observed that addition of ceramide induced lateral phase separation of fluid phospholipid bilayers into regions of gel and liquid-crystalline (fluid) phases, with ceramide partitioning largely into the gel phase of d62-DPPC.

Holopainen and coworkers were perhaps the first to examine ceramide/glycerophospholipid mixtures by DSC (Holopainen et al. 1997). They used liposomes consisting of dimyristoylphosphatidylcholine (DMPC) and natural ceramide. Calorimetry revealed that ceramide increased the pretransition temperature till at $X_{cer} > 0.07$ this transition was no

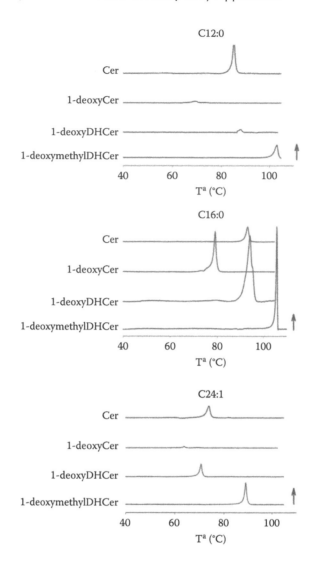

FIGURE 7.3 DSC thermograms of pure 1-deoxyceramides. Chain lengths are indicated above each plot. Results (third scans) of two very similar measurements are represented. The arrow indicates 2000 J/mol°C. The amount of sample may not be the same in all three cases. (From Jiménez-Rojo, N. et al., *Biophys. J.* 107, 2850–2859, 2014.)

longer evident. Also, increasing X_{cer} up to 0.05 increased T_m from 23.9 to 24.6°C. ΔH of the main transition decreased progressively on increase in X_{cer} up to approximately 0.10. Above this ceramide concentration, a new endotherm became evident at 22.5°C, and above $X_{cer} = 0.14$, the latter endotherm became dominant. The authors interpreted these results as microdomain formation concomitant with the presence of the ceramide-enriched phase at $X_{cer} < 0.10$.

Domain formation by ceramides was also described by a combination of DSC and infrared (IR) spectroscopy, using natural ceramides (brain, egg) and several synthetic phospholipids (Veiga et al. 1999). Calorimetry was used to detect gel–fluid transitions. Different domains, when formed, "melt" at different temperatures, so that they can be easily detected. Veiga et al. (1999) found lateral separation of ceramide-rich domains with as little as 5 mol% ceramide. Carrer and Maggio (1999) have studied mixtures of bovine brain ceramide, which contains mostly C18:0 and C24:1 fatty acids, with DPPC, both by DSC and in lipid monolayers extended at the air–water interface. The calorimetric results are essentially coincident with those of Veiga and coworkers (Veiga et al. 1999), hinting at in-plane phase separation with only 1% ceramide.

Holopainen and coworkers refined their 1997 results using mixtures of DMPC with C16:0 ceramide instead of a mixture of ceramides of natural origin (Holopainen et al. 2000). The authors constructed a partial phase diagram on the basis of DSC and x-ray scattering data. DSC heating scans showed that with increased X_{cer}, the pretransition temperature T_p first increased, till at $X_{cer} > 0.06$ it could no longer be resolved. The main transition enthalpy ΔH remained practically unaltered, while its width increased significantly, and the upper phase boundary temperature of the mixture shifted to approximately 63°C at $X_{cer} = 0.30$. On cooling, a profound phase separation was evident, and for all of the studied compositions there was an endotherm in the region close to the T_m for DMPC. At $X_{cer} \geq 0.03$, a second endotherm was evident at higher temperatures, starting at 32.1°C and reaching 54.6°C at $X_{cer} = 0.30$.

Under equilibrium conditions, the disposition of lipids in cell membranes resembles that of the lamellar phases formed by some pure phospholipids—for example, egg phosphatidylcholine—in water. However, the idea that transient nonlamellar structures may form in membranes under nonequilibrium conditions is now widespread (Goñi 2014). Such nonlamellar structures are, indeed, formed under equilibrium conditions by certain phospholipids. One of them, dielaidoyl phosphatidylethanolamine (DEPE), exhibits a reversible lamellar-to-inverted hexagonal phase transition at 65°C. Inverted hexagonal and inverted cubic phases are believed to be involved in processes of membrane fusion and fission (Nieva et al. 1995 and references therein). Veiga and coworkers have analyzed the lamellar-to-inverted hexagonal phase transition of DEPE, either pure or in mixtures with natural long-chain ceramides, by a combination of DSC and ^{31}P-NMR spectroscopy (Veiga et al. 1999). Ceramides decrease the transition temperature without significantly modifying the transition enthalpy, thus facilitating inverted hexagonal phase formation. Additionally, ^{31}P-NMR indicates that, under certain conditions, lamellar and hexagonal phases may coexist.

The different effects of long- and short-chain ceramides on the gel–fluid and lamellar–hexagonal transitions of DEPE were examined by Sot and coworkers (Sot et al. 2005) using a combination of DSC, x-ray scattering, and ^{31}P-NMR (Figure 7.4). N-palmitoyl (Cer16), N-hexanoyl (Cer6), and N-acetyl (Cer2) sphingosines were used in the 0–25 mol% concentration range. Pure hydrated ceramides exhibited cooperative endothermic order–disorder transitions at 93°C (Cer16), 60°C (Cer6), and 54°C (Cer2). In DEPE bilayers, Cer16 did not mix with the phospholipid in the gel phase, giving rise to high-melting ceramide-rich domains. Cer16 favored the lamellar–hexagonal transition of DEPE, decreasing the transition temperature. Cer2, on the other hand, was soluble in the gel phase of DEPE, decreasing the gel–fluid and increasing the lamellar–hexagonal transition temperature, thus effectively stabilizing the lamellar fluid phase. In addition, Cer2 was peculiar in that no equilibrium could be reached for the Cer2–DEPE mixture above 60°C, the lamellar–hexagonal

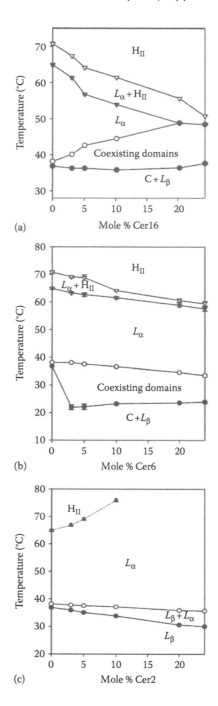

FIGURE 7.4 Temperature-composition diagrams for (a) DEPE/Cer16, (b) DEPE/Cer6, and (c) DEPE/Cer2 mixtures in excess water. Transition temperatures were derived from DSC thermograms. Phase structures were obtained from x-ray diffraction and ^{31}P-NMR. C: ceramide-rich gel phase; L_β: DEPE, or DEPE-rich, gel phase; L_α: liquid-crystalline or fluid phase; H_{II}, inverted hexagonal phase. (●, ○) Respectively, onset and completion temperatures of the gel–fluid phase transition. (▼, ▽) Respectively, onset and completion temperatures of the lamellar–hexagonal phase transition. (Modified from Sot, J. et al., *Biophys. J.* 88, 3368–3380, 2005.)

transition shifting with time to temperatures beyond the instrumental range. The properties of Cer6 are intermediate between those of the other two, this ceramide decreasing both the gel–fluid and lamellar–hexagonal transition temperatures. Temperature-composition diagrams were constructed for the mixtures of DEPE with each of the three ceramides. The different behavior of the long- and short-chain ceramides could be rationalized in terms of their different molecular geometries, Cer16 favoring negative curvature in the monolayers and thus inverted phases, and the opposite being true of the micelle-forming Cer2. Carrer and Maggio (1999) were also able to construct a phase diagram (temperature-composition diagram) for the C8:0 ceramide/DMPC mixture. In their case, C8:0 ceramide constitutes, as C6:0 did in Sot et al. (2005), an asymmetric-chain-length ceramide, and the authors suggested a partial interdigitation of C8:0 ceramide chains in the gel phase. Ceramide chain-length asymmetry was also explored by Pinto and coworkers using C24:1 ceramide and palmitoyl-oleoyl-phosphatidylcholine (POPC) binary mixtures (Pinto et al. 2008). Note that in this case, the asymmetry arises from a long N-acyl chain. The solubility of NCer in the fluid POPC was low, driving gel–fluid phase separation, and the binary phase diagram was characterized by multiple and large coexistence regions between the interdigitated gel phases and the fluid phase. At 37°C, the relevant phases were the fluid and the partially interdigitated gel. Moreover, the formation of NCer interdigitated gel phases led to strong morphological alterations in the lipid vesicles, driving the formation of cochleate-type tubular structures.

7.3.2 Ceramide in Mixtures with Sphingomyelins

Ceramide-enriched domains in sphingomyelin/ceramide mixtures were described by Sot and coworkers using mainly DSC and fluorescence microscopy (Sot et al. 2006). DSC data showed for pure egg sphingomyelin a rather narrow transition centered at 39°C. Egg ceramide, even at low proportions (5 mol%), had the effect of widening the phase transition, shifting it to higher temperatures. More important, the endotherms of ceramide-containing samples had a clearly asymmetric shape, indicating formation of high-T melting ceramide/sphingomyelin domains. In fact, the observed overall endotherms could be fitted to three-component endotherms, corresponding to the same number of coexisting domains with different compositions. The corresponding phase diagram includes a wide temperature/composition region of coexisting domains (Sot et al. 2006). Essentially similar results were obtained for DPPC/brain Cer mixtures (Carrer and Maggio 1999), an example of chemically different molecules behaving in the same way due to structural similarities. Interestingly, Sot et al. found that sphingomyelin/ceramide bilayers were resistant to solubilization by Triton X-100 at 4°C, the nonsolubilized residue being enriched in ceramide (Sot et al. 2006). The DSC results were refined by Busto and coworkers using C16:0 ceramide and C16:0 SM (Busto et al. 2009). DSC of multilamellar vesicles revealed a widening of the C16 SM main gel–fluid phase transition (41°C) on C16 Cer incorporation, with formation of a further endotherm at higher temperatures that could be deconvoluted into two components. DSC data reflected the presence of C16 Cer-enriched domains coexisting, in different proportions, with a C16 SM-enriched phase. The C16 SM-enriched phase was no longer detected in DSC thermograms containing >30 mol% C16 Cer. The same system was studied by Leung and coworkers using ^{2}H-NMR (Leung et al. 2012). The alternate use of either perdeuterated or proton-based N-acyl chain palmitoylsphingomy- elin (PSM) and PCer in our ^{2}H NMR studies allowed the separate observation of gel–fluid transitions in each lipid in the presence of the other one, and this in turn provided direct information on the lipids' miscibility over a wide temperature range. The results confirmed the stabilization of the C16 SM gel state by C16 Cer. Moreover, overlapping NMR and DSC data revealed that the DSC signals paralleled the melting of the major component (C16 SM) except at intermediate (20 and 30 mol%) fractions of C16 Cer. In such cases, the DSC endotherm reported on the presumably highly cooperative melting of C16 Cer. Up to at least 50 mol% C16 Cer, C16 SM, and C16 Cer mixed ideally in the liquid crystalline phase; in the gel phase, C16 Cer became incorporated into C16 SM:C16 Cer membranes with no evidence of pure solid C16 Cer.

The influence of ceramide N-acyl chain length on its miscibility with SM has been stud- ied by Westerlund et al. (2010) and by Jiménez-Rojo and coworkers (Jiménez-Rojo et al. 2014a). The former authors investigated the effect of acyl chain composition in ceramides (C4–C24:1) on their miscibility with N-palmitoyl-sphingomyelin (C16 SM) using DSC. They found that short-chain (C4 and C8) ceramides induced phase separation and lowered the T_m and enthalpy of the C16 SM endotherm. They concluded that short-chain cerami- des were more miscible in the fluid-phase than in the gel-phase SM bilayers. Long-chain ceramides induced apparent heterogeneity in the bilayers. The main C16 SM endotherm decreased in cooperativity and enthalpy with increasing ceramide concentration. New ceramide-enriched components could be seen in the thermograms at all ceramide con- centrations above $X_{cer}=0.05$. These broad components had higher T_m values than pure C16 SM. C24:1 ceramide exhibited complex behavior in the C16 SM bilayers. The miscibility of C24:1 ceramide with C16 SM at low ceramide concentrations ($X_{cer}<0.10$) concentrations was exceptionally good according to the cooperativity of the transition. At higher concen- trations, multiple components were detected, which might have arisen from interdigitated gel phases formed by this highly asymmetric ceramide. The results of this study indicate that short-chain and long-chain ceramides have very different effects on the sphingomy- elin bilayers. In turn, Jiménez-Rojo and coworkers varied the acyl chain length both in ceramides and in SM (Jiménez-Rojo et al. 2014a). Most SM:ceramide mixtures (at a 9:1 molar ratio) gave rise to bilayers containing separate SM-rich and Cer-rich domains, but the particular thermotropic properties of the mixtures were often unrelated to the charges in N-acyl chain length.

In the studies reviewed in the previous two paragraphs, the main variable was the length of the ceramide N-acyl chain. However, Maula and coworkers studied the effect of vary- ing the sphingoid base length (Maula et al. 2012). They synthesized ceramides consisting of a C12-, C14-, C16-, C18-, or C20-sphing-4-enin derivative coupled to palmitic acid. The ceramides were studied in mixtures with PSM. All of the analogs were able to thermally stabilize PSM, and a chain-length-dependent increase in the main phase transition tem- perature of equimolar PSM/Cer bilayers was revealed by DSC.

The same authors performed remarkable synthetic chemistry work, preparing a series of ceramide analogs. Maula and coworkers synthesized palmitoylceramide analogues in which a methyl group was introduced to the amide-nitrogen or the C3-oxygen of the sphingosine backbone (Maula et al. 2011). A DSC analysis of equimolar mixtures of palmitoylsphingomyelin (C16 SM) and C16 ceramide showed that C16 Cer analogs formed gel phases with PSM but that they melted at lower temperatures compared with the system with canonical PCer. Taken together, these results show that 2N-methylation weakened the ceramide–SM interactions, whereas the 3O-methylated ceramide behaved more like C16 Cer in interactions with C16 SM. These findings are compatible with the view that interlipid interactions between the amide-nitrogen and neighboring lipids are important for the cohesive properties of sphingolipids in membranes.

Because ceramide lacks a headgroup that could shield its hydrophobic body from unfavorable interactions with water, accommodation of ceramide under the larger phosphocholine headgroup of SM could contribute to their favorable interactions. To elucidate the role of the SM headgroup in SM/ceramide interactions, Artetxe and coworkers explored the effects of reducing the size of the phosphocholine headgroup (removing one, two, or three methyls on the choline moiety, or the choline moiety itself) (Artetxe et al. 2013). Using DSC and fluorescence spectroscopy, they found that the size of the SM headgroup had no marked effect on the thermal stability of ordered domains formed by SM analog/palmitoyl ceramide (C16 Cer) interactions. In more complex bilayers composed of a fluid glycerophospholipid, an SM analog, and C16 Cer, the thermal stability and molecular order of the laterally segregated gel domains were roughly identical despite variation in SM headgroup size. The data suggested that the association between PCer and SM analogs was stabilized by ceramide's aversion to disordered phospholipids, by interfacial hydrogen bonding between C16 Cer and the SM analogs, and by attractive van der Waals' forces between saturated chains of C16 Cer and SM analogs.

An interesting step forward in this type of study was made by Boulgaropoulos and coworkers, who explored ternary mixtures of ceramide, POPC, and SM using DSC together with x-ray scattering and IR spectroscopy (Boulgaropoulos et al. 2011). They found that the POPC/SM equimolar mixture showed a single broad main transition at 21.3°C, with a half-width of ~8°C. In the presence of 5 mol% ceramide, however, the single transition peak split into two broad peaks at 19°C and 44°C, and at higher ceramide concentrations (up to 35 mol%), the two transition peaks separated further, one of them approaching the T_m of pure POPC (−33°C) and the second shifting more and more toward temperatures above that of pure SM (39°C). At and above 35 mol% ceramide, exotherms and endotherms were detected at temperatures compatible with the presence of pure ceramide domains.

7.4 MIXTURES WITH CHOLESTEROL (AND OTHER LIPIDS)

Souza and coworkers were able to derive the thermotropic binary phase diagram of mixtures of C16:0 ceramide and cholesterol in excess water, using differential scanning calorimetry and x-ray diffraction (Souza et al. 2009). These mixtures were self-organized in lamellar mesostructures, which, among other particularities, showed two ceramide-to-cholesterol crystalline phases with molar proportions that approached 2:3 and 1:3. The 2:3 phase crystallized in a tetragonal arrangement with a lamellar repeat distance of 3.50 nm, which indicates an unusual lipid stacking, probably unilamellar. The uncommon mesostructures formed by ceramides with cholesterol should be considered in view of their proposed interaction in phospholipid:ceramide:cholesterol mixtures.

In fact, cholesterol introduces a further degree of complexity into phospholipid: ceramide binary mixtures. Fidorra and coworkers provided a fine example of this in their work in which binary (POPC/ceramide) and ternary (POPC/ceramide/cholesterol) mixtures were studied with a combination of calorimetric, spectroscopic, and microscopic techniques (Fidorra et al. 2006). The binary mixture exhibited gel–fluid phase coexistence over a broad composition and temperature span. Addition of cholesterol causes a gradual ordering effect of the fluid phase, with smaller changes in the ceramide-rich gel phase, up to a certain cholesterol concentration. Above that point (e.g., 22 mol% cholesterol in POPC/ceramide 5:1 molar ratio), a complex situation occurs in which three regions coexist in the same giant unilamellar vesicle (GUV), which the authors assign tentatively to one gel and two ordered phases. Interestingly, no coexistence of L_0 and L_α phases was observed in the POPC/ceramide/cholesterol mixture, unlike systems in which SM (i.e., ceramide phosphorylcholine) was used instead of Cer (de Almeida et al. 2005; Dietrich et al. 2001).

The observation by Megha and coworkers (Megha 2004) that ceramide displaces cholesterol from L_0 domains containing PC, or SM, and cholesterol is very interesting, and it elicited a whole series of studies, in which, however, calorimetric techniques were not preponderant. A study from our laboratory (Sot et al. 2008) provided morphological and DSC evidence for the generation of ceramide-enriched gel domains within L_0 domains in GUV composed of PC:SM:PE:cholesterol on addition of ceramide.

Alonso and coworkers have studied the phenomenon of cholesterol C16 ceramide displacement in ternary bilayers with C16 SM, that is, in the absence of an L_0 phase (Busto et al. 2010). DSC of multilamellar vesicles and confocal fluorescence microscopy of GUVs concurred in showing immiscibility, but no displacement, between L_0 cholesterol-enriched and gel-like ceramide-enriched phases at high SM/(cholesterol + ceramide) ratios. At higher cholesterol content, C16 Cer was unable to displace cholesterol to any extent, even at $X_{Chol} < 0.25$. It is interesting that an opposite strong cholesterol-mediated Cer displacement from its tight packing with SM was clearly detected, completely abolishing the ability of Cer to generate large microdomains and giving rise instead to a single ternary phase. These observations in model membranes in the absence of the lipids commonly used to form a liquid-disordered phase support the role of cholesterol as the key determinant in controlling its own displacement from L_0 domains by ceramide. As a consequence of this study, two novel gel lamellar phases of ternary composition have been described, containing ceramide and cholesterol, and either DPPC or C16 SM (Busto et al. 2014). These phases are stabilized by direct cholesterol–ceramide interactions and present intermediate properties between phospholipid–cholesterol L_0 and phospholipid–ceramide gel phases. The calorimetric midpoint gel–fluid transition temperatures are 56°C and 54°C for the SM- and PC-based mixtures, respectively.

DSC has also been applied to the study of ceramide in more complex mixtures, such as cholesteryl oleate/fatty acid/cholesterol/ceramide (Souza et al. 2011) or lysophospholipid/1,3-butanediol/ceramide (Konno et al. 2014).

ACKNOWLEDGMENTS

The authors thank the Spanish and Basque Governments for long-standing support to their laboratory.

REFERENCES

Artetxe, I., Sergelius, C., Kurita, M., Yamaguchi, S., Katsumura, S., Slotte, J. P., and Maula, T. (2013), Effects of sphingomyelin headgroup size on interactions with ceramide, *Biophys. J.* 104, 604–612.

Boulgaropoulos, B., Arsov, Z., Laggner, P., and Pabst, G. (2011), Stable and unstable lipid domains in ceramide-containing membranes, *Biophys. J.* 100, 2160–2168.

Bouwstra, J. A., and Ponec, M. (2006), The skin barrier in healthy and diseased state, *Biochim. Biophys. Acta* 1758, 2080–2095.

Busto, J. V., Fanani, M. L., De Tullio, L., Sot, J., Maggio, B., Goñi, F. M., and Alonso, A. (2009), Coexistence of immiscible mixtures of palmitoylsphingomyelin and palmitoylceramide in monolayers and bilayers, *Biophys. J.* 97, 2717–2726.

Busto, J. V., García-Arribas, A. B., Sot, J., Torrecillas, A., Gómez-Fernández, J. C., Goñi, F. M., and Alonso, A. (2014), Lamellar gel (lβ) phases of ternary lipid composition containing ceramide and cholesterol, *Biophys. J.* 106, 621–630.

Busto, J. V., Sot, J., Requejo-Isidro, J., Goñi, F. M., and Alonso, A. (2010), Cholesterol displaces palmitoylceramide from its tight packing with palmitoylsphingomyelin in the absence of a liquid-disordered phase, *Biophys. J.* 99, 1119–1128.

Carrer, D. C., and Maggio, B. (1999), Phase behavior and molecular interactions in mixtures of ceramide with dipalmitoylphosphatidylcholine, *J. Lipid Res.* 40, 1978–1989.

de Almeida, R. F. M., Loura, L. M. S., Fedorov, A., and Prieto, M. (2005), Lipid rafts have different sizes depending on membrane composition: A time-resolved fluorescence resonance energy transfer study, *J. Mol. Biol.* 346, 1109–1120.

Dietrich, C., Bagatolli, L. A., Volovyk, Z. N., Thompson, N. L., Levi, M., Jacobson, K., and Gratton, E. (2001), Lipid rafts reconstituted in model membranes, *Biophys. J.* 80, 1417–1428.

Fidorra, M., Duelund, L., Leidy, C., Simonsen, A. C., and Bagatolli, L. A. (2006), Absence of fluid-ordered/fluid-disordered phase coexistence in ceramide/POPC mixtures containing cholesterol, *Biophys. J.* 90, 4437–4451.

Goñi, F. M. (2014), The basic structure and dynamics of cell membranes: An update of the Singer–Nicolson model, *Biochim. Biophys. Acta* 1838, 1467–1476.

Goñi, F. M., and Alonso, A. (2006a), Biophysics of sphingolipids I. Membrane properties of sphingosine, ceramides and other simple sphingolipids, *Biochim. Biophys. Acta* 1758, 1902–1921.

Goñi, F. M., and Alonso, A. (2006b), Differential scanning calorimetry in the study of lipid structures. In B. Larijani, C. A. Rosser and R. Woscholski (Eds.), *Chemical Biology. Techniques and Applications*, 47–66, John Wiley: Chichester.

Goñi, F. M., and Alonso, A. (2009), Effects of ceramide and other simple sphingolipids on membrane lateral structure, *Biochim. Biophys. Acta* 1788, 169–177.

Han, C. H., Sanftleben, R., and Wiedmann, T. S. (1995), Phase properties of mixtures of ceramides, *Lipids* 30, 121–128.

Hannun, Y. A., Greenberg, C. S., and Bell, R. M. (1987), Sphingosine inhibition of agonist-dependent secretion and activation of human platelets implies that protein kinase C is a necessary and common event of the signal transduction pathways, *J. Biol. Chem.* 262, 13620–13626.

Holopainen, J. M., Lehtonen, J. Y., and Kinnunen, P. K. (1997), Lipid microdomains in dimyristoylphosphatidylcholine-ceramide liposomes, *Chem. Phys. Lipids* 88, 1–13.

Holopainen, J. M., Lemmich, J., Richter, F., Mouritsen, O. G., Rapp, G., and Kinnunen, P. K. (2000), Dimyristoylphosphatidylcholine/C16:0-ceramide binary liposomes studied by differential scanning calorimetry and wide- and small-angle x-ray scattering, *Biophys. J.* 78, 2459–2469.

Huang, H. W., Goldberg, E. M., and Zidovetzki, R. (1996), Ceramide induces structural defects into phosphatidylcholine bilayers and activates phospholipase A2, *Biochem. Biophys. Res. Commun.* 220, 834–838.

Jiménez-Rojo, N., García-Arribas, A. B., Sot, J., Alonso, A., and Goñi, F. M. (2014a), Lipid bilayers containing sphingomyelins and ceramides of varying N-acyl lengths: A glimpse into sphingolipid complexity, *Biochim. Biophys. Acta* 1838, 456–464.

Jiménez-Rojo, N., Sot, J., Busto, J. V., Shaw, W. A., Duan, J., Merrill, A. H., Alonso, A., and Goñi, F. M. (2014b), Biophysical properties of novel 1-deoxy-(dihydro)ceramides occurring in mammalian cells, *Biophys. J.* 107, 2850–2859.

Kolesnick, R. N. (1987), 1,2-Diacylglycerols but not phorbol esters stimulate sphingomyelin hydrolysis in GH3 pituitary cells, *J. Biol. Chem.* 262, 16759–16762.

Konno, Y., Naito, N., Yoshimura, A., and Aramaki, K. (2014), A study on the formation of liquid ordered phase in lysophospholipid/cholesterol/1,3-butanediol/ water and lysophospholipid/ceramide/1,3-butanediol/water systems, *J. Oleo Sci.* 63, 823–828.

Leung, S. S. W., Busto, J. V., Keyvanloo, A., Goñi, F. M., and Thewalt, J. (2012), Insights into sphingolipid miscibility: Separate observation of sphingomyelin and ceramide N-acyl chain melting, *Biophys. J.* 103, 2465–2474.

Maggio, B. (1994), The surface behavior of glycosphingolipids in biomembranes: A new frontier of molecular ecology, *Prog. Biophys. Mol. Biol.* 62, 55–117.

Maggio, B., Ariga, T., Sturtevant, J. M., and Yu, R. K. (1985), Thermotropic behavior of binary mixtures of dipalmitoylphosphatidylcholine and glycosphingolipids in aqueous dispersions, *Biochim. Biophys. Acta* 818, 1–12.

Maula, T., Artetxe, I., Grandell, P.-M., and Slotte, J. P. (2012), Importance of the sphingoid base length for the membrane properties of ceramides, *Biophys. J.* 103, 1870–1879.

Maula, T., Kurita, M., Yamaguchi, S., Yamamoto, T., Katsumura, S, and Slotte, J. P. (2011), Effects of sphingosine 2N- and 3O-methylation on palmitoyl ceramide properties in bilayer membranes, *Biophys. J.* 101, 2948–2956.

Megha, L. E. (2004), Ceramide selectively displaces cholesterol from ordered lipid domains (rafts): Implications for lipid raft structure and function, *J. Biol. Chem.* 279, 9997–10004.

Nieva, J. L., Alonso, A., Basánez, G., Goñi, F. M., Gulik, A., Vargas, R., and Luzzati, V. (1995), Topological properties of two cubic phases of a phospholipid:cholesterol:diacylglycerol aqueous system and their possible implications in the phospholipase C-induced liposome fusion, *FEBS Lett.* 368, 143–147.

Okazaki, T., Bell, R. M., and Hannun, Y. A. (1989), Sphingomyelin turnover induced by vitamin D3 in HL-60 cells. Role in cell differentiation, *J. Biol. Chem.* 264, 19076–19080.

Pinto, S. N., Silva, L. C., de Almeida, R. F. M., and Prieto, M. (2008), Membrane domain formation, interdigitation, and morphological alterations induced by the very long chain asymmetric C24:1 ceramide, *Biophys. J.* 95, 2867–2879.

Ruiz-Arguello, M. B., Basanez, G., Goñi, F. M., and Alonso, A. (1996), Different effects of enzyme-generated ceramides and diacylglycerols in phospholipid membrane fusion and leakage, *J. Biol. Chem.* 271, 26616–26621.

Shah, J., Atienza, J. M., Duclos, R. I. J., Rawlings, A. V., Dong, Z., and Shipley, G. G. (1995a), Structural and thermotropic properties of synthetic C16:0 (palmitoyl) ceramide: Effect of hydration, *J. Lipid Res.* 36, 1936–1944.

Shah, J., Atienza, J. M., Rawlings, A. V., and Shipley, G. G. (1995b), Physical properties of ceramides: Effect of fatty acid hydroxylation, *J. Lipid Res.* 36, 1945–1955.

Simon, C. G., and Gear, A. R. L. (1998), Membrane-destabilizing properties of C2-ceramide may be responsible for its ability to inhibit platelet aggregation, *Biochemistry* 37, 2059–2069.

Small, D. M. (1970), Surface and bulk interactions of lipids and water with a classification of biologically active lipids based on these interactions, *Fed. Proc.* 29, 1320–1326.

Sot, J., Aranda, F. J., Collado, M. I., Goñi, F. M., and Alonso, A. (2005), Different effects of long- and short-chain ceramides on the gel-fluid and lamellar-hexagonal transitions of phospholipids: A calorimetric, NMR, and x-ray diffraction study, *Biophys. J.* 88, 3368–3380.

Sot, J., Bagatolli, L. A., Goñi, F. M., and Alonso, A. (2006), Detergent-resistant, ceramide-enriched domains in sphingomyelin/ceramide bilayers, *Biophys. J.* 90, 903–914.

Sot, J., Ibarguren, M., Busto, J. V., Montes, L. R., Goñi, F. M., and Alonso, A. (2008), Cholesterol displacement by ceramide in sphingomyelin-containing liquid-ordered domains, and generation of gel regions in giant lipidic vesicles, *FEBS Lett.* 582, 3230–3236.

Souza, S. L., Capitán, M. J., Álvarez, J., Funari, S. S., Lameiro, M. H., and Melo, E. (2009), Phase behavior of aqueous dispersions of mixtures of n-palmitoyl ceramide and cholesterol: A lipid system with ceramide–cholesterol crystalline lamellar phases, *J. Phys. Chem. B* 113, 1367–1375.

Souza, S. L., Hallock, K. J., Funari, S. S., Vaz, W. L. C., Hamilton, J. A., and Melo, E. (2011), Study of the miscibility of cholesteryl oleate in a matrix of ceramide, cholesterol and fatty acid, *Chem. Phys. Lipids* 164, 664–671.

Stoffel, W. (1971), Sphingolipids, *Annu. Rev. Biochem.* 40, 57–82.

Veiga, M. P., Arrondo, J. L., Goñi, F. M., and Alonso, A. (1999), Ceramides in phospholipid membranes: Effects on bilayer stability and transition to nonlamellar phases, *Biophys. J.* 76, 342–350.

Wertz, P. W., and Downing, D. T. (1989), Integral lipids of mammalian hair, *Comp. Biochem. Physiol. B* 92, 759–761.

Westerlund, B., Grandell, P.-M., Isaksson, Y. J. E., and Slotte, J. P. (2010), Ceramide acyl chain length markedly influences miscibility with palmitoyl sphingomyelin in bilayer membranes, *Eur. Biophys. J.* 39, 1117–1128.

CHAPTER 8

Peptide–Membrane Interactions Studied by Isothermal Titration Calorimetry

Johannes Klingler and Sandro Keller

CONTENTS

8.1 INTRODUCTION

Many amphipathic peptides interact with biological membranes by partitioning into the interfacial region of the lipid bilayer, that is, the chemically heterogeneous interface between the hydrocarbon core and the aqueous phase (Bechinger 2009). Numerous antimicrobial, fusogenic, and cell-penetrating peptides (CPPs) also destabilize the bilayer and compromise its barrier function, adopt a transbilayer orientation, cause vesicle (hemi)fusion or agglom-eration, and so on (Bechinger 2009; Li et al. 2004). In the extreme case, membrane interac-tions of peptides possessing detergent-like properties may lead to membrane solubilization (Bechinger and Lohner 2006; Keller et al. 2005). For all these reasons, a detailed, quantita-tive characterization of peptide–membrane interactions is of great importance in both basic biological and applied pharmaceutical research. In most cases, such interactions cannot be described by a stoichiometric binding model but, rather, involve a surface partition equilib-rium, in which the peptide partitions between the aqueous phase (usually containing buffer and additional salt) and the lipid bilayer phase (Seelig 2004).

The membrane affinity of an amphipathic peptide or protein is determined not only by its overall hydrophobicity but also by its ability to form secondary structure on membrane association, oftentimes in the form of amphipathic helices (Fernandez-Vidal et al. 2007), and by its ability to engage in polar interactions with lipid headgroups (Broecker et al. 2014). A critical parameter is the amphipathicity of a peptide (Fernandez-Vidal et al. 2007) as given by its hydrophobic moment (Eisenberg et al. 1984). In the case of charged peptides, the surface partition equilibrium is further modulated by Coulombic interactions, that is, long-range electrostatic repulsion between membrane-bound and free peptide molecules and, in the presence of anionic lipids, attraction between cationic peptides and the mem-brane lipids (Seelig 2004). Despite the complex interplay among different free-energy con-tributions to membrane partitioning, the interaction of a peptide with a membrane can conceptually be parsed into three thermodynamic steps (Seelig 2004): first, Coulombic attraction between a cationic peptide and anionic lipids leads to an increase in peptide concentration in the interfacial aqueous phase near the membrane surface; second, direct contact between peptide and membrane is established, as driven by the hydrophobic effect and short-range electrostatic interactions, such as monopole–dipole, dipole–dipole, or, more generally, multipolar interactions; third, the bound peptide may assume or change secondary structure to form hydrogen bonds or engage in short-range electrostatic inter-actions (Broecker et al. 2014). The first step can be accounted for by simple electrostatics as given by Gouy–Chapman theory (Aveyard and Haydon 1973; McLaughlin 1977, 1989), while the second and third steps are the determinants of the surface partition equilib-rium. Additional effects resulting from or accompanying peptide–membrane interactions, such as those impairing membrane integrity in the case of pore-forming (Wenk and Seelig 1998) or detergent-like peptides (Bechinger and Lohner 2006; Keller et al. 2005), are ther-modynamically more complex and will not be discussed in this chapter.

High-sensitivity isothermal titration calorimetry (ITC; Wiseman et al. 1989) constitutes a particularly useful method for the study of peptide–membrane interactions. A combina-tion of experiments known as *uptake* and *release* titrations can be performed and analyzed to assess the thermodynamics of such interactions by means of a surface partition equi-librium (Heerklotz et al. 1999) that is modulated by electrostatics (Keller et al. 2006). This chapter explains the theory of surface partition equilibria and Coulombic membrane effects and demonstrates its implementation in the form of fitting equations for the simultaneous analysis of uptake and release experiments, for which the experimental background is also provided and discussed. The theoretical and experimental approach is exemplified using the CPP penetratin (Keller et al. 2007), and the driving forces underlying its association with lipid membranes are discussed comprehensively.

8.2 THEORY

8.2.1 Bilayer Partitioning without Surface Electrostatics

Assuming the physically simple case of ideal mixing in both the aqueous (aq) phase and the bilayer (b) phase, membrane partitioning can be quantified in terms of a mole fraction partition coefficient (Heerklotz et al. 1996; Tanford 1973), which is defined as

$$K_P^{b/aq} \equiv \frac{X_P^b}{X_P^{aq}} = \frac{c_P^b(c_W + c_P^{aq})}{c_P^{aq}(c_L + c_P^b)} \tag{8.1}$$

Here, $X_P^b \equiv c_P^b / (c_L + c_P^b)$ and $X_P^{aq} \equiv c_P^{aq} / (c_W + c_P^{aq})$ are the mole fractions of the peptide (P) in the bilayer and in the aqueous phase, respectively, where c_P^b, c_P^{aq}, c_L, and c_W denote the molar concentrations of membrane-bound and free peptide, lipid (L), and water (W), respectively. Using the approximation $c_W + c_P^{aq} \approx c_W$ in the numerator and the substitution $c_P^{aq} = c_P - c_P^b$ in the denominator of Equation 8.1, one obtains

$$K_P^{b/aq} = \frac{c_P^b c_W}{\left(c_P - c_P^b\right)\left(c_L - c_P^b\right)} \tag{8.2}$$

Rearranging Equation 8.2 yields the following expression for the molar concentration of membrane-bound peptide:

$$c_P^b = \frac{1}{2K_P^{b/aq}}\left(K_P^{b/aq}(c_P - c_L) - c_W + \sqrt{K_P^{b/aq^2}(c_P + c_L)^2 - 2K_P^{b/aq}(c_P - c_L)c_W + c_W^2}\right) \tag{8.3}$$

Alternatively, a mole ratio partition coefficient can be employed to account for nonideality of mixing, as detailed elsewhere (Heerklotz and Seelig 2000; Keller et al. 2006).

8.2.2 Transbilayer Distribution

Under conditions in which no transbilayer equilibration of the peptide occurs during the time needed to complete an ITC experiment, which is typically on the order of several dozen minutes to a few hours, the concentration of lipid interacting with the peptide has to be corrected to

$$c_L \rightarrow \gamma c_L \tag{8.4}$$

where the so-called lipid accessibility factor, γ (Heerklotz and Seelig 2000), denotes the fraction of lipid accessible for interaction with the peptide. This factor amounts to 1.0 if the peptide equilibrates rapidly across the membrane and to 0.5 or 0.6 for large unilamellar vesicles (LUVs) or small unilamellar vesicles (SUVs), respectively, if no transbilayer movement is observed in the experimental time frame. The value of 0.6 for SUVs stems from the fact that, because of their small diameter and resulting high curvature, 60% of the lipid molecules reside in the outer leaflet. Intermediate values of γ hint at partial peptide translocation during the experiment and, thus, to a nonequilibrium situation (Keller et al. 2006), which precludes rigorous, quantitative interpretation in terms of the presented model. A suitable surface partitioning model, which explicitly accounts for the kinetics of membrane translocation, has been developed by Martins et al. (2012), as detailed in Chapter 10 (Martins and Moreno 2015).

8.2.3 Coulombic Effects

In the case of a charged peptide, the partition equilibrium between the bulk aqueous phase and the bilayer phase, as described by $K_P^{b/aq}$, is modulated by electrostatic repulsion between bound and free peptide molecules and, in the case of oppositely charged lipid headgroups, by electrostatic attraction of peptide molecules to the membrane (Beschiaschvili and Seelig 1990). A partition equilibrium corrected for these electrostatic effects is then established between the interfacial aqueous (i) phase, that is, the region of the aqueous phase in the vicinity of the membrane surface, and the bilayer phase. The corresponding *intrinsic* partition coefficient is defined as

$$K_P^{b/i} \equiv \frac{X_P^b}{X_P^i} \tag{8.5}$$

where X_P^i is the mole fraction of the peptide in the interfacial aqueous phase. Peptide partitioning between the bulk and the interfacial aqueous phases is described by a Boltzmann term according to

$$K_P^{i/aq} \equiv \frac{X_P^i}{X_P^{aq}} = \exp\left(\frac{-z_P^{eff} e \Delta\phi^{i/aq}}{kT}\right) \tag{8.6}$$

where:
   z_P^{eff} is the effective signed charge number of the peptide
   e is the elementary charge
   $\Delta\phi^{i/aq}$ is the electrostatic potential at the membrane surface with respect to the bulk aqueous phase
   k is the Boltzmann constant
   T is the absolute temperature

The apparent partition coefficient then becomes

$$K_P^{b/aq} = K_P^{b/i} K_P^{i/aq} = K_P^{b/i} \exp\left(\frac{-z_P^{eff} e \Delta\phi^{i/aq}}{kT}\right) \tag{8.7}$$

Gouy–Chapman theory (Aveyard and Haydon 1973; McLaughlin 1977, 1989), which is based on the Poisson equation and the Boltzmann distribution, relates $\Delta\phi^{i/aq}$ to the surface charge density of the membrane, σ, resulting in

$$\sigma = \text{sgn}(\Delta\phi^{i/aq}) \sqrt{2000 R T \varepsilon_0 \varepsilon_r \sum_I c_I^{aq} \left(\exp\left(\frac{-z_I e \Delta\phi^{i/aq}}{kT} \right) - 1 \right)} \tag{8.8}$$

where:

R is the universal gas constant
ε_0 is the permittivity of free space
z_I is the signed charge number of ionic species (I) in the sample (corresponding to z_P^{eff} for the peptide)

The relative permittivity of the aqueous phase, ε_r, is calculated according to $174 - 0.32\ T/K$ (Lide 2005). The summation in Equation 8.8 goes over the bulk aqueous concentrations, c_I^{aq}, of all ionic species, which include the peptide, the buffer, and additional salt. A second, independent expression (Beschiaschvili and Seelig 1990) for the surface charge density is given by its definition:

$$\sigma = \frac{z_P^{eff} R_P^b}{A_L + R_P^b A_P} e \tag{8.9}$$

where $R_P^b \equiv c_P^b / c_L$ is the peptide/lipid mole ratio in the bilayer phase. Here, A_L denotes the surface area requirement of one lipid molecule, which is 0.68 nm² for 1-palmitoyl-2-oleoyl-*sn*-glycero-3-phosphocholine (POPC) (Altenbach and Seelig 1984), and A_P is the surface area requirement of the peptide, which, to a good approximation, may be assumed to be 0 in many cases of shallow interaction with the membrane. Note, however, that this assumption does not apply to peptides that lead to significant expansion or contraction of the bilayer on insertion into the headgroup region. In such cases, A_P has to be estimated or, better, taken from independent experimental data such as surface pressure measurements (Maget-Dana 1999). $\Delta\phi^{i/aq}$ is implicitly given by the equality of Equations 8.8 and 8.9, which can be established by standard computational iteration methods. If the bilayer contains anionic lipids, Equation 8.9 has to be expanded to

$$\sigma = \frac{z_P^{eff} R_P^b - (1 - \nu) X_{L^-}^0}{A_L + R_P^b A_P} e \tag{8.10}$$

where:

$X_{L^-}^0$ is the mole fraction of anionic lipid in the peptide-free membrane
ν is the fraction of negatively charged lipid headgroups neutralized by binding of monovalent counterions (e.g., Na⁺, K⁺)

The latter can be described by a Langmuir binding isotherm (Beschiaschvili and Seelig 1990) according to

$$\nu = \frac{K_{I^+}^{L^-} c_{I^+}^i}{1 + K_{I^+}^{L^-} c_{I^+}^i} \tag{8.11}$$

with $K_{I^+}^{L^-}$ denoting the association constant of a monovalent cation to an anionic lipid headgroup. Values of this binding constant for Na^+ and K^+ to anionic phospholipid headgroups are $K_{Na^+}^{L^-} = 0.6$ L mol^{-1} and $K_{K^+}^{L^-} = 0.15$ L mol^{-1}, respectively (Eisenberg et al. 1979). To relate the interfacial ion concentration, $c_{I^+}^i$, to the corresponding value in the bulk aqueous solution, $c_{I^+}^{aq}$, another Boltzmann term is employed:

$$c_{I^+}^i = c_{I^+}^{aq} \exp\left(\frac{-e\Delta\phi^{i/aq}}{kT}\right) \tag{8.12}$$

8.3 METHODOLOGY

8.3.1 Experimental Design

Assessment of membrane partitioning of peptides and other solutes by means of ITC is usually accomplished by uptake experiments (Heerklotz et al. 1996). Here, a peptide solution is titrated with lipid vesicles, leading to the *uptake* of the peptide into the lipid bilayer. If the peptide is able to translocate across the bilayer, *uptake* occurs into both membrane leaflets, otherwise into the outer leaflet only. The resulting binding isotherm can then be fitted with the theoretical model described in Section 8.2 and the ITC fitting equations given in Section 8.3.2. To obtain correct best-fit values of the fitting parameters, knowledge of the transbilayer distribution of the peptide is required. However, uptake experiments alone cannot provide this information. To this end, so-called release titrations (Heerklotz et al. 1999; Heerklotz and Seelig 2000) can be performed. Here, lipid vesicles are first preloaded in a homogeneous manner, that is, on both the inner and the outer leaflets, with peptide and then titrated into buffer, leading to the *release* of the peptide into the aqueous phase. If the peptide is able to translocate across the bilayer, *release* occurs from both membrane leaflets, otherwise the fraction preloaded on the inner leaflet remains bound. A simultaneous fit of the binding isotherms obtained from a set of uptake and release experiments using the model described in Section 8.2 and the ITC fitting equations given in Section 8.3.2 thus allows an assessment of both membrane partitioning and translocation of a peptide. A schematic representation of uptake and release experiments is given in Figure 8.1.

8.3.2 Fitting Equations

The following fitting equation for ITC uptake and release titrations relies on finite concentration differences (Keller et al. 2006). In contrast with approaches based on cumulative heats (Heerklotz et al. 1996) or partial derivatives (Heerklotz et al. 1999; Heerklotz and Seelig 2000) derived for nonionic solutes, this method allows an assessment of the

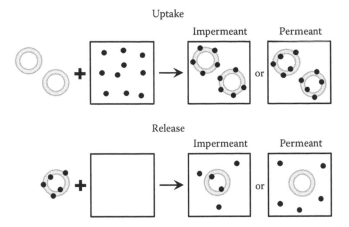

FIGURE 8.1 Principle of uptake and release experiments to assess membrane partitioning and translocation of a peptide. (Modified from Keller, S. et al., *ChemBioChem*, 8, 546–552, 2007.)

membrane-partitioning thermodynamics and membrane translocation of charged solutes such as peptides (Keller et al. 2006). The heats produced or consumed during an uptake or release experiment, normalized with respect to the molar amount of injected lipid, Q_L or Q_{L+P}, respectively, are described by

$$Q_{L;L+P} = \Delta n_P^b \frac{\Delta H_P^{b/aq}}{\Delta n_L} + Q_{L;L+P,dil} = V \Delta c_P^b \frac{\Delta H_P^{b/aq}}{\Delta n_L} + Q_{L;L+P,dil} \tag{8.13}$$

where:

$\Delta H_P^{b/aq}$ is the molar transfer enthalpy of the peptide from the bulk aqueous phase to the bilayer phase

$Q_{L;L+P,dil}$ is the heat of dilution in uptake and release experiments, respectively

Δn_L and Δn_P^b are the changes in the molar amounts of lipid and membrane-bound peptide, respectively, after each injection

V is the volume of the calorimeter cell

Δc_P^b is the change in the concentration of membrane-bound peptide

Δc_P^b can be expressed as the volume-weighted difference between the equilibrium concentrations of bound peptide before ($\hat{c}_P^b$) and after (c_P^b) an injection of volume ΔV, respectively, according to

$$\Delta c_P^b = c_P^b - \left(1 - \frac{\Delta V}{V}\right)\hat{c}_P^b - \frac{\Delta V}{V} c_P^{b,s} \tag{8.14}$$

Here, $c_P^{b,s}$ is the concentration of membrane-bound peptide in the ITC syringe (s) and thus amounts to $c_P^{b,s} = 0$ for an uptake experiment. In deriving the above equation, it was assumed that the volume that is injected from the syringe into the calorimeter cell expels the corresponding volume of cell content into the calorimetrically inert access tube so quickly that this volume no longer contributes to the measured heat of reaction. The actual ITC fit function for uptake and release experiments is obtained by inserting Equation 8.14 into Equation 8.13, yielding

$$Q_{L;L+P} = V\left(c_P^b - \left(1 - \frac{\Delta V}{V}\right)\hat{c}_P^b - \frac{\Delta V}{V}c_P^{b,s}\right)\frac{\Delta H_P^{b/aq}}{\Delta n_L} + Q_{L;L+P,dil} \tag{8.15}$$

Explicit expressions for c_P^b, $\hat{c}_P^b$, and $c_P^{b,s}$ are given by Equation 8.3. If the peptide is not able to translocate across the bilayer, the release experiment requires a correction of c_P appearing in Equation 8.3 according to

$$c_P \rightarrow \left(\frac{c_P^{b,s}}{c_P^s}(\gamma-1)+1\right)c_P \tag{8.16}$$

with c_P^s being the total peptide concentration in the ITC syringe. Furthermore, before substitution into Equation 8.15, $c_P^{b,s}$ has to be corrected to

$$c_P^{b,s} \rightarrow \gamma c_P^{b,s} \tag{8.17}$$

8.3.3 Fitting Procedure

Membrane partitioning and translocation on the basis of ideal mixing can be assessed by fitting a set of uptake and release experiments with Equations 8.3, 8.4, and 8.15 through 8.17. In the case of a charged peptide, Equation 8.7 has to be inserted into Equation 8.3, while Equations 8.8 and 8.10 yield $\Delta\phi^{i/aq}$. During the fitting procedure, $\Delta H_P^{b/aq}$, $Q_{L,dil}$, $Q_{L+P,dil}$, $K_P^{b/i}$, z_P^{eff}, and γ are adjusted by iterative numerical methods to yield the closest agreement between theoretical and experimental data. A step-by-step description of how to perform such fitting procedures by nonlinear least-squares fitting using an Excel spreadsheet with the Solver add-in is provided elsewhere (Vargas et al. 2013). To obtain information on the reliability of the best-fit values of the adjustable parameters, it is advisable to perform confidence interval analysis using Fisher's F distribution, as detailed elsewhere (Kemmer and Keller 2010). To furnish a comprehensive thermodynamic picture of the peptide–membrane interactions under investigation, the following standard relationships need to be employed. The change in standard molar Gibbs free energy on partitioning of the peptide from the interfacial aqueous phase into the bilayer phase, $\Delta G_P^{o,b/i}$, is obtained as

$$\Delta G_P^{o,b/i} = -RT \ln K_P^{b/i} \tag{8.18}$$

The molar enthalpy change is given by using the approximation $\Delta H_P^{b/aq} = \Delta H_P^{b/i} = \Delta H_P^{o,b/i}$, and the corresponding change in the entropic term is

$$-T\Delta S_P^{o,b/i} = \Delta G_P^{o,b/i} - \Delta H_P^{o,b/i} \tag{8.19}$$

In summary, the surface partition equilibrium reflecting the *intrinsic*, that is, non-Coulombic, membrane affinity of the peptide is characterized by $K_P^{b/i}$ or, equivalently, $\Delta G_P^{o,b/i}$, as well as $\Delta H_P^{o,b/i}$ and $-T\Delta S_P^{o,b/i}$. The determinant of Coulombic attraction between the peptide and the membrane is z_P^{eff}, that is, the *effective* peptide valence sensed by the membrane, which is often smaller in magnitude than the formal valence, z (Ladokhin and White 2001). Finally, membrane permeability as reflected in the transbilayer distribution of the peptide is manifested in the lipid accessibility factor, γ.

8.4 EXAMPLE

Penetratin is an archetypical representative of the class of CPPs. It is a 16-residue cationic peptide with a formal net charge of $z = +7$ derived from the third helix of the Antennapedia homeodomain of *Drosophila* (Derossi et al. 1994). Penetratin is soluble in aqueous solutions (Derossi et al. 1994) and interacts with lipid bilayers, on which the largely unstructured peptide adopts an α-helical conformation (Magzoub et al. 2002). At high peptide/lipid ratios in the membrane, it also forms β-sheets (Magzoub et al. 2002). The thermodynamics of membrane partitioning and membrane translocation of penetratin has been characterized by fluorescence spectroscopy and ITC (Keller et al. 2007). Binding of penetratin to SUVs composed of a mixture of POPC and the anionic phospholipid 1-palmitoyl-2-oleoyl-sn-glycero-3-[phospho-rac-(1-glycerol)] (POPG) is characterized by an intrinsic partition coefficient of $K_P^{b/i} = 1.2 \times 10^3$ and an effective peptide charge of $z_P^{eff} = 4.9$ (Keller et al. 2007). Furthermore, it has been shown that penetratin does not cross the bilayer by passive diffusion (Bárány-Wallje et al. 2005; Keller et al. 2007), as reflected in $\gamma = 0.6$ (Keller et al. 2007).

We chose uptake and release titrations of penetratin interacting with SUVs composed of POPC and POPG at a molar ratio of 3:1 in 10 mM phosphate buffer, 154 mM NaF, pH 7.4 at 37°C (Keller et al. 2007) as an example to demonstrate the implementation of the analysis presented in Sections 8.2 and 8.3. A detailed description of the sample preparation and experimental details are given elsewhere (Vargas et al. 2013). Figure 8.2a shows the resulting thermograms, which were baseline-corrected and integrated using the public-domain software NITPIC (Keller et al. 2012), and Figure 8.2b depicts the corresponding reaction heats and fits. The best-fit values and 95% confidence intervals of the thermodynamic parameters thus retrieved are displayed in the top row of Table 8.1. The fitting procedure largely corresponds to that applied previously (Keller et al. 2007) but, additionally, includes γ as a fitting parameter. The best-fit values of $K_P^{b/i}$ and z_P^{eff} are in accord with those reported earlier (Keller et al. 2007). The best-fit value found for γ, being close to 0.6, also confirms the inability of penetratin to translocate across the bilayer.

8.4.1 Nonadditivity of Intrinsic and Coulombic Contributions

In an alternative fitting procedure, z_P^{eff} was excluded as a fitting parameter but, instead, was related to $\Delta G_P^{o,b/i}$ according to

$$z_P^{eff} = z\left(1 + 0.2\frac{\Delta G_P^{o,b/i}}{12.5 \text{ kJ/mol}}\right) \qquad (8.20)$$

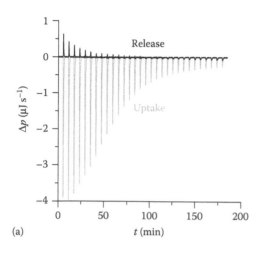
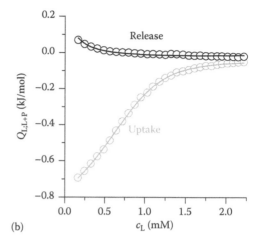

(a) (b)

FIGURE 8.2 Interactions of penetratin with POPC/POPG SUVs at 37°C in 10 mM phosphate buffer containing 154 mM NaF at pH 7.4. Experimental data (circles) and global fit (lines) of a set of uptake and release experiments: uptake: 10 μL aliquots of 20 mM vesicles were injected into 20 μM penetratin; release: 10 μL aliquots of 20 mM lipid in the form of SUVs preloaded with 300 μM penetratin were injected into buffer. (a) Differential heating power, Δp, vs. time, t. (Experimental data from Keller, S. et al., *ChemBioChem* 8, 546–552, 2007.) (b) Reaction heats normalized with respect to the molar amount of lipid injected, $Q_{L;L+P}$, (circles) and fits (lines) vs. lipid concentration in the ITC sample cell, c_L.

TABLE 8.1 Thermodynamic Parameters Characterizing the Membrane Association of Penetratin

$K_P^{b/i}$ (10^3)	$\Delta G_P^{o,b/i}$ (kJ/mol)	$\Delta H_P^{o,b/i}$ (kJ/mol)	$-T\Delta S_P^{o,b/i}$ (kJ/mol)	z_P^{eff}	γ
1.3	−18.5	−50.2	31.7	4.8	0.58
(0.7 to 2.2)	(−19.9 to −17.0)	(−52.4 to −48.0)	(28.1 to 35.4)	(4.4 to 5.4)	(0.51 to 0.66)
1.1	−18.1	−49.7	31.6	5.0	0.60
(0.9 to 1.4)	(−18.7 to −17.6)	(−50.7 to −48.6)	(29.9 to 33.1)	(−)	(0.57 to 0.63)

Note: Top row: Best-fit values and 95% confidence intervals (in parentheses) of thermodynamic parameters as obtained from global analysis of a set of uptake and release experiments monitored by ITC: changes on partitioning from the interfacial aqueous phase into the bilayer as reflected in the intrinsic partition coefficient, $K_P^{b/i}$; the change in standard molar Gibbs free energy, $\Delta G_P^{o,b/i}$, standard molar enthalpy, $\Delta H_P^{o,b/i}$, and the entropic term, $-T\Delta S_P^{o,b/i}$; the effective charge number, z_P^{eff}; and the lipid accessibility factor, γ. Bottom row: Best-fit values and 95% confidence intervals of the other fitting parameters when z_P^{eff} is excluded as an adjustable parameter and, instead, made dependent on $\Delta G_P^{o,b/i}$ according to Equation 8.20.

This rule of thumb suggested by Ladokhin and White (2001) considers the nonadditivity of Coulombic and intrinsic contributions to peptide–membrane interactions. Accordingly, each increment in the magnitude of $\Delta G_P^{o,b/i}$ of 12.5 kJ/mol reduces z_P^{eff} by 20% as compared with z. The best-fit values and 95% confidence intervals of the thermodynamic parameters obtained on the basis of this rule of thumb are shown in the bottom row of Table 8.1.

The results from the two fitting approaches are in excellent agreement, demonstrating that z_P^{eff} is appropriately described by Equation 8.20. The second fitting approach has a reduced number of adjustable parameters, thus resulting in narrower confidence intervals (cf. Table 8.1), which is particularly advantageous for the analysis of datasets with low signal-to-noise ratios.

8.4.2 Influence of Secondary-Structure Formation

As outlined in the introduction, the thermodynamics of a surface partition equilibrium is determined by the establishment of a close contact between the peptide and the membrane and the formation of, or a change in, secondary structure. Thus, one may parse $\Delta G_P^{o,b/i}$ into two contributions according to

$$\Delta G_P^{o,b/i} = \Delta G_P^{o,b/i,uf} + \Delta G_P^{o,ss} \tag{8.21}$$

where $\Delta G_P^{o,b/i,uf}$ and $\Delta G_P^{o,ss}$ denote the energetic contributions of membrane partitioning of a (hypothetical) unfolded (uf) peptide and of secondary-structure (ss) formation at the membrane interface, respectively. If the peptide is located at the membrane interface, as is generally the case for amphipathic peptides such as penetratin, $\Delta G_P^{o,b/i,uf}$ can be estimated using the Wimley–White interfacial hydrophobicity scale (Wimley and White 1996; Hristova and White 2005).

For the penetratin sequence, values of $\Delta G_P^{o,b/i,uf} = -1.7$ kJ/mol and $\Delta G_P^{o,b/i,uf} = -3.6$ kJ/mol are obtained for interfacial partitioning of all residues, the anionic form of the C-terminus, and, respectively, the neutral and cationic forms of the N-terminus. Note that, at pH 7.4 in the bulk aqueous phase, considerable fractions of both the neutral and cationic forms of the N-terminus can be present in the aqueous phase near the membrane interface, whereas the C-terminus can be safely assumed to be fully deprotonated (Wimley and White 1996; Hristova and White 2005). Protection of both termini, that is, acetylation of the N-terminus and amidation of the C-terminus, results in a value of $\Delta G_P^{o,b/i,uf} = -9.1$ kJ/mol. Accordingly, both with and without protection of the termini, a highly favorable $\Delta G_P^{o,ss}$ is required to arrive at the experimental net value of $\Delta G_P^{o,b/i} = -18.5$ kJ/mol, thus highlighting the importance of secondary-structure formation in determining the energetics of peptide–membrane interactions.

As seen in Table 8.1, membrane partitioning of penetratin is driven by enthalpy and opposed by entropy. Traditionally, one would expect the opposite scenario for a process governed by hydrophobicity (Tanford 1973). However, as demonstrated earlier in this section, on the basis of its hydrophobicity alone, penetratin would barely be able to interact with membranes. The observation of membrane partitioning that is driven by enthalpy and opposed by entropy is in line with exothermic contributions such as secondary-structure formation and short-range electrostatic interactions between the peptide and the membrane (Broecker et al. 2014). Additional key parameters describing the membrane interactions of penetratin with regard to electrostatics and thermodynamics are discussed in the following section.

8.4.3 Coulombic Contributions to Membrane Partitioning

Membrane electrostatics in terms of the surface charge density, σ, as obtained from Equations 8.8 and 8.10, is summarized in Figure 8.3a. The membrane surface potential, $\Delta\phi^{i/aq}$, which is shown in Figure 8.3b, can be obtained by equating the σ values calculated with the aid of Equations 8.8 and 8.10. σ continuously decreases in magnitude with increasing X_P^b because the negative charges of the POPG headgroups become increasingly neutralized by adsorption of the cationic peptide. Note that X_P^b in the ITC sample cell decreases in the course of an uptake experiment but increases in the course of a release experiment. The apparent mole fraction partition coefficient, $K_P^{b/aq}$, as given by Equation 8.7, is shown in Figure 8.4a. In response to the changes in the membrane surface potential, it increases by a factor of 10 but decreases by a factor of 2.5 during uptake and release titrations, respectively. From Equation 8.7 and a standard relationship analogous to Equation 8.18, a decrease in $\Delta G_P^{o,i/aq}$ from –10.3 kJ/mol to –16.5 kJ/mol in the course of the uptake experiment is obtained, underlining the considerable contribution of Coulombic effects to the membrane association equilibrium of penetratin. Figure 8.4b depicts the fraction of peptide that is bound to the membrane, F_b, as obtained by the ratio of c_P^b, given by Equation 8.3, to the total peptide concentration, c_p. The mole fraction of penetratin in the bilayer, X_P^b, in dependence on the bulk aqueous concentration of the peptide, c_P^{aq}, is provided in Figure 8.5a. While this curve bends downward with increasing c_P^{aq}, a straight line with a slope of $K_P^{b/i}/c_W$ is obtained when X_P^b is plotted as a function of the interfacial aqueous peptide concentration, c_P^i, as shown in Figure 8.5b. This confirms that the interactions of an amphipathic cationic peptide such as penetratin with anionic phospholipid membranes are properly described by the present combination of a surface partition equilibrium with a simple electrostatic model of the membrane–water interfacial region.

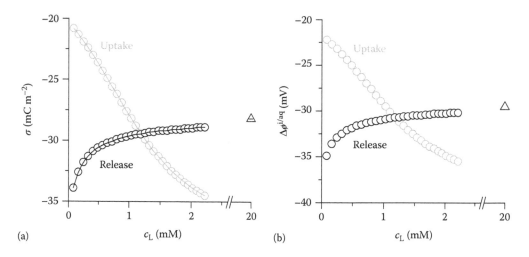

FIGURE 8.3 Membrane electrostatics of the interactions of penetratin with POPC/POPG SUVs at 37°C. (a) Membrane surface charge density, σ, according to Equation 8.8 (lines) and Equation 8.10 (symbols) vs. lipid concentration in the ITC sample cell (circles) and the syringe (triangle), c_L. (b) Electrostatic membrane potential, $\Delta\phi^{i/aq}$, as given by the equality of Equations 8.8 and 8.10, vs. lipid concentration in the ITC sample cell (circles) and the syringe (triangle), c_L.

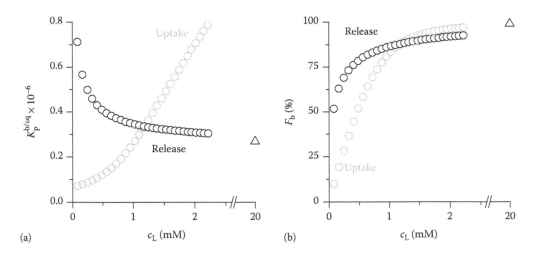

FIGURE 8.4 Parameters characterizing the interactions of penetratin with POPC/POPG SUVs at 37°C. (a) Apparent mole fraction partition coefficient, $K_P^{b/aq}$, as defined by Equation 8.1 and given by Equation 8.7, vs. lipid concentration in the ITC sample cell (circles) and the syringe (triangle), c_L. (b) Fraction of membrane-bound peptide, F_b, as obtained by the ratio of c_P^b, given by Equation 8.3, and c_P vs. lipid concentration in the ITC sample cell (circles) and the syringe (triangle), c_L.

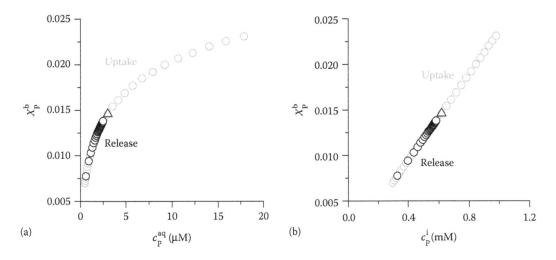

FIGURE 8.5 Partitioning of penetratin between the aqueous phase and POPC/POPG SUVs at 37°C. (a) Mole fraction of peptide in the membrane, X_P^b, vs. bulk aqueous peptide concentration in the ITC sample cell (circles) and the syringe (triangle), c_P^{aq}. (b) X_P^b vs. interfacial aqueous peptide concentration in the ITC sample cell (circles) and the syringe (triangle), c_P^i.

8.5 CONCLUSIONS

In conclusion, this chapter summarizes the theoretical and methodological framework required for assessing the interactions of amphipathic peptides with phospholipid membranes by means of ITC, as exemplified for the CPP penetratin. The protocol is generally applicable to amphipathic peptides whose membrane interactions, under the experimental conditions applied, preserve membrane integrity and are not subject to slow transmembrane equilibration. Alternative approaches exist for more complex cases such as membrane solubilization (Keller et al. 2005) and slow translocation rates (Martins et al. 2012; Martins and Moreno 2015).

ACKNOWLEDGMENTS

We thank Martin Textor (University of Kaiserslautern) for helpful comments on the manuscript. This work was supported by the Phospholipid Research Center (PRC; Heidelberg, Germany).

REFERENCES

Altenbach, C. and Seelig, J. (1984), Ca^{2+} binding to phosphatidylcholine bilayers as studied by deuterium magnetic resonance. Evidence for the formation of a Ca^{2+} complex with two phospholipid molecules, *Biochemistry* 23, 3913–3920.

Aveyard, R. and Haydon, D. A. (1973), *An Introduction to the Principles of Surface Chemistry*, Cambridge University Press: Cambridge.

Bárány-Wallje, E., Keller, S., Serowy, S., Geibel, S., Pohl, P., Bienert, M., and Dathe, M. (2005), A critical reassessment of penetratin translocation across lipid membranes, *Biophys. J.* 89, 2513–2521.

Bechinger, B. (2009), Rationalizing the membrane interactions of cationic amphipathic antimicrobial peptides by their molecular shape, *Curr. Opin. Colloid Interface Sci.* 14, 349–355.

Bechinger, B. and Lohner, K. (2006), Detergent-like actions of linear amphipathic cationic antimicrobial peptides, *Biochim. Biophys. Acta* 1758, 1529–1539.

Beschiaschvili, G. and Seelig, J. (1990), Melittin binding to mixed phosphatidylglycerol/phosphatidylcholine membranes, *Biochemistry* 29, 52–58.

Broecker, J., Fiedler, S., Gimpl, K., and Keller, S. (2014), Polar interactions trump hydrophobicity in stabilizing the self-inserting membrane protein Mistic, *J. Am. Chem. Soc.* 136, 13761–13768.

Derossi, D., Joliot, A. H., Chassaing, G., and Prochiantz, A. (1994), The third helix of the Antennapedia homeodomain translocates through biological membranes, *J. Biol. Chem.* 269, 10444–10450.

Eisenberg, D., Weiss, R. M., and Terwilliger, T. C. (1984), The hydrophobic moment detects periodicity in protein hydrophobicity, *Proc. Natl. Acad. Sci. USA* 81, 140–144.

Eisenberg, M., Gresalfi, T., Riccio, T., and McLaughlin, S. (1979), Adsorption of monovalent cations to bilayer membranes containing negative phospholipids, *Biochemistry* 18, 5213–5223.

Fernandez-Vidal, M., Jayasinghe, S., Ladokhin, A. S., and White, S. H. (2007), Folding amphipathic helices into membranes: Amphiphilicity trumps hydrophobicity, *J. Mol. Biol.* 370, 459–470.

Heerklotz, H. H., Binder, H., and Epand, R. M. (1999), A "release" protocol for isothermal titration calorimetry, *Biophys. J.* 76, 2606–2613.

Heerklotz, H., Lantzsch, G., Binder, H., Klose, G., and Blume, A. (1996), Thermodynamic characterization of dilute aqueous lipid/detergent mixtures of POPC and $C_{12}EO_8$ by means of isothermal titration calorimetry, *J. Phys. Chem.* 100, 6764–6774.

Heerklotz, H. and Seelig, J. (2000), Titration calorimetry surfactant–membrane partitioning and membrane solubilization, *Biochim. Biophys. Acta* 1508, 69–85.

Hristova, K. and White, S. H. (2005), An experiment-based algorithm for predicting the partitioning of unfolded peptides into phosphatidylcholine bilayer interfaces, *Biochemistry* 44, 12614–12619.

Keller, S., Böthe, M., Bienert, M., Dathe, M., and Blume, A. (2007), A simple fluorescence-spectroscopic membrane translocation assay, *ChemBioChem* 8, 546–552.

Keller, S., Heerklotz, H., and Blume, A. (2006), Monitoring lipid membrane translocation of sodium dodecyl sulfate by isothermal titration calorimetry, *J. Am. Chem. Soc.* 128, 1279–1286.

Keller, S., Sauer, I., Strauss, H., Gast, K., Dathe, M., and Bienert, M. (2005), Membrane-mimetic nanocarriers formed by a dipalmitoylated cell-penetrating peptide, *Angew. Chem. Int. Ed.* 44, 5252–5255.

Keller, S., Vargas, C., Zhao, H., Piszczek, G., Brautigam, C. A., and Schuck, P. (2012), High-precision isothermal titration calorimetry with automated peak-shape analysis, *Anal. Chem.* 84, 5066–5073.

Kemmer, G. and Keller, S. (2010), Nonlinear least-squares data fitting in Excel spreadsheets, *Nat. Protoc.* 5, 267–281.

Ladokhin, A. S. and White, S. H. (2001), Protein chemistry at membrane interfaces: Non-additivity of electrostatic and hydrophobic interactions, *J. Mol. Biol.* 309, 543–552.

Li, W., Nicol, F., and Szoka Jr., F. C. (2004), GALA: A designed synthetic pH-responsive amphipathic peptide with applications in drug and gene delivery, *Adv. Drug Delivery Rev.* 56, 967–985.

Lide, D. R. (2005), *CRC Handbook of Chemistry and Physics*, 86th edition, CRC: Boca Raton, FL.

Maget-Dana, R. (1999), The monolayer technique: A potent tool for studying the interfacial properties of antimicrobial and membrane-lytic peptides and their interactions with lipid membranes, *Biochim. Biophys. Acta* 1462, 109–140.

Magzoub, M., Eriksson, L. E. G., and Gräslund, A. (2002), Conformational states of the cell-penetrating peptide penetratin when interacting with phospholipid vesicles: Effects of surface charge and peptide concentration, *Biochim. Biophys. Acta* 1563, 53–63.

Martins, P. A. T. and Moreno, M. J. (2015), Kinetics of the interaction of amphiphiles with lipid bilayers using ITC. In M. Bastos (Ed.), *Biocalorimetry: Foundations and Contemporary Approaches*, Taylor and Francis: Boca Raton, FL.

Martins, P. T., Velazquez-Campoy, A., Vaz, W. L. C., Cardoso, R. M. S., Valério, J., and Moreno, M. J. (2012), Kinetics and thermodynamics of chlorpromazine interaction with lipid bilayers: Effect of charge and cholesterol, *J. Am. Chem. Soc.* 134, 4184–4195.

McLaughlin, S. (1977), Electrostatic potentials at membrane-solution interfaces, *Curr. Top. Membr.* 9, 71–144.

McLaughlin, S. (1989), The electrostatic properties of membranes, *Annu. Rev. Biophys. Biophys. Chem.* 18, 113–136.

Seelig, J. (2004), Thermodynamics of lipid–peptide interactions, *Biochim. Biophys. Acta* 1666, 40–50.

Tanford, C. (1973), *The Hydrophobic Effect*, Wiley: New York.

Vargas, C., Klingler, J., and Keller, S. (2013), Membrane partitioning and translocation studied by isothermal titration calorimetry, *Methods Mol. Biol.* 1033, 253–271.

Wenk, M. R. and Seelig, J. (1998), Magainin 2 amide interaction with lipid membranes: Calorimetric detection of peptide binding and pore formation, *Biochemistry* 37, 3909–3916.

Wimley, W. C. and White, S. H. (1996), Experimentally determined hydrophobicity scale for proteins at membrane interfaces, *Nat. Struct. Biol.* 3, 842–848.

Wiseman, T., Williston, S., Brandts, J. F., and Lin, L. N. (1989), Rapid measurement of binding constants and heats of binding using a new titration calorimeter, *Anal. Biochem.* 179, 131–137.

CHAPTER 9

DSC Studies on the Modulation of Membrane Lipid Polymorphism and Domain Organization by Antimicrobial Peptides

Karl Lohner

CONTENTS

9.1 WHY STUDY THE INTERACTIONS OF ANTIMICROBIAL PEPTIDES (AMPs) WITH LIPIDS USING MICROCALORIMETRY?

The reemergence of infectious diseases and, in particular, the rapid increase of pathogenic multidrug-resistant bacterial strains have become a significant problem for the treatment of various infections, especially in intensive care units, which often results in the need to switch the treatment to more costly second-line antibiotics. Unfortunately, these alternative drugs may also become ineffective over time, as observed, for example, for vancomycin, an antibiotic of last resort (Chang et al. 2003). Moreover, the efficacy of antibiotics is steadily declining, while bacteria becoming resistant to conventional antibiotics continue to proliferate (Hancock 2014). Consequently, the World Health Organization has ranked antibiotic resistance as a priority and urged the development of alternative antibiotic agents with completely novel mechanisms of action. This should counteract the decline in approved antibiotics since the early 1980s, when only two new classes (oxazolidinones and daptomycin) were introduced onto the market (Nordberg et al. 2005).

One alternative strategy is based on antimicrobial peptides (AMPs), effector molecules of innate immunity that provide a first line of defense against a substantial array of pathogenic microorganisms (Hancock 2001; Zasloff 2002). Currently, more than 2500 natural and synthetic AMPs are listed in the Antimicrobial Peptide Database (http://aps.unmc.edu/AP) (Wang et al. 2009). In contrast to conventional antibiotics, these cationic amphipathic peptides composed of 10–40 amino acid residues predominantly act without specific receptors, but interfere with the lipid matrix of bacterial cell membranes (Lohner and Blondelle 2005). There is consensus that the positive charge of the peptide is essential for initial binding to the negatively charged bacterial membrane surface, which allows discrimination between bacterial and host cell membranes, and its hydrophobicity is needed for insertion into and disruption of the membrane (e.g., Henderson and Lee 2013; Lohner and Blondelle 2005). These two parameters determine the window of activity (Findlay et al. 2010). Thus, presumably, the amphipathic topology of AMPs, that is, their physicochemical properties rather than a specific amino acid sequence, is responsible for their biological activities.

Considerable efforts have been made to elucidate the molecular mechanism(s) of action of AMPs in model and *in vitro* studies, to provide a sound basis for rational peptide design (Blondelle and Lohner 2010; Hilpert et al. 2005; Rathinakumar et al. 2009). Numerous studies have demonstrated that AMPs interfere with the integrity of bacterial membranes via diverse mechanisms; for reviews, see, for example, Bechinger and Lohner (2006); Lohner (2009); Wimley (2010). Therefore, the role of membrane lipid composition in the activity and target cell specificity of AMPs has been investigated using liposomes that mimic the more complex biological membrane and applying a wide variety of biophysical techniques. Within this plethora of approaches, differential scanning calorimetry (DSC) has proved to be a powerful nonperturbing technique to analyze the modulation of the thermotropic behavior as well as the domain organization of membrane mimetic systems in the presence of AMPs (Chiu and Prenner 2011; Lohner and Prenner 1999; McElhaney 1986).

9.2 CHOICE OF PROPER MEMBRANE MIMETIC SYSTEMS

As indicated in Section 9.1, AMPs interact with cell membranes to different extents; that is, they show different cell specificity and membrane activity, which has been related to the different membrane architecture and lipid composition of bacterial and mammalian cells (Lohner 2001; Lohner and Blondelle 2005). Briefly, while Gram-positive bacteria have a simple lipid bilayer membrane protected by a thick lipoteichoic acid-peptidoglycan layer, the cell envelope of Gram-negative bacteria consists of an inner and a unique outer membrane separated by an intervening layer of peptidoglycan in the periplasmic space. The outer membrane of Gram-negative bacteria has a distinctive, highly asymmetric lipid distribution with lipopolysaccharides (LPS) located exclusively in its outer leaflet. Negatively charged lipids such as phosphatidylglycerol (PG) and diphosphatidylglycerol or cardiolipin (CL) and the zwitterionic phosphatidylethanolamine (PE) are major components of the cytoplasmic membrane of both bacterial classes (Lohner 2001 and references therein). Although the detailed phospholipid composition varies among different species and growth conditions, it seems that in general, Gram-negative bacteria display a higher content of PE and Gram-positive bacteria a higher content of PG. Further, there is also a difference in the fatty acid composition between Gram-negative and Gram-positive bacteria. In the former, a saturated fatty acid is often linked to position sn-1 and an unsaturated fatty acid to position sn-2, whereas in the latter, saturated branched chains are frequently found. The archetype of mammalian cell membranes is represented by the plasma membrane of red blood cells, characterized by an asymmetric distribution of phospholipids between the outer and inner lipid leaflet of the bilayers exposing the neutral phosphatidylcholine (PC) and sphingomyelin (SM) to the extracellular side (Rothman and Lenard 1977; Bevers et al. 1999). In addition, this membrane contains about 25 wt% of cholesterol. Normally, no sterols are found in bacterial cytoplasmic membranes. Given the distinct lipid distribution in cell membranes, it follows that they display different physicochemical properties, contributing to the discrimination by AMPs between bacterial and mammalian cell membranes (Latal et al. 1997; Lohner et al. 1997; Lohner 2001), which is a prerequisite for development as an antibiotic. Thus, liposomes composed of the respective lipid components are well suited as model systems for studying the effect of AMPs on lipid bilayers to elucidate the molecular mechanism(s) of action. Table 9.1 gives an overview of the lipid components to be preferentially chosen to mimic bacterial and mammalian cell membranes.

Especially in the time when DSC was being pioneered, thermograms were recorded from natural cell membranes and lipid extracts from bacteria (Steim et al. 1969; Jackson and Sturtevant 1977; Brandenburg and Blume 1987). A main conclusion of these experiments was that the growth temperature of the organisms was always several degrees above the gel–fluid phase transition of their cytoplasmic membrane, and hence, the fluid liquid-crystalline phase is the viable phase (for more details, see Chapter 6 in this volume). These early experiments also demonstrated that DSC is the method of choice to study the

TABLE 9.1 Typical Lipid Composition for Simple Membrane Mimics for Bacterial and Mammalian Cell Membranes

		Phospholipid	**Other Lipids**
	Headgroup	Preferred Hydrocarbon Chains	
Gram-positive bacteria	PG ≫ CL,PE	(Branched) saturated	LTA[a]
Gram-negative bacteria	PE ≫ CL,PG	Mixed (saturated and unsaturated)	LPS[b]
Red blood cell	PC~SM ≫ PE[c]	Mixed	Cholesterol[c]

[a] Lipoteichoic acid (LTA) to mimic the effect of cell wall components.
[b] Lipopolysaccharides (LPS) to mimic the outer leaflet of the outer membrane.
[c] To mimic the outer leaflet of the cytoplasmic membrane.

thermotropic phase transitions of membranes, as well as of their lipid extracts or synthetically prepared lipids, and is able to provide thermodynamic parameters to characterize the thermotropic behavior of such systems. However, detailed data analysis becomes very difficult for natural membranes, which are composed of a large number of lipids, as well as for lipid mixtures consisting of more than two components, as discussed, for example, by A. Blume (Chapter 6 in this volume) and Raudino et al. (2011). Therefore, many model studies are performed using either single-component systems or binary lipid mixtures. A quantitative characterization of the thermotropic phase behavior of the membrane model systems is indispensable for the interpretation of the effects observed in the presence of AMPs. An extensive overview of phase diagrams of lipid mixtures and their analysis and interpretation is given in Chapter 6 in this volume. In addition, further data on individual lipids and lipid mixtures can be found in a compilation by Koynova and Caffrey (2002). Contributions regarding model systems mimicking bacterial and mammalian cytoplasmic membranes have also come from our group. For example, we characterized binary mixtures of both disaturated (Lohner et al. 2001) and mixed-chain PG/PE (Pozo et al. 2005), disaturated PG/CL (Prossnigg et al. 2010), and PC/SM (Degovics et al. 1997).

Besides simplicity, there is also another reason why initial studies are often performed using monocomponent liposomes: the fact that one of the major determinants in discrimination between bacterial and mammalian cell membranes is the electrostatically driven interaction between cationic AMPs and the anionic bacterial membrane, which largely triggers membrane affinity and hence, membrane selectivity (Lohner and Blondelle 2005). Therefore, liposomes composed of PG and PC are used as simple mimics for bacterial and mammalian cytoplasmic membranes, respectively. Both lipids exhibit a very similar thermotropic phase behavior, and hence, it is argued that in addition, these phospholipids represent a good model system to study the influence of charge and, further, to assess in a first set of experiments the impact of electrostatic interaction and consequently membrane specificity. The disaturated myristoyl and palmitoyl species are commonly used, because their gel–fluid phase transition, also termed the *main* or *chain melting transition*, occurs at around 24°C and 41°C, respectively, which is an easily accessible temperature range by DSC (Koynova and Caffrey 2002). Note, however, that the behavior of PGs is highly dependent on ionic strength and pH (e.g., Degovics et al. 2000; Garidel et al. 1997).

In terms of binary lipid mixtures, PG/PC bilayers are more frequently studied as bacterial membrane mimic than PG/PE bilayers, although PG/PC mixtures are not strictly a good mimic for cytoplasmic bacterial membranes, which contain little if any PC. Moreover, it has been shown that the mixing behavior of PC/PG and PE/PG differs (see Chapter 6 in this volume), which may be due in part to the different molecular shapes of PE and PC as well as to the hydrogen-bonding capacity of PE (Sevcsik et al. 2008). This will result in different lipid packing and, in turn, different overall physicochemical properties, which will influence the interaction and hence the mode of action with AMPs. For example, using DSC, we have shown that the human cathelicidin LL-37 interacts differently with equimolar mixtures of PG/PE and PG/PC having identical hydrocarbon chains (Sevcsik et al. 2008). As can be deduced from Figure 9.1, the phase behavior of dipalmitoyl-PG/dipalmitoyl-PC (DPPG/DPPC) in the presence of LL-37 closely resembled the features of

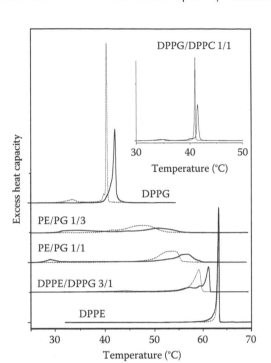

FIGURE 9.1 Interaction of liposomes composed of DPPG/DPPE at various molar ratios as indicated in the panel with the human antimicrobial peptide cathelicidin LL-37 (lipid-to-peptide molar ratio 25/1). Insert shows the thermograms for an equimolar mixture of DPPG/DPPC in the absence and presence of LL-37. Thermograms obtained for pure lipids are represented by dotted lines. Data were recorded at a scan rate of 0.5°C/min in PBS buffer, pH 7.4. (From Sevcsik, E., Hemolytic and antimicrobial activity of LL-37 is based on diverse modes of membrane perturbation, PhD Thesis, Graz University of Technology, Austria, 2006.)

pure DPPG, showing a shift of the phase transition to higher temperatures. In contrast, two well-separated phase transitions were observed for the DPPG/DPPE mixture at the same molar ratio, which were attributed to peptide-poor and peptide-rich domains due to the preferential interaction of the cationic peptide with the negatively charged DPPG. This will lead to an enrichment of PE in the peptide-poor domains, shifting the phase transition of these domains closer to that of pure dipalmitoyl-PE, which is around 64°C (Sevcsik et al. 2008). Similarly, the cyclic antimicrobial hexapeptide cyclo-RRWWRF induced phase separation in mixtures of DPPG/DPPE, but only induced a shift of the transition temperature in mixtures of dipalmitoyl-PG/dimyristoyl-PC (Arouri et al. 2009). These examples emphasize that care has to be taken over selection of the model system, especially when such findings are correlated with *in vivo* data.

9.2.1 Thermodynamic Information Gained from a DSC Thermogram

As described in detail in other chapters of this book, DSC is widely used for the study of phase transitions and conformational changes in biological systems, including proteins, nucleic acids, and lipid assemblies. In the case of a reversible transition, it is possible, in principle, to determine the following thermodynamic information from a single differential calorimetric scan: (1) transition temperature, (2) transition enthalpy, and (3) cooperativity (Hinz and Schwarz 2001). It should be stressed that the thermodynamic and kinetic properties of various lipid thermotropic phase transitions can vary considerably, and thus the process observed by DSC may not always occur under equilibrium conditions (Lewis et al. 2007). Thus, for the proper interpretation of the experimental data as well as the theoretical aspects of applying DSC in studying thermotropic phase transitions under both equilibrium and nonequilibrium conditions, the reader is referred to an excellent review by Lewis et al. (2007), which also contains data acquisition and data analysis protocols. Nevertheless, on interaction with lipid bilayers, AMPs will affect these parameters to a varying extent, from which some overall conclusions in terms of lipid-peptide interaction can be deduced immediately (Chiu and Prenner 2011; Lohner and Prenner 1999; McElhaney 1986; Raudino et al. 2011):

1. Phase transition temperature: A shift of the main phase transition temperature (T_m) gives information on the membrane (dis)ordering effect of AMPs. On insertion, AMPs mostly disorder the lipid bilayer, which favors the fluid phase over the gel phase, resulting in a decrease of T_m. In contrast, an increase of T_m indicates a stabilization of the gel phase due to tighter lipid packing. This can be caused by several factors, such as shielding of surface charge, dehydration of the lipid headgroups, or induction of specific gel phases, such as an interdigitated phase, discussed in Section 9.3.3. Similarly, monitoring the transition temperature for the transition from lamellar fluid to inverse hexagonal phase (T_{HII}) yields information if planar bilayer or highly curved nonlamellar structures are preferentially formed in the presence of AMPs, which confirms their effect on membrane curvature strain, described in Section 9.3.5.

2. Phase transition enthalpy: The absolute value of the phase transition enthalpy (ΔH_{cal}) allows assumptions to be made on the type of phase transition. Enthalpy values in the range of 20–50 kJ/mol lipid are indicative of a chain melting transition, while lower values, mostly below 5 kJ/mol lipid, are typical for fluid–fluid phase transitions, such as the transition from lamellar fluid to inverse hexagonal phase or from lamellar to cubic phase (Koynova and Caffrey 2002). Both transitions are characterized by low cooperativity, resulting in a broad transition range, and therefore are not always easily discernible from the heat capacity profile. Moreover, the enthalpy of the so-called *pretransition* observed in disaturated PGs and PCs is in the same order of magnitude and corresponds to a rearrangement of the lipids from a tilted $L_{\beta'}$-phase to the ripple $P_{\beta'}$-phase. This transition is particularly sensitive to impurities (Lohner 1991) and hence can be taken as a first indication of incorporation of peptides in bilayers. Further, if AMPs disorder the lipid bilayer, the enthalpy of the chain melting transition will be reduced due to the induced number of *trans*-gauche isomerization in the hydrocarbon chains, eliminating this fraction of phospholipids from undergoing the gel–fluid transition. Increased ΔH_{cal} can be expected, for example, when lipid ordering occurs. In addition, in the case of phase separation deconvolution of the

thermogram, yielding ΔH_{cal} for the individual transitions allows the fraction of lipids undergoing a respective phase transition to be deduced.

3. Cooperativity: A point of considerable interest is the average number of lipid molecules acting as a cooperative unit (n) at the transition temperature. It should be stressed, however, that reality is probably better described by a cluster size distribution, in which the size of clusters (the number of lipid molecules undergoing the phase transition at the same point) varies with temperature, and hence, the cooperative unit represents an average size characteristic for the state around T_m (Freire and Biltonen 1978; Marsh et al. 1977). n can be calculated from the ratio

$$n = \frac{\Delta H_{vH}}{\Delta H_{cal}} \tag{9.1}$$

where ΔH_{vH} is the van't Hoff enthalpy. The larger the value of n, the more cooperative is the phase transition. For a single symmetric reversible transition, ΔH_{vH} can be calculated from the experimentally determined parameters according to

$$\Delta H_{vH} = \frac{4RT_m^2 \Delta C_{p,max}}{\Delta H_{cal}} \tag{9.2}$$

where:
  R is the gas constant
  $\Delta C_{p,max}$ is the maximum excess heat capacity

The cooperative unit can also be estimated from the half-width of the transition ($\Delta T_{1/2}$) (Mabrey and Sturtevant 1978; Privalov 1979). Delineation for the alternative approximation of ΔH_{vH} can be found in Lewis et al. (2007):

$$\Delta H_{vH} = \frac{3.5255 \, RT_m^2}{\Delta T_{1/2}} \tag{9.3}$$

However, based on this relationship, the transition half-width $\Delta T_{1/2}$ is often just used as a qualitative measure of cooperativity. Therefore, many investigators compare only $\Delta T_{1/2}$ in the presence and absence of AMPs, whereby a broadening of the phase transition indicates loss of cooperativity, that is, formation of lipid clusters with reduced average size. It should be mentioned that care has to be taken in the case of asymmetric or overlapping transitions, when a cautious fitting procedure of the individual peaks has to be performed.

Finally, the shape of the thermogram also contains information on the kind of interaction of AMPs and lipids, as the miscibility of the peptide with the two lipid phases, gel and fluid, respectively, will influence the heat capacity profile. If a peptide mixes well with the fluid phase (low nearest neighbor interaction energy) and does not mix well with the gel phase (high nearest neighbor interaction energy), the peptide will homogeneously distribute in the fluid phase but aggregate in the gel state (Raudino et al. 2011). Consequently, the corresponding heat capacity profile displays an asymmetric broadening at the low-temperature side of the transition and, in addition, may be shifted to lower temperatures (Grabitz et al. 2002). In the following sections, typical patterns of DSC thermograms related to certain modes of interaction will be described. However, it must be stressed that analysis of DSC data does not *per se* allow conclusions on the phase state, and therefore, structural changes induced by AMPs must be linked to complementary methods, preferably structural techniques, in particular, x-ray scattering (Pabst et al. 2012).

9.3 BACTERIAL KILLING BASED ON MEMBRANE PERTURBATION/DISRUPTION BY AMPs

AMPs can kill bacteria via diverse molecular mechanisms (Hancock 2001; Hancock and Rozek 2002; Lohner 2009; Lohner and Blondelle 2005). Here, we focus on those that interfere with the integrity of the cell membrane, which can be investigated by DSC using membrane model systems. In this respect, the most frequently discussed modes of action include the carpet model (Shai 2002) and the formation of toroidal pores (Ludtke et al. 1996; Matsuzaki et al. 1996). In the latter model, AMPs together with the lipid form a water-filled pore, as shown for a number of α-helical- and β-sheet-type peptides under certain experimental conditions (Huang 2006). In the carpet model, AMPs accumulate at the membrane, being aligned parallel to the bilayer surface, and insert into the membrane above a certain threshold concentration, resulting in membrane permeabilization and eventually disruption (Shai 2002). At the molecular level, different processes may apply that can lead to loss of membrane integrity and can be detected by monitoring the thermotropic phase behavior of liposomes in the absence and presence of AMPs. They are briefly listed here and will be discussed in more detail in Sections 9.3.1 through 9.3.5:

1. Interfacial activity model (Rathinakumar and Wimley 2010; Wimley 2010)

2. Phase separation (Arouri et al. 2009; Epand et al. 2010; Epand and Epand 2009, 2011; Latal et al. 1997; Lohner 2009; Zweytick et al. 2011)

3. Free volume model (Huang 2000; Lohner 2009; Sevcsik et al. 2007)

4. Detergent-like action (Bechinger and Lohner 2006; Hristova et al. 1997)

5. Modifying membrane curvature strain (Koller and Lohner 2014; Lohner and Blondelle 2005)

9.3.1 Interfacial Activity Model: Perturbation of Lipid Packing

The interfacial activity model is based on the considerations that the hydrocarbon core (about 30 Å) of an unperturbed lipid bilayer membrane is one of the most hydrophobic microenvironments found in nature and is normally a strict permeability barrier to polar or charged molecules (Wiener and White 1992). Thus, the property defined as *interfacial activity* "is the ability of a peptide to perturb the permeability barrier imposed by the hydrocarbon core by partitioning into the interfacial region of the bilayer and driving local rearrangements in vertical lipid packing (i.e., normal to the bilayer plane) that alter the segregation between the hydrocarbon core and the interfacial groups" (Wimley 2010). It has been suggested that peptides characterized by imperfect segregation of polar/charged and hydrophobic amino acids will be most efficient at deforming the bilayer and disrupting the hydrocarbon chain packing to simultaneously accommodate the hydrophobic and polar/charged groups of the peptide (Rathinakumar et al. 2009). Partitioning itself depends on the appropriate balance of electrostatic and hydrophobic interactions between peptides, water, and lipids. Isothermal titration calorimetry (ITC) experiments (see Chapter 6 in this volume) revealed that the driving force for binding is not solely the electrostatic attraction of the cationic peptide to the negatively charged membrane surface; hydrophobic effects also have to be considered. The influence of these competing effects can be elucidated by ITC, as shown for a short model peptide (Hoernke et al. 2012). While electrostatic binding alone would only increase the phase transition temperature due to dehydration processes in the headgroup, total or partial incorporation of a peptide in the hydrophobic core of a bilayer will significantly perturb the hydrocarbon chain packing. This will decrease the cooperativity of the phase transition and, in addition, favor the fluid phase, in which, due to increased *trans*-gauche isomerization of the hydrocarbon chains, the amino acid residues can be accommodated in such a way as to maximize hydrophobic interactions. As a result, the phase transition will be broadened and concomitantly shifted to lower phase transition temperatures. Such thermograms can be expected especially for natural AMPs, which frequently do not exhibit ideal amphipathicity, and hence are more prone to perturb membranes in a concentration-dependent manner in the proposed way than synthetic AMPs, which are frequently designed to exhibit ideal amphipathicity (Wang et al. 2009).

9.3.2 Lipid Phase Separation Induced by AMPs: Charge Clustering

For a number of AMPs, it has been shown that these cationic amphipathic peptides can induce lateral phase separation in bilayers composed of anionic and zwitterionic lipids because of preferential binding to the negatively charged lipid component (Epand et al. 2010; Epand and Epand 2009, 2011; Lohner et al. 1997; Lohner and Prenner 1999). These laterally heterogeneous domains with a preferred location of AMPs in one domain will differ significantly in their properties in respect of membrane thickness, fluidity, and curvature strain. The melting properties of the segregated lipid domains will be different from each other and, in turn, will give rise to changed thermotropic properties expressed in different phase transition temperatures, enthalpies, and cooperativity. Thus, DSC is a useful technique to monitor such laterally heterogeneous domains within a bilayer, if the domains are stable on the time scale of the lipid phase transition (Raudino et al. 2011). As well as no requirement for molecular probes, another advantage is the detection of domains that are too small to be imaged by microscopic techniques (Epand 2007). Promotion of lateral phase separation has been investigated using mainly PE as a matrix lipid mixed with either PG or CL at various molar ratios to mimic bacterial cytoplasmic membranes. On preferential interaction of AMPs with the anionic lipid, this domain will segregate, leading to enrichment of the lipid bulk phase with PE. Hence, the melting temperature of each domain will be similar to that of the more abundant component, which leads to the splitting of the phase transition in proportion to the composition and properties of the domains. In particular, it can be expected that the PE-enriched domain will become more cooperative and will exhibit a phase transition temperature closer to the pure component (see also Figure 9.1). For example, such behavior was found for liposomes composed of PE/PG in the presence of PGLa, human defensin, or human LL-37 (Lohner et al. 1997; Lohner and Prenner 1999; Sevcsik et al. 2008). The respective thermograms (Figure 9.2) are characterized by two phase transitions, which can be related to a domain enriched in both AMPs and PG melting at a lower temperature, while a domain enriched in PE and depleted of AMPs melts at a higher temperature. Phase separation for monocomponent systems, that is, formation of peptide-poor and peptide-rich domains, has been also described (see the following section). Moreover, DSC studies on monocomponent systems clearly showed that a given AMP can affect the phase behavior of different species of anionic lipids to a different extent (Arouri et al. 2009; Jing et al. 2005), emphasizing the value of DSC for gaining information on peptides' lipid specificity.

The importance of induced lateral phase separation has been specifically proposed as a mechanism contributing to the antimicrobial activity of a designed α/β peptide (Epand et al. 2006) and an oligo-acyl-lysine (OAK) compound (Epand et al. 2008; Radzishevsky et al. 2007). In the presence of OAKs, similar thermograms were obtained for 1-palmitoyl-2-oleoyl-sn-glycero-3-phosphoethanolamine (POPE)/tetraoleoyl-cardiolipin (TOCL) and POPE/ palmitoyl-oleoyl-phosphatidylglycerol (POPG) mixtures, indicating that segregation of anionic from zwitterionic lipids is not primarily dependent on the structure of the anionic headgroup of the lipid (Epand et al. 2008). Nor does it seem to depend on the nature of the hydrocarbon chains, as similar observations were obtained with disaturated phospholipids (Arouri et al. 2009; Lohner et al. 1997; Sevcsik et al. 2008). These results emphasize that electrostatic interaction favoring the binding of the cationic peptides to the

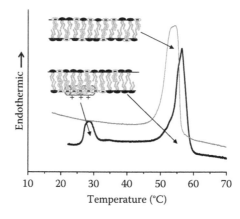

FIGURE 9.2 Thermogram of an equimolar mixture of DPPG/DPPE (thin dark gray line) exhibiting a broad phase transition around 54°C and in the presence of LL-37 (thick black line). Lipid-to-peptide molar ratio was 25/1. Experimental conditions as in Figure 9.1. The insert shows a scheme of the binary lipid mixture (top) as well as of the mixture in the presence of the peptide, indicating the segregation of membrane domains (bottom).

anionic lipids is the driving force in the creation of segregated domains. However, clustering of anionic lipids was most effective for substances with sequential positive charges contained within a flexible molecule that can adapt to the arrangement of charged groups on the surface of the bacterial cell membrane (Epand and Epand 2009; Epand et al. 2010). Further, it was proposed to use the ability to induce phase separation for the prediction of relative sensitivity of bacteria toward a respective AMP (Epand et al. 2008). Because of the higher amount of PE present in membranes of Gram-negative bacteria, it can be expected that this mechanism will be more relevant for bacteria of this kind. Finally, it is tempting to speculate that the specific interaction of AMPs with anionic lipids may have important implications for the structure and integrity of membranes and, in turn, also influence the function of membrane proteins, adversely affecting bacteria.

9.3.3 Free Volume Model: Chain Interdigitation and Bilayer Thinning

Early DSC experiments using liposomes composed of DPPG to study the effect of the frog skin antimicrobial peptide PGLa on bacterial membrane mimics revealed a concentration-dependent behavior, in which a shoulder at the high-temperature side of T_m was observed on increasing the concentration of the peptide (Latal et al. 1997). Concomitantly, the pretransition of DPPG decreased in a similar manner to the main transition of DPPG. The presence of a pretransition was taken as an indication that unperturbed lipid domains must be still present, since the pretransition has been shown to respond even to small perturbations of the lipid matrix (Lohner 1991). This conclusion was supported by the fact that the peptide did not affect the half-width of the main transition, indicative of the cooperativity of the transition, which, like the pretransition, is a very sensitive parameter for impurities. The concomitant decrease of enthalpy of the pure DPPG fraction and increase of enthalpy of the PGLa-affected domains indicates phase separation resulting in pure lipid or, at least, peptide-poor and peptide-rich domains. Initially, it was proposed that the temperature shift observed for the peptide-enriched domain was due to electrostatic shielding resulting from nonspecific screening of the negative surface charges by the cationic AMP, as described for monovalent ions (Cevc et al. 1980). X-ray data indicated that the tilted $L_{\beta'}$-phase of DPPG coexists with an untilted gel phase, which was later identified as a quasi-interdigitated phase (Sevcsik et al. 2007), as discussed in the following paragraph.

Similar observations were made with other AMPs and PG liposomes; LL-37 was more efficient than PGLa at inducing the stabilized gel phase (Sevcsik et al. 2007). At a lipid-to-peptide molar ratio of 25:1, this was the only phase observed, which allowed an unambiguous structural analysis by small- and wide-angle x-ray scattering (SAX and WAX). Using a global data analysis (Pabst 2006), a drastically reduced (by 13.5 Å) bilayer thickness was calculated from the SAX data in association with a characteristic electron density profile lacking the trough for the terminal methyl group of the hydrocarbon chains. Furthermore, the area per lipid was significantly reduced (40 Å² as compared with 46 Å² of the $L_{\beta'}$ phase of DPPG), which indicated that the headgroup tilts away from an orientation parallel to the membrane surface, allowing a tighter hydrocarbon chain packing. This was in accordance with the changed wide-angle x-ray pattern, indicating that in the presence of the peptide, the hydrocarbon chains are aligned perpendicular to the bilayer plane and not tilted as observed for pure DPPG. Based on these facts, the formation of an interdigitated bilayer structure could be deduced, in which the peptide shields the tail ends of the lipids from the interfacial water, and which was therefore called a quasi-interdigitated ($L_{\beta Iq}$) phase (Figure 9.3). This example illustrates the need for complementary structural techniques to assign the phase unequivocally, which would not be possible by DSC experiments alone.

The formation of this phase was explained by the alignment of the peptide parallel to the bilayer plane inducing free volume (*voids*) in the hydrophobic core of the membrane. The extent of this perturbation will depend on a number of parameters, such as the concentration, molecular size, and aggregation state of the peptide, which determine the lateral area occupied by the peptide in the membrane interface, as well as on the membrane penetration depth of the peptide. The effect of penetration depth was tested by performing DSC and x-ray experiments on PGs and PCs of various chain lengths (Sevcsik et al. 2007). Indeed, decreasing the chain length of PG abolished the formation of the interdigitated bilayer structure, while interdigitation was observed to some extent for very-long-chain PCs. Therefore, the fraction of $L_{\beta Iq}$, which cannot be deduced accurately from x-ray scattering data, can be easily calculated from the deconvolution of the DSC thermogram. This behavior was found not only for LL-37, but also for PGLa and melittin, as well as for a shorter synthetic derivative of LL-37, OP-145 (Malanovic et al. 2015). Of these, LL-37 was most effective, demonstrating that there is a tight coupling between the peptide properties and those of the lipid bilayer. Hence, a general phase diagram was proposed for lipid/peptide mixtures as long as the amphipathic peptides adopt a spatial orientation parallel to the membrane plane, while the phase boundary for a quasi-interdigitated phase of PC/peptide mixtures is shifted toward longer chain lengths due to the deeper penetration of such peptides into PC bilayers, while electrostatic interactions between the cationic peptide and the anionic lipid head groups inhibit deeper insertion into the acyl chain region of PG membranes (Dathe et al. 2002; Sevcsik et al. 2007). Finally, it is interesting to note that the cationic antibiotic peptide polymyxin B (Ranck and Tocanne 1982b) and large,

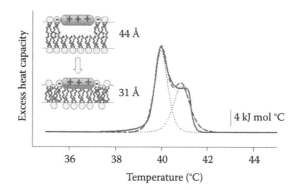

FIGURE 9.3 Thermogram of a DPPG/PGLa mixture at a lipid-to-peptide molar ratio of 25/1. Experimental conditions as in Figure 9.1. The insert shows the incorporation of the peptide in the bilayer interface creating a void, which is compensated by moving the inner monolayer toward the hydrophobic side of the peptide. The phosphate-to-phosphate distance determined by SAX scattering for the pure lipid gel phase and the lipid/peptide mixture is indicated.

amphipathic cations such as acetylcholine (Ranck and Tocanne 1982a) have been shown to give rise to hydrocarbon chain interdigitation in DPPG vesicles as well. In all these cases, the resulting fluid phase was significantly thinner than the pure lipid L_α-phase.

9.3.4 Detergent-Like Behavior: Micellization of Lipid Bilayers

DSC experiments in combination with dilatometry, x-ray scattering and nuclear magnetic resonance (NMR) techniques gave rise to the idea of gradual membrane disintegration (Bechinger and Lohner 2006). For example, a strongly concentration-dependent behavior of disaturated PCs was observed for the lytic peptides melittin (Monette and Lafleur 1995; Posch et al. 1983) and δ-lysin from *Staphylococcus aureus* (Lohner et al. 1986, 1999). At very low melittin concentrations (lipid-to-peptide molar ratio > 500:1), the main phase transition temperature was slightly but significantly increased. Since these effects could not be accounted for by local perturbations induced by the peptide around the site of interaction, it was suggested that the peptide-affected domains create line defects in an ordered lipid lattice (Colotto et al. 1993). In contrast, at very high peptide concentrations (lipid-to-peptide molar ratio of 15:1), which better reflect the *in vivo* situation of full coverage of the bacterial surface (Roversi et al. 2014), a completely different heat capacity profile was recorded, characterized by a very low-enthalpic, broad asymmetric transition. The fraction of this component depended on the concentration of the peptide. Similar thermograms were also obtained for other systems at high peptide concentrations, such as δ-lysin (Lohner et al. 1999), LL-37 (Sevcsik et al. 2007), and OP-145 (Malanovic et al. 2015). For these samples, SAX experiments were performed, showing a diffuse particle scattering in the gel phase. Modeling of the x-ray data revealed disk-shaped lipid–peptide aggregates surrounded by a peptide ring about 10 Å thick (Figure 9.4) (Lohner et al. 1999; Malanovic et al. 2015; Sevcsik et al. 2007). Interestingly, the diameter of the disk-like micelles varied between the AMPs (Table 9.2), which may be due to the flexibility of the helical peptides to

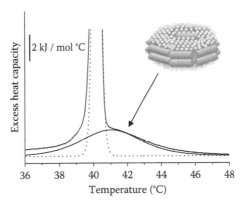

FIGURE 9.4 Disintegration of DPPC multilamellar vesicles on addition of 4 mol% of OP-145 derived from a fragment of LL-37. A broad transition underlies the highly cooperative chain melting transition of DPPC, which arises from disk-like lipid–peptide aggregates (scheme shown in insert) induced on incubation with the peptide. Peptide-to-lipid molar ratio was 25/1. Experimental conditions as in Figure 9.1. Deconvolution of the thermogram indicated that about one-quarter of the multilamellar vesicle was disintegrated.

TABLE 9.2 Structural Parameters of Disk-Like Lipid–Peptide Aggregates Estimated from SAX Experiments in the Gel Phase

Model System	Peptide Length[a]	Disk Diameter (Å)	Bilayer Thickness (Å)	References
DPPC/OP-145	24	480	60	(Malanovic et al. 2015)
DPPC/LL-37	37	270	55	(Sevcsik et al. 2007)
DMPC/LL-37	37	270	49	(Sevcsik et al. 2007)
DMPC/δ-lysin	26	140	52	(Lohner et al. 1999)
DMPC/melittin	26	235	n.d.[b]	(Dufourcq et al. 1986)

[a] Number of amino acids constituting the respective peptide.

[b] The study was performed using quasi-elastic light scattering, gel filtration, and freeze-fracture electron microscopy; the exact value was not determined, but was assumed to correspond to regular bilayer thickness.

adopt certain curvatures. The appearance of disk-like micelles has also been described for DMPC in the presence of melittin (Dufourcq et al. 1986). Although these disk-like lipid–peptide aggregates fused into large bilayer sheets when heated above T_m, these experiments still indicate that at high concentrations, penetration of AMPs into neutral lipid bilayers may lead to disintegration, similarly to a detergent-like disruption of a membrane.

There has been some debate about the peptide arrangement that surrounds the hydrophobic core of the disk-like micelle. A perpendicular orientation to the bilayer plane was proposed for δ-lysin (Lohner et al. 1999), while a parallel orientation was suggested for LL-37, considering that the double diameter of an α-helix is comparable to the thickness of the hydrophobic core of the membrane (~38 Å for DPPC and ~33 Å for DMPC as compared with an estimated helix diameter of ~16 Å) (Sevcsik et al. 2007). This would be analogous to the "bicycle-tire" model proposed for melittin (Dempsey 1990) and apolipoprotein E (Raussens et al. 1998). If such an arrangement is the case, disk formation should be more difficult in bilayers of longer-chain PCs, as parallel arranged peptides no longer match the hydrophobic core of the bilayer. Indeed, for the longer-chain distearoyl-PC (DSPC) and diarachidoyl-PC (DAPC) in the presence of LL-37, no low-enthalpic broad transition, characteristic of micelle formation, was observed by DSC, but a second rather sharp transition at higher temperatures appeared. X-ray data analysis confirmed that no disk-like lipid–peptide aggregates were formed, but demonstrated a coexistence of interdigitated and noninterdigitated lipid domains, quite similar to the situation observed for mixtures of DPPG and LL-37 (Sevcsik et al. 2007).

9.3.5 AMPs Affecting Membrane Curvature

Gram-negative bacteria contain substantial amounts of so-called nonlamellar-phase-forming lipids, such as PE and CL, and precisely regulate their lipid composition in a narrow window close to a lamellar to nonlamellar phase boundary (McElhaney 1992; Morein et al. 1996; Rilfors et al. 1993). Due to their molecular shape (an inverted cone), PE and CL lead to an increase of the lateral pressure in the center of the bilayer and, in turn, to membrane curvature strain (Koller and Lohner 2014). Thus, on dispersion in aqueous media, these lipids adopt primarily the inverted hexagonal (H_{II}) phase, but also cubic phases (Rappolt et al. 2003; Seddon and Templer 1995; Siegel and Banschbach 1990). Therefore, it has been proposed that AMPs may interfere with the balance of lamellar versus nonlamellar forming forces, shifting it toward nonlamellar structures, which in turn may lead to membrane disintegration (Lohner 2009). A correlation between antimicrobial activity and cubic phase formation of *Escherichia coli* lipid extracts was reported for cationic amphipathic peptides derived from a fragment of human lactoferrin (Zweytick et al. 2008, 2011). Cubic phase formation was also found for lipid extracts from *E. coli* and *Acholeplasma laidlawii* in the presence of gramicidin S, a cyclic AMP (Staudegger et al. 2000). The promotion of this non-lamellar phase was explained by an increase of spontaneous curvature on interaction of the AMPs with the lipid bilayer. Fluid–cubic phase transitions are not easily detected by DSC, because they may be metastable and exhibit low cooperativity and very low enthalpy change. For example, using very slow scan rates, the formation of thermodynamically stable cubic phases of methylated 1,2-dioleoyl-sn-glycero-3-phosphoethanolamine (DOPE) was detected by DSC, exhibiting an enthalpy of about 1 kJ/mol (Siegel and Banschbach 1990).

However, PEs that adopt the H_{II} phase are good models to test the effect on spontaneous curvature, as the lamellar to inverse hexagonal phase transition is accompanied by a change in curvature from zero (bilayer) to negative values (inverse hexagonal). Changes in T_{HII} can be easily measured by DSC and present the most common parameter describing the curvature behavior of AMPs (Haney et al. 2010; Lohner and Prenner 1999; Matsuzaki et al. 1998). Hence, AMPs that induce negative spontaneous curvature, that is, that favor the H_{II} phase, will decrease the lamellar to inverse hexagonal phase transition temperature (T_{HII}), while AMPs that induce positive spontaneous curvature, that is, that stabilize the bilayer, will lead to an increase of T_{HII}. A number of examples are listed in Table 9.3, which, however, does not represent a complete list.

In a recent review, Haney et al. (2010) tried to relate changes in T_{HII} to the proposed mode of action of AMPs. In brief, an increase of T_{HII} was correlated with an increase in positive curvature, which was suggested to play a role in toroidal pore formation (Matsuzaki et al. 1998). The structure of such a phospholipid–peptide pore was solved by neutron scattering (Ludtke et al. 1996), which showed that the phospholipids bend from the membrane interface toward the hydrophobic interior (positive curvature) and are intercalated along the long axis between the peptide molecules, which are oriented perpendicular to the bilayer plane. Furthermore, induction of a positive curvature by AMPs was suggested to be possibly also involved in the micellization of a bilayer (Haney et al. 2010). In this case, the peptide would cover a rather spherical lipid particle, with the hydrocarbon chains packed into the interior of this aggregate. The existing models to explain the significance of enhancing negative curvature strain (decrease of T_{HII}) are less clear. One idea is that when a sufficient number of peptides aggregate, they induce negative curvature stress followed by the formation of a non-bilayer intermediate redistributing the peptide between the two leaflets (Powers et al. 2005). Another model to explain this peptide redistribution is the sinking raft model (Pokorny and Almeida 2004). As suggested for polyphemusin, this could also be a mechanism to translocate AMPs across the membrane to act on an intracellular target (Powers et al. 2005).

The impact on membrane function of membrane curvature strain and its changes on incorporation of AMPs has been discussed (Koller and Lohner 2014; Lohner and Blondelle 2005). It was suggested earlier that the high lateral hydrocarbon chain pressure exhibited by

TABLE 9.3   Effect on the Fluid Lamellar to Inverse Hexagonal Phase Transition Temperature (T_{HII}) of Selected Membrane-Active Peptides in Different Lipid Model Systems

Peptide	Lipid System	References
Increase of T_{HII}		
Acylated/nonacylated variants of LF11	POPE	(Zweytick et al. 2011)
Melittin	POPE	(Hickel et al. 2008)
Magainin 2	DiPoPE[a]	(Matsuzaki et al. 1998)
Trichogin GA IV and variants	DiPoPE	(Epand et al. 1999)
MSI-78	DiPoPE	(Hallock et al. 2003)
MSI-843	DiPoPE	(Thennarasu et al. 2005)
MSI-367	DiPoPE	(Thennarasu et al. 2010)
RL16	DiPoPE	(Alves et al. 2008)
Oxyopinins (Oxki1/2)[a]	DEPE	(Nomura and Corzo 2006)
Decrease of T_{HII}		
NK-2	POPE	(Willumeit et al. 2005)
Nisin[b]	POPE/DOPE	(El Jastimi and Lafleur 1999)
Beta-17	DiPoPE	(Epand et al. 2003)
Polyphemusin I	DiPoPE	(Powers et al. 2005)
Penetratin	DiPoPE	(Alves et al. 2008)
Oritavancin	CL/POPE	(Domenech et al. 2009)
RMAF4, R/K-RMAD4, Crp4[c]	DOPS[d]/DOPE/DOPC[e]	(Schmidt et al. 2012)

[a] Dipalmitoleoyl-phosphatidylethanolamine (DiPoPE).
[b] Determined by P^{31}-NMR.
[c] Determined by SAX.
[d] 1,2-dioleoyl-sn-glycero-3-phospho-L-serine (DOPS).
[e] 1,2-dioleoyl-sn-glycero-3-phosphocholine (DOPC).

nonlamellar-phase-preferring lipids such as PE and CL controls the conformation of integral membrane proteins (de Kruijff 1997). Therefore, AMPs that affect the lateral hydrocarbon chain pressure on insertion (note, both positive and negative curvature induction) may lead as a secondary effect to conformational changes of integral membrane proteins and hence to impairment of membrane function.

9.4 CONCLUDING REMARKS

As outlined in a review by Richard and Raquel Epand (Epand and Epand 2003), the chemical composition of biological membranes is complex, including a number of other constituents, such as proteins, as well as phospholipids, and it would therefore be too difficult to completely mimic this environment with liposomes. In terms of bacterial membranes, cell wall components have to be considered too, such as lipoteichoic acid (LTA) and peptidoglycan in the case of Gram-positive bacteria or lipopolysaccharides in the case of Gram-negative bacteria (Freire et al. 2015). While a wealth of data exists on the interaction of AMPs with LPS, particularly on their activity to neutralize LPS, a trigger of sepsis (e.g., Heinbockel et al. 2015 and references therein), little is known about LTA (Gutberlet et al. 1994; Malanovic et al. 2015). Moreover, transversal membrane asymmetry is another aspect that has to be taken into account. In this respect, recent developments (Marquardt et al. 2015 and references therein) may open new avenues to expand our range of membrane mimetic systems, which would fit with the current trends toward preparing increasingly complex systems. Nevertheless, liposomes composed of phospholipids have been extensively studied by DSC (Biltonen and Lichtenberg 1993; Heerklotz 2004), whereby the use of synthetic lipids allowed systematic investigation of the impact of the nature of headgroups and hydrocarbon chains, as well as environmental factors such as pH and ionic strength, on their polymorphic phase behavior. This has yielded invaluable knowledge toward a better understanding of membrane properties and function, as well as on the interaction of membrane-active molecules with membranes, as described in this chapter.

ACKNOWLEDGMENTS

A vast amount of the data described in this chapter was obtained by former and current members of my group, whom I thank for their discussions and contributions. This research was supported by the Austrian Science Fund FWF, Project No. I 1763-B21 (to K.L.).

REFERENCES

Alves, I. D., Goasdoue, N., Correia, I., Aubry, S., Galanth, C., Sagan, S., Lavielle, S., and Chassaing, G. (2008), Membrane interaction and perturbation mechanisms induced by two cationic cell penetrating peptides with distinct charge distribution, *Biochim. Biophys. Acta* 1780, 948–959.

Arouri, A., Dathe M., and Blume, A. (2009), Peptide induced demixing in PG/PE lipid mixtures: A mechanism for the specificity of antimicrobial peptides towards bacterial membranes? *Biochim. Biophys. Acta* 1788, 650–659.

Bechinger, B. and Lohner, K. (2006), Detergent-like actions of linear amphipathic cationic antimicrobial peptides, *Biochim. Biophys. Acta* 1758, 1529–1539.

Bevers, E. M., Comfurius, P., Dekkers, D. W., and Zwaal, R. F. (1999), Lipid translocation across the plasma membrane of mammalian cells, *Biochim. Biophys. Acta* 1439, 317–330.

Biltonen, R. L. and Lichtenberg, D. (1993), The use of differential scanning calorimetry as a tool to characterize liposome preparations, *Chem. Phys. Lipids* 64, 129–142.

Blondelle, S. E. and Lohner, K. (2010), Optimization and high-throughput screening of antimicrobial peptides, *Curr. Pharm. Des.* 16, 3204–3211.

Brandenburg, K. and Blume, A. (1987), Investigations into the thermotropic phase behaviour of natural membranes extracted from gram-negative bacteria and artificial membrane systems made from lipopolysaccharides and free lipid A, *Thermochim. Acta* 119, 127–142.

Cevc, G., Watts, A., and Marsh, D. (1980), Non-electrostatic contribution to the titration of the ordered-fluid phase transition of phosphatidylglycerol bilayers, *FEBS Lett.* 120, 267–270.

Chang, S., Sievert, D. M., Hageman, J. C., Boulton, M. L., Tenover, F. C., Downes, F. P., Shah, S., et al. (2003), Infection with vancomycin-resistant *Staphylococcus aureus* containing the vanA resistance gene, *N. Engl. J. Med.* 348, 1342–1347.

Chiu, M. H. and Prenner, E. J. (2011), Differential scanning calorimetry: An invaluable tool for a detailed thermodynamic characterization of macromolecules and their interactions, *J. Pharm. Bioallied Sci.* 3, 39–59.

Colotto, A., Kharakoz, D. P., Lohner, K., and Laggner, P. (1993), Ultrasonic study of melittin effects on phospholipid model membranes, *Biophys. J.* 65, 2360–2367.

Dathe, M., Meyer, J., Beyermann, M., Maul, B., Hoischen, C., and Bienert, M. (2002), General aspects of peptide selectivity towards lipid bilayers and cell membranes studied by variation of the structural parameters of amphipathic helical model peptides, *Biochim. Biophys. Acta* 1558, 171–186.

de Kruijff, B. (1997), Lipid polymorphism and biomembrane function, *Curr. Opin. Chem. Biol.* 1, 564–569.

Degovics, G., Latal, A., and Lohner, K. (2000), X-ray studies on aqueous dispersions of dipalmitoylphosphatidylglycerol in the presence of salt, *J. Appl. Crystallogr.* 33, 544–547.

Degovics, G., Latal, A., Prenner, E., Kriechbaum, M., and Lohner, K. (1997), Structure and thermotropic behaviour of mixed choline phospholipid model membranes, *J. Appl. Crystallogr.* 30, 776–780.

Dempsey, C. E. (1990), The actions of melittin on membranes, *Biochim. Biophys. Acta* 1031, 143–161.

Domenech, O., Francius, G., Tulkens, P. M., Van Bambeke, F., Dufrene, Y., and Mingeot-Leclercq, M. P. (2009), Interactions of oritavancin, a new lipoglycopeptide derived from vancomycin, with phospholipid bilayers: Effect on membrane permeability and nanoscale lipid membrane organization, *Biochim. Biophys. Acta* 1788, 1832–1840.

Dufourcq, J., Faucon, J.-F., Fourche, G., Dasseux, J.-L., le Maire, M., and Gulik-Krzywicki, T. (1986), Morphological changes of phosphatidylcholine bilayers induced by melittin: Vesicularization, fusion, discoidal particles, *Biochim. Biophys. Acta* 859, 33–48.

El Jastimi, R. and Lafleur, M. (1999), Nisin promotes the formation of non-lamellar inverted phases in unsaturated phosphatidylethanolamines, *Biochim. Biophys. Acta* 1418, 97–105.

Epand, R. F., Epand, R. M., Monaco, V., Stoia, S., Formaggio, F., Crisma, M., and Toniolo, C. (1999), The antimicrobial peptide trichogin and its interaction with phospholipid membranes, *Eur. J. Biochem.* 266, 1021–1028.

Epand, R. F., Maloy, W. L., Ramamoorthy, A., and Epand, R. M. (2010), Probing the "charge cluster mechanism" in amphipathic helical cationic antimicrobial peptides, *Biochemistry* 49, 4076–4084.

Epand, R. F., Schmitt, M. A., Gellman, S. H., and Epand, R. M. (2006), Role of membrane lipids in the mechanism of bacterial species selective toxicity by two alpha/beta-antimicrobial peptides, *Biochim. Biophys. Acta* 1758, 1343–1350.

Epand, R. F., Umezawa, N., Porter, E. A., Gellman, S. H., and Epand, R. M. (2003), Interactions of the antimicrobial beta-peptide beta-17 with phospholipid vesicles differ from membrane interactions of magainins, *Eur. J. Biochem.* 270, 1240–1248.

Epand, R. M. (2007), Detecting the presence of membrane domains using DSC, *Biophys. Chem.* 126, 197–200.

Epand, R. M. and Epand, R. F. (2003), Liposomes as models for antimicrobial peptides, *Methods Enzymol.* 372, 124–133.

Epand, R. M. and Epand, R. F. (2009), Lipid domains in bacterial membranes and the action of antimicrobial agents, *Biochim. Biophys. Acta* 1788, 289–294.

Epand, R. M. and Epand, R. F. (2011), Bacterial membrane lipids in the action of antimicrobial agents, *J. Pept. Sci.* 17, 298–305.

Epand, R. M., Rotem, S., Mor, A., Berno, B., and Epand, R. F. (2008), Bacterial membranes as predictors of antimicrobial potency, *J. Am. Chem. Soc.* 130, 14346–14352.

Findlay, B., Zhanel, G. G., and Schweizer, F. (2010), Cationic amphiphiles, a new generation of antimicrobials inspired by the natural antimicrobial peptide scaffold, *Antimicrob. Agents Chemother.* 54, 4049–4058.

Freire, E. and Biltonen, R. (1978), Estimation of molecular averages and equilibrium fluctuations in lipid bilayer systems from the excess heat capacity function, *Biochim. Biophys. Acta* 514, 54–68.

Freire, J. M., Gaspar, D., Veiga, A. S., and Castanho, M. A. (2015), Shifting gear in antimicrobial and anticancer peptides biophysical studies: From vesicles to cells, *J. Pept. Sci.* 21, 178–185.

Garidel, P., Johann, C., Mennicke, L., and Blume, A. (1997), The mixing behavior of pseudobinary phosphatidylcholine-phosphatidylglycerol mixtures as a function of pH and acyl chain length, *Eur. Biophys. J.* 26, 447–459.

Grabitz, P., Ivanova, V. P., and Heimburg, T. (2002), Relaxation kinetics of lipid membranes and its relation to the heat capacity, *Biophys. J.* 82, 299–309.

Gutberlet, T., Milde, K., Bradaczek, H., Haas, H., and Mohwald, H. (1994), Miscibility of lipotei-choic acid in dipalmitoylphosphatidylcholine studied by monofilm investigations and fluorescence microscopy, *Chem. Phys. Lipids* 69, 151–159.

Hallock, K. J., Lee, D. K., and Ramamoorthy, A. (2003), MSI-78, an analogue of the magainin antimicrobial peptides, disrupts lipid bilayer structure via positive curvature strain, *Biophys. J.* 84, 3052–3060.

Hancock, R. E. (2001), Cationic peptides: Effectors in innate immunity and novel antimicrobials, *Lancet Infect. Dis.* 1, 156–164.

Hancock, R. E. (2014), Collateral damage, *Nat. Biotechnol.* 32, 66–68.

Hancock, R. E. W. and Rozek, A. (2002), Role of membranes in the activities of antimicrobial cationic peptides, *FEMS Microbiol. Lett.* 206, 143–149.

Haney, E. F., Nathoo, S., Vogel, H. J., and Prenner, E. J. (2010), Induction of non-lamellar lipid phases by antimicrobial peptides: A potential link to mode of action, *Chem. Phys. Lipids* 163, 82–93.

Heerklotz, H. (2004), The microcalorimetry of lipid membranes, *J. Phys. Condens. Matter* 16, 441–467.

Heinbockel, L., Marwitz, S., Barcena Varela, S., Ferrer-Espada, R., Reiling, N., Goldmann, T., Gutsmann, T., et al. (2015), Therapeutic administration of peptide pep19-2.5 and ibuprofen reduces inflammation and prevents lethal sepsis, *PLoS One* 10(7), e0133291.

Henderson, J. M. and Lee, K. Y. C. (2013), Promising antimicrobial agents designed from natural peptide templates, *Curr. Opin. Solid State Mater. Sci.* 17, 175–192.

Hickel, A., Danner-Pongratz, S., Amenitsch, H., Degovics, G., Rappolt, M., Lohner, K., and Pabst, G. (2008), Influence of antimicrobial peptides on the formation of nonlamellar lipid mesophases, *Biochim. Biophys. Acta* 1778, 2325–2333.

Hilpert, K., Volkmer-Engert, R., Walter, T., and Hancock, R. E. (2005), High-throughput generation of small antibacterial peptides with improved activity, *Nat. Biotechnol.* 23, 1008–1012.

Hinz, H.-J. and Schwarz, F. P. (2001), Measurements and analysis of results obtained on biological substances with differential scanning calorimetry, *Pure Appl. Chem.* 73, 745–759.

Hoernke, M., Schwieger, C., Kerth, A., and Blume, A. (2012), Binding of cationic pentapeptides with modified side chain lengths to negatively charged lipid membranes: Complex interplay of electrostatic and hydrophobic interactions, *Biochim. Biophys. Acta* 1818, 1663–1672.

Hristova, K., Selsted, M. E., and White, S. H. (1997), Critical role of lipid composition in membrane permeabilization by rabbit neutrophil defensins, *J. Biol. Chem.* 272, 24224–24233.

Huang, H. W. (2000), Action of antimicrobial peptides: Two-state model, *Biochemistry* 39, 8347–8352.

Huang, H. W. (2006), Molecular mechanism of antimicrobial peptides: The origin of cooperativity, *Biochim. Biophys. Acta* 1758, 1292–1302.

Jackson, M. B. and Sturtevant, J. M. (1977), Studies of the lipid phase transitions of Escherichia coli by high sensitivity differential scanning calorimetry, *J. Biol. Chem.* 252, 4749–4751.

Jing, W., Prenner, E. J., Vogel, H. J., Waring, A. J., Lehrer, R. I., and Lohner, K. (2005), Headgroup structure and fatty acid chain length of the acidic phospholipids modulate the interaction of membrane mimetic vesicles with the antimicrobial peptide protegrin-1, *J. Pept. Sci.* 11, 735–743.

Koller, D. and Lohner, K. (2014), The role of spontaneous lipid curvature in the interaction of interfacially active peptides with membranes, *Biochim. Biophys. Acta* 1838, 2250–2259.

Koynova, R. and Caffrey, M. (2002), An index of lipid phase diagrams, *Chem. Phys. Lipids* 115, 107–219.

Latal, A., Degovics, G., Epand, R. F., Epand, R. M., and Lohner, K. (1997), Structural aspects of the interaction of peptidyl-glycylleucine-carboxyamide, a highly potent antimicrobial peptide from frog skin, with lipids, *Eur. J. Biochem.* 248, 938–946.

Lewis, R. N., Mannock, D. A., and McElhaney, R. N. (2007), Differential scanning calorimetry in the study of lipid phase transitions in model and biological membranes: Practical considerations, *Methods Mol. Biol.* 400, 171–195.

Lohner, K. (1991), Effects of small organic molecules on phospholipid phase transitions, *Chem. Phys. Lipids* 57, 341–362.

Lohner, K. (2001), The role of membrane lipid composition in cell targeting and their mechanism of action. In K. Lohner (Ed.), *Development of Novel Antimicrobial Agents: Emerging Strategies*, 149–165, Horizon Scientific: Norfolk.

Lohner, K. (2009), New strategies for novel antibiotics: Peptides targeting bacterial cell membranes, *Gen. Physiol. Biophys.* 28, 105–116.

Lohner, K. and Blondelle, S. E. (2005), Molecular mechanisms of membrane perturbation by antimicrobial peptides and the use of biophysical studies in the design of novel peptide antibiotics, *Comb. Chem. High Throughput Screen.* 8, 241–256.

Lohner, K., Laggner, P., and Freer, J. H. (1986), Dilatometric and calorimetric studies of the effect of *Staphylococcus aureus* delta-lysin on the phospholipid phase transition, *J. Solution Chem.* 15, 189–198.

Lohner, K., Latal, A., Degovics, G., and Garidel, P. (2001), Packing characteristics of a model system mimicking cytoplasmic bacterial membranes, *Chem. Phys. Lipids* 111, 177–192.

Lohner, K., Latal, A., Lehrer, R. I., and Ganz, T. (1997), Differential scanning microcalorimetry indicates that human defensin, HNP-2, interacts specifically with biomembrane mimetic systems, *Biochemistry* 36, 1525–1531.

Lohner, K. and Prenner, E. J. (1999), Differential scanning calorimetry and X-ray diffraction studies of the specificity of the interaction of antimicrobial peptides with membrane-mimetic systems, *Biochim. Biophys. Acta* 1462, 141–156.

Lohner, K., Staudegger, E., Prenner, E. J., Lewis, R. N., Kriechbaum, M., Degovics, G., and McElhaney, R. N. (1999), Effect of staphylococcal delta-lysin on the thermotropic phase behavior and vesicle morphology of dimyristoylphosphatidylcholine lipid bilayer model membranes. Differential scanning calorimetric, ^{31}P nuclear magnetic resonance and Fourier transform infrared spectroscopic, and X-ray diffraction studies, *Biochemistry* 38, 16514–16528.

Ludtke, S. J., He, K., Heller, W. T., Harroun, T. A., Yang, L., and Huang, H. W. (1996), Membrane pores induced by magainin, *Biochemistry* 35, 13723–13728.

Mabrey, S. and Sturtevant, J. M. (1978), High-sensitivity differential scanning calorimetry in the study of biomembranes and related model systems. In E. D. Korn (Ed.), *Methods in Membrane Biology*, vol. 9, 237–274, Plenum: New York.

Malanovic, N., Leber, R., Schmuck, M., Kriechbaum, M., Cordfunke, R. A., Drijfhout, J. W., de Breij, A., Nibbering, P. H., Kolb, D., and Lohner, K. (2015), Phospholipid-driven differences determine the action of the synthetic antimicrobial peptide OP-145 on Gram-positive bacterial and mammalian membrane model systems, *Biochim. Biophys. Acta* 1848, 2437–2447.

Marquardt, D., Geier, B., and Pabst, G. (2015), Asymmetric lipid membranes: Towards more realistic model systems, *Membranes* 5, 180–196.

Marsh, D., Watts, A., and Knowles, P. F. (1977), Cooperativity of the phase transition in single- and multibilayer lipid vesicles, *Biochim. Biophys. Acta* 465, 500–514.

Matsuzaki, K., Murase, O., Fujii, N., and Miyajima, K. (1996), An antimicrobial peptide, magainin 2, induced rapid flip-flop of phospholipids coupled with pore formation and peptide translocation, *Biochemistry* 35, 11361–11368.

Matsuzaki, K., Sugishita, K., Ishibe, N., Ueha, M., Nakata, S., Miyajima, K., and Epand, R. M. (1998), Relationship of membrane curvature to the formation of pores by magainin 2, *Biochemistry* 37, 11856–11863.

McElhaney, R. N. (1986), Differential scanning calorimetric studies of lipid-protein interactions in model membrane systems, *Biochim. Biophys. Acta* 864, 361–421.

McElhaney, R. N. (1992), Membrane structure. In J. Maniloff, R. N. McElhaney, L. R. Finch, and J. B. Baseman (Eds.), *Mycoplasma: Molecular Biology and Pathogenesis*, 113–155, American Society for Microbiology: Washington DC.

Monette, M. and Lafleur, M. (1995), Modulation of melittin-induced lysis by surface charge density of membranes, *Biophys. J.* 68, 187–195.

Morein, S., Andersson, A., Rilfors, L., and Lindblom, G. (1996), Wild-type *Escherichia coli* cells regulate the membrane lipid composition in a "window" between gel and non-lamellar structures, *J. Biol. Chem.* 271, 6801–6809.

Nomura, K. and Corzo, G. (2006), The effect of binding of spider-derived antimicrobial peptides, oxyopinins, on lipid membranes, *Biochim. Biophys. Acta* 1758, 1475–1482.

Nordberg, P., Monnet, D. L., and Cars, O. (2005), *Antibacterial Drug Resistance: Options for Concerned Action*, WHO, Department of Medicines Policy and Standards: Geneva.

Pabst, G. (2006), Global properties of biomimetic membranes: Perspectives on molecular features, *Biophys. Rev. Lett.* 1, 57–84.

Pabst, G., Zweytick, D., Prassl, R., and Lohner, K. (2012), Use of X-ray scattering to aid the design and delivery of membrane-active drugs, *Eur. Biophys. J* 41, 915–929.

Pokorny, A. and Almeida, P. F. F. (2004), Kinetics of dye efflux and lipid flip-flop induced by δ-lysin in phosphatidylcholine vesicles and the mechanism of graded release by amphipathic, α-helical peptides, *Biochemistry* 43, 8846–8857.

Posch, M., Rakusch, U., Mollay, C., and Laggner, P. (1983), Cooperative effects in the interaction between melittin and phosphatidylcholine model membranes. Studies by temperature scanning densitometry, *J. Biol. Chem.* 258, 1761–1766.

Powers, J. P., Tan, A., Ramamoorthy, A., and Hancock, R. E. (2005), Solution structure and interaction of the antimicrobial polyphemusins with lipid membranes, *Biochemistry* 44, 15504–15513.

Pozo, N. B., Lohner, K., Deutsch, G., Sevcsik, E., Riske, K. A., Dimova, R., Garidel, P., and Pabst, G. (2005), Composition dependence of vesicle morphology and mixing properties in a bacterial model membrane system, *Biochim. Biophys. Acta* 1716, 40–48.

Privalov, P. L. (1979), Stability of proteins: Small globular proteins, *Adv. Protein Chem.* 33, 167–241.

Prossnigg, F., Hickel, A., Pabst, G., and Lohner, K. (2010), Packing behaviour of two predominant anionic phospholipids of bacterial cytoplasmic membranes, *Biophys. Chem.* 150, 129–135.

Radzishevsky, I. S., Rotem, S., Bourdetsky, D., Navon-Venezia, S., Carmeli, Y., and Mor, A. (2007), Improved antimicrobial peptides based on acyl-lysine oligomers, *Nat. Biotechnol.* 25, 657–659.

Ranck, J. L. and Tocanne, J. F. (1982a), Choline and acetylcholine induce interdigitation of hydrocarbon chains in dipalmitoylphosphatidylglycerol lamellar phase with stiff chains, *FEBS Lett.* 143, 171–174.

Ranck, J. L. and Tocanne, J. F. (1982b), Polymyxin B induces interdigitation in dipalmitoylphosphatidylglycerol lamellar phase with stiff hydrocarbon chains, *FEBS Lett.* 143, 175–178.

Rappolt, M., Hickel, A., Bringezu, F., and Lohner, K. (2003), Mechanism of the lamellar/inverse hexagonal phase transition examined by high resolution X-ray diffraction, *Biophys. J.* 84, 3111–3122.

Rathinakumar, R., Walkenhorst, W. F., and Wimley, W. C. (2009), Broad-spectrum antimicrobial peptides by rational combinatorial design and high-throughput screening: The importance of interfacial activity, *J. Am. Chem. Soc.* 131, 7609–7617.

Rathinakumar, R. and Wimley, W. C. (2010), High-throughput discovery of broad-spectrum peptide antibiotics, *FASEB J.* 24, 3232–3238.

Raudino, A., Sarpietro, M. G., and Pannuzzo, M. (2011), The thermodynamics of simple biomembrane mimetic systems, *J. Pharm. Bioallied Sci.* 3, 15–38.

Raussens, V., Fisher, C. A., Goormaghtigh, E., Ryan, R. O., and Ruysschaert, J. M. (1998), The low density lipoprotein receptor active conformation of apolipoprotein E. Helix organization in n-terminal domain-phospholipid disc particles, *J. Biol. Chem.* 273, 25825–25830.

Rilfors, L., Wieslander, A., and Lindblom, G. (1993), Regulation and physicochemical properties of the polar lipids in *Acholeplasma laidlawii, Subcell. Biochem.* 20, 109–166.

Rothman, J. E. and Lenard, J. (1977), Membrane asymmetry, *Science* 195, 743–753.

Roversi, D., Luca, V., Aureli, S., Park, Y., Mangoni, M. L., and Stella, L. (2014), How many antimicrobial peptide molecules kill a bacterium? The case of PMAP-23, *ACS Chem. Biol.* 9, 2003–2007.

Schmidt, N. W., Tai, K. P., Kamdar, K., Mishra, A., Lai, G. H., Zhao, K., Ouellette, A. J., and Wong, G. C. (2012), Arginine in alpha-defensins: Differential effects on bactericidal activity correspond to geometry of membrane curvature generation and peptide-lipid phase behavior, *J. Biol. Chem.* 287, 21866–21872.

Seddon, J. M. and Templer, R. H., (1995), Polymorphism of lipid water systems. In R. Lipowsky and E. Sackmann (Eds.), *Structure and Dynamics of Membranes*, 97–160, North-Holland: Amsterdam.

Sevcsik, E. (2006), Hemolytic and antimicrobial activity of LL-37 is based on diverse modes of membrane perturbation, PhD Thesis, Graz University of Technology, Austria.

Sevcsik, E., Pabst, G., Jilek, A., and Lohner, K. (2007), How lipids influence the mode of action of membrane-active peptides, *Biochim. Biophys. Acta* 1768, 2586–2595.

Sevcsik, E., Pabst, G., Richter, W., Danner, S., Amenitsch, H., and Lohner, K. (2008), Interaction of LL-37 with model membrane systems of different complexity: Influence of the lipid matrix, *Biophys. J.* 94, 4688–4699.

Shai, Y. (2002), Mode of action of membrane active antimicrobial peptides, *Biopolymers* 66, 236–248.

Siegel, D. P. and Banschbach, J. (1990), Lamellar/inverted cubic (Lalpha/QII) phase transition in N-methylated dioleoylphosphatidylethanolamine, *Biochemistry* 29, 5975–5981.

Staudegger, E., Prenner, E. J., Kriechbaum, M., Degovics, G., Lewis, R. N., McElhaney, R. N., and Lohner, K. (2000), X-ray studies on the interaction of the antimicrobial peptide gramicidin S with microbial lipid extracts: Evidence for cubic phase formation, *Biochim. Biophys. Acta* 1468, 213–230.

Steim, J. M., Tourtellotte, M. E., Reinert, J. C., McElhaney, R. N., and Rader, R. L. (1969), Calorimetric evidence for the liquid-crystalline state of lipids in a biomembrane, *Proc. Natl. Acad. Sci. USA* 63, 104–109.

Thennarasu, S., Huang, R., Lee, D. K., Yang, P., Maloy, L., Chen, Z., and Ramamoorthy, A. (2010), Limiting an antimicrobial peptide to the lipid-water interface enhances its bacterial membrane selectivity: A case study of MSI-367, *Biochemistry* 49, 10595–10605.

Thennarasu, S., Lee, D. K., Tan, A., Prasad, K. U., and Ramamoorthy, A. (2005), Antimicrobial activity and membrane selective interactions of a synthetic lipopeptide MSI-843, *Biochim. Biophys. Acta* 1711, 49–58.

Wang, G., Li, X., and Wang, Z. (2009), APD2: The updated antimicrobial peptide database and its application in peptide design, *Nucleic Acids Res.*, 37, D933–D937.

Wiener, M. C. and White, S. H. (1992), Structure of a fluid dioleoylphosphatidylcholine bilayer determined by joint refinement of X-ray and neutron diffraction data. III. Complete structure, *Biophys. J.* 61, 437–447.

Willumeit, R., Kumpugdee, M., Funari, S. S., Lohner, K., Navas, B. P., Brandenburg, K., Linser, S., and Andra, J. (2005), Structural rearrangement of model membranes by the peptide antibiotic NK-2, *Biochim. Biophys. Acta* 1669, 125–134.

Wimley, W. C. (2010), Describing the mechanism of antimicrobial peptide action with the interfacial activity model, *ACS Chem. Biol.* 5, 905–917.

Zasloff, M. (2002), Antimicrobial peptides of multicellular organisms, *Nature* 415, 389–395.

Zweytick, D., Deutsch, G., Andra, J., Blondelle, S. E., Vollmer, E., Jerala, R., and Lohner, K. (2011), Studies on lactoferricin-derived *Escherichia coli* membrane-active peptides reveal differences in the mechanism of N-acylated versus nonacylated peptides, *J. Biol. Chem.* 286, 21266–21276.

Zweytick, D., Tumer, S., Blondelle, S. E., and Lohner, K. (2008), Membrane curvature stress and antibacterial activity of lactoferricin derivatives, *Biochem. Biophys. Res. Commun.* 369, 395–400.

CHAPTER 10

Kinetics of the Interaction of Amphiphiles with Lipid Bilayers Using ITC

Patrícia A. T. Martins and Maria João Moreno

CONTENTS

10.1 INTRODUCTION

10.1.1 Topology of the Lipidic Phase

Liposomes are closed bilayers that encapsulate small volumes of aqueous phase. Depending on their size and method of preparation, they may be unilamellar or may present several bilayers with thin layers of aqueous media between them.

In studies of the interaction of solutes with lipid bilayers by isothermal titration calorimetry (ITC), preformed liposomes are usually added to the solute, and only the lipid in the outer leaflet of the external bilayer is directly accessible to the solute. Depending on the rates of insertion/desorption and translocation, the solute may or may not equilibrate with all the lipidic phase. The relevant parameter to quantitatively characterize the affinity of the solute to the lipid bilayers is, therefore, the concentration of accessible lipid, not the total lipid concentration.

For unilamellar vesicles with a diameter equal to or larger than 100 nm (large unilamellar vesicles [LUVs]), the fraction of lipids in the outer leaflet is nearly 0.5. When the diameter is smaller (small unilamellar vesicles [SUVs]), the surface of the outer leaflet is significantly larger than that of the inner leaflet, and therefore the fraction of lipids in contact with the external aqueous phase is larger than 0.5. The fraction of lipids in the outer leaflet of SUVs may be estimated from Equation 10.1; SUVs with a radius of 20 nm have typically 60% of the lipid in the outer leaflet.

$$S^{out} = 4\pi r^2; \quad S^{in} = 4\pi (r - h)^2 \tag{10.1}$$

where:

S^{out} is the surface of the outer leaflet

S^{in} is the surface of the inner leaflet

h is the thickness of the bilayer (typically 4 nm)

Depending on the method of preparation and on the lipids used, liposomes may present some degree of multilamellarity. In this case, before using Equation 10.1, it is necessary to characterize the fraction of lipid in the inner layers (fraction of multilamellarity). Typically, 100 nm LUVs prepared by extrusion are essentially unilamellar when they contain significant fractions of charged lipids but have nearly 10% of the lipid in inner bilayers when prepared from neutral lipids (Cardoso et al. 2011; Martins et al. 2012). The degree of multilamellarity tends to increase with the total lipid concentration and with the increase in the pore size used at the extrusion step. There are several distinct protocols followed by different groups to overcome the problem of multilamellarity, including cycles of freeze and thaw and extrusion through progressively smaller pore sizes, but a systematic evaluation of this problem and its possible solutions is not available in the literature. The fraction of lipid in inner bilayers may be evaluated through the reaction of dithionite with lipids bearing the fluorescent group 7-nitrobenz-2-oxa-1,3-diazol-4-yl (NBD) (McIntyre and Sleight 1991) incorporated in the liposomes at small molar fractions (typically NBD-1,2-dimyristoyl-sn-glycero-3-phosphoethanolamine (NBD-DMPE) at 0.1 mol%) (Martins et al. 2012). This lipid needs to be added to the liposomes in the early stages of the preparation, and this method cannot, therefore, be followed to characterize previously prepared liposomes. A related method has been developed recently by us that uses the fluorescent amphiphile NBD-C_{16} incorporated in the liposomes at a molar percentage of 0.1% (Martins et al. 2012). This fluorophore is significantly more soluble in the aqueous phase and equilibrates with all the lipid compartments during overnight incubation at temperatures above the lipid phase transition. It may therefore be added to preformed liposomes, allowing their characterization *a posteriori*. When dithionite is added to preformed liposomes containing NBD-C_{16}, there is a three-phase decrease in fluorescence: (i) fast reaction with the fluorophore in the outer leaflet of the external bilayer; (ii) translocation of NBD-C_{16} from the inner into the outer leaflet of the external bilayer with concomitant reaction with dithionite; and (iii) a much slower fluorescence decrease due to reaction with the fluorophore initially in internal bilayers, resulting from either permeation of dithionite through the external bilayer (Moreno et al. 2006) or equilibration of the fluorophore with the external bilayer (Martins et al. 2012).

10.1.2 Interaction of Solutes with Membranes: Insertion/ Desorption and Translocation

Lipid bilayers are very heterogeneous media with regions of very distinct polarity and density. A polar surface is in contact with the aqueous phase, while the bilayer interior is formed by nonpolar hydrocarbon. The density at the polar interface formed by the hydrated lipid headgroup is higher than that of bulk water, while it is lower at the center of the lipid bilayer and intermediate in the polymer-like region between the two (Marrink and Berendsen 1994). In addition to this transversal heterogeneity, membranes composed of several lipid types may show lateral phase separation, where the size, dynamics, and properties of the distinct phases depend on the lipids present (Simons and Vaz 2004). When considering the interaction of solutes with lipid bilayers, it is necessary to take this heterogeneity into account. At the equilibrium position in the membrane, the solute may be adsorbed at the bilayer/water interface (charged and very polar solutes), embedded in the nonpolar center of the bilayer (nonpolar solutes), or inserted in the bilayer at the interface between the polar headgroups and the hydrocarbons (amphiphilic or easily polarized solutes). The kinetics of interaction of the solutes with the membranes depends on their equilibrium position and on the energy barriers in the pathway. The general kinetic scheme for interaction with LUVs is shown in Equation 10.2:

$$S_W \underset{k_-}{\overset{k_+^{LUV}[LUV]}{\rightleftarrows}} S_L^o \overset{k_f}{\longleftrightarrow} S_L^i \tag{10.2}$$

where:

S_W is the solute in the aqueous phase
S_L^o is the solute interacting with the outer leaflet of the lipid bilayer
S_L^i is the solute interacting with the inner leaflet of the lipid bilayer
k_+ is the rate constant for association with the lipid bilayer
k_- is the rate constant for dissociation from the lipid bilayer
k_f is the rate constant for translocation through the lipid bilayer
$[LUV]$ is the concentration of liposomes

This scheme is not valid for SUVs, as it assumes the same rate of translocation in both directions, or for giant unilamellar vesicles (GUVs), because it neglects the aqueous media inside the liposomes.

The rate constant for the equilibration of the solute with the outer leaflet of the liposome (k) is dependent on the rates of both association and dissociation (Equation 10.3)

$$k = k_- + k_+^{LUV}[LUV] \tag{10.3}$$

while the equilibration between both leaflets is only dependent on k_f (Cardoso et al. 2011). The overall equilibration with the lipid bilayer depends on all three rate constants and may be calculated through the numerical integration of the differential equations obtained from the kinetic scheme (Equation 10.4):

$$\frac{d[S]_w}{dt} = k_- [S]_L^o - k_+^{LUV} [LUV][S]_w$$

$$\frac{d[S]_L^o}{dt} = k_+^{LUV} [LUV][S]_w - k_- [S]_L^o + k_f \left([S]_L^i - [S]_L^o\right)$$

$$\frac{d[S]_L^i}{dt} = k_f \left([S]_L^o - [S]_L^i\right) \tag{10.4}$$

When the rates of insertion/desorption and translocation differ by at least one order of magnitude, the set of differential equations may be simplified, and the association of the solute with the membrane is well described by a monoexponential variation with the characteristic rate constant (β) given by the analytical solutions shown in Equation 10.5:

$$\beta = \frac{k_-}{2} + k_+^{LUV}[LUV] \quad \text{if } k_f \gg k \quad \text{(fast translocation)}$$

$$\beta = k_- + k_+^{LUV}[LUV] \quad \text{if } k_f \ll k \quad \text{(slow translocation)} \tag{10.5}$$

The lipidic phase in Equation 10.5 is considered with its actual topology, fragmented in liposomes (LUVs). This is important when characterizing the rate of association, because if the rate is close to diffusion controlled, it is affected by the state of division of the lipid phase. It has, however, been shown that the rate of association of amphiphiles with lipid bilayers is much slower than diffusion controlled, especially for the case of membranes in the liquid-ordered phase (Abreu et al. 2004; Cardoso et al. 2011; Estronca et al. 2007; Sampaio et al. 2005). When this is the case, the state of division of the lipidic phase does not affect the rate of association and it is only necessary to consider the total amount of lipidic phase accessible to the solute (Estronca et al. 2014; Filipe et al. 2014). In this case, the bimolecular rate constant for association with the membrane may be replaced by

$$k_+^{LUV}[LUV] = k_+^L [L]^* \tag{10.6}$$

where:

k_+^L is the rate of association calculated with respect to the concentration of lipid in direct contact with the aqueous phase where the solute is added

$[L]^*$ is half the total lipid for the case of LUVs

The equilibrium association with the LUVs is given by the ratio of the association and dissociation rate constants (Equation 10.7):

$$K_{LUV} = \frac{k_+^{LUV}}{k_-} \tag{10.7}$$

This equilibrium constant is related to the constant expressed in terms of concentration of accessible lipid (K_L) and to the partition coefficient (K_P) by Equation 10.8:

$$K_L = \frac{k_+^L}{k_-} = \frac{K_{LUV}}{n_{LUV}^{L^*}}$$

$$K_P = \frac{k_+^{V_L}}{k_-} = \frac{K_L}{\overline{V_L}} \tag{10.8}$$

where:

$n_{LUV}^{L^*}$ is the number of lipid molecules accessible to the external aqueous media per LUV

$k_+^{V_L}$ is the pseudo-first-order rate constant for insertion (expressed in terms of fractional volume of the lipidic phase accessible to the solute in the aqueous phase)

$\overline{V_L}$ is the molar volume of the lipids in the bilayer

The association of very polar and charged solutes with lipid bilayers is usually a fast process that is beyond the time resolution of the ITC technique. Their translocation through the bilayer involves the dissolution of the polar groups in the center of the bilayer (or the formation of pores) and occurs very slowly, often being outside the useful time window of the ITC as well. Nonpolar solutes diffuse very quickly through the hydrocarbon portion of the bilayer. High energy barriers are also not expected at the bilayer interface, as those solutes must have a small size in order to be soluble in the aqueous phase. Amphiphilic and strongly polarizable solutes are, however expected, to have intermediate rates of interaction with lipid bilayers, and this may be characterized directly by ITC. This chapter shows examples of the use of the ITC technique to characterize the rate of translocation of amphiphilic solutes through lipid bilayers in the liquid-disordered state.

In the kinetic schemes and equations considered in this section, it is assumed that the association with the membranes follows a simple partition between two phases (aqueous media and membrane). This is a good approximation for situations in which the local concentration of solute in the membrane is low (less than a few percent), and therefore there is little perturbation of the lipidic phase due to the presence of the solute. If this is not the case, the properties of the membrane change as it is loaded with the solute, and the rate constants for the association/dissociation and translocation steps become dependent on the local concentration of the solute, leading to complicated kinetics.

10.1.3 Effects of Solute Concentration

The kinetics of interaction of solutes with lipid membranes and the distribution attained at equilibrium may depend significantly on the local concentration of solute, and low solute concentrations are therefore recommended. On the other hand, a good signal-to-noise ratio is necessary to follow kinetics with the ITC. To optimize the conditions to use in the ITC, it is therefore important to know the maximum solute concentration in the membrane that does not lead to significant perturbation.

For charged solutes, the first perturbation observed is due to the effects on the electrostatic surface potential of the lipid bilayer. The effect on the surface potential of the membrane and consequently on the observed partition coefficient may be calculated using the Gouy–Chapman theory (Moreno et al. 2010 and references cited therein) and is given in Equations 10.9 and 10.10. At low values of ionic strength (10 mM NaCl), a local concentration of 1 mol% of solutes with a global charge equal to −1 and occupying a surface area similar to that of the phospholipids (0.63 nm² [Smaby et al. 1997]) leads to a surface potential (ψ_0) equal to −10 mV. At this surface potential, the observed partition coefficient $\left(K_P^{obs}\right)$ is decreased by 34% as compared with the intrinsic partition coefficient (K_P). A local concentration of 5 mol% leads to $\psi_0 = -46$ mV and to a reduction of 84% in the observed partition coefficient: $K_P^{obs} = 0.16 K_P$.

$$\sigma = \frac{\psi_0}{|\psi_0|} \sqrt{2000 \varepsilon RT \sum_i C_i \left(e^{-z_i F \psi_0 / RT} - 1 \right)}$$

$$\sigma = e_0 \frac{\sum\limits_j z_j \left(n_j / n_L \right)}{\sum\limits_j A_j \left(n_j / n_L \right)} \tag{10.9}$$

where:

C_i is the concentration of the specie i in the aqueous phase, with charge z_i

n_j is the number of moles of specie j in the membrane, with charge z_j and molar area A_j.

$$K_P^{obs} = K_P e^{-z F \psi_0 / RT} \tag{10.10}$$

These electrostatic effects are much less pronounced at the ionic strength usually considered as physiologic (0.15 M univalent salts). In these conditions, 5 mol% charged solute leads to $\psi_0 = -13$ mV and $K_P^{obs} = 0.58 K_P$.

The rate of desorption is expected to be faster when the electrostatic repulsion between the lipid bilayer and the solute is significant. The rate of insertion may also be affected, but this depends strongly on whether or not the insertion is close to diffusion controlled. The presence of an electrostatic potential on the surface may also influence the rate of translocation, as it leads to an increase in the area per lipid and could affect the equilibrium location of the solute associated with the membrane (Martins et al. 2012).

At local concentrations above 5 mol%, most solutes significantly affect the properties of the lipid bilayers irrespective of their effects on the electrostatic properties (Martins et al. 2012; Moreno et al. 2010). This effect depends on the properties of the solute and the lipid bilayer, and may lead to a decrease or an increase in the equilibrium association and rate constants. A well-studied example is the association of bile salts with lipid bilayers. At low local concentrations (below 5 mol%), the partition coefficient decreases as the concentration of bile salt increases due to the negative charge imposed by the solute (Schubert and Schmidt 1988). At very high local concentrations, the observed partition coefficient increases despite the increase in the electrostatic potential, reflecting strong perturbations in the topology and properties of the bile salt/membrane system (Schubert and Schmidt 1988).

To characterize the kinetics of the interaction of solutes with lipid bilayers, it is therefore very important to use the smallest concentration of solute that leads to an acceptable signal-to-noise ratio. As a reference value for small amphiphiles in the neutral form or with a single charge at an ionic strength of 0.15 M, one may consider that a lipid to bound solute ratio of 25 does not lead to significant changes in the properties of the lipid bilayer and that above a ratio of 50, electrostatic effects do not have to be considered (Martins et al. 2012; Moreno et al. 2010). After the characterization of the interaction with the unperturbed membrane, it may be relevant to increase the solute concentration to evaluate its effects on the equilibrium and kinetic parameters.

10.2 HOW TO OBTAIN THE RATE OF TRANSLOCATION USING THE ITC

When unilamellar liposomes in the liquid-disordered phase are added to an aqueous solution containing small amphiphilic solutes in the monomeric form, the equilibration with the outer leaflet occurs within seconds or less, being followed by solute translocation and equilibration with the inner leaflet. This additional available lipidic phase leads to an increase in the amount of solute associated with the liposomes, which is seen in the thermogram as a slow variation in the heat evolved with the same sign as the first fast component (endothermic or exothermic).

The additional heat generated by equilibration with the inner leaflet depends on the extent of partition previously established with the outer leaflet (Figure 10.1).

At very small volumes of the lipidic phase, the amount of amphiphile that partitions into the bilayer is linearly dependent on the volume of this phase, and therefore the heat variation associated with the second (slower) step is equal to that associated with the first (fast) step. However, in this region of the titration curve, the amount of amphiphile that partitions

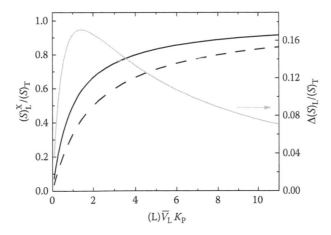

FIGURE 10.1 Variation of the fraction of amphiphile in the outer leaflet (- - -) and in both leaflets (black line) with the fractional volume of the lipid phase ($[L]\,\overline{V_L}$) multiplied by the partition coefficient (K_P). The variation in the fraction of amphiphile associated with the lipid bilayer due to partition into the inner leaflet ($\Delta[S]_L/[S]_T$) is also shown (gray line).

into the bilayer is very small, and therefore the heat evolved is small, with a poor signal-to-noise ratio. As the volume of the lipidic phase is increased, the fractional increase due to association with the inner leaflet decreases, because the titration curve is leaving the linear region. However, because the total amount of amphiphile that partitions into the membrane is increasing, the heat evolved due to partition into the inner leaflet is still increasing. At very large volumes of the lipidic phase, all amphiphile partitions into the outer leaflet of the bilayer, and, consequently, equilibration with the inner leaflet does not lead to any additional heat variation. The heat generated during the slow step is maximal at intermediate lipid concentrations that correspond to $\left([L]\overline{V_L}\right)K_P = \sqrt{2}$. The optimal lipid concentration to characterize translocation with this protocol $\left([L]_{trans}^{optim}\right)$ is therefore given by

$$[L]_{trans}^{optim} = \frac{\sqrt{2}}{K_P \overline{V}_L} \tag{10.11}$$

At this optimal lipid concentration, 60% of the amphiphile associates with the lipid bilayer: 40% in the first equilibration step (with the outer leaflet) and 20% additional amphiphile in the second step (due to equilibration with the inner leaflet).

An example of the application of this protocol to characterize the rate of translocation is given in Figure 10.2 for chlorpromazine (CPZ) in bilayers with 1-palmitoyl-2-oleoyl-sn-glycero-3-phosphocoline (POPC) and 1-palmitoyl-2-oleoyl-sn-glycero-3-phosphoserine (POPS) at the molar ratio of 9:1.

The time dependence of the heat evolved is given by the time derivative of the total amphiphile associated with the lipid bilayer (outer plus inner leaflet; differential equations given in Equation 10.4 now adapted for the use of lipid volume instead of liposome concentration) multiplied by the enthalpy variation of partition (Equation 10.12):

$$\frac{dQ}{dt} = \Delta H \frac{dnS_L}{dt} = \Delta H \left(\frac{dnS_L^o}{dt} + \frac{dnS_L^i}{dt}\right)$$

with

$$\frac{dnS_L^o}{dt} = K_P k_- [L]^* \overline{V}_L nS_W - k_- nS_L^o - k_f \left(nS_L^o - nS_L^i\right)$$

$$\frac{dnS_L^i}{dt} = k_f \left(nS_L^o - nS_L^i\right)$$

$$nS_W = nS_T - nS_L^o - nS_L^i \tag{10.12}$$

The rate constant of translocation may be obtained from the best fit of Equation 10.12 after the numerical integration of the differential equations for nS_L^o and nS_L^i. In most situations, the rate of desorption obtained from this fit does not reflect the kinetics of desorption, because it is limited by the time response of the equipment. This may be evaluated through comparison

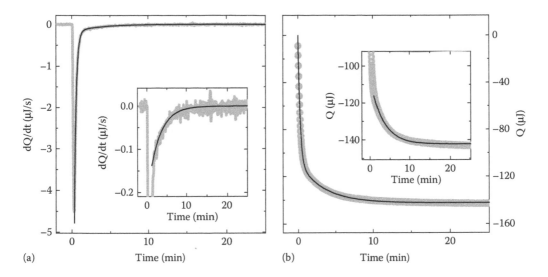

FIGURE 10.2 Time dependence of the heat evolved due to the addition of 10 μL of 15 mM POPC:POPS 9:1 (final [L] = 0.1 mM) to 10 μM CPZ, HEPES buffer 10 mM pH = 7.4 with 150 mM NaCl at 33°C. (a) Differential heat; (b) integrated heat. The experimental signal is in gray and the best fit in black. The inset shows a closer look at longer times and the best fit with the equations considering fast equilibration of CPZ between the aqueous phase and the outer leaflet of the liposomes.

with the rate constant obtained for a fast reaction such as NaOH/HCl neutralization or Ca^{2+}/EDTA binding, which leads to $k \sim 0.07$ s^{-1} for the VP-ITC instrument from MicroCal.

Although the addition of liposomes is not an instantaneous process—typically the addition of 10 μL of a liposome suspension takes 20 s (a rate of 0.5 μL s^{-1})—it is only possible to describe the kinetics of the interaction after the injection is finished. This is because the temporal evolution of the lipid concentration is affected by the efficiency of mixing and because the equilibration with the lipid added in the distinct steps during the injection is not independent. The best fit shown in Figure 10.2a was started at the end of the injection and considered the kinetics of interaction with the total amount of lipid injected.

The fitting of the total differential heat evolved is quite complicated, requires several assumptions, and may lead to a misinterpretation of the insertion/desorption rate constants obtained. One may also consider the analysis of only the slow component of the heat evolution ($t > 60$ s, four times the equipment characteristic time corresponding to 98% completion of the fast process if limited by the time response of the equipment) and assume fast equilibration of solute between the aqueous media and the outer leaflet of the liposomes (Equation 10.13).

$$nS_L^o = \left(nS_T - nS_L^i\right) \frac{K_P[L]^* \overline{V_L}}{1 + K_P[L]^* \overline{V_L}}$$

(10.13)

To perform the best fit of the time dependence of the differential heat evolved, it is necessary to know the total amount of solute associated with the lipidic phase at a given time, nS_L, which is given by

$$nS_L = nS_L^o + nS_L^i$$

$$= nS_T \frac{K_P [L]^* \overline{V_L}}{1 + K_P [L]^* \overline{V_L}} + nS_L^i \frac{1}{1 + K_P [L]^* \overline{V_L}} \tag{10.14}$$

The time dependence of the differential heat evolved is calculated using Equation 10.12, with the time derivative of nS_L obtained from Equation 10.14, and is given by

$$\frac{dQ}{dt} = \Delta H \frac{1}{1 + K_P [L]^* \overline{V_L}} \frac{dnS_L^i}{dt} \tag{10.15}$$

which is equal to

$$\frac{dQ}{dt} = \Delta H n S_T \frac{K_P [L]^* \overline{V_L}}{1 + K_P [L]^* \overline{V_L}} \frac{k_f}{1 + K_P [L]^* \overline{V_L}} \exp\left(-\frac{t}{\tau_f}\right); \quad \frac{1}{\tau_f} = k_f \frac{1 + K_P [L] \overline{V_L}}{1 + K_P [L]^* \overline{V_L}} \tag{10.16}$$

where τ_f is the observed characteristic time for the translocation step.

The quality of the best fit of Equation 10.16 to the slow component of dQ/dt is insensitive to the value of K_P, leading, however, to distinct values of k_f, because the relation between this parameter and the observed characteristic time is a function of K_P.

Alternatively, the heat evolved may be integrated and the rate of translocation obtained from the best fit of the numerical integration of Equation 10.12 (to describe the complete curve) or using the analytical solution for integration of Equation 10.16 (to describe only the slow component). The analytical solution for the heat evolved due to the slow translocation step is given by Equation 10.17 with τ_f as in 10.16:

$$Q(t) = \Delta H n S_L(t)$$

$$= \Delta H \left\{ nS_L(\infty) + \left[nS_L(0) - nS_L(\infty) \right] \right\} \exp\left(-\frac{t}{\tau_f}\right)$$

with

$$nS_L(\infty) = nS_T \frac{K_P [L] \overline{V_L}}{1 + K_P [L] \overline{V_L}}$$

$$nS_L(0) = nS_T \frac{K_P [L]^* \overline{V_L}}{1 + K_P [L]^* \overline{V_L}} \tag{10.17}$$

This procedure was followed to perform the best fit shown in Figure 10.2b, numerical integration of Equations 10.12 in the main panel and Equation 10.17 in the inset.

The four different approaches lead to similar, although not identical, values of the rate constants. In the example given in Figure 10.2, the value obtained globally for k_f was $3.5(\pm 0.3) \times 10^{-3} s^{-1}$. The rate of desorption obtained from the best fit of the complete curve was $3.1 \times 10^{-2} s^{-1}$ and 2.4×10^{-2} s^{-1} for the differential and integrated heat, respectively. This corresponds to a rate constant for equilibration between the aqueous phase and the outer monolayer near 0.07 s^{-1} $\left(k = k_- + \kappa_+^{V_L} [L] \overline{V_L} = k_- + K_p k_- [L] \overline{V_L} \right)$, which is the time resolution of the equipment.

The best fit of the integrated heat has the advantage that the relative weight of the heat evolved due to interaction with the outer and inner leaflets is more easily seen. From the two possible approaches, the complete curve or only the slow component, the numerical integration of Equation 10.12 to describe the complete curve may be preferable, because this solution is only valid when translocation is much slower than equilibration with the outer leaflet.

This protocol requires a very stable baseline to allow the time dependence of the heat evolved to be obtained with minimal baseline correction. In the experiment shown in Figure 10.2, the baseline was allowed to equilibrate for 1 h before the lipid was injected. It is also important to use the first injection for the analysis, because this leads to the maximal additional heat evolved due to equilibration with the inner leaflet. The diffusion of ligand into the syringe, or lipid into the cell, during the baseline equilibration is usually not significant due to the large size of the liposomes and the relatively large viscosity of the liposome suspension (the heat evolved in the first injection of a typical ligand/liposome titration is usually in agreement with the whole titration). Some exchange between the cell and syringe contents may, however, take place, leading to partition coefficients and enthalpy variations slightly different from those obtained by the usual complete titration. If the parameters are very different, however, this may indicate that the model is not adequately describing the system. The use of different temperatures, pH values, and concentrations (ligand and lipid) may be helpful to clarify the adequacy of the model and give confidence in the parameters obtained.

The quantitative characterization of the translocation rate constant given in this section can only be followed when the enthalpy variation on partition is large and the rate constant is intermediate. Considering $\Delta H = -22$ kJ/mol and $K_p = 1.3 \times 10^4$, which is typical for the interaction of CPZ with POPC:POPS 9:1 bilayers (Martins et al. 2012), the optimal lipid concentration in the cell is 0.11 mM (15 mM in the syringe for a 10 μL injection and a cell volume of 1,410 μL), and up to 10 μM solute in the cell may be used, leading to a ratio of 20 lipids per solute in the membrane (Moreno et al. 2010). The differential heat evolved at the observed characteristic time of translocation (τ_f) would be -205 nJ s^{-1} for $k_f = 5 \times 10^{-3} s^{-1}$ ($\tau_f = 2.5$ min), well above the noise level of 8 nJ s^{-1}, but only -42 nJ s^{-1} for $k_f = 1 \times 10^{-3}$ s^{-1} ($\tau_f = 12$ min), which corresponds to the lower limit accessible by this method. The time resolution of the equipment (15 s) dictates the upper limit for the characteristic time of translocation, which should be 10 times larger. The accessible range of k_f for typical solute/membrane systems is therefore $[10^{-3} \text{ s}^{-1}, 10^{-2} \text{ s}^{-1}]$. For rates of translocation outside this interval, the ITC equipment and protocols currently available allow only qualitative information to be obtained. One such method is the uptake and release protocol, developed by Heerklotz and coworkers (Tsamaloukas et al. 2007), which is explained in the next section together with some precautions required to avoid artifacts.

10.3 QUALITATIVE INFORMATION ON THE RATE OF TRANSLOCATION: UPTAKE AND RELEASE PROTOCOL

This protocol is based on the comparison of the heats evolved within the time of the ITC experiment when liposomes are added to the solute (uptake) and when liposomes previously equilibrated with solute are diluted (release). If the rate of translocation is fast, all lipid is available for equilibration during the duration of each step in the titration (typically 2–3 min), and the two titration curves should be symmetric. On the other hand, if translocation is very slow, only half of the lipid (outer leaflet) is available for solute equilibration in the uptake experiment, and half of the solute associated with the liposomes will equilibrate with the aqueous media outside the liposomes in the release experiment, leading to nonsymmetric titration curves. The results from both uptake and release titrations must be adjusted with the same set of parameters (K_p and ΔH) to obtain the equilibration factor γ, which is 1 for fast translocation and 0.5 for slow translocation with large unilamellar liposomes (Martins et al. 2012; Tsamaloukas et al. 2007). This factor is used to calculate the concentration of lipid available in the uptake and release protocol ($[L]^*$) and the concentration of solute available in the release protocol (nS_T^*) (Equation 10.18):

$$[L]^*(i) = [L]^*(i-1)\left\{1 - \frac{V(i)}{V_{cell}}\right\} + \gamma\left\{[L]^{Syr}\frac{V(i)}{V_{cell}}\right\}$$

$$nS_T^*(i) = nS_T^*(i-1)\left\{1 - \frac{V(i)}{V_{cell}}\right\} + \left\{\gamma[S]_L^{Syr} + [S]_W^{Syr}\right\}V(i) \qquad (10.18)$$

An example of the application of this protocol is shown in Figure 10.3 for CPZ in POPC:Chol:POPS (PcCPs) 6:3:1 (Figure 10.3a) and pure POPC bilayers (Figure 10.3b) at 25°C. The value obtained from the best global fit for the equilibration factor is 0.5 (or less) for the more ordered PcCPs membrane and slightly larger than 0.5 for the fluid POPC membrane. This indicates that only the lipid (solute) in the outer leaflet is available to equilibrate with the aqueous media outside the liposomes (slow translocation), although some translocation occurs during the ITC experiment for the more fluid POPC membrane.

It is important to note that the local concentration of solute in the membrane changes throughout the titration in the uptake protocol, while it is fixed in the release protocol. One should first optimize the conditions of the uptake experiments and perform the analysis assuming both slow and fast translocation ($\gamma = 0.5$ and 1). The concentration of lipid used in the syringe for the release experiment should lead to 90%–99% solute binding,

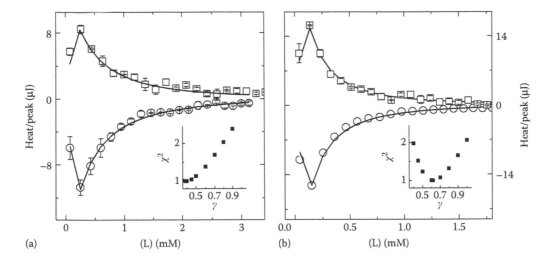

FIGURE 10.3 Uptake (circles) and release (squares) experiments at 25°C with membranes composed of (a) POPC:Chol:POPS (PcCPs, 6:3:1) and (b) POPC. The lines are the best fit of a simple partition with $\gamma = 0.5$. The inserts show the dependence of the square deviation between the global best fit and the experimental results (χ^2) as a function of γ. The concentration of CPZ in the uptake experiments was 25 and 5 μM, and the lipid concentration in the syringe was 25 and 15 mM, for PcCPs and POPC, respectively. The ratio of lipid to CPZ in the release experiments was 50:1 for both membranes.

and a ratio of lipid to bound solute (L:S) close to that obtained at 50% solute binding in the uptake experiment should be used (Coreta-Gomes et al. 2015). The results shown in Figure 10.3 following the uptake protocol lead to an L:S ratio near 40 and 80 when considering $\gamma = 0.5$ and 1, respectively. An L:S ratio equal to 50 was therefore selected for the release experiments. The lipid concentration in the syringe was 25 mM for the ternary lipid mixture and 15 mM for POPC, leading to 98 and 99% CPZ binding, respectively, in the release experiment.

Another important limitation of this protocol results from the fact that the temperature in the syringe is not controlled in the ITC equipment available. The tip of the syringe is inserted into the cell, and its temperature is equal to that of the cell, but the remaining content of the syringe is at a noncontrolled temperature, between that of the cell and room temperature. In most ITC applications, this is not a limitation, because the content of the syringe for the next injection is at the required temperature and equilibrates during the time between injections. However, when the syringe contains liposomes and solutes, the equilibration may take minutes to hours, and the system is halfway between the equilibrium distribution at room temperature and that attained at the temperature of the experiment. The results obtained at temperatures far from room temperature may therefore be misleading.

If all these precautions have been followed, the equilibration factor obtained gives a semiquantitative estimate of the rate of translocation. Values of γ near 0.5 indicate that negligible translocation occurred during the experiment (2–3 min for each injection, nearly 1 h for the whole titration), leading to a rate of translocation lower than $5 \times 10^{-4} s^{-1}$. If the value obtained for γ is much smaller than 0.5, this is an indication of a high degree of multilamellarity in the liposome sample. On the other hand, γ near 1 indicates that all lipid is available, and therefore k_f must be larger than $10^{-2} s^{-1}$. Intermediate values of γ indicate that some translocation occurs during the titration experiment. In this situation, it may be possible to obtain the rate of translocation using the methodology indicated in the previous section. Quantitative interpretations of intermediate values of γ should not be done, because under those conditions, the model is not expected to describe the titration (the extent of translocation varies as the titration proceeds, with heats being collected at injection i that reflect equilibrations started in previous injections).

The protocols described in this and the previous section may lead to the quantitative characterization of the rate of translocation or, at least, to a good estimate of this parameter. Together with the partition coefficient, this information may be used to calculate the rate of permeation through lipid bilayers and therefore to evaluate the contribution of passive processes to bioavailability and pharmacokinetics (Martins et al. 2012).

REFERENCES

Abreu, M. S. C., Moreno, M. J., and Vaz, W. L. C. 2004. Kinetics and thermodynamics of association of a phospholipid derivative with lipid bilayers in liquid-disordered and liquid-ordered phases. *Biophys. J.* 87:353–365.

Cardoso, R. M. S., Martins, P. A. T., Gomes, F., Doktorovova, S., Vaz, W. L. C., and Moreno, M. J. 2011. Chain-length dependence of insertion, desorption, and translocation of a homologous series of 7-nitrobenz-2-oxa-1,3-diazol-4-yl-labeled aliphatic amines in membranes. *J. Phys. Chem. B* 115:10098–10108.

Coreta-Gomes, F. M., Martins, P. A. T., Velazquez-Campoy, A., Vaz, W. L. C., Geraldes, C. F. G., and Moreno, M. J. 2015. Interaction of bile salts with model membranes mimicking the gastrointestinal epithelium: A study by isothermal titration calorimetry. *Langmuir* 31:9097–9104.

Estronca, L. M. B. B., Filipe, H. A. L., Salvador, A., Moreno, M. J., and Vaz, W. L. C. 2014. Homeostasis of free cholesterol in the blood: A preliminary evaluation and modeling of its passive transport. *J. Lipid Res.* 55:1033–1043.

Estronca, L. M. B. B., Moreno, M. J., and Vaz, W. L. C. 2007. Kinetics and thermodynamics of the association of dehydroergosterol with lipid bilayer membranes. *Biophys. J.* 93:4244–4253.

Filipe, H. A. L., Salvador, A., Silvestre, J. M., Vaz, W. L. C., and Moreno, M. J. 2014. Beyond Overton's rule: Quantitative modeling of passive permeation through tight cell monolayers. *Mol. Pharm.* 11:3696–3706.

Marrink, S. J., and Berendsen, H. J. C. 1994. Simulation of water transport through a lipid-membrane. *J. Phys. Chem.* 98:4155–4168.

Martins, P. T., Velazquez-Campoy, A., Vaz, W. L. C., Cardoso, R. M. S., Valerio, J., and Moreno, M. J. 2012. Kinetics and thermodynamics of chlorpromazine interaction with lipid bilayers: Effect of charge and cholesterol. *J. Am. Chem. Soc.* 134:4184–4195.

McIntyre, J. C., and Sleight, R. G. 1991. Fluorescence assay for phospholipid membrane asymmetry. *Biochemistry* 30:11819–11827.

Moreno, M. J., Bastos, M., and Velazquez-Campoy, A. 2010. Partition of amphiphilic molecules to lipid bilayers by isothermal titration calorimetry. *Anal. Biochem.* 399:44–47.

Moreno, M. J., Estronca, L. M. B. B., and Vaz, W. L. C. 2006. Translocation of phospholipids and dithionite permeability in liquid-ordered and liquid-disordered membranes. *Biophys. J.* 91:873–881.

Sampaio, J. L., Moreno, M. J., and Vaz, W. L. C. 2005. Kinetics and thermodynamics of association of a fluorescent lysophospholipid derivative with lipid bilayers in liquid-ordered and liquid-disordered phases. *Biophys. J.* 88:4064–4071.

Schubert, R., and Schmidt, K. H. 1988. Structural-changes in vesicle membranes and mixed micelles of various lipid compositions after binding of different bile-salts. *Biochemistry* 27:8787–8794.

Simons, K., and Vaz, W. L. C. 2004. Model systems, lipid rafts, and cell membranes. *Annu. Rev. Biophys. Biomol. Struct.* 33:269–295.

Smaby, J. M., Momsen, M. M., Brockman, H. L., and Brown, R. E. 1997. Phosphatidylcholine acyl unsaturation modulates the decrease in interfacial elasticity induced by cholesterol. *Biophys. J.* 73:1492–1505.

Tsamaloukas, A. D., Keller, S., and Heerklotz, H. 2007. Uptake and release protocol for assessing membrane binding and permeation by way of isothermal titration calorimetry. *Nat. Protoc.* 2:695–704.

III

Nucleic Acids and Proteins: Stability and Their Interactions with Ligands

CHAPTER 11

Calorimetry of DNA

Energetic Basis of the Double Helix

Peter L. Privalov

CONTENTS

11.1 INTRODUCTION

The discovery that all genetic information is stored in large polymeric molecules, deoxyribonucleic acids (DNA), was one of the greatest achievements of twentieth-century science (Avery et al. 1944). This polymer is arranged from the four simple bases, which appeared to be in surprising proportions in all DNAs (Chargaff 1950):

$$\left[\text{Adenine}\right]=\left[\text{Thymine}\right]\text{and}\left[\text{Guanine}\right]=\left[\text{Cytosine}\right] \tag{11.1}$$

The pairing of these bases suggests that DNA might be formed by two complementary polynucleotides. Based on the unpublished crystallographic data of Rosalind Franklin, Watson and Crick (1953) suggested that the two complementary strands of DNA form a double helix (Figure 11.1a).

The idea that the DNA structure represents a double helix gained wide acceptance, firstly, because it explained the mechanism of coding genetic information in the sequence of AT and GC base pairs, as a digital Morse code. Moreover, this explained the mechanism of genetic information replication: the two complementary strands dissociate, and new complementary strands are synthesized along each. This immediately raised a number

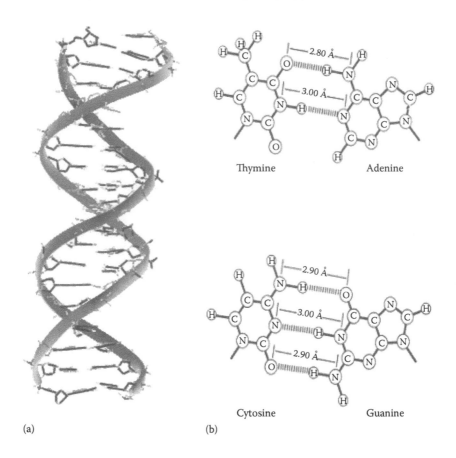

FIGURE 11.1   (a) The DNA double helix. (b) Hydrogen bonding of the complementary bases.

of questions: How much work is needed to separate the complementary strands of DNA? How stable is the DNA double helix? What are the forces stabilizing this construction?

According to the Watson–Crick model, an essential role in stabilization of the DNA double helix is played by the hydrogen bonds between complementary bases: two between adenine and thymine and three between cytosine and guanine (Figure 11.1b). One would expect, therefore, that an increase of the CG base pair content should lead to increased stability of the double helix (Watson et al. 1987). This is just what was observed: the thermal stability of the DNA duplex does indeed increase with an increase in CG content (Marmur and Doty 1962; Wartell and Benight 1985). This experimental fact was considered a strong argument for the correctness of the double helix Watson–Crick DNA model, according to which the hydrogen bonds between the complementary bases play a dominant role in stabilization of the double helix. However, thermodynamic verification was still required, that is, direct determination of the enthalpy and entropy of formation of various DNA duplexes differing in their content of CG and AT base pairs. This required calorimetric measurement of the heat effect of DNA duplex dissociation, particularly on heating, or measurement at fixed temperatures of the heats of complementary strand association into a double helix. However, realization of these experiments appeared to be far from simple, because (a) pure DNA was available in very limited amounts and (b) the process of dissociation–association of these polymeric molecules needs to be studied in highly dilute solutions to eliminate side effects resulting from their nonspecific interactions. Calorimetric study of DNA thermodynamics, therefore, required calorimeters significantly more sensitive than those existing at that time. Such an instrument, now called the differential scanning calorimeter (DSC), was specially designed for measuring the heat of DNA melting (Privalov et al. 1965). The isothermal titration calorimeter (ITC) was subsequently developed to an instrument of higher sensitivity, permitting measurements of the heat of association of the DNA complementary strands. (McKinnon et al. 1984). The latest modifications of these instruments, the Nano-DSC and the Nano-ITC, designed at the Johns Hopkins University, are now manufactured by TA Instruments (for the evolution of microcalorimetry, see Privalov 2012).

11.2 HEAT OF DNA MELTING

Figure 11.2 presents one of the first DSC recordings of the heat effect observed on heating the natural DNA of phage T_2 (Privalov et al. 1965). It shows extensive excess heat absorption in the temperature range from 75°C to 85°C, in which, according to optical studies, the DNA double helix breaks down and the complementary strands dissociate. Linear extrapolation of the initial heat capacity to higher temperatures appeared to show that the DNA dissociation proceeds without noticeable change of the heat capacity. According to Kirchoff's relation ($\Delta C_p = \partial \Delta H / \partial T$), this meant that the enthalpy of DNA unfolding should not depend on temperature. On the other hand, studies of DNA melting in solutions of different NaCl concentration showed that with increased ionic strength, the DNA melting temperature and the enthalpy of melting increase (Privalov et al. 1969). However, at that time, this change of enthalpy was attributed to the effect of salt, and it was widely accepted that the enthalpy of double helix formation does not depend on temperature.

With progress in the synthesis of polynucleotides, calorimetric studies of DNA thermodynamics have shifted completely to studying the melting of synthetic DNA and RNA duplexes. The main advantage of using long complementary synthetic polynucleotides made up of identical bases was that such duplexes melt on heating in a rather short temperature range (Figure 11.3). The sharp excess heat effect associated with dissociation of the complementary strands could, therefore, be easily recorded. However, an essential

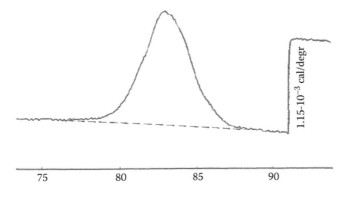

FIGURE 11.2 The very first DSC record of the heat effect at melting of 0.64 mg phage T_2 DNA in a 0.7 mL calorimetric cell. (From Privalov, P.L. et al., *Soviet Physics Journal*, 20, 1393–1396, 1965.)

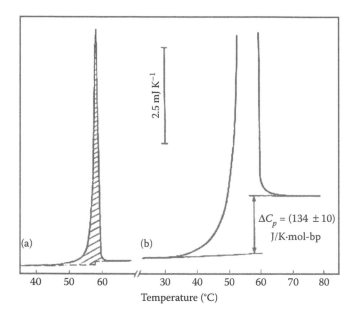

FIGURE 11.3 (a) DSC recording of poly(A)poly(U) melting at a 0.3 mM concentration and (b) a fragment of recording at 5.0 mM. (From Filimonov, V.V. and Privalov, P.L., *J. Mol. Biol.*, 122, 465–470, 1978.)

disadvantage of long synthetic polynucleotides was that they were not of unique length and did not necessarily form exact one-to-one duplexes. Furthermore, long polynucleotides could not be synthesized with all possible base sequences. Interest has, therefore, shifted to short synthetic oligonucleotides, which can be synthesized with any defined sequence of base pairs. However, such short DNA duplexes dissociate cooperatively on heating with smaller molar enthalpies and, correspondingly, over a broader temperature range, that is, their melting heat effect is much less pronounced (for reviews, see Breslauer 1978; Record et al. 1981; Filimonov 1986). In consequence, truly quantitative study of DNA thermodynamics became possible only with the appearance of the Nano-DSC and Nano-ITC instruments (Jelesarov et al. 1999).

11.3 SHORT DNA DUPLEXES

Figure 11.4a shows the Nano-DSC recordings of the heat effects on heating and subsequent cooling of a 12 bp DNA duplex, arranged from four (CGC/GCG) triplets, demonstrating the perfect reversibility of the temperature-induced processes of its dissociation/association: the excess heats of these two processes appear as mirror images. It is notable that linear extrapolation of the initial apparent heat capacity function to the higher temperatures implies that, although dissociation of strands proceeds with extensive excess heat absorption, it does not result in a noticeable heat capacity increment, that is, $\Delta C_p = 0$. It appears, therefore, that the heat effect of duplex unfolding/dissociation, determined as the area of the excess heat above the extrapolated initial heat capacity line, equals about 420 kJ/mol and, according to Kirchoff's relation ($\Delta C_p = \partial \Delta H / \partial T$), it does not depend on temperature. Thus, the enthalpy of formation of this duplex at room temperature should also be about −420 kJ/mol.

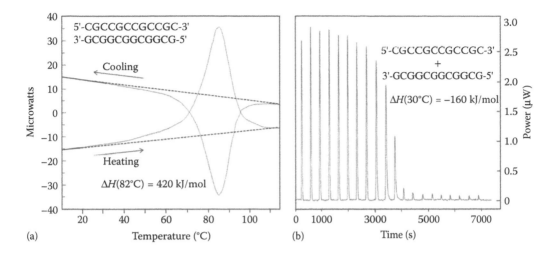

FIGURE 11.4 Original DSC recordings of (a) the heat effect on heating and subsequent cooling at a constant rate of 1 K/min of the 230 μM 12 CG DNA duplex and (b) ITC titration at 30°C of the 5′-CGCCGCCGCGC-3′ strand into the 3′-GCGGCGGCGGCG-5′ complementary strand by injection of 10 μL portions into the 1 mL cell at 30°C. All experiments in 150 mM NaCl, 5 mM Na phosphate, pH 7.4 solutions.

Figure 11.4b presents an original Nano-ITC recording of the heat effects on titration of one of the strands of the 12 bp CG duplex by its complementary strand at 30°C. According to this experiment, the enthalpy of duplex formation at that temperature is only −160 kJ/mol. Thus, the magnitude of this enthalpy measured by ITC at 30°C is in sharp contrast to the DSC-measured enthalpy of temperature-induced dissociation of this duplex on heating (or to its association on consequent cooling), taking place at around 85°C.

There could be several reasons for the observed discrepancy between the DSC- and the ITC-measured enthalpies: (a) the melting enthalpy of the duplex does, in fact, depend on temperature, that is, the assumption that DNA melting proceeds without heat capacity increment is incorrect; (b) the duplex formed at low temperature is not completely folded; (c) the separated strands have residual structure, so that to associate they must first unfold, and the heat of their unfolding contributes significantly to the observed heat effect of duplex formation.

11.4 TEMPERATURE-INDUCED DISSOCIATION OF DNA DUPLEXES

Considering the apparent heat capacity function of the DNA duplex (Figure 11.4a), one notes that it starts to increase from the very beginning of heating, in the temperature range from 0°C to 45°C, over which the duplex is generally regarded as still being fully folded. This increase of the heat capacity might result from increasing fluctuations of the duplex, particularly fraying of its ends, with a rise in temperature. If so, one would expect that the specific heat capacity of the DNA duplex, calculated per base pair, should depend on its length, increasing with a reduction in the number of base pairs in the duplex.

As shown in Figure 11.5, the initial partial molar heat capacity functions of three CG duplexes of different size (9, 12, and 15 base pairs) are different: the initial partial molar

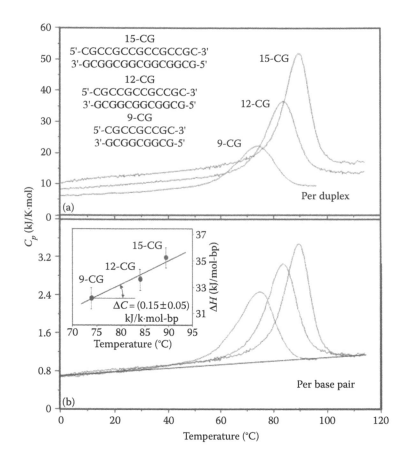

FIGURE 11.5 The partial heat capacity functions of the three considered CG DNA duplexes calculated (a) per mole of duplex (molar heat capacity) and (b) per mole of base pair (specific molar heat capacity), all measured at the same molarity, 230 μM, of the duplexes in 150 mM NaCl, 5 mM Na phosphate, pH 7.4 solutions. Inset: the dependence of the excess enthalpy on the transition temperature, the slope of which gives an estimate of ΔC_p.

heat capacity increases with the number of base pairs. However, recalculated per mole of base pair, the specific partial molar heat capacity functions of these three duplexes at temperatures below and above the heat absorption peaks appear to be very similar. It follows, therefore, that *contributions of the base pairs to the DNA heat capacities are additive*. Thus, one can conclude that *the increase of heat capacity of the duplex on heating results from intensified fluctuations of the whole of its body and not only of its ends*.

Figure 11.5 also shows that an increase in the number of base pairs results in increased duplex stability: longer duplexes dissociate at higher temperatures with larger excess heat effects. The calculated specific unfolding enthalpy per base pair plotted against the transition temperatures (inset in Figure 11.5) suggests that the averaged slope of this temperature dependence appears to be $\partial\Delta H/\partial T = (150 \pm 20)$ J/K mol bp for these CG duplexes. Bearing in mind Kirchoff's relation, $\partial\Delta H/\partial T = \Delta C_p$, one can conclude that *unfolding of these DNA duplexes proceeds with a heat capacity increment of about this magnitude*.

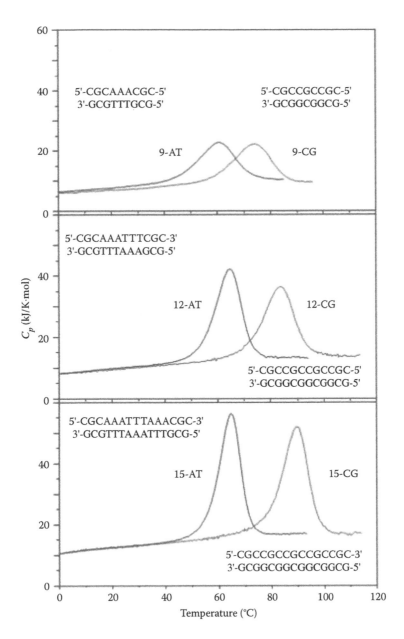

FIGURE 11.6 Comparison of the partial molar heat capacities of the 9, 12, and 15 base pair CG and AT duplexes, all at the identical molar concentration of 230 μM in 150 mM NaCl, 5 mM Na phosphate, pH 7.4 solutions.

A similar situation is observed with DNA duplexes containing (AAA/TTT) triplets in their middle part (Figure 11.6). Comparison of their heat capacity profiles with those of duplexes arranged from only CG base pairs shows that the initial and final partial molar heat capacities of all duplexes containing the same number of base pairs are indistinguishable, notwithstanding their very different thermostability. It follows that *the AT and GC base pair contributions to the total heat capacity of the duplex are similar and additive.* As expected, the presence of the AT base pair decreases the stability of the duplex significantly; however, the observation that *unfolding/dissociation of the less thermostable AT duplexes proceeds with a noticeably larger heat effect* was absolutely unexpected.

Since the specific enthalpy of duplex unfolding increases with an increase of its transition temperature (inset in Figure 11.5), one might be surprised by why the heat capacity increment on duplex unfolding is not observed in its heat capacity profile. It appears that the heat capacity increment on duplex unfolding is screened by the gradual effect of thermal energy accumulation by the duplex on heating. Thus, the apparent heat capacity function of the duplex at temperatures below the extensive heat absorption peak (associated with the cooperative separation of its strands) cannot be regarded simply as an intrinsic heat capacity of the fully folded duplex. However, precise determination of the excess enthalpy of duplex melting requires knowledge of the partial heat capacity of the fully folded duplex over the whole considered temperature range. Such a standard heat capacity function can be constructed using the heat capacity increment of duplex unfolding. Subtracting this heat capacity increment from the heat capacity of the 12 bp DNA duplex above 110°C (where it is completely unfolded) and connecting this point to the heat capacity of the duplex at 0°C (where it is assumed to be completely folded), we obtain the hypothetical heat capacity function for the fully folded duplex (Figure 11.7). Thus, all the area above this standard heat capacity function should be regarded as the *total heat of DNA melting.* Deconvolution analysis of the excess heat absorption profile shows that the total heat effect includes the heat accumulated in thermal fluctuations of the duplex, which gradually intensify with temperature, and the heat associated with cooperative dissociation of the DNA strands.

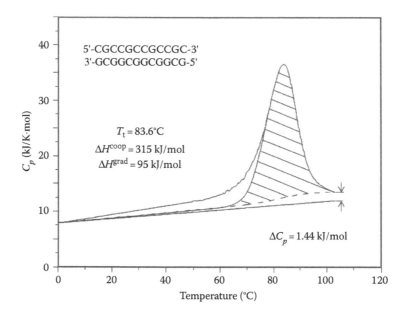

FIGURE 11.7 Deconvolution of the heat capacity profile of the 12 bp CG duplex. The hatched area shows the excess heat association with cooperative dissociation of the duplex. The open area between 0°C and ~70°C represents the gradual accumulation of heat.

11.5 HEAT OF COMPLEMENTARY STRAND ASSOCIATION AT FIXED TEMPERATURES

The dependence of the enthalpy of DNA duplex formation on temperature can also be studied by measuring the heats of association of the complementary strands at various fixed temperatures using the Nano-ITC. The main problem in this case is that the oligonucleotides possess residual structure. Therefore, to associate into a duplex, the complementary oligonucleotides must first unfold, and the heat of their unfolding, which can be determined by DSC melting of the separated strands, should be taken into account. Figure 11.8 illustrates this procedure of correcting the ITC data for the contribution of residual structures in the separated oligonucleotides and also for premelting of the duplex (for details, see Privalov 2012; Jelesarov et al. 1999).

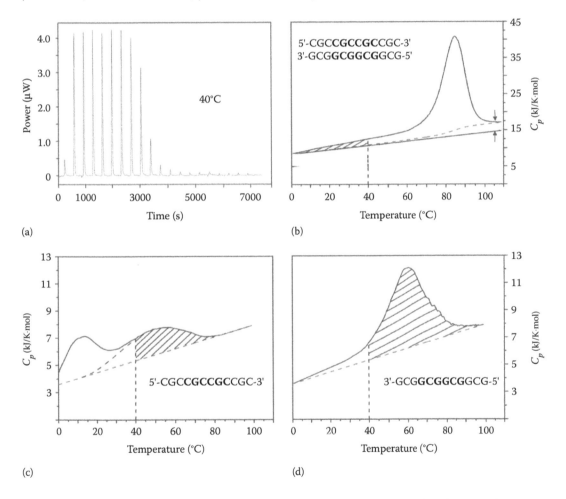

FIGURE 11.8 (a) ITC titration of the 5′-CGCCGCCGCCGC-3′ strand by the complementary 3′-GCGGCGGCGGCG-5′ strand at 40°C. (b) The DSC-measured partial molar heat capacity of the 12 bp CG duplex; the hatched area represents the enthalpy of the duplex premelting on heating to 40°C. (c) and (d) The partial heat capacities of the two isolated oligonucleotides; the hatched areas show the enthalpy of the residual structures in the single strands remaining at 40°C. All in 150 mM NaCl, 5 mM Na phosphate, pH 7.4 solutions.

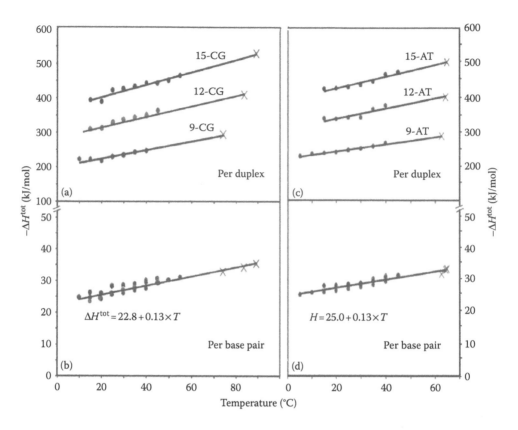

FIGURE 11.9   (a) and (c) Molar enthalpies of formation of the three CG and AT duplexes differing in number of base pairs, measured by ITC and corrected for the residual structures; (b) and (d) Specific molar (per base pair) enthalpies of these duplexes. Crosses indicate the total enthalpy values of formation of the considered duplexes estimated from the DSC-measured excess heat of duplex melting and attributed to the transition temperatures, T_t. All in 150 mM NaCl, 5 mM Na phosphate, pH 7.4 solutions. (From Vaitiekunas, P., et al., *Nucl. Acids Res.*, 43, 8577–8589, 2015.)

Following correction for residual structure in the two strands and for duplex premelting, the ITC-measured *total* enthalpies of formation of the 9, 12, and 15 base pair CG DNA duplexes are plotted against temperature in Figure 11.9. The crosses represent the DSC-measured *total* enthalpies at their melting temperatures. It is notable that the ITC-measured total enthalpies of association of the complementary strands, following the two corrections, lie precisely on lines that project to the DSC-measured total enthalpies of temperature-induced dissociation/association of the corresponding duplexes (indicated by crosses). Thus, calculated per base pair, all these enthalpies are expressed by the same linear function of temperature. The slope of these specific enthalpy functions for all the CG and AT duplexes is $\partial \Delta H / \partial T = \Delta C_p = (130 \pm 5)$ J/K mol bp, which is very close to the heat capacity increment determined to a first approximation from the DSC experiments (Figure 11.5). The latter value, (130 ± 5) J/K mol bp, is certainly more accurate, since it is determined using enthalpy values measured over a much wider temperature range. It is striking that this DNA heat capacity increment value is very close to the previously measured heat capacity increment on unfolding polymeric RNA, poly(AU)$_n$: $\Delta C_p = 134$ J/K mol bp (Figure 11.3).

11.6 CONTRIBUTION OF THE CG AND AT BASE PAIRS TO DNA DUPLEX STABILIZATION

As follows from Figure 11.9, knowing the heat capacity increment, ΔC_p, one can determine the total enthalpy of duplex dissociation/association at any other temperature:

$$\Delta H(T) = \Delta H(T_t) - \Delta C_p \times (T_t - T) \tag{11.2}$$

Using the thermodynamic characteristics of the cooperative unfolding of duplexes consisting only of CG base pairs, and assuming that contributions of all base pairs are additive, one can determine the enthalpy contribution of the CG base pair by dividing the molar values by the number of base pairs. In the case of the AT base pair, its enthalpic contributions to the duplex formation can be determined from the ITC-measured enthalpy of duplex formation with mixed content of GC and AT base pairs by firstly excluding the expected contribution of the CG base pairs:

$$\Delta H_{AT}(T) = \frac{\Delta H_{duplex}(T) - N_{CG} \cdot \Delta H_{CG}(T)}{N_{AT}} \tag{11.3}$$

Analysis of the enthalpies of formation of various duplexes presented in Figure 11.9 showed that the contribution of CG base pairing to the total enthalpy of duplex formation at standard temperature 25°C is $\Delta H_{CG}(\text{tot}) = -(27 \pm 1)$ kJ/mol and $\Delta H_{AT}(\text{tot}) = -(32 \pm 2)$ kJ/mol. It appears, therefore, that the larger stabilizing effect of CG base pairs results not from its larger enthalpy contribution, as was believed, but, perhaps, from its lower entropy contribution.

Unfortunately, the entropy contributions of base pairing cannot be determined from the ITC data on the heats of association of complementary oligonucleotides, because this process is not simple, as it includes unfolding of the oligonucleotide's residual structures. However, the entropy of duplex formation can be determined from the DSC data on the temperature-induced cooperative dissociation/association of the DNA duplex, which is a simple, reversible reaction, since the separated oligonucleotides do not form residual structures at temperatures of duplex dissociation. The entropy of cooperative dissociation of a heterodimer is

$$\Delta S^{coop}(H_t) = \frac{\Delta H_m^{tcoop}}{H_t} + R\ln\left(\frac{[N]}{2}\right) \tag{11.4}$$

Extrapolating this to standard temperature (25°C), one finds that the entropic contribution of CG base pairing to the cooperative phase of DNA dissociation is, indeed, significantly lower than that of AT base pairing: $\Delta S^{CG}(\text{coop}) = (40 \pm 1)$ J/K mol bp, while $\Delta S^{AT}(\text{coop}) = (80 \pm 3)$ J/K mol bp. The contributions of these base pairs to the enthalpy of cooperative phase transition are $\Delta H^{CG}(\text{coop}) = (18 \pm 1)$ J/K mol bp and $\Delta H^{AT}(\text{coop}) = (28 \pm 2)$ J/K mol bp, respectively. Bearing in mind that $\Delta G = \Delta H - T\Delta S$, one finds that the contribution of the CG base pair to duplex stabilization at 25°C is $\Delta G_{CG} = 6.4$ kJ/mol bp, while $\Delta G_{AT} = 4.7$ kJ/mol bp. Thus, *the CG-rich DNA duplex is more stable than the AT-rich duplex not because the enthalpy of CG dissociation is larger than that of ATs, but because the entropy of its dissociation is lower* (for details, see Vaitiekunas et al. 2015).

The question is then: What is the reason for such dramatic dominance of the contribution of the AT base pair over the CG, especially in entropy? It certainly cannot be a difference in hydrogen bonding between the complementary bases, since the AT pair has fewer such bonds than CG. It cannot be a difference in stacking interactions of the bases packed in the double helix, since these are not very different for these two pairs. Thus, it can only be a component external to the DNA: the water specifically bound by the AT base pair.

11.7 DNA HYDRATION

The larger negative enthalpy and especially the larger entropy of AT base pairing, in comparison with that of GC pairing, can be caused only by the water hydrating the AT pair: the water molecules that are fixed by this pair are observed crystallographically as a spine of ordered water molecules in the minor groove of AT-rich DNA (Drew and Dickerson 1981; Shui et al. 1998). Comparing the excess entropy contribution of the AT base pair (40 J/K mol bp) with the entropy of ice melting (22 J/K mol), one can conclude that the AT-fixed water molecule affects the state of the neighboring water molecules.

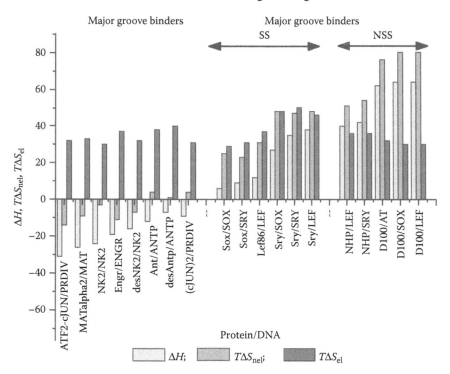

FIGURE 11.10 Enthalpies (ΔH) and entropy factors ($T\Delta S_{nel}$: nonelectrostatic; $T\Delta S_{el}$: electrostatic) of binding proteins to the minor and major grooves of their optimal and suboptimal DNAs. SS: sequence-specific; NSS: non-sequence-specific DNA binding domains. Each dataset is labeled with the name of the protein followed by the DNA designation. (From Privalov, P.L., et al., *Nucl. Acids Res.* 39, 2483–2491, 2011.)

Water ordering in the minor groove of AT-rich DNA also explains our earlier observation that protein binding to the AT-rich minor groove is an entropy-driven process, in contrast to protein binding to the major groove, which is usually enthalpy driven (Figure 11.10). The entropies of specific binding of various transcription factors, particularly the HMG boxes, to the minor groove of DNA at sites formed by five–seven AT base pairs amount to 170–240 J/K mol (Dragan et al. 2004; Privalov et al. 2007, 2011). Since the entropy of association of two kinetic units is negative, this suggests that the water that is removed from the minor groove of the DNA must be in an ordered state in this groove.

REFERENCES

Avery, O., MacLeod, C., and McCarty, M. 1944. Studies on the chemical nature of the substance inducing transformation of pneumococcal types. *J. Exp. Med.* 79:137–158.

Breslauer, K. J. 1986. Methods for obtaining thermodynamic data on oligonucleotide transitions. In Hinz, H.-J. (Ed.), *Thermodynamic Data for Biochemistry and Biotechnology*, 402–427. Springer: Berlin.

Chargaff, E. 1950. Chemical specificity of nucleic acids and mechanism of their enzymatic degradation. *Experientia* 6:201–209.

Dragan, A. I., Read, C. M., Makeyeva, E. N., Milgotina, E. I., Churchill, M. E. A., Crane-Robinson, C. and Privalov, P. L. 2004. DNA binding and bending by HMG boxes: Energetic determinants of specificity. *J. Mol. Biol.* 343:371–393.

Drew, H. R., and Dickerson, R. E. 1981. Structure of a B-DNA dodecamer. III. Geometry of hydration. *J. Mol. Biol.* 151:535–556.

Filimonov, V. V. 1986. The thermodynamics of conformational transitions in polynucleotides. In Hinz, H.-J. (Ed.), *Thermodynamic Data for Biochemistry and Biotechnology*, 377–401. Springer: Berlin.

Filimonov, V. V., and Privalov, P. L. 1978. Thermodynamics of base interaction in (A)n and (A.U)n. *J. Mol. Biol.* 122:465–470.

Jelesarov, I., Crane-Robinson, C., and Privalov, P. L. 1999. The energetics of HMG box interactions with DNA: Thermodynamic description of the target DNA duplexes. *J. Mol. Biol.* 294:981–995.

Marmur, J., and Doty, P. 1962. Determination of the base composition of deoxyribonucleic acid from its thermal melting temperature. *J. Mol. Biol.* 5:109–118.

McKinnon, I. R., Parody-Morreale, A., and Gill, S. J. 1984. A twin titration microcalorimeter for the study of biochemical reactions. *Anal. Biochem.* 139:134–139.

Privalov, P. L. 2012. *Microcalorimetry of Macromolecules*. John Wiley: New Jersey.

Privalov, P. L., Dragan, A. I., and Crane-Robinson, C. 2011. Interpreting protein/DNA interactions: Distinguishing specific from non-specific and electrostatic from non-electrostatic. *Nucleic Acids Res.* 39:2483–2491.

Privalov, P. L., Dragan, A. I., Crane-Robinson, C., Breslauer, K. J., Remeta, D. P., and Minetti, C. A. S. 2007. What drives proteins into the major and minor grooves of DNA? *J. Mol. Biol.* 365:1–9.

Privalov, P. L., Monaselidze, D. R., Mrevlishvili, G. M., and Magaldadze, V. A. 1965. Intramolecular heat of fusion of macromolecules. *Soviet Physics Journal* 20:1393–1396.

Privalov, P. L., Pritsyn, O. B., and Birshtein, T. M. 1969. Determination of stability of DNA double helix in an aqueous media. *Biopolymers* 8:559–571.

Record, M. T., Mazur, S. J., Melancon, P., Roe, J. H., Shaner, S. L., and Unger, L. 1981. Double helical DNA: Conformations, physical properties, and interactions with ligands. *Annu. Rev. Biochem.* 30:997–1024.

Shui, X., Sines, C. C., McFail-Isom, L., VanDerveer, D., and Williams, L. D. 1998. Structure of the potassium form of CGCGAATTCGCG: DNA deformation by electrostatic collapse around inorganic cations. *Biochemistry* 37:16877–16887.

Vaitiekunas, P., Crane-Robinson, C., and Privalov, P. L. 2015. The energetic basis of the DNA double helix: A combined microcalorimetric approach. *Nucl. Acids Res.* 43:8577–8589.

Wartell, R. M., and Benight, A. S. 1985. Thermal denaturation of DNA molecules: A comparison of theory with experiment. *Phys. Rep.* 126:67–107.

Watson, J. D., and Crick, F. H. C. 1953. A structure for deoxyribonucleic acid. *Nature* 171:964–967.

Watson, J. D., Hopkins, N. H., Roberts, J. W., Steiz, J. A., and Weiner A. M. 1987. *Molecular Biology of the Gene*, 4th Edition. Benjamin Cummings: Menlo Park, CA.

CHAPTER 12

Allostery and Cooperative Interactions in Proteins Assessed by Isothermal Titration Calorimetry

Adrian Velazquez-Campoy

CONTENTS

12.1 ALLOSTERY AND COOPERATIVITY IN PROTEINS

The traditional definition of *allostery*, the cooperative phenomenon in which the binding of a given ligand to a macromolecule is influenced by the binding of another ligand, is somewhat imprecise and vague. It is clear that in the absence of direct ligand–ligand interaction, the long-range cooperativity effect is made possible through conformational changes in the macromolecule. On the other hand, ligand binding does not induce such conformational change as a result of the macromolecule–ligand collision, but ligand binding shifts the conformational equilibrium toward certain conformational states. Therefore, the definition of *allostery* must include the conformational equilibrium of the macromolecule.

Proteins are highly dynamic molecules exhibiting a complex conformational landscape modulated by their interaction with the solvent and different effectors. That conformational landscape is constituted by an ensemble of states with populations or molar fractions dependent on their conformational Gibbs energy. The conformational Gibbs energy can be modulated by extrinsic factors such as temperature, pressure, pH, ionic strength, and ligands. Different conformational states of a protein interact with a given ligand with different binding affinities, and thus the binding of the ligand will reduce their overall Gibbs energy to different extents. The conformational equilibrium is then redistributed toward those conformational states able to bind that ligand. In addition, the different conformational states possess different biological activities (i.e., different abilities to interact with other biological partners), and this phenomenon may lead to increased or decreased activity, or even to the acquisition of a new biological activity. Therefore, the interaction of a protein with a ligand will determine its biological activity by influencing its ability to interact with other ligands. Thus, in a broad sense, allosterism is the modulation of the protein conformational equilibrium by ligand binding (Wyman 1948, 1963, 1964; Wyman and Allen 1951; Del Sol et al. 2009). For example, as illustrated in Figure 12.1, if, in the absence of ligand *A*, a protein is able to interact with ligand *B*, but ligand *A* binding displaces the protein conformational equilibrium by populating a certain protein conformational state that does not interact (or show a very low binding affinity) with ligand *B*, then the two ligands exhibit negative cooperativity; on the contrary, if ligand *A* binding displaces the protein conformational equilibrium by populating a certain protein conformational state that interacts better with ligand *B*, then the two ligands exhibit positive cooperativity. If ligands *A* and *B* are identical, it is called homotropic cooperativity; if ligands *A* and *B* are different, it is called heterotropic cooperativity. Although traditionally homotropic cooperativity has been directly associated with the oligomeric nature of proteins whose quaternary structure represents the structural and energetic basis for the cooperative effect, the fact is that many monomeric proteins display homotropic behavior (e.g., calmodulin, transferrin, and lectins).

A remarkable corollary to the previous discussion is that all proteins can be considered allosteric, because all proteins have a conformational landscape modulated by ligand binding (Gunasekaran and Nussinov 2004). Furthermore, all proteins show homotropic and heterotropic cooperativity. The most common case of homotropy is that of the influence of pH on the protein conformational equilibrium. Proteins contain functional ionizable groups that may be protonated or deprotonated. Protons are ligands binding to specific protein binding sites, and their binding or dissociation modulates the protein conformational equilibrium. The most common case of heterotropy is that of the influence of pH on the interaction of a protein with a ligand (e.g., substrate or cofactor) (Eftink and Biltonen 1980; Baker and Murphy 1996). In addition to possible direct interaction with the ligand, proton binding or dissociation modulates the protein conformational equilibrium by shifting the equilibrium toward certain states that may be interacting better or worse with the ligand.

An important point is that allosterism does not imply cooperativity, but cooperativity usually implies allosterism. A protein with a single ligand-binding site displaying a simplified conformational equilibrium with two conformational states having different ligand-binding affinities is an allosteric system, but there is no cooperative behavior, since cooperativity requires more than one binding site. Another important point is that binding

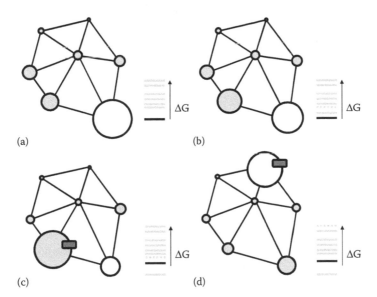

FIGURE 12.1 Schematic depiction of the conformational equilibrium of a protein and its modulation by ligand binding (allostery). (a) Conformational states are populated (size of the circle) according to their Gibbs energy. (b) A mutation in the protein may redistribute the conformational equilibrium by lowering the Gibbs energy of a certain partially folded conformation and increasing its relative population. This phenomenon may lead to increased or decreased biological activity or to the acquisition of a new biological activity in the protein if the conformational states differ in their biological activities. Similarly, the binding of a ligand redistributes the conformational equilibrium by lowering the Gibbs energy (in an amount equal to $-RT\ln(1 + \beta[L])$, where β is the binding affinity and $[L]$ is the free ligand concentration) of a conformational state that preferentially binds that ligand following a conformational selection (c) or an induced fit (d) scheme. The main difference between these two models is related to whether the states coexist in equilibrium (i.e., whether they are significantly populated according to the Gibbs energy gap between them); in conformational selection, the energy gap is small and both states are significantly populated, whereas in induced fit, the energy gap is large and the binding competent state is hardly populated. The redistribution of conformational equilibrium by ligand binding constitutes the structural and energetic basis for allostery.

cooperativity is not a conserved property associated with protein homology or structural or topological similarity. Homologous proteins sharing a common fold do not necessarily show the same cooperative behavior regarding ligand binding (Jobichen et al. 2009). On the other hand, different ligands binding to a given protein do not necessarily elicit the same cooperative effect (Williams et al. 2006; Schön et al. 2011; Liu et al. 2013; Sahun-Roncero et al. 2013; Courter et al. 2014).

In what follows, a review of the general formalism for homotropic and heterotropic cooperativity underlying allosteric interactions in proteins based on the binding polynomial will be given. Finally, two examples of complex systems, a pentameric protein exhibiting negative homotropy and a molecule exhibiting coupled homotropic and heterotropic effects, will be discussed.

12.2 BINDING POLYNOMIAL IN INTERMOLECULAR INTERACTIONS

A macromolecule interacting with several ligands is the paradigmatic case for protein function regulation. The mathematical description of such a system based on the binding partition function or binding polynomial was developed by Jeffries Wyman several decades ago (Wyman and Gill 1990 and references therein). Despite its simplicity, its easy generalization from the simplest scenario to the more complex ones, and its beauty, this formalism has been overlooked (mainly due to its apparent complexity) as the most appropriate framework for approaching any ligand-binding system, and in many cases, the mathematical description of a given system has been accomplished by elaborating an *ad hoc* approach within a limited scope. In addition, such formalism not only allows a straightforward description of the system on solid foundations, but also allows the application of general and powerful thermodynamic relationships and concepts conducive to physical, allosteric, and polysteric linkage. In this chapter, the formalism has been simplified; a thorough description can be found elsewhere (Gill 1989; Vega et al. 2015).

A macromolecule, P, with n binding sites for a given ligand, L, can be represented by a set of overall association constants $\{\beta_i, i = 0, \ldots, n\}$, each one associated with a different liganded state $\{PL_i, i = 0, \ldots, n\}$:

$$\beta_i = \frac{[PL_i]}{[P][L]^i} \tag{12.1}$$

The constant β_i is not related to specific binding sites, but with a subset of $\binom{n}{i}$ macromolecule–ligand complexes. Hence, β_i reflects the binding of i ligand molecules to any i binding sites among the n possible binding sites (no matter their binding affinity and the specific location of the binding sites) to form the PL_i complex. These association constants constitute a set of macroscopic phenomenological model-free parameters governing the binding equilibrium. Therefore, no mechanistic interpretation should be inferred from them. However, based on these association constants, appropriate binding models based on site-specific microscopic association constants can be derived.

The partition function of the system, Z, is operationally defined as

$$Z = \sum_{i=0}^{n} \frac{[PL_i]}{[P]} = \sum_{i=0}^{n} \beta_i [L]^i = \sum_{i=0}^{n} \mathcal{P}_i([L]) \tag{12.2}$$

an n-degree polynomial in $[L]$ (i.e., the binding polynomial), with positive coefficients equal to the overall association constants. The molar fraction of each liganded state, χ_i, is given by

$$\chi_i = \frac{[PL_i]}{[P]_T} = \frac{\beta_i [L]^i}{Z} = \frac{\mathcal{P}_i([L])}{Z} \tag{12.3}$$

where $[P]_T$ is the total concentration of the macromolecule.

The partition function contains all relevant information related to the binding; in particular, it may be used to calculate the average number of ligand molecules bound per macromolecule, $n_{LB} = [L]_{bound}/[P]_T$, the average excess molar enthalpy of the system, $\langle \Delta H \rangle$, and the average excess molar Gibbs energy of the system, $\langle \Delta G \rangle$ (Wyman 1964; Schellman 1975; Wyman and Gill 1990):

$$n_{LB} = \langle i \rangle = \left(\frac{\partial lnZ}{\partial ln[L]} \right)_{T,p,\dots} = -\frac{1}{R} \frac{\partial \left(\frac{\langle \Delta G \rangle}{T} \right)}{\partial ln[L]} \Bigg|_{[L],p,\dots} = \frac{\sum_{i=0}^{n} i\beta_i [L]^i}{\sum_{i=0}^{n} \beta_i [L]^i} = \sum_{i=0}^{n} i\chi_i$$

$$\langle \Delta H \rangle = -R \left(\frac{\partial lnZ}{\partial \frac{1}{T}} \right)_{[L],p,\dots} = \frac{\partial \left(\frac{\langle \Delta G \rangle}{T} \right)}{\partial \frac{1}{T}} \Bigg|_{[L],p,\dots} = \frac{\sum_{i=0}^{n} \Delta H_i \beta_i [L]^i}{\sum_{i=0}^{n} \beta_i [L]^i} = \sum_{i=0}^{n} \Delta H_i \chi_i$$

$$\langle \Delta G \rangle = -RTlnZ = -RTln \left(\sum_{i=0}^{n} \beta_i [L]^i \right) \tag{12.4}$$

where:
  ΔH_i is the overall enthalpy associated with the formation of complex PL_i
  T is the absolute temperature
  p is the pressure

Similar to the heat capacity (capability for storing and delivering thermal energy with the environment upon changes in temperature), which is defined by a homotropic second derivative of the partition function,

$$\langle \Delta C_P \rangle = \left(\frac{\partial \langle \Delta H \rangle}{\partial T} \right)_{[L],p,\dots} = -\frac{1}{T^2} \frac{\partial^2 \left(\frac{\langle \Delta G \rangle}{T} \right)}{\partial \left(\frac{1}{T} \right)^2} \Bigg|_{[L],p,\dots} = \frac{R}{T^2} \frac{\partial^2 lnZ}{\partial \left(\frac{1}{T} \right)^2} \Bigg|_{[L],p,\dots}$$

$$= \langle \Delta C_P \rangle_{int} + \frac{\langle \Delta H^2 \rangle - \langle \Delta H \rangle^2}{RT^2} \tag{12.5}$$

the binding capacity B_L is defined by a homotropic second derivative of the binding polynomial (Di Cera et al. 1988; Schellman 1990):

$$B_L = \left(\frac{\partial n_{LB}}{\partial ln[L]} \right)_{T,p,\dots} = \left(\frac{\partial^2 lnZ}{\partial ln[L]^2} \right)_{T,p,\dots} = -\frac{1}{R} \frac{\partial^2 \left(\frac{\langle \Delta G \rangle}{T} \right)}{\partial ln[L]^2} \Bigg|_{T,p,\dots} = \langle i^2 \rangle - \langle i \rangle^2 \tag{12.6}$$

and it reflects the capability of the macromolecule for storing and delivering chemical energy (ligand molecules) with the environment upon changes in the chemical potential of the ligand. A positive heat capacity guarantees thermodynamic stability in a system (from the second law of thermodynamics), and similarly, the binding capacity is also positive. The binding capacity is an index for the binding cooperativity, since it is a reflection of the statistical weight of intermediate ligation states in the equilibrium distribution, and it is related to the Hill coefficient, n_H (slope of the Hill plot), another index for quantifying binding cooperativity (Wyman and Gill 1990; Schellman 1990):

$$n_H = \left(\frac{\partial \ln \left(\frac{n_{LB}}{n-n_{LB}} \right)}{\partial \ln[L]} \right)_{T,p,\ldots} = \frac{\left(\frac{\partial n_{LB}}{\partial \ln[L]} \right)_{T,p,\ldots}}{n_{LB} \left(1 - \frac{n_{LB}}{n} \right)} = \frac{B_L}{B_L^0} \qquad (12.7)$$

which indicates that the Hill coefficient is the ratio between the binding capacity of the system, B_L, and the binding capacity for a reference macromolecule with n identical and independent binding sites, B_L^0. The maximal cooperativity ($n_H = n$) occurs in a macromolecule with n identical binding sites and no intermediate ligation states ($\beta_i = 0$ for $i = 1, \ldots, n - 1$), which corresponds to a binding polynomial given by

$$Z = 1 + \beta_n [L]^n \qquad (12.8)$$

Factorability of the binding polynomial is possible if the following relationship is fulfilled for every i ($i = 2, \ldots, n$) (Wyman and Philipson 1974; Wyman and Gill 1990):

$$\rho_i = \frac{\left(\dfrac{\beta_i}{\binom{n}{i}} \right)}{\left(\dfrac{\beta_{i-1}}{\binom{n}{i-1}} \right)^{i/i-1}} \leq 1 \qquad (12.9)$$

If all $\rho_i = 1$, the binding sites are identical and independent; if all $\rho_i \leq 1$, the binding sites are either nonidentical and independent or exhibit negative cooperativity (negative homotropy); and if some $\rho_i > 1$, some of the binding sites exhibit positive cooperativity (positive homotropy) (Wyman and Philipson 1974; Wyman and Gill 1990). Once the system is ascribed to any of these different scenarios using the model-free general formalism, particular models can be defined in terms of site-specific microscopic or intrinsic association constants and cooperativity constants.

The binding polynomial can be generalized for a macromolecule able to bind two ligands, A with n_A sites and B with n_B sites (Wyman and Gill 1990):

$$Z = \sum_{t=0}^{n_B} \sum_{s=0}^{n_A} \frac{[PA_s B_t]}{[P]} = \sum_{t=0}^{n_B} \sum_{s=0}^{n_A} \beta_{st} [A]^s [B]^t \tag{12.10}$$

where β_{st} is the overall association constant for the complex $PA_s B_t$. The constant β_{st} can be split into two factors: $\beta_{st} = \beta_{s0} \beta_{t/s}$, where β_{s0} is the overall association constant for $P + A_s \leftrightarrow PA_s$, and $\beta_{t/s}$ is the overall association constant for $PA_s + B_t \leftrightarrow PA_s B_t$. Thus, from the point of view of ligand A, the binding polynomial can be written as

$$Z = \sum_{s=0}^{n_A} \beta_{s0} [A]^s \sum_{t=0}^{n_B} \beta_{t/s} [B]^t = \sum_{s=0}^{n_A} \beta_{s0} Z_{s,B} [A]^s \tag{12.11}$$

where $Z_{S,B}$ is the binding subpolynomial for ligand B restricted to the subensemble of sA-ligated species. Furthermore, the binding polynomial can be renormalized by grouping all terms for complexes without A ligands (i.e., $Z_{0,B}$):

$$Z = \frac{\sum_{s=0}^{n_A} \beta_{s0} [A]^s \sum_{t=0}^{n_B} \beta_{t/s} [B]^t}{\sum_{t=0}^{n_B} \beta_{t/0} [B]^t} = \sum_{s=0}^{n_A} \beta_{s0} \frac{Z_{s,B}}{Z_{0,B}} [A]^s = \sum_{s=0}^{n_A} \beta_s^{\text{app}} [A]^s \tag{12.12}$$

where β_s^{app} is an apparent association constant for the subensemble of sA-ligated species. Renormalizing the binding polynomial implies changing the reference state P for the subensemble of states contained within the renormalization factor. From this,

$$\beta_s^{\text{app}} = \beta_{s0} \frac{Z_{s,B}}{Z_{0,B}}$$

$$\left(\frac{\partial \ln \beta_s^{\text{app}}}{\partial \ln [B]} \right)_{T, p, \ldots} = n_{s,BB} - n_{0,BB} = \Delta n_{s,BB}$$

$$RT^2 \left(\frac{\partial \ln \beta_s^{\text{app}}}{\partial T} \right)_{[B] p, \ldots} = \Delta H_s^{\text{app}} = \Delta H_{s0} + \langle H_{s,B} \rangle - \langle H_{0,B} \rangle \tag{12.13}$$

where:

$n_{s,BB}$ is the average number of ligand B molecules bound per macromolecule in the subensemble of sA-ligated species

$\Delta H_{s,0}$ is the binding enthalpy for the complex PA_s

$\langle \Delta H_{s,B} \rangle$ is the average binding enthalpy for ligand B in the subensemble of sA-ligated species

These are well-known chemical linkage relationships (e.g., pH dependency of ligand-binding parameters due to coupling of proton exchange at ionizable groups, or the effect of a ligand on the binding of another ligand) (Eftink and Biltonen 1980; Baker and Murphy 1996; Du et al. 2000; Velazquez-Campoy et al. 2006). From the binding polynomial average quantities can be calculated:

$$n_{AB} = \left(\frac{\partial \ln Z}{\partial \ln [A]} \right)_{T, p, [B], \ldots}$$

$$n_{BB} = \left(\frac{\partial \ln Z}{\partial \ln [B]} \right)_{T, p, [A], \ldots}$$

$$\langle \Delta H \rangle = RT^2 \left(\frac{\partial \ln Z}{\partial T} \right)_{p, [A], [B] \ldots}$$

$$\langle \Delta G \rangle = -RT \ln Z \tag{12.14}$$

The binding polynomial can also be generalized for a macromolecule exhibiting different conformations with different ligand-binding affinities, within the general allosteric model (Wyman and Gill 1990). According to this general allosteric model, the binding polynomial for a macromolecule with $m + 1$ conformations and n ligand-binding sites is given by

$$Z = \sum\nolimits_{s=0}^{m} \sum\nolimits_{i=0}^{n} \frac{[P_s L_i]}{[P]} = \sum\nolimits_{s=0}^{m} \sum\nolimits_{i=0}^{n} \frac{\beta_{si} [P_s][L]^i}{[P]} = \sum\nolimits_{s=0}^{m} \sum\nolimits_{i=0}^{n} \beta_{si} \gamma_s [L]^i \tag{12.15}$$

where:

β_{si} is the overall association constant for the complex $P_s L_i$

γ_s is the conformational equilibrium constant for the P_s conformation

Regrouping terms, the binding polynomial can be written as

$$Z = \sum_{s=0}^{m} \gamma_s Z_s$$

$$Z = \sum\nolimits_{i=0}^{n} \left(\sum\nolimits_{s=0}^{m} \beta_{si} \gamma_s \right) [L]^i \tag{12.16}$$

where Z_s is the binding subpolynomial restricted to the P_s conformation. Again, we can renormalize the binding polynomial:

$$Z = \frac{\sum_{s=0}^{m}\sum_{i=0}^{n}\beta_{si}\gamma_s[L]^i}{\sum_{i=0}^{n}\beta_{0i}[L]^i} = \frac{\sum_{s=0}^{m}\gamma_s\sum_{i=0}^{n}\beta_{si}[L]^i}{\sum_{i=0}^{n}\beta_{0i}[L]^i} = \sum_{s=0}^{m}\gamma_s\frac{Z_s}{Z_0} = \sum_{s=0}^{m}\gamma_s^{app}$$

$$Z = \frac{\sum_{s=0}^{m}\sum_{i=0}^{n}\beta_{si}\gamma_s[L]^i}{\sum_{s=0}^{m}\gamma_s} = \sum_{i=0}^{n}\frac{\sum_{s=0}^{m}\beta_{si}\gamma_s}{\sum_{s=0}^{m}\gamma_s}[L]^i = \sum_{i=0}^{n}\beta_i^{app}[L]^i$$

$$Z = \frac{\sum_{s=0}^{m}\sum_{i=0}^{n}\beta_{si}\gamma_s[L]^i}{\sum_{s=0}^{m}\gamma_s} = \frac{\sum_{s=0}^{m}\gamma_s\sum_{i=0}^{n}\beta_{si}[L]^i}{\sum_{s=0}^{m}\gamma_s} = \frac{\sum_{s=0}^{m}\gamma_s Z_s}{\sum_{s=0}^{m}\gamma_s} \tag{12.17}$$

where:

γ_s^{app} is the apparent conformational equilibrium constant for the P_s conformation

β_i^{app} is the apparent overall association constant in the subensemble of iL-ligated species

From this,

$$\gamma_s^{app} = \gamma_s\frac{Z_s}{Z_0}$$

$$\Delta G_s^{app} = \Delta G_s + \langle\Delta G_{s,L}\rangle - \langle\Delta G_{0,L}\rangle$$

$$\left(\frac{\partial\ln\gamma_s^{app}}{\partial\ln[L]}\right)_{T,p,\ldots} = n_{s,LB} - n_{0,LB} = \Delta n_{s,LB}$$

$$RT^2\left(\frac{\partial\ln\gamma_s^{app}}{\partial T}\right)_{[L],p,\ldots} = \Delta H_s^{app} = \Delta H_s + \langle\Delta H_{s,L}\rangle - \langle\Delta H_{0,L}\rangle \tag{12.18}$$

where:

$n_{s,LB}$ is the average number of ligand molecules bound per macromolecule in the subensemble of the P_s conformation

ΔH_s is the conformational enthalpy for the P_s conformation

$\langle\Delta G_{s,L}\rangle$ and $\langle\Delta H_{s,L}\rangle$ are the average ligand-binding Gibbs energy and enthalpy in the subensemble of the P_s conformation

Equations 12.17 and 12.18 contain well-known linkage relationships between conformational equilibrium and ligand binding (Wyman and Gill 1990; Straume and Freire 1992). Similarly,

$$\beta_i^{\mathrm{app}} = \frac{\sum_{s=0}^{m} \beta_{si}\gamma_s}{\sum_{s=0}^{m} \gamma_s}$$

$$RT^2 \left(\frac{\partial \ln \beta_i^{\mathrm{app}}}{\partial T} \right)_{[L],p,\dots} = \Delta H_i^{\mathrm{app}} = \left\langle \Delta H_{i,si} \right\rangle + \left\langle \Delta H_{s,si} \right\rangle - \left\langle \Delta H_{s,s0} \right\rangle \qquad (12.19)$$

where:

$\left\langle \Delta H_{i,si} \right\rangle$ is the average ligand-binding enthalpy in the conformational subensemble of iL-ligated species

$\left\langle \Delta H_{s,si} \right\rangle$ is the average conformational enthalpy in the conformational subensemble of iL-ligated species

By imposing certain constraints, the general allosteric model can be significantly simplified through elimination of many of the possible macromolecular states (Wyman 1972). The Monod–Wyman–Changeux (MWC) model corresponds to an oligomeric macromolecule consisting of n identical subunits behaving independently regarding ligand binding and undergoing a simultaneous concerted conformational change (i.e., no mixed conformational species) between two possible conformations, R and T, upon ligand binding, and the binding polynomial is given by (see Equations 12.16 and 12.17)

$$Z = \frac{1}{1+\gamma}\left(1+k_R[L]\right)^n + \frac{\gamma}{1+\gamma}\left(1+k_T[L]\right)^n = \sum_{i=0}^{n} \binom{n}{i}\left(\frac{1}{1+\gamma}k_R^i + \frac{\gamma}{1+\gamma}k_T^i \right)[L]^i$$

$$= \sum_{i=0}^{n} \beta_i^{\mathrm{app}}[L]^i \qquad (12.20)$$

where $\gamma = [T]/[R]$, and k_R and k_T are the site-specific microscopic association constants for the subunits in R and T conformation, respectively (Monod et al. 1965). Although the conformations R and T are not intrinsically cooperative regarding ligand binding, ligand binding drives a cooperative transition between the two conformations.

The Pauling–Koshland–Nemethy–Filmer (PKNF) model corresponds to an oligomeric macromolecule consisting of n subunits undergoing individual, sequential conformational changes between two conformations, R and T (with $k_T = 0$), upon ligand binding (with mixed conformational species, and $s = i$), and the binding polynomial is given by (see Equations 12.16 and 12.17)

$$Z = \sum_{i=0}^{n} \beta_{ii}\gamma_i[L]^i = \sum_{i=0}^{n} \beta_i^{\mathrm{app}}[L]^i \qquad (12.21)$$

where β_{ii} is the overall association constant for i ligand molecules to the macromolecule with i subunits in R conformation (Kohsland et al. 1966). The apparent association constant β_i^{app} contains the intrinsic association constant and several factors accounting for binding cooperative interactions and conformational changes. Unlike the MWC model,

PKNF model can describe negative cooperativity. In the case of a monomeric protein with a single ligand-binding site ($n = 1$), the system is still allosteric, because the macromolecule exhibits equilibrium between two conformational states with different ligand-binding affinities, but there is no cooperativity (see Equations 12.20 and 12.21):

$$Z = 1 + \frac{k_R + \gamma k_T}{1 + \gamma}[L] = 1 + \beta_i^{app}[L] \tag{12.22}$$

The binding equations are obtained from the binding polynomial, resulting in a combination of the chemical equilibrium equations and the mass conservation equations for both the macromolecule and the ligand:

$$[P]_T = [P]Z$$
$$[L]_T = [L] + [P]_T \frac{\partial \ln Z}{\partial \ln [L]} \tag{12.23}$$

or alternatively,

$$[P]_T = [P]Z$$
$$[L]_T = [L] + [P][L]\frac{\partial Z}{\partial [L]} \tag{12.24}$$

In the first method, Equation 12.23, a polynomial equation in $[L]$, is obtained:

$$[L] + [P]_T \frac{\partial \ln Z}{\partial \ln [L]} - [L]_T = 0 \tag{12.25}$$

whereas in the second method, Equation 12.24, a set of nonlinear equations in $[P]$ and $[L]$, is obtained:

$$[P]Z - [P]_T = 0$$
$$[L] + [P][L]\frac{\partial Z}{\partial [L]} - [L]_T = 0 \tag{12.26}$$

These equations can be solved analytically for simple systems (e.g., one or two classes of independent binding sites) or numerically for any system (e.g., using the Newton–Raphson or the bisection) (Press et al. 2007; Vega et al. 2015). Thus, solving for $[P]$ and $[L]$ for given values of $[P]_T$, $[L]_T$, and $\{\beta_i\}$, the concentration of each complex can be calculated (Equations 12.1 and 12.3):

$$[PL_i] = [P]_T \frac{\beta_i [L]^i}{Z} = \beta_i [P][L]^i \tag{12.27}$$

Both methods are equivalent; however, method 2 is more easily generalized (in particular, for polysteric systems, where ligand-binding effects are linked to association and dissociation of the macromolecule, and systems with more than one ligand).

12.3 ISOTHERMAL TITRATION CALORIMETRY: CHARACTERIZING COOPERATIVE INTERACTIONS

Considered a minor technique in the applied chemistry field some decades ago, isothermal titration calorimetry (ITC) enjoys nowadays a golden age with widespread use in molecular and structural biology labs, facilitated by technical developments and the availability of user-friendly commercial calorimeters. ITC has become the preferred technique for characterizing binding interactions. ITC is unique among the many tools intended to study binding in that it allows the simultaneous determination of the association equilibrium constant, the binding enthalpy, and the binding stoichiometry. The high sensitivity of modern calorimeters (absolute heat smaller than 1 μcal) makes possible estimating dissociation constants in the millimolar to nanomolar range. In addition, displacement assays offer the possibility to extend that range by several orders of magnitude. The practical use of ITC ranges from the study of the interaction between two natural binding partners, or the elucidation of binding cooperativity phenomena underlying allosteric regulation, to the optimization of lead compounds in drug discovery and development.

ITC is especially appropriate for studying allosteric and cooperative interactions because any alteration in the system (as it occurs in homotropic and heterotropic effects) will have an impact on the ligand-binding parameters. Usually, the binding Gibbs energy is rather insensitive to changes in the system, a reflection of the enthalpy–entropy compensation phenomenon (Cooper et al. 2001; Breiten et al. 2013); however, its first derivatives (enthalpy and entropy of binding) are quite sensitive to those changes (Hinz 1983). Thus, it is not unusual to find cooperativity Gibbs energies smaller than 5 kJ/mol, while the magnitude of their associated cooperativity enthalpies and entropies is much larger, in the order of tens of kilojoules per mole. Fortunately, in ITC it is possible to obtain information about those three potentials in a single experiment, and to better discriminate between different possible scenarios.

The direct observable in an ITC experiment is the heat per injection, which reflects the formation or dissociation of macromolecule–ligand complexes elicited by the addition of titrant solution. Once the concentrations of macromolecule–ligand complexes after each injection j, $[PL_i]_j$, are calculated, the heat effect associated with each injection, Q_j, can be calculated as a finite-difference or incremental quotient:

$$q_j = V_0\left([P]_{T,j}\langle\Delta H\rangle_j - [P]_{T,j-1}\langle\Delta H\rangle_{j-1}\left(1-\frac{v_j}{V_0}\right)\right)$$

$$Q_j = \frac{q_j}{v_j[S]_0} + q_d \tag{12.28}$$

where the heat effect has been normalized by the amount of titrant injected ($\Delta n_{LT}=v_j[S]_0$) and an adjustable parameter q_d accounting for a nonzero background injection heat due to titrant dilution or injection has been added. Finally, the experimental data (normalized heat per injection) are compared to the calculated values and the thermodynamic binding parameters ($\{\beta_i\}$, $\{\Delta H_i\}$) can be estimated through iterative nonlinear least-squares regression analysis.

From Equation 12.28, it is clear that, except in the case of a macromolecule with a single binding site or with a set of identical and independent binding sites, ITC does not provide a direct measurement of n_{LB} or B_L (Equations 12.4 and 12.6), since each binding site will contribute differently to the global heat signal (with an intrinsic interaction enthalpy and with a cooperativity enthalpy). However, from the estimated binding parameters, both n_{LB} and B_L can be calculated. There are other techniques for binding studies (e.g., dialysis and thin-layer optical method) that allow the direct determination of n_{LB} and B_L.

12.4 COOPERATIVE HOMOTROPIC INTERACTIONS: THE CASE OF NUCLEOPLASMIN BINDING HISTONES

Nucleoplasmin (NP) is a nuclear DNA chaperone participating in different activities: chromatin remodeling, nucleosome assembly, genome stability, ribosome biogenesis, DNA duplication, and transcriptional regulation (Laskey et al. 1978; Akey and Luger 2003; Prado et al. 2004). NP is a thermostable acidic protein with a pentameric structure; each of the five identical subunits shows an ordered region involved in the formation of the pentameric ring and a disordered region (tail) that is important for histone interaction (Dutta et al. 2001). NP is able to interact with different histones (e.g., H5 linker histone and H2A/H2B histone dimer). Experimental evidence provided by ITC and small-angle x-ray scattering (SAXS) (Taneva et al. 2009) and electron microscopy (Ramos et al. 2010) made it possible to establish the expected 1:5 NP–histone stoichiometry.

NP can be represented by five overall association constants, β_i, and five overall binding enthalpies, ΔH_i. Therefore, the binding polynomial is given by

$$Z = 1 + \beta_1 [L] + \beta_2 [L]^2 + \beta_3 [L]^3 + \beta_4 [L]^4 + \beta_5 [L]^5 = \sum_{i=0}^{5} \beta_i [L]^i \qquad (12.29)$$

The scheme corresponding to this binding polynomial is shown in Figure 12.2. According to the first method (Equations 12.23 and 12.25), the binding equations are expressed as follows:

$$[P]_T = [P]Z = [P] \sum_{i=0}^{5} \beta_i [L]^i$$

$$[L]_T = [L] + [P]_T \frac{\partial \ln Z}{\partial \ln [L]} = [L] + [P]_T \frac{\sum_{i=0}^{5} i\beta_i [L]^i}{\sum_{i=0}^{5} \beta_i [L]^i} \qquad (12.30)$$

from which a six-degree polynomial equation in $[L]$ is obtained, which can be solved numerically. On the other hand, according to the second method (Equations 12.24 and 12.26), the binding equations are expressed as follows:

	P	1	P	1
	PL	$5K[L]$	PL	$\beta_1[L]$
	$PL_{2,A}$	$5K^2 k_1 [L]^2$	PL_2	$\beta_2[L]^2$
	$PL_{2,B}$	$5K^2 k_1 k_2 [L]^2$		
	$PL_{3,A}$	$5K^3 k_1^2 k_2 [L]^3$	PL_3	$\beta_3[L]^3$
	$PL_{3,B}$	$5K^3 k_1^2 k_2^2 k_3 [L]^3$		
	PL_4	$5K^4 k_1^3 k_2^3 k_3^2 [L]^4$	PL_4	$\beta_4[L]^4$
	PL_5	$5K^5 k_1^4 k_2^5 k_3^5 [L]^5$	PL_5	$\beta_5[L]^5$

FIGURE 12.2 Scheme corresponding to the binding polynomial for a pentameric protein with five ligand-binding sites according to the general model-free formalism (in terms of the overall association constants, β_i [Equation 12.29]) or the PKNF cooperativity model (in terms of the intrinsic association constant, K, and cooperative interaction constants [Equation 12.34]). The interpretation of the cooperative interaction constants is the following: k_1 reflects the cooperative effect due to binding of an additional ligand, k_2 reflects the cooperative effect due to site–site interaction between two adjacently bound ligands, and k_3 reflects the cooperative effect due to site–site interaction between three adjacently bound ligands.

$$[P]_T = [P]Z = [P]\sum_{i=0}^{5} \beta_i [L]^i$$

$$[L]_T = [L] + [P][L]\frac{\partial Z}{\partial [L]} = [L] + [P]\sum_{i=0}^{5} i\beta_i [L]^i \qquad (12.31)$$

and this set of nonlinear equations in [P] and [L] can also be solved numerically. The concentration of the macromolecule–ligand complexes can be readily calculated from the free macromolecule and ligand concentrations, [P] and [L]:

$$[PL_i] = [P]_T \chi_i = [P]_T \frac{\beta_i [L]^i}{\sum_{i=0}^{5} \beta_i [L]^i} = \beta_i [P][L]^i \qquad (12.32)$$

If these calculations are performed for each injection j, then the normalized heat associated with injection j along the calorimetric titration is given by

$$Q_j = \frac{1}{v_j [L]_0}\left(V_0\left(\sum_{i=0}^{5} \Delta H_i \left([PL_i]_j - [PL_i]_{j-1}\left(1 - \frac{v_j}{V_0}\right)\right)\right) + q_d \right) \qquad (12.33)$$

and the nonlinear least-squares regression analysis allows estimating the thermodynamic binding parameters. Figure 12.3a illustrates a typical titration obtained by injecting histone H5 into NP solution.

Nonlinear regression with the general model is a nontrivial task, since in this case there are 10 parameters to be determined from a single titration. In order to reduce potential degeneracy and parameter correlation, direct and reverse titrations can be globally analyzed (Taneva et al. 2009). Once the overall thermodynamic parameters are determined

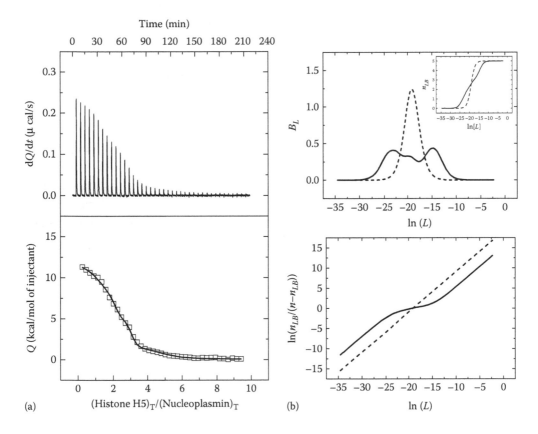

FIGURE 12.3 (a) Calorimetric titration corresponding to the interaction of NP with histone H5 (20 mM PIPES, 150 mM NaCl, pH 7.4, 25°C). Raw thermogram (thermal power as a function of time; top panel) and binding isotherm (normalized heat effect per injection as a function of molar ratio; bottom panel) are shown. (b) Top panel: Binding capacity for NP interacting with histone H5 (continuous line) compared to that of a reference protein with five identical and independent binding sites (dashed line). Inset: Binding isotherm representing n_{LB} as a function of ligand concentration for NP interacting with histone H5 (continuous line) compared to that of a reference protein with five identical and independent binding sites (dashed line). Although the maximal binding capacity for NP is lower than that of the reference protein, it is clear that NP is able to respond over a wide range of histone concentrations (seven orders of magnitude in ligand concentration, compared to three orders of magnitude for the reference protein) with a more uniform binding capacity. Bottom panel: Hill plot for NP interacting with histone H5 (continuous line) compared to that of a reference protein with five identical and independent binding sites (dashed line). The marked decrease in the slope of the Hill plot in the nanomolar region provides a distinctive feature of negative cooperativity.

using the general model based on the overall association constants, appropriate specific models can be selected according to the value of the ρ parameters (see Equation 12.9). Direct titrations provided evidence that some ρ parameters (in particular, ρ_2 and ρ_3) are smaller than 1, and therefore the binding sites are nonidentical or exhibit negative cooperativity, or both (Taneva et al. 2009). Because the symmetry in the homopentameric NP imposes that the binding sites must be identical, there must be negative homotropic cooperativity. Consequently, the system can be then described applying the PKNF model, which, contrary to the MWC model, is able to account for negative cooperativity, by using site-specific association constants and cooperativity constants (see Equation 12.21):

$$Z = 1 + 5K[L] + 5K^2 k_1 [L]^2 + 5K^2 k_1 k_2 [L]^2 + 5K^3 k_1^2 k_2 [L]^3 + 5K^3 k_1^2 k_2^2 k_3 [L]^3$$

$$+ 5K^4 k_1^3 k_2^3 k_3^2 [L]^4 + K^5 k_1^4 k_2^5 k_3^5 [L]^5 \tag{12.34}$$

The scheme corresponding to this binding polynomial is shown in Figure 12.2, and it represents an extension of a previous model employed for studying the binding of cholera toxin (a pentameric protein) to the oligosaccharide fraction of its cell surface receptor (Schön and Freire 1989). This model allowed reducing the number of parameters to four equilibrium constants and four enthalpies, and solving the parameter correlation problem during regression analysis. The normalized heat associated with injection j along the calorimetric titration is given by

$$Q_j = \frac{1}{v_j [L]_0} \left(V_0 \left[\Delta H \left([PL]_j - [PL]_{j-1} \left(1 - \frac{v_j}{V_0} \right) \right) + (2\Delta H + \Delta h_1) \right. \right.$$

$$\left([PL_{2,A}]_j - [PL_{2,A}]_{j-1} \left(1 - \frac{v_j}{V_0} \right) \right)$$

$$+ (2\Delta H + \Delta h_1 + \Delta h_2) \left([PL_{2,B}]_j - [PL_{2,B}]_{j-1} \left(1 - \frac{v_j}{V_0} \right) \right)$$

$$+ (3\Delta H + 2\Delta h_1 + \Delta h_2) \left([PL_{3,A}]_j - [PL_{3,A}]_{j-1} \left(1 - \frac{v_j}{V_0} \right) \right) + (3\Delta H + 2\Delta h_1 + 2\Delta h_2 + \Delta h_3)$$

$$\left([PL_{3,B}]_j - [PL_{3,B}]_{j-1} \left(1 - \frac{v_j}{V_0} \right) \right)$$

$$+ (4\Delta H + 3\Delta h_1 + 3\Delta h_2 + 2\Delta h_3) \left([PL_4]_j - [PL_4]_{j-1} \left(1 - \frac{v_j}{V_0} \right) \right)$$

$$\left. \left. + (5\Delta H + 4\Delta h_1 + 5\Delta h_2 + 5\Delta h_3) \left([PL_5]_j - [PL_5]_{j-1} \left(1 - \frac{v_j}{V_0} \right) \right) \right] + q_d \right)$$

$$\tag{12.35}$$

where ΔH and Δh_i are the intrinsic binding enthalpy and the cooperative binding enthalpies.

According to the experimental data (Taneva et al. 2009), histone H5 binding was characterized by a strong negative cooperativity with a 2000-fold reduction in binding affinity for the fifth histone bound to NP compared to the first one (dissociation constants of 0.1 and 210 nM for the first and fifth histones bound to NP, respectively). A similar study performed with H2A/H2B dimers also indicated negative cooperativity but significantly lower affinity (dissociation constants of 65 nM and 1 μM for the first and fifth histone dimers bound to NP, respectively). Thus, H5 and H2A/H2B interacting with NP are characterized by Hill coefficients at half saturation of 0.26 and 0.56, respectively. Figure 12.3b shows the binding capacity for NP/H5 interaction, compared to a reference system consisting of identical and independent binding sites. It is noticeable that NP presents lower binding capacity than the reference system (protein with five identical and independent binding sites), but it is able to elicit a biochemical response (load and unload histones) over a broader histone concentration range. In addition, the data revealed a different structural source for negative cooperativity for H5 and H2A/H2B dimers. Because $k_1 < 1$ for H5, the cooperativity observed between nonadjacent ligands points to a conformational change as the main source for negative cooperativity in H5 binding; however, direct histone–histone interaction cannot be excluded as an additional source of cooperativity. Because $k_1 = 1$ for H2A/H2B dimers, cooperativity is observed only when histone dimers bind adjacently to NP, and therefore ligand–ligand interaction must be the main source of cooperative interaction in the binding of H2A/H2B dimers.

It has been suggested that the difference in affinity of NP for the H5 and H2A/H2B, the negative cooperativity, and the common binding sites shared on NP are related to the biological activity of NP (Taneva et al. 2009). In particular, NP is involved in the decondensation of sperm DNA during fertilization. Thus, H2A/H2B-loaded NP in contact with condensed sperm chromatin may promote the dissociation of linker histone from DNA and the exchange of basic DNA-associated proteins by H2A/H2B dimers.

Positive homotropic cooperativity acts as an on–off switch where it is important to activate or deactivate (or load or unload) a given protein within a narrow range of ligand concentration. Negative homotropic cooperativity provides a means to render proteins operative over a wider ligand concentration range, while exhibiting high binding affinity and a statistically significant population of intermediate ligation states with biological relevance. This is important for NP, which is located in the nucleus at high concentration and surrounded by potentially interacting histones (and other proteins) at fairly high concentrations as well.

12.5 HETEROTROPIC COOPERATIVE INTERACTIONS: THE CASE OF β-SUBUNIT FROM F$_1$-ATPASE BINDING NUCLEOTIDES

F-ATP synthase or F-ATPase is responsible for the synthesis of ATP in bacteria, chloroplast, and mitochondria (Boyer 1993, 1997; Abrahams et al. 1994). This membrane-embedded supramolecular assembly couples the interconversion of chemical and mechanical energy using the proton electrochemical gradient across the membrane in order to store energy in the form of high-energy-content phosphoanhydride bonds in ATP. The soluble fraction of the F-ATPase, F_1, which is directly responsible for ATP synthesis, contains the catalytic core, a heterohexamer with alternant regulatory α- and catalytic β-subunits. Proton flow through the F_O transmembrane fraction causes a rotary motion within F_O while keeping F_1 stationary, driving ATP synthesis in the F_1 fraction. As the internal F-ATPase axis rotates, the conformation of the α- and β-subunits in the F_1 fraction is altered in such a way that it switches from an "open" state (ADP and phosphate can enter the active site) to a "closed" state (ADP and phosphate are bound loosely) to a "tight" state (ADP and phosphate covalently bind to form ATP) (Boyer 1993). A further conformational change results in ATP release and reversion to the open state. Although this machinery has been the focus of many studies, several aspects of its functioning are still uncertain; in particular, the role of magnesium in the interaction of α- and β-subunits with nucleotides is somewhat unclear.

In order to go deeper into the potential nucleotide–magnesium heterotropic cooperative interaction, the independent subunits were isolated (Perez-Hernandez et al. 2002; Pulido et al. 2010; Salcedo et al. 2014). The characterization of the two independent subunits interacting with nucleotides, as well as the interaction of nucleotides with magnesium, by ITC binary titrations (i.e., those involving two reactants) is somewhat trivial. However, the characterization of the interaction of the two independent subunits with nucleotides in the presence of magnesium represents a difficult challenge. No experimental setup directly provides the expected binding parameters, because in the ternary titration simultaneous association and dissociation processes involving the three reactants (protein subunit, nucleotide, and magnesium) will always occur. For example, titrating a nucleotide–magnesium solution into a protein subunit–magnesium solution will reflect several simultaneous coupled processes (e.g., nucleotide–magnesium complex dissociation and association, nucleotide binding to protein subunit, and nucleotide–magnesium complex binding to protein subunit). If, instead of plain apparent binding parameters, a precise determination of the binding parameters and cooperativity parameters is intended, the model must explicitly account for all different equilibria between the three reactants.

The ternary system constituted by the α- or β-subunit (from now on, $F_1\beta(\alpha)$), ATP or ADP (from now on, AT(D)P), and magnesium has a central pivotal element: the nucleotide. AT(D)P may bind one or two magnesium ions (homotropy regarding magnesium binding to the nucleotide) (Nakamura et al. 2013) and may also bind to the $F_1\beta(\alpha)$ subunit (heterotropy regarding protein subunit and magnesium binding to the nucleotide). The global scheme for the different interactions taking place in this system is shown in Figure 12.4. The binding polynomial for this system, taking AT(D)P as the reference species, is given by

$$Z = 1 + \beta_1 \left[Mg \right] + \beta_2 \left[Mg \right]^2 + \beta_P \left[F_1\beta(\alpha) \right] + \beta_P' \beta_1 \left[F_1\beta(\alpha) \right] \left[Mg \right] \tag{12.36}$$

where:

β_1 and β_2	are the overall association constants for the nucleotide–magnesium interaction
β_P	is the association constant for the nucleotide–subunit complex
β_P'	is the association constant for the nucleotide–magnesium–subunit complex

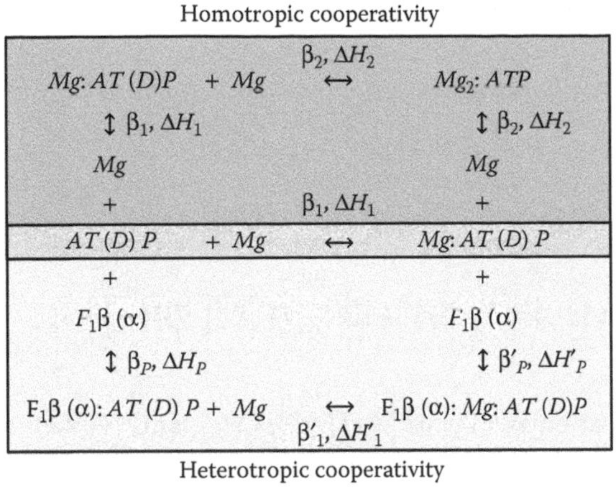

Homotropic cooperativity

Heterotropic cooperativity

FIGURE 12.4   Schematic depiction of the ternary equilibrium between F_1-ATPase α- or β-subunits ($F_1\beta(\alpha)$), ATP or ADP ($AT(D)P$), and magnesium (Mg). Nucleotides may bind one or two (not for ADP) magnesium atoms, and nucleotides may bind to α- or β-subunits with or without bound magnesium. Therefore, it is apparent that there are two cooperativity cycles, a homotropic cycle for nucleotide–magnesium interaction and a heterotropic cycle for nucleotide–magnesium–subunit interaction, coupled by a common element. Association constants and enthalpies are indicated for each binding event. There will be homotropic cooperativity if $\kappa_{hom} = 4\beta_2/\beta_1^2 \neq 1$; in addition, there will be heterotropic cooperativity if $\kappa_{het} = \beta'_P/\beta_P = \beta'_1/\beta_1 \neq 1$.

The homotropic cooperativity for magnesium binding to ATP (it is assumed that ADP only binds one magnesium atom) is quantified through the cooperativity constant, κ_{hom}, and the cooperativity enthalpy, Δh_{hom}:

$$\kappa_{hom} = \frac{4\beta_2}{\beta_1^2}$$
$$\Delta h_{hom} = \Delta H_2 - 2\Delta H_1 \tag{12.37}$$

where κ_{hom} is equal to ρ_2 (see Equation 12.9). The heterotropic cooperativity is quantified through the cooperativity constant, κ_{het}, and the cooperativity enthalpy, Δh_{het}:

$$\kappa_{het} = \frac{\beta'_P}{\beta_P} = \frac{\beta'_1}{\beta_1}$$
$$\Delta h_{het} = \Delta H'_P - \Delta H_P = \Delta H'_1 - \Delta H_1 \tag{12.38}$$

where it is reflected that the effect of the heterotropic cooperativity is reciprocal: if ligand A increases (decreases) the affinity of ligand B by a given factor, then ligand B increases (decreases) the affinity of ligand A by the same factor. Also, it is evident that the four association constants appearing in the heterotropic cycle (see Figure 12.4) are not independent (which is a consequence of the energy conservation principle).

According to the second method (Equations 12.24 and 12.26), the binding equations are expressed as follows:

$$\left[AT(D)P \right]_T = \left[AT(D)P \right] Z$$

$$\left[Mg \right]_T = \left[Mg \right] + \left[AT(D)P \right]\left[Mg \right] \frac{\partial Z}{\partial \left[Mg \right]}$$

$$\left[F_i\beta(\alpha) \right]_T = \left[F_i\beta(\alpha) \right] + \left[AT(D)P \right]\left[F_i\beta(\alpha) \right] \frac{\partial Z}{\partial \left[F_i\beta(\alpha) \right]} \tag{12.39}$$

and this set of nonlinear equations in $[AT(D)P]$, $[Mg]$, and $[F_i\beta(\alpha)]$ can be solved numerically. The concentration of the different complexes can be readily calculated from the free concentrations:

$$\left[Mg{:}\,AT(D)P \right] = \beta_1 \left[AT(D)P \right]\left[Mg \right]$$

$$\left[Mg_2{:}\,AT(D)P \right] = \beta_2 \left[AT(D)P \right]\left[Mg \right]^2$$

$$\left[F_i\beta(\alpha){:}\,AT(D)P \right] = \beta_P \left[F_i\beta(\alpha) \right]\left[AT(D)P \right]$$

$$\left[F_i\beta(\alpha){:}\,Mg{:}\,AT(D)P \right] = \beta_P'\beta_1 \left[F_i\beta(\alpha) \right]\left[AT(D)P \right]\left[Mg \right] \tag{12.40}$$

If these calculations are performed for each injection j, then the normalized heat associated with injection j along the ternary calorimetric titration (injecting a nucleotide–magnesium solution into a protein subunit–magnesium solution) is given by

$$
\begin{aligned}
Q_j = \frac{1}{v_j[L]_0} \Bigg(& V_0 \Bigg(\Delta H_P \left(\left[F_i\beta(\alpha){:}\,AT(D)P \right]_j - \left[F_i\beta(\alpha){:}\,AT(D)P \right]_{j-1}\left(1 - \frac{v_j}{V_0} \right) \right) \\
& + \left(\Delta H_P' + \Delta H_1 \right)\left(\left[F_i\beta(\alpha){:}\,Mg{:}\,AT(D)P \right]_j - \left[F_i\beta(\alpha){:}\,Mg{:}\,AT(D)P \right]_{j-1}\left(1 - \frac{v_j}{V_0} \right) \right) \\
& + \Delta H_1 \left([Mg{:}\,AT(D)P] - [Mg{:}\,AT(D)P]_{j-1}\left(1 - \frac{v_j}{V_0} \right) - [Mg{:}\,AT(D)P]_{\mathrm{syr}}\frac{v_j}{V_0} \right) \\
& + \Delta H_2 \left(\left[Mg_2{:}\,ATP \right]_j - \left[Mg_2{:}\,ATP \right]_{j-1}\left(1 - \frac{v_j}{V_0} \right) - \left[Mg_2{:}\,ATP \right]_{\mathrm{syr}}\frac{v_j}{V_0} + q_d \right)
\end{aligned}
\tag{12.41}
$$

where $[Mg{:}\,AT(D)P]_{\mathrm{syr}}$ and $[Mg_2{:}\,ATP]_{\mathrm{syr}}$ are the concentrations of $Mg{:}\,AT(D)P$ and $Mg_2{:}\,ATP$ complexes in the injection syringe, and the additional term in the last two terms is a required correction accounting for the increase in nucleotide–magnesium complexes in the calorimetric cell due to direct titrant transference from the syringe upon injection

(Houtman et al. 2007). Nonlinear least-squares regression analysis allows estimating the thermodynamic binding parameters. It is obvious that β_1, β_2, and β_P (with their respective enthalpies) can be obtained through binary titrations (titrating nucleotide into α- or β-subunit, and titrating magnesium into nucleotide), while β'_P (with its respective enthalpy) must be obtained through the ternary titration (titrating nucleotide into α- or β-subunit in the presence of magnesium).

According to the experimental data (Perez-Hernandez et al. 2002; Pulido et al. 2010; Salcedo et al. 2014), there is negative homotropic cooperativity for ATP interacting with Mg ($\kappa_{hom} = 0.01$; that is, the second magnesium ion binds to ATP with a 100-fold lower binding affinity), which corresponds to a Hill coefficient at half saturation of 0.18. On the other hand, α- and β-subunits bind both nucleotides with enthalpically driven affinities in the micromolar range (see Figure 12.5). While there is positive heterotropic cooperativity for ATP and Mg binding to α- and β-subunits ($\kappa_{het} = 300$ and 8, respectively; that is, binding of magnesium to ATP increases its binding affinity to α- and β-subunits by factors of 300 and 8, respectively), there is no significant heterotropic cooperativity for ADP and Mg binding to α- and β-subunits ($\kappa_{het} = 1.1$ and 2, respectively).

The β-subunit binds Mg-ATP and Mg-ADP with similar affinity, while the α-subunit binds Mg-ATP with higher affinity. Thus, the α- and β-subunits' interaction with ATP is metal dependent, while their interaction with ADP is metal independent. Or, in other words, Mg interaction with ATP is dependent on the "environment" (i.e., ATP free, bound to α-subunit, or bound to β-subunit), contrary to Mg interaction with ADP. Comparing the binding of Mg-ATP and Mg-ADP to the β-subunit, the only significant difference is the binding heat capacity (ΔC_p values of –0.15 and –0.63 kJ/K·mol for Mg-ADP and

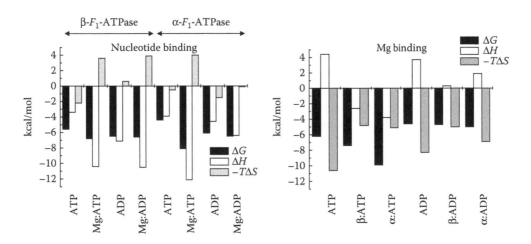

FIGURE 12.5 Binding parameters (ΔG, ΔH, and $-T\Delta S$) corresponding to the interaction of free nucleotides and nucleotide–magnesium complexes with α- and β-subunits from F_1-ATPase (right panel) and to the interaction of magnesium with free nucleotides and nucleotides complexed with α- and β-subunits from F_1-ATPase (right panel), derived from the analysis of calorimetric titrations employing the ternary model shown in Figure 12.4. Differences in the binding parameters taking the interaction with free nucleotide as a reference reflect the heterotropic cooperativity.

Mg-ATP, respectively), values obtained by performing titrations at different temperatures between 15°C and 30°C, which suggests a larger conformational change upon Mg-ATP binding where the β-subunit adopts a more closed conformation in that ternary complex, in agreement with experimental evidence obtained by nuclear magnetic resonance (NMR) (Yagi et al. 2009) and with the proposed model for ATP synthesis that couples conformational changes within the $\alpha_3\beta_3$ hexamer and ADP association and ATP dissociation (Boyer 1993).

12.6 CONCLUDING REMARKS

The binding polynomial was introduced several decades ago for describing binding interactions in biological systems. This concept allows employing the powerful structure and formalism from thermodynamics, as well as handling systems of different complexity levels in a simple and general manner (Gill et al. 1985; Gill 1989). However, its implementation for ITC data analysis (and other binding techniques) has been considered somewhat difficult and cumbersome, and different system-specific approaches have been developed instead, which results in a lack of uniformity in modeling and data analysis, and a serious drawback for the untrained researcher. As it has been shown with two examples, histones binding to nucleoplasmin (homotropic cooperativity) and nucleotides and magnesium binding to F_1-ATPase α- and β-subunits (coupled homotropic and heterotropic cooperativity), the mathematical framework based on the binding polynomial can be easily generalized to many scenarios and applied to a great diversity of biological systems, providing precise and valuable information on allosteric interactions and cooperativity.

ACKNOWLEDGMENTS

This work was supported by the Spanish Ministry of Science and Innovation (BFU2010-19451), the Spanish Ministry of Economy and Competitiveness (BFU2013-47064-P), and Diputacion General de Aragon (Protein Targets Group B89).

REFERENCES

Abrahams, J. P., Leslie, A. G. W., Lutter, R., and Walker, J. E. 1994. Structure at 2.8 Å resolution of F1-ATPase from bovine heart mitochondria. *Nature* 370:621–628.

Akey, C. W. and Luger, K. 2003. Histone chaperones and nucleosome assembly. *Curr. Opin. Struct. Biol.* 13:6–14.

Baker, B. M. and Murphy, K. P. 1996. Evaluation of linked protonation effects in protein binding reactions using isothermal titration calorimetry. *Biophys. J.* 71:2049–2055.

Boyer, P. D. 1993. The binding change mechanism for ATP synthase. Some probabilities and possibilities. *Biochim. Biophys. Acta* 1140:215–250.

Boyer, P. D. 1997. The ATP synthase: A splendid molecular machine. *Annu. Rev. Biochem.* 66:717–749.

Breiten, B., Lockett, M. R., Sherman, W., Fujita, S., Al-Sayah, M., Lange, H., Bowers, C. M., Heroux, A., Krilov, G., and Whitesides, G. M. 2013. Water networks contribute to enthalpy/entropy compensation in protein–ligand binding. *J. Am. Chem. Soc.* 135:15579–15584.

Cooper, A., Johnson, C. M., Lakey, J. H., and Nollmann, M. 2001. Heat does not come in different colours: Entropy-enthalpy compensation, free-energy windows, quantum confinement, pressure perturbation calorimetry, solvation and the multiple causes of heat capacity effects in biomolecular interactions. *Biophys. Chem.* 93:215–230.

Courter, J. R., Madani, N., Sodroski, J., Schön, A., Freire, E., Kwong, P. D., Hendrickson, W. A., Chaiken, I. M., LaLonde, J. M., and Smith 3rd, A. B. 2014. Structure-based design, synthesis and validation of CD4-mimetic small molecule inhibitors of HIV-1 entry: Conversion of a viral entry agonist to an antagonist. *Acc. Chem. Res.* 47:1228–1237.

Del Sol, A., Tsai, C. J., Ma, B., and Nussinov, R. 2009. The origin of allosteric functional modulation: Multiple pre-existing pathways. *Structure* 17:1042–1050.

Di Cera, E., Gill, S. J., and Wyman, J. 1988. Binding capacity: Cooperativity and buffering in biopolymers. *Proc. Natl. Acad. Sci. USA* 85:449–452.

Du, W., Liu, W. S., Payne, D. J., and Doyle, M. L. 2000. Synergistic inhibitor binding to *Streptococcus pneumoniae* 5-enolpyruvylshikimate-3-phosphate synthase with both monovalent cations and substrate. *Biochemistry* 39:10140–10146.

Dutta, S., Akey, I. V., Dingwall, C., Hartman, K. L., Laue, T., Nolte, R. T., Head, J. F., and Akey, C. W. 2001. The crystal structure of nucleoplasmin-core: Implications for histone binding and nucleosome assembly. *Mol. Cell* 8:841–853.

Eftink, M, and Biltonen, R. 1980. Thermodynamics of interacting biological systems. In Beezer, A. E. (ed.), *Biological Calorimetry*, 343–412. Academic Press: London.

Gill, S. J. 1989. Thermodynamics of ligand binding to proteins. *Pure Appl. Chem.* 61:1009–1020.

Gill, S. J., Richey, B., Bishop, G., and Wyman, J. 1985. Generalized binding phenomena in an allosteric macromolecule. *Biophys. Chem.* 21:1–14.

Gunasekaran, K., Ma, B., and Nussinov, R. 2004. Is allostery an intrinsic property of all dynamic proteins? *Proteins* 57:433–443.

Hinz, H. J. 1983. Thermodynamics of protein-ligand interactions: Calorimetric approaches. *Annu. Rev. Biophys. Bioeng.* 12:285–317.

Houtman, J. C., Brown, P. H., Bowden, B., Yamaguchi, H., Apella, E., Samelson, L. E., and Schuck, P. 2007. Studying multisite binary and ternary protein interactions by global analysis of isothermal titration calorimetry data in SEDPHAT: Application to adaptor protein complexes in cell signaling. *Protein Sci.* 16:30–42.

Jobichen, C., Fernandis, A. Z., Velazquez-Campoy, A., Leung, K. Y., Mok, Y. K., Wenk, M. R., and Sivaraman, J. 2009. Identification and characterization of the lipid-binding property of GrlR, a locus of enterocyte effacement regulator. *Biochem. J.* 420:191–199.

Koshland Jr., D. E., Nemethy, G., and Filmer, D. 1966. Comparison of experimental binding data and theoretical models in proteins containing subunits. *Biochemistry* 5:365–385.

Laskey, R. A., Honda, B. M., Mills, A. D., and Finch, J. T. 1978. Nucleosomes are assembled by an acidic protein which binds histones and transfers them to DNA. *Nature* 275:416–420.

Liu, Y., Schön, A., and Freire, E. 2013. Optimization of CD4/gp120 inhibitors by thermodynamic-guided alanine-scanning mutagenesis. *Chem. Biol. Drug Des.* 81:72–78.

Monod, J., Wyman, J., and Changeux, J. P. 1965. On the nature of allosteric transitions: A plausible model. *J. Mol. Biol.* 12:88–118.

Nakamura, S., Koga, S., Shibuya, N., Seo, K., and Kidokoro S. 2013. A new multi-binding model for isothermal titration calorimetry analysis of the interaction between adenosine 5′-triphosphate and magnesium ion. *Thermochim. Acta* 563:82–89.

Perez-Hernandez, G., Garcia-Hernandez, E., Zubillaga, R. A., and de Gomez-Puyou, M. T. 2002. Structural energetics of MgADP binding to the isolated beta subunit of F1-ATPase from thermophilic *Bacillus* PS3. *Arch. Biochem. Biophys.* 408:177–183.

Prado, A., Ramos, I., Frehlick, L. J., Muga, A., and Ausio, J. 2004. Nucleoplasmin: A nuclear chaperone. *Biochem. Cell Biol.* 82:437–445.

Press, W. H., Teukolsky, S. A., Vetterling, W. T., and Flannery, B. P. 2007. *Numerical Recipes: The Art of Scientific Computing*, 3rd ed. Cambridge University Press: New York.

Pulido, N. O., Salcedo, G., Perez-Hernandez, G., Jose-Nuñez, C., Velazquez-Campoy, A., and Garcia-Hernandez, E. 2010. Energetic effects of magnesium in the recognition of adenosine nucleotides by the F(1)-ATPase beta subunit. *Biochemistry* 49:5258–5268.

Ramos, I., Martin-Benito, J., Finn, R., Bretaña, L., Aloria, K., Arizmendi, J. M., Ausio, J., Muga, A., Valpuesta, J. M., and Prado, A. 2010. Nucleoplasmin binds histone H2A-H2B dimers through its distal face. *J. Biol. Chem.* 285:33771–33778.

Sahun-Roncero, M., Rubio-Ruiz, B., Conejo-Garcia, A., Velazquez-Campoy, A., Entrena, A., and Hurtado-Guerrero, R. 2013 Determination of potential scaffolds for human choline kinase α1 by chemical deconvolution studies. *ChemBioChem* 2013 14:1291–1295.

Salcedo, G., Cano-Sanchez, P., de Gomez-Puyou, M. T., Velazquez-Campoy, A., and Garcia-Hernandez, E. 2014. Isolated noncatalytic and catalytic subunits of F1-ATPase exhibit similar, albeit not identical, energetic strategies for recognizing adenosine nucleotides. *Biochim. Biophys. Acta.* 1837:44–50.

Schellman, J. 1975. Macromolecular binding. *Biopolymers* 14:999–1018.

Schellman, J. 1990. Fluctuation and linkage relations in macromolecular solution. *Biopolymers* 29:215–224.

Schön, A. and Freire, E. 1989. Thermodynamics of intersubunit interactions in cholera toxin upon binding to the oligosaccharide portion of its cell surface receptor, ganglioside GM1. *Biochemistry* 28:5019–5024.

Schön, A., Lam, S. Y., and Freire, E. 2011. Thermodynamics-based drug design: Strategies for inhibiting protein-protein interactions. *Future Med. Chem.* 3:1129–1137.

Straume, M. and Freire, E. 1992. Two-dimensional differential scanning calorimetry: Simultaneous resolution of intrinsic protein structural energetics and ligand binding interactions by global linkage analysis. *Anal. Biochem.* 203:259–268.

Taneva, S. G., Bañuelos, S., Falces, J., Arregi, I., Muga, A., Konarev, P. V., Svergun, D. I., Velazquez-Campoy, A., and Urbaneja, M. A. 2009. A mechanism for histone chaperoning activity of nucleoplasmin: Thermodynamic and structural models. *J. Mol. Biol.* 393:448–463.

Vega, S., Abian, O., and Velazquez-Campoy, A. 2015. A unified framework based on the binding polynomial for characterizing biological systems by isothermal titration calorimetry. *Methods* 76:99–115.

Velazquez-Campoy, A., Goñi, G., Peregrina, J. R., and Medina, M. 2006. Exact analysis of heterotropic interactions in proteins: Characterization of cooperative ligand binding by isothermal titration calorimetry. *Biophys. J.* 91:1887–1904.

Williams, R., Holyoak, T., McDonald, G., Gui, C., and Fenton, A. W. 2006 Differentiating a ligand's chemical requirements for allosteric interactions from those for protein binding. Phenylalanine inhibition of pyruvate kinase. *Biochemistry* 45:5421–5429.

Wyman, J. 1948. Heme proteins. *Adv. Protein Chem.* 4:407–531.

Wyman, J. 1963. Allosteric effects in hemoglobin. *Cold Spring Harb. Symp. Quant. Biol.* 28:483–489.

Wyman, J. 1964. Linked functions and reciprocal effects in hemoglobin: A second look. *Adv. Protein Chem.* 19:223–286.

Wyman, J. 1972. On allosteric models. *Curr. Topics Cell. Reg.* 6:207–223.

Wyman, J. and Allen, D. W. 1951. The problem of the heme interactions in hemoglobin and the basis of the Bohr effect. *J. Polym. Sci.* 7:499–518.

Wyman, J. and Gill, S. J. 1990. *Binding and Linkage: Functional Chemistry of Biological Macromolecules.* University Science Books: Mill Valley, CA.

Wyman, J. and Philipson, P. 1974. A probabilistic approach to cooperativity of ligand binding by a polyvalent molecule. *Proc. Natl. Acad. Sci. USA* 71:3431–3434.

Yagi, H., Kajirawa, N., Iwabuchi, T., Izumi, K., Yoshida, M., and Akutsu, H. 2009. Stepwise propagation of the ATP-induced conformational change of the F1-ATPase β subunit revealed by NMR. *J. Biol. Chem.* 284:2374–2382.

CHAPTER 13

Protein Stability in Crowded Environments

Michael Senske, Simon Ebbinghaus, and Christian Herrmann

CONTENTS

13.1 INTRODUCTION

Proteins have evolved to function in the presence of high concentrations of cosolutes. The cellular and extracellular environments are crowded by heterogeneous types of molecules. Those molecules include macromolecules such as proteins and nucleic acids, as well as small organic compounds such as osmolytes and inorganic salts. This phenomenon is known as the (macro)molecular crowding effect (Minton 1981). The overall concentration of macromolecules can be as high as 400 g/L (Boersma et al. 2015; Gnutt et al. 2015; Zimmerman and Trach 1991). In addition, cellular organelles and compartments can lead to confinement phenomena (Zhou et al. 2008).

In contrast to the natural environment, proteins are commonly overexpressed and then isolated for experiments *in vitro* in dilute buffer solutions. It remains an intriguing question to what extent the natural crowding environment influences protein function, structure, and stability. In this chapter, we analyze the impact of various cosolutes (dextran, poly[ethylene glycol] [PEG], glucose, urea, salts, and aqueous ionic liquids) on the protein stability of two model proteins: ubiquitin and RNase A. We probe the thermal stability of both proteins in these artificial crowding environments by differential scanning calorimetry (DSC). The difference in free energy of unfolding of both proteins in the presence of cosolutes to the dilute solution, $\Delta\Delta G_u$, is then dissected into its enthalpic and entropic contributions ($\Delta\Delta H_u$ and $\Delta\Delta S_u$) in order to get mechanistic insights into the stabilizing or destabilizing mechanism of each cosolute (Senske et al. 2014). Intriguingly, even though cosolutes follow often simple size-dependent linear concentration dependencies on the free energy level, their underlying enthalpic and entropic components ($\Delta\Delta H_u$ and $\Delta\Delta S_u$) can be manifold (Sukenik et al. 2013b).

13.2 THERMODYNAMIC CHARACTERIZATION OF PROTEIN FOLDING PROCESS BY DIFFERENTIAL SCANNING CALORIMETRY

DSC is one of the most common methods to study the thermal unfolding of proteins. In theory, DSC allows a full thermodynamic characterization of a protein folding process (Lopez and Makhatadze 2002). The melting temperature, T_m, the calorimetric and van't Hoff enthalpy of unfolding, $\Delta H_{u,cal}$ and $\Delta H_{u,vH}$, respectively, and the heat capacity change upon unfolding, ΔC_p, can be obtained as direct-fit parameters from the thermograms. At the melting temperature $\Delta G_u = 0$, this means that the changes in enthalpy and entropy exactly compensate each other: $\Delta H_u = T\Delta S_u$. Using Kirchoff's laws (Equations 13.1 and 13.2), ΔH_u and ΔS_u can be calculated for all temperatures if $\Delta C_p(T)$ is known. Kirchhoff's laws describe the temperature dependence of the standard enthalpy change $\Delta H°$ and the standard entropy change $\Delta S°$. The temperature dependence of both parameters depends on the standard heat capacity change, $\Delta C_p°$. Since the experimental results discussed in this chapter were not obtained at standard conditions, Kirchhoff's laws were applied assuming that $\Delta H° \approx \Delta H_u$, $\Delta S° \approx \Delta S_u$, and $\Delta C_p° = \Delta C_p$. Substituting Equations 13.1 and 13.2 into $\Delta G_u = \Delta H_u - T\Delta S_u$ results in Equation 13.3 (at T_m, $\Delta G_u = 0$, resulting in $\Delta S_u = \Delta H_u/T_m$). Thereby, a full thermodynamic characterization of the protein folding process is obtained. The most crucial parameter in the extrapolation using Kirchhoff's laws is the heat capacity change ΔC_p. However, ΔC_p is the parameter that can be measured with least accuracy. In Section 13.4, we discuss two different ways to estimate ΔC_p.

$$\Delta H_u\left(T\right) = \Delta H_u\left(T_m\right) + \int_{T_m}^{T} \Delta C_p dT \tag{13.1}$$

$$\Delta S_u\left(T\right) = \Delta S_u\left(T_m\right) + \int_{T_m}^{T} \frac{\Delta C_p}{T} dT \tag{13.2}$$

$$\Delta G_u\left(T\right) = \Delta H_u\left(T_m\right)\left(1 - \frac{T}{T_m}\right) + \int_{T_m}^{T} \Delta C_p dT - T\int_{T_m}^{T} \frac{\Delta C_p}{T} dT \tag{13.3}$$

13.3 THERMAL UNFOLDING OF UBIQUITIN IN CROWDED ENVIRONMENTS

In the following two sections, the macromolecular crowding effect (dextran and PEG) is compared to a molecular crowding effect (glucose), as well as to the impact of KCl, an inorganic salt. We have chosen ubiquitin as a model protein to analyze the mechanistic principles of the crowding effect. Dextran and PEG serve as artificial crowding agents. Both polymers are commonly used crowders and are believed to act via their excluded volumes on the protein folding equilibrium (Christiansen and Wittung-Stafshede 2014; Sasahara et al. 2003). This excluded volume effect is thought to lead to an entropic protein stabilization since the conformational entropy of the expanded and flexible denatured state is more strongly reduced in the presence of a crowding agent than the conformational entropy of the compact and rigid native state of the protein (Minton 1981; Zhou et al. 2008).

Glucose is used to compare the macromolecular crowding effect to the molecular crowding effect, since the excluded volume theory predicts a dependence of the crowding effect on the crowding radius (Cheung et al. 2005; Sakaue and Raphael 2006; Waegele and Gai 2011). Furthermore, KCl gives mechanistic insights into the Hofmeister effect on proteins, which will be analyzed in more detail in Section 13.6. Figure 13.1 shows the DSC traces of ubiquitin at pH 2 in dilute solution in comparison to the crowding solutions. The experiments needed to be conducted at low pH (pH 2–4) since ubiquitin tends to aggregate at higher pH upon thermal unfolding (Wintrode et al. 1994). Ubiquitin is known to fold in a two-state manner (Wintrode et al. 1994). The estimation of a ratio of $\Delta H_{u,cal}/\Delta H_{u,vH}$ near unity (Table 13.1) for all measurements justifies a two-state folding mechanism (Privalov 1979).

Two different two-state models were used to describe the experimental DSC data (Lopez and Makhatadze 2002; Senske et al. 2014). The difference between both models is the treatment of the baseline before and after the thermal transition. The model by Lopez and Makhatadze describes the heat capacity of the folded and unfolded protein by two linear equations. The model used by Senske et al. uses a sum of one linear function and the heat

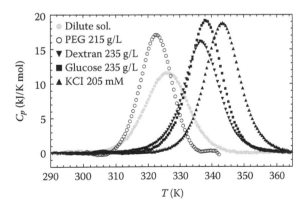

FIGURE 13.1 Excess heat capacity profiles of the thermal unfolding of ubiquitin at pH 2 in dilute solution and in the presence of different cosolutes.

TABLE 13.1 Thermodynamic Data for the Thermal Unfolding of Ubiquitin at pH 2 in Dilute Solution and in the Presence of Different Cosolutes

	T_M/K	$\Delta H_{U,FIT}$/KJ MOL^{-1}	$\Delta H_{U,CAL}$/KJ MOL^{-1}	$\Delta H_{U,VH}$/KJ MOL^{-1}
Dilute	324.02 ± 0.25	177.0 ± 5.0	167.8 ± 7.1	207.9 ± 1.8
Dextran (235 g/L)	335.79 ± 0.57	243.9 ± 7.5	234 ± 12	266.5 ± 7.5
PEG (105 g/L)	321.600 ± 0.011	196.06 ± 0.19	188	223
PEG (215 g/L)	320.363 ± 0.045	226.61 ± 1.0	210	277
Glucose (120 g/L)	330.44 ± 0.42	220.9 ± 4.6	221 ± 15	236 ± 10
Glucose (235 g/L)	336.44 ± 0.64	257.3 ± 4.6	250.6 ± 5.9	278.4 ± 1.8
KCL (205 MM)	342.82 ± 0.40	265.0 ± 1.7	261.8 ± 2.5	275.27 ± 0.75

Note: The DSC data were analyzed according to the two-state transition model described by Lopez and Makhatadze 2002. The uncertainties are provided as the 68% confidence interval of the mean, derived from repeated measurements, using a Student's t-distribution. In the case of PEG, the uncertainty is the error of the fit of a single thermogram.

capacity change weighted by the fraction of the unfolded proteins to describe the baseline before and after the transition. While the first model has the advantage to fit DSC curves reliably even with slight uncertainties in the baselines (which is often the case in solutions of high concentrations of cosolutes), it overestimates ΔC_p. Both models give the same enthalpy values.

As can be seen from Figure 13.1, glucose, dextran, and KCl lead to a stabilization of ubiquitin (increase of the melting temperature), while PEG causes a destabilization (decrease of the melting temperature). Table 13.1 lists the thermodynamic parameters obtained from the fit of the two-state unfolding model (Lopez and Makhatadze 2002) to the experimental data points. The evaluation of ΔC_p is discussed in the following section.

13.4 HEAT CAPACITY CHANGE OF UBIQUITIN

The heat capacity change ΔC_p describes the difference between the heat capacity of the unfolded protein and the heat capacity of the folded protein. ΔC_p is temperature dependent and starts to decrease at elevated temperatures (>70°C) (Gomez et al. 1995). At low temperatures, it is often treated as temperature independent. There are two main methods to experimentally determine ΔC_p: (1) direct determination via fitting of the DSC curves and (2) indirect determination via measuring ΔH at different melting temperatures (by pH variation) and calculating $\partial \Delta H_u/\partial T_m$. Figure 13.2 shows the results of the estimation of $\Delta C_p(T)$ by both methods. The ΔH_u–T_m data were recorded by variation of the pH (2–4). The data were described by a second-order polynomial. $\Delta C_p(T)$ is then obtained as the first derivative of the fitted function. In Figure 13.2b, the $\Delta C_p(T)$ data from Figure 13.2a are compared to the ΔC_p data obtained by the fit of the DSC thermograms with the model used by Senske et al. (2014). Both methods give similar results for $\Delta C_p(T)$. However, the ΔH_u–T_m method cannot resolve the nonlinear temperature dependence of ΔC_p (Gomez et al. 1995).

The estimation of $\Delta C_p(T)$ by ΔH_u versus T_m recorded by variation of pH is only valid if the pH has no influence on the heat capacities of the folded and unfolded proteins. In

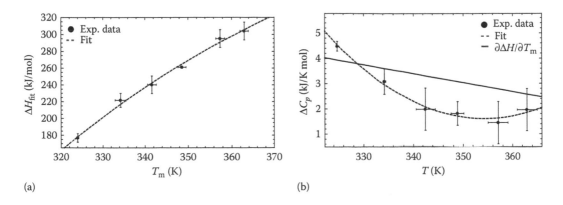

(a)   (b)

FIGURE 13.2   (a) $\Delta H_{u,fit}$ vs. T_m. A second-order polynomial was fitted to the data points in order to estimate the temperature dependence of ΔC_p. (b) ΔC_p obtained as a direct-fit parameter of the DSC thermograms and ΔC_p estimated as $\partial \Delta H_{u,fit}/\partial T_m$ from the plot in (a). The two-state model used by Senske et al. (2014) was used to extract ΔC_p by the direct fit of the DSC traces. A second-order polynomial was fitted to the ΔC_p data.

Figure 13.3, the partial molar heat capacity of ubiquitin is plotted as a function of temperature for various pH values. The heat capacity of the folded and unfolded proteins remains unchanged with increasing pH, justifying the pH variation as a suitable method to determine $\Delta C_p(T)$.

For the following calculations of the excess enthalpy $\Delta \Delta H_u$ and excess entropy $T \Delta \Delta S_u$, the assumption $\Delta C_{p,dilute} = \Delta C_{p,cosolute}$ was used, since ΔC_p of ubiquitin in the presence of glucose, dextran, and PEG could not be determined by a direct fit of the DSC traces reliably, because of small baseline innacuracies in the presence of these cosolutes. However, in a previous study, we could show by pH variation and estimating $\Delta C_{p,cosolute}(T)$ from the $\Delta H_u - T_m$ curve that the cosolute has only a small effect on $\Delta C_p(T)$ (Senske et al. 2014).

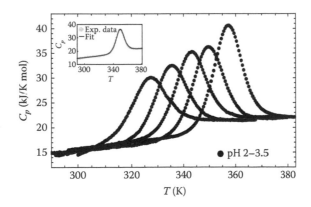

FIGURE 13.3   Partial molar heat capacity of ubiquitin as a function of temperature and pH (2–3.5). The inset shows the quality of the fit of the two-state model used by Senske et al. (2014) to the data points. The ΔC_p data in Figure 13.2b were obtained from these fits.

13.5 THERMODYNAMIC ANALYSIS OF COSOLUTE EFFECT ON UBIQUITIN

The determination of T_m, $\Delta H_u(T_m)$, and $\Delta C_p(T)$ allows a full thermodynamic character-ization according to Equations 13.1 through 13.3. Since ΔG, ΔH, and ΔS are temperature dependent, enthalpies obtained from DSC curves in the presence of cosolute cannot be directly compared to enthalpies obtained in the dilute solution. Thus, using Equations 13.1 through 13.3, the respective thermodynamic parameter in the presence of a cosolute was extrapolated to the T_m in dilute solution. Calculating the respective excess parameter $\Delta\Delta X_u = \Delta X_{u,cosolute} - \Delta X_{u,dilute}$ ($X = G, H, S$) gives an expression for the underlying thermo-dynamic mechanism of the cosolute effect.

In Figure 13.4, the excess entropy $T\Delta\Delta S_u$ and the excess enthalpy $\Delta\Delta H_u$ are compared to each other. Figure 13.4a shows the results of the data extrapolation using $\Delta C_p(T)$ obtained from the ΔH_u–T_m curve, and Figure 13.4b shows the results of the data extrapolation using $\Delta C_p(T)$ obtained from the fit of the thermograms. Both $\Delta C_p(T)$ functions give equivalent results. The dashed diagonal line represents a full enthalpy–entropy compensation result-ing in $\Delta\Delta G_u = 0$. While all stabilizing cosolutes (glucose, dextran, and KCl) are located within region II on the right side of the diagonal ($\Delta\Delta G_u > 0$), PEG, which is destabilizing, is located on the left side within region I ($\Delta\Delta G_u < 0$). The area I corresponds to an overall entropic destabilization that is offset by an enthalpic stabilization. The adjacent area II describes a cosolute effect that is enthalpically stabilizing with a minor contribution of an entropic destabilization. Therefore, glucose, dextran, and KCl stabilize ubiquitin via an enthalpic pathway, while PEG leads to an entropic destabilization that is counteracted by an enthalpic stabilization. The thermodynamic excess parameters of PEG and dextran are surprising since the excluded volume theory predicts an entropic protein stabilization with an unaffected enthalpic contribution (Zhou 2013). An ideal volume crowder would be located on the negative part of the y-axis in the entropy–enthalpy diagram. However, real crowder molecules—even though they are mainly excluded from the protein surface—tend

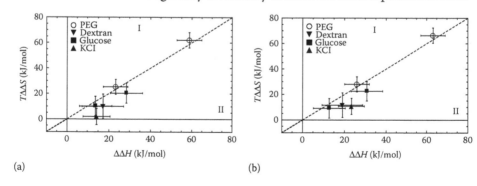

(a) (b)

FIGURE 13.4 Entropy–enthalpy plot showing the respective excess parameters of ubiquitin in the presence of the different cosolutes. For glucose and PEG, both measured concentrations are depicted. The dashed line represents full entropy–enthalpy compensation. Region I indicates coso-lutes inducing an entropic destabilization that is offset by a minor enthalpic stabilization. Data in area II are characterized by an enthalpic stabilization with an additional minor destabilizing entro-pic contribution. (a) Data obtained using $\Delta C_p(T)$ obtained from $\partial\Delta H_{u,fit}/\partial T_m$. (b) Data obtained using $\Delta C_p(T)$ obtained from direct fits of the DSC curves.

also to attractively interact with the protein via unspecific interactions. Attractive interactions are thought to be enthalpically destabilizing, since the larger solvent-exposed surface area in the unfolded state provides more binding sites for the cosolute molecules (Minton 2013; Wang et al. 2012; Zhou 2013).

PEG and dextran do not fit at all in the category of an excluded volume crowder. Their entropic contribution is destabilizing and the enthalpic component is stabilizing. While small osmolytes are known to mainly stabilize proteins enthalpically (Attri et al. 2010; Politi and Harries 2010), an enthalpic stabilization by macromolecular crowding agents was unknown until recent experimental studies provided first evidence for such an osmolyte-like thermodynamic mechanism (Benton et al. 2012; Senske et al. 2014).

The enthalpic stabilization by osmolytes is thought to be water mediated (Bruzdziak et al. 2013; Gilman-Politi and Harries 2011; Heyden et al. 2008; Hunger et al. 2012). The Harries group could show that the origin of the enthalpic stabilization is distortions in the hydration bond network of water due to the formation of nonoptimal hydrogen bonds between the cosolute and the water molecules. The cosolute is thereby excluded from the direct contact to the protein. However, the water molecules optimize their remaining hydrogen bonds. This optimization of the H-bonds is mediated toward the protein, in which also internal hydrogen bonds are strengthened (Gilman-Politi and Harries 2011; Sapir and Harries 2014).

We propose a similar water-mediated enthalpic stabilization of ubiquitin in the presence of glucose and dextran. On a chemical basis, this might not be surprising, since glucose is the monomeric equivalent of dextran. PEG can be classified as a cosolute for which two main effects are overlaid. On the one hand, attractive unspecific interactions lead to a protein destabilization; on the other hand, PEG also seems to be able to induce a water-mediated enthalpic stabilization. Both pathways have opposite enthalpic contributions causing together an entropic destabilization with a minor enthalpic stabilization. The pH dependence of the cosolute effect of PEG on ubiquitin strengthens this hypothesis (Senske et al. 2014). In the case of salts, electrostatic Poisson–Boltzmann effects could dominate at small concentrations (Pegram et al. 2010; Sukenik et al. 2013a). However, a solvent-mediated enthalpic contribution arising from nonelectrostatic salt–peptide interactions could also play a major role in the cosolute effect of KCl on ubiquitin (Senske et al. 2014; Sukenik et al. 2013a).

13.6 STABILITY OF RNASE A IN AQUEOUS SOLUTIONS OF IONIC LIQUIDS

Ionic liquids (ILs) are a special class of cosolutes. They are molten salts at room temperature and become of interest as cosolvents for water, in biphasic systems, and as neat solvents (Fujita et al. 2006; van Rantwijk et al. 2003). Here, we focus on the cosolute effect of IL compounds on the thermal stability of RNase A, which is a well-characterized two-state folder (Pace et al. 1999). Constantinescu et al. (2007) studied the effect of various ILs on the melting temperature of RNase A by DSC. With the exception of [choline][dhp], all ILs lead to a nonlinear decrease of the melting temperature. In this section, we analyze the underlying thermodynamic mechanism of the disparate changes of the thermal stability of RNase A. The DSC scans of RNase A in all aqueous solutions of ILs were highly reversible.

This enables the extraction of thermodynamic data from the DSC curves (Sanchez-Ruiz 1992). The following ILs were analyzed: [bmim][BF$_4$], [bmim]Br, [C$_{4,4,4,4}$N]Br, [emim] [SCN], [hmim]Br, [bmpyrr]Br, [emim]Cl, [emim][N(CN)$_2$], [emim]Br, [emim][EtSO$_4$], and [choline][dhp]. The ILs studied consist of the following ions: 1-ethyl-3-methylimidazo-lium ([emim]$^+$), 1-butyl-3-methylimidazolium ([bmim]$^+$), 1-hexyl-3-methylimidazolium ([hmim]$^+$), 1-butyl-1-methylpyrrolidinium ([bmpyrr]$^+$), tetrabutylammonium ([C$_{4,4,4,4}$N]$^+$), 2-hydroxy-N,N,N-trimethylethanammonium ([choline]$^+$), thiocyanate ([SCN]$^-$), ethylsul-fate ([EtSO$_4$]), dicyanimide ([N(CN)$_2$]), tetrafluoroborate ([BF$_4$]), and dihydrogen phos-phate ([dhp]).

Figure 13.5 shows representative DSC scans in the presence of ILs in comparison to RNase A in 10 mM Na$_2$HPO$_4$ at pH 5.5. As can be seen from the figure, [hmim]Br desta-bilizes RNase A, while [choline][dhp] stabilizes RNase A. In the case of [hmim]Br, ΔH_u decreases monotonically, while [choline][dhp] has a nonmonotonical effect on ΔH_u. At small concentrations of [choline][dhp], ΔH_u increases, while a decrease of ΔH_u can be observed at concentrations higher than 1.5 mol dm^{-3}.

Using the same procedure as described for ubiquitin, the excess thermodynamic parameters were calculated for RNase A in the presence of ILs. In this case, $\Delta C_p(T)$ could not be reliably determined from the DSC curves. Therefore, we used the experimental values of ΔC_p at different temperatures from the Privalov lab (Privalov 1990). For the calculations of $\Delta\Delta H_u$ and $\Delta\Delta S_u$, it was assumed that $\Delta C_{p,\text{buffer}} = \Delta C_{p,\text{IL}}$. Even though it was not possible to extract absolute ΔC_p values from the DSC traces, the ΔC_p values obtained from the fit of the DSC curves do not change for most of the IL measurements. However, in some cases, a slight increase could be observed. This increase—which was not consid-ered in the calculation of the excess parameter—reinforces the qualitative outcome of this analysis.

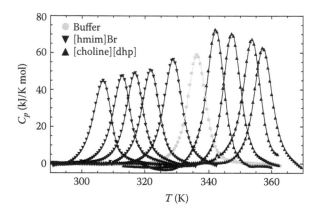

FIGURE 13.5 Excess heat capacities of RNase A in the presence of different concentrations of [hmim]Br and [choline][dhp] in comparison to the dilute aqueous buffer solution. The data points are joined to guide the eye. The concentrations from left to right are 1.5 mol dm^{-3}, 1.0 mol dm^{-3}, 0.75 mol dm^{-3}, 0.5 mol dm^{-3}, and 0.25 mol dm^{-3} ([hmim]Br) and 0.5 mol dm^{-3}, 1.5 mol dm^{-3}, 3 mol dm^{-3}, 4 mol dm^{-3} ([choline][dhp]).

Figure 13.6 shows the results of the thermodynamic analysis in a similar entropy-enthalpy diagram as presented for ubiquitin in Section 13.5. Intriguingly, nearly all destabilizing ILs can be found in region I of the diagram, indicating an entropic destabilization that is counteracted by a minor enthalpic stabilization. This is in a sharp contrast to guanidine hydrochloride (GdmCl), which causes a clear enthalpic destabilization (region IV). It is widely accepted that GdmCl attractively interacts with proteins and thereby reduces the enthalpy of the protein unfolding transition (Attri et al. 2010; Canchi and Garcia 2013; Makhatadze and Privalov 1992).

A similar discrepancy between entropically destabilizing salts and GdmCl was found by Sukenik and Harries (Sukenik et al. 2013a). In that study, the authors assigned a destabilizing enthalpic contribution to the interaction of ions with the charged surface of the protein, while the nonelectrostatic salt–peptide interactions are enthalpically stabilizing. The measured net effect falls within the same category in the entropy–enthalpy diagram as the results for RNase A in the presence of the destabilizing ILs. It remains to be determined whether the enthalpic stabilizing contributions seen for the destabilizing ILs, and for glucose, dextran, PEG, and KCl in the case study of ubiquitin, have the same origin. Intriguingly, urea and GdmCl seem to destabilize proteins via a different thermodynamic mechanism than ILs, even though attractive interactions are thought to be the underlying mechanism of both cosolute classes (Constantinescu et al. 2007; Makhatadze and Privalov 1992). Spectroscopic results indicate that urea and GdmCl fit rather well into the hydrogen bond network of water and have no long-range effect on the water dynamics (Cooper et al. 2014; Funkner et al. 2012; van der Post et al. 2013). This might rule out distortions in the hydrogen bond network of water as an origin of stabilizing enthalpic effects in solutions of urea and GdmCl. It seems that for the ILs—similar to PEG—several molecular

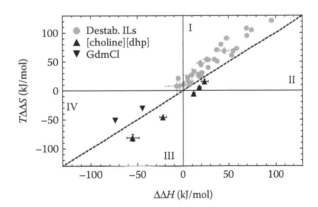

FIGURE 13.6 Thermodynamic excess parameters $\Delta\Delta H$ and $T\Delta\Delta S$ of RNase A in the presence of all destabilizing ILs and the stabilizing IL [choline][dhp] in comparison to GdmCl. (The data of GdmCl are taken from Makhatadze, G. I. and Privalov, P. L., *J. Mol. Biol.*, 226:491–505, 1992.) Data in region I are characterized by an entropic destabilizing that is offset by an enthalpic stabilization, data in area II describe an enthalpic stabilization, data in area III describe an entropic stabilization, and data in area IV indicate an enthalpic destabilization as the main thermodynamic driving force.

mechanisms come into play. This could also explain the nonlinear decrease of T_m of RNase A (Constantinescu et al. 2007). In contrast, GdmCl shows a nearly linear decrease in T_m with concentration.

The stabilizing IL [choline][dhp] is located in area II (enthalpic stabilization) at 0.5 mol dm^{-3} and changes into region III of the diagram at high concentrations (>3 mol dm^{-3}). This indicates an entropic stabilization that is offset by an enthalpic destabilization. As discussed in the previous section, this phenomenon is commonly ascribed to the cosolute effect of macromolecular crowding agents. However, in the case of RNase A—in sharp contrast to ubiquitin—even osmolytes tend to stabilize the protein entropically (Kaushik and Bhat 1998, 2003; O'Connor et al. 2004, 2007; Santoro et al. 1992; Xie and Timasheff 1997). Therefore, the depletion force in the case of RNase A might be dominated by entropic contributions. It remains an open question why the destabilizing ILs have a stabilizing enthalpic contribution while the only stabilizing IL has only a small stabilizing enthalpic effect at low concentration (up to 0.5 mol dm^{-3}) and even a pronounced destabilizing enthalpic contribution at high concentrations (>3 mol dm^{-3}).

13.7 THERMODYNAMIC FINGERPRINT OF PROTEIN FOLDING IN COSOLUTE SOLUTIONS

In summary, DSC has been shown to be the method of choice to extract a full thermodynamic characterization of cosolute effects on the thermal stability of proteins. We analyzed the effect of disparate cosolutes on two different proteins. Even though the various cosolutes have different physical properties, some thermodynamic pathways were similar. PEG and all tested ILs except [choline][dhp] are entropically destabilizing with a counteracting stabilizing enthalpic contribution. In contrast, the opposing thermodynamic pathways of the stabilizing cosolutes in the case of ubiquitin and the stabilizing IL [choline][dhp] in the case of RNase A indicate the variety of cosolute effects on proteins that can be found. In a recent study of Dandage and coworkers (2015), the enthalpic and entropic cosolute effects were extracted from kinetic data. The authors found that the underlying thermodynamic mechanism of a certain cosolute can differ from one protein to another, indicating the complexity of these effects. Therefore, the thermodynamic driving force induced by a cosolute on a protein can be considered the thermodynamic fingerprint of this particular cosolute–protein pair. In addition to sophisticated computational studies, systematic variations of cosolute properties and protein properties such as surface area, charge, and hydrophobicity are needed to explain this heterogeneity. So far, it remains speculative why the same cosolute can interact with different proteins via different thermodynamic pathways and why cosolutes with different thermodynamic pathways often become indistinguishable on the level of free energy.

ACKNOWLEDGMENTS

We acknowledge funding from the Cluster of Excellence RESOLV (EXC 1069) funded by the German Research Foundation (DFG). M.S. is supported by the Verband der Chemischen Industrie. S.E. acknowledges funding from the Ministry of Innovation, Science and Research of the state of North Rhine-Westphalia (Rückkehrerprogramm).

REFERENCES

Attri, P., Venkatesu, P., and Lee, M. J. 2010. Influence of osmolytes and denaturants on the structure and enzyme activity of alpha-chymotrypsin. *J. Phys. Chem. B* 114:1471–1478.

Benton, L. A., Smith, A. E., Young, G. B., and Pielak, G. J. 2012. Unexpected effects of macromolecular crowding on protein stability. *Biochemistry* 51:9773–9775.

Boersma, A. J., Zuhorn, I. S., and Poolman, B. 2015. A sensor for quantification of macromolecular crowding in living cells. *Nat. Methods* 12:227–229.

Bruzdziak, P., Panuszko, A., and Stangret, J. 2013. Influence of osmolytes on protein and water structure: A step to understanding the mechanism of protein stabilization. *J. Phys. Chem. B* 117:11502–11508.

Canchi, D. R. and Garcia, A. E. 2013. Cosolvent effects on protein stability. *Annu. Rev. Phys. Chem.* 64:273–293.

Cheung, M. S., Klimov, D., and Thirumalai, D. 2005. Molecular crowding enhances native state stability and refolding rates of globular proteins. *Proc. Natl. Acad. Sci. U.S.A.* 102:4753–4758.

Christiansen, A. and Wittung-Stafshede, P. 2014. Synthetic crowding agent dextran causes excluded volume interactions exclusively to tracer protein apoazurin. *FEBS Lett.* 588:811–814.

Constantinescu, D., Weingärtner, H., and Herrmann, C. 2007. Protein denaturation by ionic liquids and the Hofmeister series: A case study of aqueous solutions of ribonuclease A. *Angew. Chem. Int. Ed. Engl.* 46:8887–8889.

Cooper, R. J., Heiles, S., DiTucci, M. J., and Williams, E. R. 2014. Hydration of guanidinium: Second shell formation at small cluster size. *J. Phys. Chem. A* 118:5657–5666.

Dandage, R., Bandyopadhyay, A., Jayaraj, G. G., Saxena, K., Dalal, V., Das, A., and Chakraborty, K. 2015. Classification of chemical chaperones based on their effect on protein folding landscapes. *ACS Chem. Biol.* 10:813–820.

Fujita, K., Forsyth, M., MacFarlane, D. R., Reid, R. W., and Elliott, G. D. 2006. Unexpected improvement in stability and utility of cytochrome c by solution in biocompatible ionic liquids. *Biotechnol. Bioeng.* 94:1209–1213.

Funkner, S., Havenith, M., and Schwaab, G. 2012. Urea, a structure breaker? Answers from THz absorption spectroscopy. *J. Phys. Chem. B* 116:13374–13380.

Gilman-Politi, R. and Harries, D. 2011. Unraveling the molecular mechanism of enthalpy driven peptide folding by polyol osmolytes. *J. Chem. Theory Comput.* 7:3816–3828.

Gnutt, D., Gao, M., Brylski, O., Heyden, M., and Ebbinghaus, S. 2015. Excluded-volume effects in living cells. *Angew. Chem. Int. Ed. Engl.* 54:2548–2551.

Gomez, J., Hilser, V. J., Xie, D., and Freire, E. 1995. The heat capacity of proteins. *Proteins* 22:404–412.

Heyden, M., Bründermann, E., Heugen, U., Niehues, G., Leitner, D. M., and Havenith, M. 2008. Long-range influence of carbohydrates on the solvation dynamics of water—answers from terahertz absorption measurements and molecular modeling simulations. *J. Am. Chem. Soc.* 130:5773–5779.

Hunger, J., Tielrooij, K. J., Buchner, R., Bonn, M., and Bakker, H. J. 2012. Complex formation in aqueous trimethylamine-N-oxide (TMAO) solutions. *J. Phys. Chem. B* 116:4783–4795.

Kaushik, J. K. and Bhat, R. 1998. Thermal stability of proteins in aqueous polyol solutions: Role of the surface tension of water in the stabilizing effect of polyols. *J. Phys. Chem. B* 102:7058–7066.

Kaushik, J. K. and Bhat, R. 2003. Why is trehalose an exceptional protein stabilizer? An analysis of the thermal stability of proteins in the presence of the compatible osmolyte trehalose. *J. Biol. Chem.* 278:26458–26465.

Lopez, M. M. and Makhatadze, G. I. 2002. *Differential Scanning Calorimetry. Calcium-Binding Protein Protocols: Volume 2: Methods and Techniques*. Chapter 9: 113–119 Totowa, NJ: Humana Press.

Makhatadze, G. I. 2001. Measuring protein thermostability by differential scanning calorimetry. *Curr. Protoc. Protein Sci.* Chapter 7: Unit 7.9.

Makhatadze, G. I. and Privalov, P. L. 1992. Protein interactions with urea and guanidinium chloride. A calorimetric study. *J. Mol. Biol.* 226:491–505.

Minton, A. P. 1981. Excluded volume as a determinant of macromolecular structure and reactivity. *Biopolymers* 20:2093–2120.

Minton, A. P. 2013. Quantitative assessment of the relative contributions of steric repulsion and chemical interactions to macromolecular crowding. *Biopolymers* 99:239–244.

O'Connor, T. F., Debenedetti, P. G., and Carbeck, J. D. 2004. Simultaneous determination of structural and thermodynamic effects of carbohydrate solutes on the thermal stability of ribonuclease A. *J. Am. Chem. Soc.* 126:11794–11795.

O'Connor, T. F., Debenedetti, P. G., and Carbeck, J. D. 2007. Stability of proteins in the presence of carbohydrates; experiments and modeling using scaled particle theory. *Biophys. Chem.* 127:51–63.

Pace, C. N., Grimsley, G. R., Thomas, S. T., and Makhatadze, G. I. 1999. Heat capacity change for ribonuclease A folding. *Protein Sci.* 8:1500–1504.

Pegram, L. M., Wendorff, T., Erdmann, R., Shkel, I., Bellissimo, D., Felitsky, D. J., and Record, M. T., Jr. 2010. Why Hofmeister effects of many salts favor protein folding but not DNA helix formation. *Proc. Natl. Acad. Sci. U.S.A.* 107:7716–7721.

Politi, R. and Harries, D. 2010. Enthalpically driven peptide stabilization by protective osmolytes. *Chem. Commun. (Camb.)* 46:6449–6451.

Privalov, P. L. 1979. Stability of proteins: Small globular proteins. *Adv. Protein Chem.* 33:167–241.

Privalov, P. L. and Makhatadze, G. I. 1990. Heat capacity of proteins II. Partial molar heat capacity of the unfolded polypeptide chain of proteins: Protein unfolding effects. *J. Mol. Biol.* 213:385–391.

Sakaue, T. and Raphael, E. 2006. Polymer chains in confined spaces and flow-injection problems: Some remarks. *Macromolecules* 39:2621–2628.

Sanchez-Ruiz, J. M. 1992. Theoretical analysis of Lumry-Eyring models in differential scanning calorimetry. *Biophys. J.* 61:921–935.

Santoro, M. M., Liu, Y. F., Khan, S. M. A., Hou, L. X., and Bolen, D. W. 1992. Increased thermal-stability of proteins in the presence of naturally-occurring osmolytes. *Biochemistry* 31:5278–5283.

Sapir, L. and Harries, D. 2014. Origin of enthalpic depletion forces. *J. Phys. Chem. Lett.* 5:1061–1065.

Sasahara, K., McPhie, P., and Minton, A. P. 2003. Effect of dextran on protein stability and conformation attributed to macromolecular crowding. *J. Mol. Biol.* 326:1227–1237.

Senske, M., Törk, L., Born, B., Havenith, M., Herrmann, C., and Ebbinghaus, S. 2014. Protein stabilization by macromolecular crowding through enthalpy rather than entropy. *J. Am. Chem. Soc.* 136:9036–9041.

Sukenik, S., Sapir, L., Gilman-Politi, R., and Harries, D. 2013a. Diversity in the mechanisms of cosolute action on biomolecular processes. *Faraday Discuss.* 160:225–237; discussion 311–327.

Sukenik, S., Sapir, L., and Harries, D. 2013b. Balance of enthalpy and entropy in depletion forces. *Curr. Opin. Colloid Interface Sci.* 18:495–501.

van der Post, S. T., Tielrooij, K. J., Hunger, J., Backus, E. H. G., and Bakker, H. J. 2013. Femtosecond study of the effects of ions and hydrophobes on the dynamics of water. *Faraday Discuss.* 160:171–189.

van Rantwijk, F., Lau, R. M., and Sheldon, R. A. 2003. Biocatalytic transformations in ionic liquids. *Trends Biotechnol.* 21:131–138.

Waegele, M. M. and Gai, F. 2011. Power-law dependence of the melting temperature of ubiquitin on the volume fraction of macromolecular crowders. *J. Chem. Phys.* 134:095104.

Wang, Y., Sarkar, M., Smith, A. E., Krois, A. S., and Pielak, G. J. 2012. Macromolecular crowding and protein stability. *J. Am. Chem. Soc.* 134:16614–16618.

Wintrode, P. L., Makhatadze, G. I., and Privalov, P. L. 1994. Thermodynamics of ubiquitin unfolding. *Proteins* 18:246–253.

Xie, G. F. and Timasheff, S. N. 1997. Mechanism of the stabilization of ribonuclease A by sorbitol: Preferential hydration is greater for the denatured than for the native protein. *Protein Sci.* 6:211–221.

Zhou, H. X. 2013. Polymer crowders and protein crowders act similarly on protein folding stability. *FEBS Lett.* 587:394–397.

Zhou, H. X., Rivas, G., and Minton, A. P. 2008. Macromolecular crowding and confinement: Biochemical, biophysical, and potential physiological consequences. *Annu. Rev. Biophys.* 37:375–397.

Zimmerman, S. B. and Trach, S. O. 1991. Estimation of macromolecule concentrations and excluded volume effects for the cytoplasm of *Escherichia coli. J. Mol. Biol.* 222:599–620.

CHAPTER 14

Isothermal Titration Calorimetry and Fluorescent Thermal and Pressure Shift Assays in Protein– Ligand Interactions

Vytautas Petrauskas, Lina Baranauskienė,

Asta Zubrienė, and Daumantas Matulis

CONTENTS

14.1 THERMODYNAMICS OF PROTEIN–LIGAND BINDING

One of the main unanswered questions in biophysics or even in life sciences is how small-molecule ligands recognize a protein and what are the exact energy contributions of all involved factors, including hydrogen bonds, hydrophobic interactions, and van der Waals and ionic forces. The goal is to be able to predict *in silico* not only whether a particular chemical would bind a particular protein, but also the exact structural arrangement of the bound ligand with an accurate estimation of all energies involved in the binding reaction. Part of the difficulty lies in our poor understanding of water structure and hydrogen bond network.

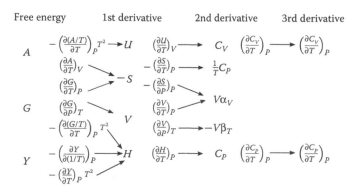

FIGURE 14.1 Energies of any reaction at equilibrium can be evaluated by one or all of the above listed thermodynamic functions (parameters). Their selection applicable in biological sciences is quite limited. Energies (A, Helmholtz energy; G, Gibbs energy; Y, Planck function [Y = G/T]) are listed in the first column, their first derivatives (U, internal energy; S, entropy; V, volume; H, enthalpy) in the second column, their second derivatives (C_V, constant-volume heat capacity; C_P, constant-pressure heat capacity; $V\alpha_V$, isochoric thermal expansion; $-V\beta_T$, isothermal compressibility) in the third column, and their third derivatives (only the ones that have been discussed or published [from Baranauskiene, L. et al., 2008]) in the last column. Arrows show the functions that would be obtained by performing the derivation as shown before the arrow. The thermodynamic functions are linked through derivative–integral relationships. Full thermodynamic characterization of any biological reaction should include all these parameters. In practice, however, most reactions are characterized only by G (e.g., affinity at constant T and P).

Let us start by reviewing the list of thermodynamic functions, that is, thermodynamic parameters, that can be directly experimentally measured and recall that the set is quite limited (Figure 14.1). Employing today's instrumentation, one can easily measure the Gibbs energy (ΔG) and enthalpy (ΔH) of protein–ligand binding. The ΔG is measured by dosing ligand at constant protein concentration and measuring a signal that is proportional to the fraction of occupied ligand-binding sites on a protein. The ΔH is easily obtained from isothermal titration calorimetry (ITC) where the heat of binding at constant pressure can be directly measured. Other functions can only be determined through derivative and integral relations between the thermodynamic functions. For example, the temperature derivative of ΔG is negative entropy (ΔS), $(\partial G/\partial T)_P = -S$; thus, plotting ΔG values determined at several temperatures as a function of temperature could yield the entropy. The temperature derivative of enthalpy is the heat capacity at constant pressure, $(\partial H/\partial T)_P = C_P$; thus, after obtaining enthalpy changes at several temperatures, one can estimate the heat capacity change of a reaction.

Several energies can be distinguished, such as the Helmholtz energy ($A = U - TS$) and the Gibbs energy ($G = U + PV - TS = H - TS$ or $G = -(\partial(AV)/\partial V)_T V^2$) and the relationships among them derived, as, for example, the Gibbs–Helmholtz equation ($(\partial(G/T)/\partial T)_p = -H/T^2$) or the Planck function $Y (Y = -G/T)$. The first derivatives of Helmholtz and Gibbs energies are U, $-S$, V, and H, and second derivatives are C_V, $V\alpha_V$

(isochoric thermal expansion), $-V\beta_T$ (isothermal compressibility), and C_P. The thermodynamic functions and their derivative and integral relationships are summarized in Figure 14.1.

Unfortunately, if the thermodynamic function is not being directly measured but estimated through derivative relationships, then some information is lost and the results could be highly misleading. For example, the ΔG of a phase change (e.g., melting of a solid pure material and protein denaturation) is zero, but the enthalpy is not zero. Obtaining the enthalpy through the van't Hoff relationship by plotting ΔG obtained at various temperatures as a function of temperature would hide the enthalpy of a phase change and would thus be essentially meaningless (Matulis 2001; Matulis and Bloomfield 2001).

14.2 FLUORESCENT THERMAL AND PRESSURE SHIFT ASSAYS VERSUS ISOTHERMAL TITRATION CALORIMETRY

ITC is an excellent method to measure protein–ligand-binding affinities, enthalpies, entropies, and indirectly heat capacities. No other method can measure experimentally the heat absorbed or released per injection, that is, the enthalpy, if the pressure is constant. These are great advantages of ITC over other available techniques.

However, ITC also has limitations that caused it to be less widely used. Unfortunately, many scientists, particularly in the biological area, are commonly not interested in thermodynamic parameters other than the Gibbs energy (affinity) of binding. They are seldom looking at enthalpies and the other above-discussed parameters. Furthermore, ITC equipment is quite expensive, it requires a significant amount of purified protein, and the time required for each experiment is about 1 h. Therefore, many give preference to quick methods that also require less material and are applicable for screening, such as enzymatic activity inhibition methods and fluorescence-based methods. However, parameters such as the enthalpy change upon binding are then never determined.

In our opinion, ITC is probably the most robust and unambiguous method to determine the dissociation constant (K_d) if the affinity falls into the right range, somewhere between micromolar and nanomolar. The window of affinities where the direct measurements of K and ΔH are the most accurate is actually rather narrow. The Wiseman parameter c, which is used to plan ITC experiments and defined as the product of K_b, stoichiometry (n), and the macromolecule concentration in the cell (M) ($c = nMK_b$), should be between 5 and 500 to give the most accurate results (Broecker et al. 2011; Wiseman et al. 1989), so the dynamic window for determination of the K_d in practice spans only two orders of magnitude. Theoretically, c can be increased to 1000, but then there is some uncertainty as to whether the binding is not stronger than it appears.

A rare but interesting observation, shown in Figure 14.2, describes a net zero enthalpy of binding. At a particular pH or in a buffer of a particular enthalpy of protonation, the observed enthalpy of binding may be nearly equal to zero. This was observed in the ITC measurement of topiramate drug binding to carbonic anhydrase (CA) II in MES buffer. The reaction performed in phosphate buffer yielded an endothermic enthalpy, while in TRIS buffer a large exothermic enthalpy was observed. A similar ligand, acetazolamide, was found to bind with highly exothermic and similar enthalpies of binding in various

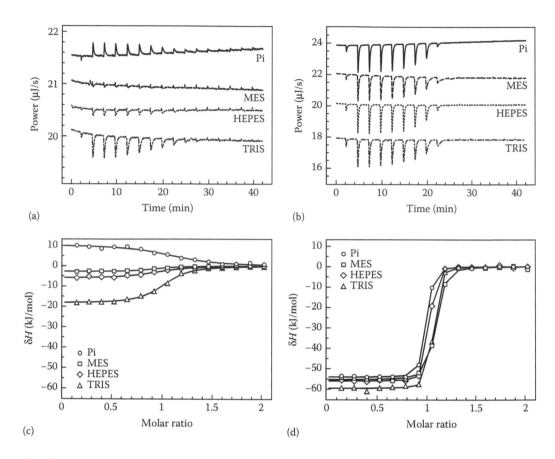

FIGURE 14.2 An example is shown of a rare but possible pitfall in ITC—disappearance of the signal due to perfect compensation of the binding and linked protonation enthalpies. The left column shows topiramate, while the right column shows acetazolamide binding to recombinant human CA II. Upper panels (a, b) show raw ITC data (shifted up or down for clarity), while the bottom panels (c, d) show integrated ITC data. Experiments were performed at pH 7.0 but in various buffers (100 mM)—sodium phosphate (Pi), MES, HEPES, and TRIS. The binding constants of topiramate and acetazolamide were 2.2×10^6 M^{-1} and 9.5×10^7 M^{-1} respectively, but the enthalpies and linked protonation reactions were quite different. Coincidentally, performing topiramate binding in MES or HEPES buffers could lead to near-zero heat signals. Therefore, for any binding reaction, results at several pH values or in several buffers should be obtained.

buffers. This effect was observed because the ligand binding to the protein is linked to several protonation reactions (Baker and Murphy 1996). The observation of zero enthalpy in some conditions illustrates how important it is to perform the binding reactions at several experimental conditions. We suggest varying either the pH or the buffer of the system in order not to miss a binding reaction that could appear not to occur if the experiment is conducted at a single particular condition.

The second best method to measure only protein–ligand affinity, in our opinion, is fluorescent thermal shift assay (FTSA), also frequently termed differential scanning fluorimetry (DSF) or, in high throughput, ThermoFluor® (Lo et al. 2004; Pantoliano et al. 2001; Todd and Salemme 2003). The method is based on the fact that ligands stabilize proteins upon binding. The result is very general and has been observed for hundreds of proteins. As long as the protein and ligand are soluble, it is possible to perform this experiment. It may take as little as 1 μl of a solution containing 1 μM protein and 1 μM ligand to determine if the protein's melting temperature is shifted upward upon adding the ligand, and thus there is binding between them. The same experiment should be repeated at several ligand concentrations to determine the K_d. Miniaturization can be applied in a robotic screening of large compound libraries to discover hits that bind to target proteins. Hundreds of compounds can be simultaneously tested in 1 h.

The FTSA method determines the linkage between two reactions—the protein unfolding reaction and the ligand-binding reaction. We are interested in determining the equilibrium constant for the binding reaction at physiological temperature, but the unfolding reaction occurs at elevated temperatures. Stabilization by the ligand usually occurs around 60°C–70°C, and the effect is then extrapolated to 37°C or even 25°C. It would seem that such a method should have large uncertainties, but in practice it gives results similar to those of ITC if performed properly.

In FTSA, we observe a fluorescence change caused by protein unfolding. Fluorescence may be intrinsic, primarily that of tryptophan, or extrinsic—a solvatochromic fluorescent compound may be added to the protein–ligand solution to report on the protein unfolding. We usually perform the experiments using 1,8-anilino-naphthalene sulfonate (ANS), which exhibits low fluorescence in water and increases fluorescence yield (Figure 14.3) upon binding to some form of the unfolded protein, thus avoiding fluorescence quenching by water (Matulis et al. 1999; Matulis and Lovrien 1998). The presence of a probe may appear as a wrong choice because the probe may bind to the active site of the protein and compete with the ligand. In fact, such competition has been observed quite rarely, and the effect can be easily subtracted by repeating the experiment at several ANS concentrations. It should be kept in mind that the probe should be molecularly as small as possible because larger hydrophobic molecules usually bind strongly to proteins. Therefore, we do not support using probes such as SYPRO Orange in FTSA (Cimmperman and Matulis 2011). Unfortunately, most fluorescence detecting equipment that have the capability to heat samples to 100°C do not have proper lasers to excite ANS and record the fluorescence at the right wavelengths.

In theory, for the FTSA to work properly, the protein unfolding transition should be perfectly reversible. However, in reality, the unfolding of most proteins is irreversible or only partially reversible. We have tested numerous proteins with varying extents of reversibility of unfolding and found that even for the proteins that exhibit a low degree of reversibility, the match between affinities determined by FTSA and ITC is still nearly perfect. This is puzzling and unexpected, but it happens most likely because there is reversibility in the narrow temperature range near the melting transition in both the absence and the presence of ligand.

FTSA has a very large dynamic window for affinity determination. It can determine affinities in the range of millimolar to picomolar, thus spanning about nine orders of magnitude for this thermodynamic parameter. The only limitation is the boiling temperature of water. Therefore, the protein melting transitions should be below 90°C to be clearly

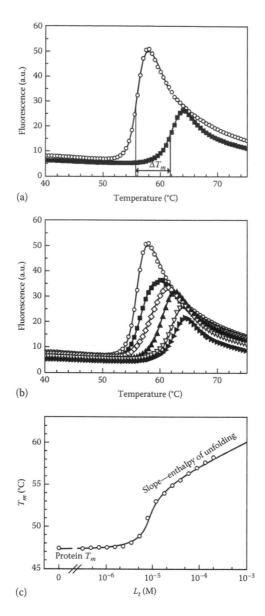

FIGURE 14.3 Typical data of FTSA. (a) Fluorescence increases upon protein thermal melting or denaturation with the midpoint equal to the melting temperature (T_m). Addition of a ligand induces the melting temperature shift due to protein stabilization upon ligand binding. (b) Increasing ligand concentrations (0, 3, 7, 17, 59, and 200 μM of 4-{2-[N-(6-methoxy-5-nitropyrimidin-4-yl)amino] ethyl}benzenesulfonamide added to 10 μM of CA II) elevated the T_m to a greater extent. (From Harding, S. E. and Chowdhry, B., *Protein-Ligand Interactions: Hydrodynamics and Calorimetry*, Oxford University Press, New York, 2001; Sudzius, J. et al., *Bioorg. Med. Chem.* 18:7413–7421, 2010.) (c) The resulting T_m values are plotted as a function of added ligand concentration, with the ligand dosing curve showing methazolamide binding to CA VII. The line is a nearly perfect fit of experimental data to Equation 14.8. The sigmoidal bend part corresponds to the binding stoichiometry and protein concentration, while the slopes of curves at transition (a and b) and the slope of the right side of the curve in (c) are proportional to the enthalpy of protein unfolding.

observable. If the stabilization by the ligand reaches a shift of 20 degrees, then the unliganded protein must melt below 70°C. Alternatively, very stable proteins may have to be destabilized by additives such as urea or guanidinium chloride or by lowering the pH. Furthermore, even very weakly binding ligands, for example, with K_d of 100 mM, can be observed as long as the concentration of added ligand exceeds 100 mM. Therefore, the method can observe in the same experimental plate both the most weakly and most tightly binding ligands. Figure 14.4a shows the FTSA dosing curves of four compounds that exhibited affinities from 1.4 µM to 0.4 nM. Figure 14.4b shows the expected dosing curves of millimolar and picomolar ligands. Similarly to ITC, FTSA can be used to determine the stoichiometry of the protein–ligand reaction, or to determine also quite accurately the concentration of the protein or ligand if the stoichiometry is known (Figure 14.4c).

The combination of both methods, ITC and FTSA, yields the most accurate results of protein–ligand affinity determination. If both methods yield K_b values that do not differ more than three times, then the affinity is determined quite accurately and can be trusted (Figure 14.5). When the affinity is tight, in the subnanomolar range, ITC provides very accurate measurement of the enthalpy, while FTSA can be used to obtain the affinity. Performing ITC at several temperatures will provide the heat capacity change upon binding, while the combination of enthalpy and Gibbs energy changes will yield the entropy change. All four thermodynamic parameters taken together provide a comprehensive picture of the binding reaction.

It is important to note that both the ITC and FTSA measurements yield so-called observed thermodynamic parameters. They should be distinguished from the intrinsic thermodynamic parameters. For example, in Figure 14.2, the enthalpy determined by ITC in various buffers for topiramate varied dramatically from exothermic to endothermic values. These enthalpies are the observed enthalpies, at particular conditions. However, the observed changes are due to the linked protonation reactions of both the protein and the ligand. In practice, it is not the observed binding enthalpy that is of interest, but rather the intrinsic enthalpy of binding (the hypothetical enthalpy change observed if there were no linked reactions) that would be independent of buffer and pH. Intrinsic values can be obtained by determining and subtracting the effect of linked reactions (Baker and Murphy 1996) that may not be limited to protonation but may involve conformational changes and other effects.

Similar to the example with enthalpy, the Gibbs energy of ethoxzolamide binding to carbonic anhydrase depends on pH (Figure 14.6a). Both ITC and FTSA show the same dependence of the affinity on pH. The affinity is greatest near neutral pH and decreases 10 times with each pH unit when going toward acidic or basic pH. The FTSA dosing curves show the decrease in affinity in the acidic and basic pH region (Figure 14.6b). In this system, the observed values by both FTSA and ITC never reach the intrinsic affinity (the affinity that would be hypothetically observed if there were no linked reactions). It is very important to dissect and determine the intrinsic values because ultimately our goal is to determine the structure–thermodynamics relationship, that is, to correlate the thermodynamic parameters with the chemical structures of the ligands and with the crystallographic structures of the protein–ligand complexes.

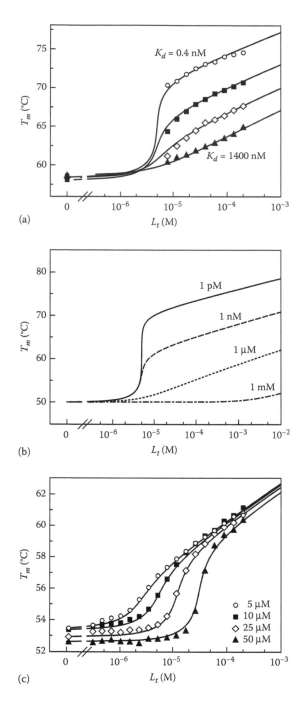

FIGURE 14.4 FTSA can determine a wide range of affinities in the same experiment. (a) Ligand dosing curves of four compounds binding to CA XIII covering affinities from micromolar to sub-nanomolar. (b) Simulated dosing curves for the K_d of 1 mM, 1 µM, 1 nM, and 1 pM. Other simulation parameters in Equation 14.8 were fixed: $T_r = 50°C$, $\Delta H_U = 450$ kJ/mol, $\Delta C_{P\_U} = 17$ kJ/(mol K), $\Delta H_b = -42$ kJ/mol, $\Delta C_{P\_b} = -0.8$ kJ/mol, and $T = 25°C$. (c) Acetazolamide dosing curves at different protein (CA II) concentrations (5, 10, 25, and 50 µM) illustrating the shift of the sigmoidal part of the curve that enables determination of the protein concentration or the stoichiometry.

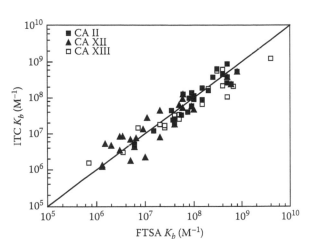

FIGURE 14.5 Comparison of K_b values determined by FTSA and ITC for several protein–ligand interactions: CA II (closed squares), CA XII (triangles), and CA XIII (open squares). Both methods determine well the affinities. However, despite numerous repetitions and great care in both experimental designs, there is a statistically significant spread of data around the perfect match line. This spread is partially due to random thermal motion (equal to $RT = 2.48$ kJ/mol at 25°C), which by itself contributes an error for K_b equal to −63% and +172%. This is equal to ±2 times in affinity (ΔG) and cannot be determined to a greater precision. This constant error in ΔG (±2.48 kJ/mol) yields a deviation of ±7.3% for $K_b = 10^6$ M^{-1} and ±5.4% for $K_b = 10^8$ M^{-1}. Therefore, any attempts to claim affinity determination with an error of less than approximately ±2 times in K_b are wrong.

The enthalpy, entropy, and heat capacity are all *temperature* derivatives of the Gibbs energy. In addition, we are trying to promote research to obtain *pressure* derivatives of the Gibbs energy, primarily the volume change upon binding. Unfortunately, experiments at high pressures are very tedious and expensive, and therefore the data are scarce and it is hard to make comparison with any other relevant data in the literature when performing high-pressure experiments.

The volume change upon binding could theoretically be obtained by measuring the total volume of the aqueous system containing free protein and free ligand versus the protein–ligand bound complex. However, such equipment does not exist and the volume change upon binding is quite small. For example, a small protein bound a ligand with the volume change of −170 mL/mol (Petrauskas et al. 2013). The volume of 1 mole of protein with molecular mass of 30 kDa, weighing 30 kg, is approximately 30 L. If 1 L of a concentrated 1 mM protein solution could be prepared, the volume of binding would be 5.7 µL. Thus, the volumes of both free and bound components should be measured with a precision of at least 1 µL—with a precision of six decimal digits. Such high precision is not yet feasible and may be prone to capillary forces that would make measurements nearly impossible.

Therefore, the volume change in protein–ligand-binding reactions is being measured indirectly by observing the affinity change upon the change in pressure (Figure 14.7). We termed the method fluorescent pressure shift assay (FPSA) or PressureFluor. It is based on the fact that proteins are expected and observed to withstand larger pressures with a bound

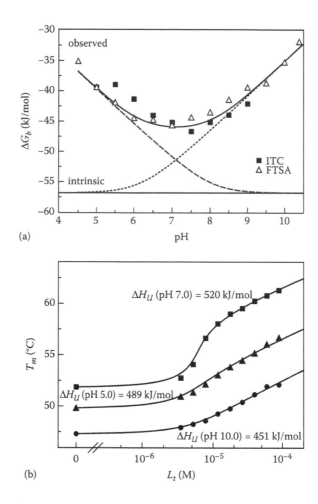

FIGURE 14.6   Any affinity determination methods, including ITC and FTSA, determine only the observed thermodynamic parameters. They do not determine the intrinsic parameters that should be independent of any linked reactions, primarily binding-linked protonation effects. (a) The data points show the affinities of ethoxzolamide binding to CA XII as a function of pH determined by ITC (squares) and FTSA (open triangles). Affinities are the same as determined by both methods and follow a U shape. Furthermore, there is no pH where observed values reach the intrinsic values. Lines show the fit of the data (solid U-shape line), the fractions of protonation forms of the protein (dotted line) and ligand (dashed line), and the pH-independent value for intrinsic affinity (straight line)—all fit to the models described in Baranauskiene and Matulis (2012), Jogaite et al. (2012), Khalifah (1973), Krishnamurthy et al. (2008), and Morkunaite et al. (2014). (b) FTSA ethoxzolamide dosing curves at various pH values (5.0, triangles; 7.0, squares; 10.0, circles). Protein (CA XII) has different T_m values and different enthalpies of unfolding (ΔH_U) at different pH values.

ligand than in an unliganded form. Thus, ligands stabilize proteins against unfolding by pressure increase.

The pressure-unfolded form of the protein is probably quite different from the temperature-unfolded form of the protein, but the method and the formulas describing the effect are similar. The structures of both unfolded forms are unknown, and it could only

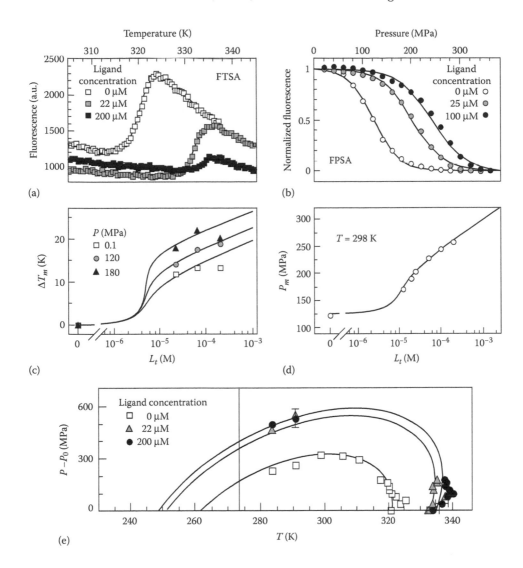

FIGURE 14.7 FPSA determines ligand affinities from their stabilization effect on proteins against denaturation by pressure instead of temperature. (a) Hsp90N protein denaturation by temperature. (b) Human serum albumin denaturation by pressure, in both the absence and the presence of ligand. (c) Ligand dosing curves by FTSA at various P values of Hsp90N. (d) Ligand dosing curve by FPSA of human serum albumin. (From Petrauskas, V. et al., *Eur. Biophys. J.* 42:355–362, 2013; Toleikis, Z. et al., *Anal. Biochem.* 413:171–178, 2011; Toleikis, Z. et al., *J. Chem. Thermodyn.* 52:24–29, 2012.) (e) Combination of FTSA and FPSA can yield a full protein stability phase diagram as a function of P and T at various ligand concentrations. The panel shows an Hsp90N stability diagram without a ligand (squares) and with 22 μM (triangles) and 200 μM (circles) of added ligand ICPD47.

be hypothesized what are the similarities and differences of the two unfolding processes. Despite the lack of significant amount of data, it can already be said that both FTSA and FPSA provide similar K_b values for the same reaction. Furthermore, the enthalpy of binding provides further information about the type of interaction, and the volume of interaction provides structural insight into the binding reaction.

14.3 DERIVATION OF EQUATIONS USED IN FLUORESCENT THERMAL SHIFT ASSAY: DERIVATION OF EQUATIONS OF LIGAND DOSING CURVES

In part, this derivation has been previously published (Cimmperman and Matulis 2011; Matulis et al. 2005). The stability of proteins is measured by shifting the equilibrium between the native (folded) and unfolded protein molecules. The equilibrium can be changed by varying environmental factors such as temperature, pressure, and chemical composition of protein solution.

The Gibbs energy (absolute) per mole of native protein can be described as the Gibbs energy under standard conditions (G_N°) plus an additional term accounting for the protein energy at a particular state:

$$G_N = G_N^\circ + RT \ln\left[N\right] \tag{14.1}$$

The same Gibbs energy formalism is valid for the unfolded protein state:

$$G_U = G_U^\circ + RT \ln\left[U\right] \tag{14.2}$$

where [N] and [U] are the molar concentrations of folded (native) and unfolded protein, respectively. Usually, the Gibbs energy of an unfolded protein is measured relative to the energy level of the folded protein, or in other words, the folded state of a protein is considered the reference state. The energy required to unfold (or melt) a protein ΔG_U is then calculated as a difference in the molar Gibbs energies G_U and G_N:

$$\Delta G_U = G_U - G_N = G_U^\circ - G_N^\circ + RT \ln[U] - RT \ln\left[N\right] \tag{14.3}$$

or with a notation of $\Delta G_U^\circ = G_U^\circ - G_N^\circ$:

$$\Delta G_U = \Delta G_U^\circ + RT \ln\left(\frac{\left[U\right]}{\left[N\right]}\right) \tag{14.4}$$

At equilibrium, $\Delta G_U = 0$ (by definition) and the change in standard Gibbs energy of unfolding is

$$\Delta G_U^\circ = -RT \ln\left(K_U\right) \tag{14.5}$$

where T is the absolute temperature, R is the molar gas constant, and

$$K_U = \frac{\left[U\right]}{\left[N\right]} \tag{14.6}$$

is the temperature-dependent equilibrium constant of protein unfolding. K_U is defined with an assumption that there are only two protein states at equilibrium, native and unfolded,

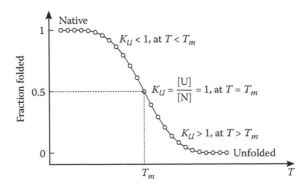

FIGURE 14.8 Typical curve of thermal unfolding of a protein, where the midpoint of the curve (when half of the protein is folded and half is unfolded) corresponds to T_m. The equilibrium unfolding constant K_U is different at different temperatures.

and the transition between states is stepwise (i.e., without any intermediate states, like "semiunfolded," the usually called "two-state process").

$$[N] \xleftarrow{\quad K_U \quad} [U] \qquad (14.7)$$

An example of protein denaturation by temperature increase is shown in Figure 14.8. A thermal unfolding curve measures the temperature dependence of the equilibrium constant of unfolding K_U.

The change in Gibbs energy of protein unfolding as a function of temperature can be expressed in terms of the changes in enthalpy ΔH_U and entropy ΔS_U:

$$\Delta G_U = \Delta H_U - T\Delta S_U \qquad (14.8)$$

Enthalpy and entropy are also temperature-dependent functions:

$$\Delta H_U = \Delta H_{U\_T_r} + \int_{T_r}^{T} \Delta C_{P\_U} dT \qquad (14.9a)$$

and

$$\Delta S_U = \Delta S_{U\_T_r} + \int_{T_r}^{T} \left(\frac{\Delta C_{P\_U}}{T} \right) dT \qquad (14.9b)$$

where ΔC_{P_U} is the change in constant-pressure heat capacity associated with protein unfolding and T_r is the reference temperature, which is usually defined as protein melting temperature without added ligand. The heat capacity weakly depends on temperature in the temperature range of protein unfolding; thus, we will consider it a temperature-independent thermodynamic parameter. This assumption enables us to integrate Equations 14.9a and 14.9b as follows:

$$\Delta H_U = \Delta H_{U\_T_r} + \Delta C_{P\_U} \int_{T_r}^{T} dT = \Delta H_{U\_T_r} + \Delta C_{P\_U}\left(T - T_r\right) \tag{14.9c}$$

$$\Delta S_U = \Delta S_{U\_T_r} + \Delta C_{P\_U} \int_{T_r}^{T} \frac{dT}{T} = \Delta S_{U\_T_r} + \Delta C_{P\_U}\left(\ln T - \ln T_r\right) = \Delta S_{U\_T_r} + \Delta C_{P\_U} \ln\left(\frac{T}{T_r}\right)$$

$$\tag{14.9d}$$

Protein stabilization by a ligand is described by an increase in protein stability against a denaturing factor. The binding equilibrium

$$\left[N\right] + \left[L\right] \xleftrightarrow{K_b} \left[NL\right] \tag{14.10}$$

is described as follows:

$$K_b = \frac{\left[NL\right]}{\left[N\right]\left[L\right]} \tag{14.11}$$

where [L] and [NL] are the concentrations of free ligand and protein–ligand complex, respectively. We will derive an equation that relates total concentrations of added protein and ligand with the equilibrium constants of protein unfolding and protein–ligand binding. Total protein and ligand concentrations P_t and L_t, respectively, are

$$P_t = \left[N\right] + \left[U\right] + \left[NL\right] \tag{14.12}$$

and

$$L_t = \left[L\right] + \left[NL\right] \tag{14.13}$$

P_t and L_t equations in combination with both equilibrium constants K_U and K_b comprise a system of four independent equations with four unknown variables [N], [U], [NL], and [L]. Substitution of [NL] in Equations 14.12 and 14.13 with [NL] = K_b[N][L] from Equation 14.11 and [U] in Equation 14.12 with [U] = K_U[N] from Equation 14.6 yields

$$\begin{cases} P_t = \left[N\right]\left(1 + K_b\left[L\right] + K_U\right) \\ L_t = \left[L\right]\left(1 + K_b\left[N\right]\right) \end{cases} \tag{14.14}$$

The elimination of [N] and [L] variables is straightforward, but the resultant algebraic expressions are highly nontrivial.

We can express L_t in another way, as a function of P_t, K_U, and K_b. Following Equation 14.11, we can write $[L] = [NL]/[N]K_b$ and put it into Equation 14.13:

$$L_t = \frac{[NL]}{[N]K_b} + [NL] = [NL]\left(\frac{1}{[N]K_b} + 1\right) \tag{14.15}$$

Let us express [N] from Equation 14.6 and [NL] from Equation 14.12: $[N] = [U]/K_U$ and $[NL] = P_t - [N] - [U] = P_t - [U]/K_U - [U]$. Substitution of [N] and [NL] in Equation 14.15 with the later expressions leads to

$$L_t = \left(P_t - \frac{[U]}{K_U} - [U]\right)\left(\frac{K_U}{[U]K_b} + 1\right) \tag{14.16}$$

We could simplify Equation 14.16 and get rid of the unknown concentration of unfolded protein [U] by sticking to a particular temperature, namely, protein melting temperature T_m. At this temperature, the protein folded and unfolded fractions are equal, which mathematically reads as

$$[U] = \frac{P_t}{2} \tag{14.17}$$

Substituting this into Equation 14.16, we get

$$L_t = \left(P_t - \frac{P_t}{2K_{U\_T_m}} - \frac{P_t}{2}\right)\left(\frac{K_{U\_T_m}}{K_{b\_T_m}}\frac{2}{P_t} + 1\right) = \frac{P_t}{2K_{U\_T_m}}\left(K_{U\_T_m} - 1\right)\left(\frac{K_{U\_T_m}}{K_{b\_T_m}}\frac{2}{P_t} + 1\right) \tag{14.18}$$

The final form of equation relating total concentrations of added protein and ligand with the equilibrium constants of protein unfolding and protein–ligand binding at temperature T_m is

$$L_t = \left(K_{U\_T_m} - 1\right)\left(\frac{P_t}{2K_{U\_T_m}} + \frac{1}{K_{b\_T_m}}\right) \tag{14.19}$$

Equilibrium constants at temperature T_m, $K_{U\_T_m}$ and $K_{b\_T_m}$, can be expressed in terms of the corresponding standard Gibbs energies:

$$K_{U\_T_m} = \exp\left(-\frac{\Delta G^\circ_{U\_T_m}}{RT}\right) \tag{14.20}$$

$$K_{b\_T_m} = \exp\left(-\frac{\Delta G^\circ_{b\_T_m}}{RT}\right) \qquad (14.21)$$

Hereafter, we continue using the thermodynamic parameters in their standard states, but for the sake of simplicity we drop the superscript ° denoting the standard state. ΔG_U and ΔG_b are temperature-dependent functions of enthalpy and entropy (see Equations 14.8, 14.9a, and 14.9b). But the enthalpy and entropy are also temperature-dependent functions; thus, we will use the following approximations of ΔH and ΔS at temperature T_m:

$$\Delta G_{U\_T_m} = \Delta H_U - T_m \Delta S_U = \Delta H_{U\_T_r} + \Delta C_{P\_U}\left(T_m - T_r\right) - T\left(\Delta S_{U\_T_r} + \Delta C_{P\_U} \ln\frac{T_m}{T_r}\right) \quad (14.22)$$

$$\Delta G_{b\_T_m} = \Delta H_b - T_m \Delta S_b = \Delta H_{b\_T_0} + \Delta C_{p\_b}\left(T_m - T_0\right) - T\left(\Delta S_{b\_T_0} + \Delta C_{p\_b} \ln\frac{T_m}{T_0}\right) \quad (14.23)$$

where T_0 is the reference temperature of ligand binding (usually 37°C or 25°C), and ΔC_{P_U} and ΔC_{P_b} are the changes in constant-pressure heat capacity of unfolding and binding. From Equations 14.19 through 14.23, we get the final form of ligand concentration L_t, which is required to reach the protein melting temperature T_m:

$$\begin{cases} L_t = \left(K_{U\_T_m} - 1\right)\left(\dfrac{Pt}{2K_{U\_T_m}} + \dfrac{1}{K_{b\_T_m}}\right) \\[2ex] K_{U\_T_m} = \exp\left(-\dfrac{\Delta H_{U\_T_r} + \Delta C_{P\_U}\left(T_m\_T_r\right) - T\left(\Delta S_{U\_T_r} + \Delta C_{P\_U} \ln T_m/T_r\right)}{RT}\right) \\[2ex] K_{b\_T_m} = \exp\left(-\dfrac{\Delta H_{b\_T_0} + \Delta C_{p\_b}\left(T_m - T_0\right) - T\left(\Delta S_{b\_T_0} + \Delta C_{p\_b} \ln T_m/T_0\right)}{RT}\right) \end{cases} \quad (14.24)$$

Several assumptions were made to derive this system of equations: (1) transitions are two state, (2) the ligand binds only the native state of the protein, (3) the constant-pressure heat capacity is temperature independent, and (4) the equilibrium constants of protein unfolding and protein–ligand binding are defined at melting temperature T_m.

14.4 DERIVATION OF EQUATIONS OF PROTEIN UNFOLDING CURVES

One of the possible ways to determine the T_m of protein unfolding is by observing the intrinsic or extrinsic fluorescence upon unfolding. The derivation of necessary equations has been previously partially published (Cimmperman and Matulis 2011; Matulis et al.

2005). The probability that protein is in the native (p_N) or unfolded (p_U) state at temperature T can be expressed as

$$p_N = \frac{1}{1+\exp\left(-\Delta G_U/RT\right)}; p_U = 1 - p_N = \frac{1}{1+\exp\left(\Delta G_U/RT\right)} \qquad (14.25)$$

The fluorescence intensity f at a particular temperature T is calculated as follows:

$$f = f_N + p_U\left(f_U - f_N\right) = f_N + \frac{f_U - f_N}{1+\exp\left(\Delta G_U/RT\right)}, \qquad (14.26)$$

where f_N and f_U are fluorescence yields at native and unfolded states, respectively. Usually, pre- and posttransition fluorescence baselines are not flat and can be approximated as linear temperature dependences:

$$f_N = f_{N\_T_m} + m_N\left(T - T_m\right) \qquad (14.27)$$

$$f_U = f_{U\_T_m} + m_U\left(T - T_m\right) \qquad (14.28)$$

where m_N and m_U are pre- and posttransition fluorescence baseline slopes, and $f_{N\_T_m}$ and $f_{U\_T_m}$ are fluorescence yields of native and unfolded states at temperature T_m, respectively. Substituting f_N and f_U in Equation 14.26 by their expressions given by Equations 14.27 and 14.28, we get

$$f = f_{N\_T_m} + m_N\left(T - T_m\right) + \frac{f_{U\_T_m} + m_U\left(T - T_m\right) - f_{N\_T_m} + m_N\left(T - T_m\right)}{1+\exp\left(\Delta G_U/RT\right)} \qquad (14.29)$$

Putting the expression of ΔG_U given by Equation 14.8 into Equation 14.29 and making use of Equations 14.9c and 14.9d at reference temperature $T = T_m$ yields

$$f = f_{N\_T_m} + m_N\left(T - T_m\right)$$

$$+ \frac{f_{U\_T_m} - f_{N\_T_m} + \left(m_U - m_N\right)\left(T - T_m\right)}{1+\exp\left(\dfrac{\Delta H_{U\_T_m} + \Delta C_{p\_U}\left(T - T_m\right) - T\left(\Delta S_{U\_T_m} + \Delta C_{p\_U}\ln T/T_m\right)}{RT}\right)} \qquad (14.30)$$

The fluorescence yield model (Equation 14.30) is fitted to the experimental fluorescence data by varying T_m, $\Delta H_{U\_T_m}$, m_N, m_U, $f_{N\_T_m}$, and $f_{U\_T_m}$.

A model where two melting transitions are discussed has been previously published (Zubriene et al. 2009). Discussion of fluorescent probes for FTSA can be found in Cimmperman and Matulis (2011).

14.5 DERIVATION OF FLUORESCENT PRESSURE SHIFT ASSAY EQUATIONS

Pressure is as important as temperature, but is a highly neglected thermodynamic parameter. In general, the Gibbs energy of unfolding is a function of both temperature and pressure. Thus, it can be described as a sum of two components: temperature and pressure-dependent terms ΔG_{U_T} and ΔG_{U_P}:

$$\Delta G_U = \Delta G_{U\_T} + \Delta G_{U\_P} \tag{14.31}$$

ΔG_{U_T} is the same as in Equation 14.8. The additional pressure-dependent term describes volumetric properties of proteins and is expressed as

$$\Delta G_{U\_P} = \Delta G_{U\_P_r} + \Delta V_P \left(P - P_r\right) \tag{14.32}$$

where P_r is reference pressure, $\Delta G_{U\_P_r}$ is the Gibbs free energy change at P_r, and ΔV_P is the pressure-dependent change in protein volume described as

$$\Delta V_P = \Delta V_{P_r} + \frac{\Delta \beta_U}{2}\left(P - P_r\right) \tag{14.33}$$

Here $\Delta \beta_U = \partial \Delta V_P / \partial P$ is the compressibility factor and ΔV_{P_r} is the change in protein volume at reference pressure. Substituting ΔV_P in Equation 14.32 with Equation 14.33, we get ΔG_{U_P} as a function of pressure at constant temperature:

$$\Delta G_{U\_P} = \Delta G_{U\_P_r} + \Delta V_{P_r}\left(P - P_r\right) + \frac{\Delta \beta_U}{2}\left(P - P_r\right)^2 \tag{14.34}$$

There is no major difference between the description of fluorescence signal in the FPSA experiment and that in FTSA, and protein fluorescence at constant temperature and elevated pressures is described using Equation 14.26 and substituting the Gibbs energy from Equation 14.34:

$$f = f_N + \frac{f_U - f_N}{1 + \exp\left(\Delta G_U / RT\right)} = f_N + \frac{f_U - f_N}{1 + \exp\left(\dfrac{\Delta G_{U\_P_r} + \Delta V_{P_r}\left(P - P_r\right) + \Delta \beta / 2\left(P - P_r\right)^2}{RT}\right)} \tag{14.35}$$

At a particular pressure $P = P_m$, the unfolding equilibrium is achieved and thus $\Delta G_{U\_P} = 0$. The protein melting pressure P_m is determined by solving the quadratic Equation 14.34:

$$P_m = P_r + \frac{-\Delta V_{P_r} - \sqrt{\left(\Delta V_{P_r}\right)^2 - 2\Delta \beta \Delta G_{U\_P_r}}}{\Delta \beta} \tag{14.36}$$

The overall Gibbs free energy as a function of temperature and pressure is described as follows:

$$\Delta G_U = \Delta H_{U_{T_r}} + \Delta C_{pU}\left(T - T_r\right) - T\left(\Delta S_{U_{T_r}} + \Delta C_{pU}\ln\frac{T}{T_r}\right) + \Delta\alpha\left(T - T_r\right)\left(P - P_r\right)$$

$$+\Delta G_{U\_P_r} + \Delta V P_r\left(P - P_r\right) + \frac{\Delta\beta_U}{2}\left(P - P_r\right)^2 \tag{14.37}$$

where $\Delta\alpha$ is the change in thermal expansion coefficient. This equation is used to fit data as in Figure 14.7e, describing the stability of the protein as a function of both temperature and pressure at various added ligand concentrations.

The above-described models are useful to analyze together both the thermal and pressure unfolding and isothermal and isobaric ligand-binding data. In our opinion, the combination of FTSA (FPSA) and ITC methods is essential to obtain high-quality data.

REFERENCES

Baker, B. M. and Murphy, K. P. 1996. Evaluation of linked protonation effects in protein binding reactions using isothermal titration calorimetry. *Biophys. J.* 71:2049–2055.

Baranauskiene, L., Matuliene, J., and Matulis, D. 2008. Determination of the thermodynamics of carbonic anhydrase acid-unfolding by titration calorimetry. *J. Biochem. Biophys. Methods* 70:1043–1047.

Baranauskiene, L. and Matulis, D. 2012. Intrinsic thermodynamics of ethoxzolamide inhibitor binding to human carbonic anhydrase XIII. *BMC Biophys.* 5:12.

Broecker, J., Vargas, C., and Keller, S. 2011. Revisiting the optimal c value for isothermal titration calorimetry. *Anal. Biochem.* 418:307–309.

Cimmperman, P. and Matulis, D. 2011. Protein thermal denaturation measurements via a fluorescent dye. In A. Podjarny, A. Deajaegere, and B. Kieffer (eds.), *Biophysical Approaches Determining Ligand Binding to Biomolecular Targets*, Chapter 8. RSC Publishing: Cambridge.

Harding, S. E. and Chowdhry, B. 2001. *Protein-Ligand Interactions: Hydrodynamics and Calorimetry*. Oxford University Press: New York.

Jogaite, V., Zubriene, A., Michailoviene, V., Gylyte, J., Morkunaite, V., and Matulis, D. 2012. Characterization of human carbonic anhydrase XII stability and inhibitor binding. *Bioorg. Med. Chem.* 21:1431–1436.

Khalifah, R. G. 1973. Carbon dioxide hydration activity of carbonic anhydrase: Paradoxical consequences of the unusually rapid catalysis. *Proc. Natl. Acad. Sci. USA* 70:1986–1989.

Krishnamurthy, V. M., Kaufman, G. K., Urbach, A. R., Gitlin, I., Gudiksen, K. L., Weibel, D. B., and Whitesides, G. M. 2008. Carbonic anhydrase as a model for biophysical and physical-organic studies of proteins and protein-ligand binding. *Chem. Rev.* 108:946–1051.

Lo, M. C., Aulabaugh, A., Jin, G., Cowling, R., Bard, J., Malamas, M., and Ellestad, G. 2004. Evaluation of fluorescence-based thermal shift assays for hit identification in drug discovery. *Anal. Biochem.* 332:153–159.

Matulis, D. 2001. Thermodynamics of the hydrophobic effect. III. Condensation and aggregation of alkanes, alcohols, and alkylamines. *Biophys. Chem.* 93:67–82.

Matulis, D., Baumann, C. G., Bloomfield, V. A., and Lovrien, R. E. 1999. 1-Anilino-8-naphthalene sulfonate as a protein conformational tightening agent. *Biopolymers* 49:451–458.

Matulis, D. and Bloomfield, V. A. 2001. Thermodynamics of the hydrophobic effect. II. Calorimetric measurement of enthalpy, entropy, and heat capacity of aggregation of alkylamines and long aliphatic chains. *Biophys. Chem.* 93:53–65.

Matulis, D., Kranz, J. K., Salemme, F. R., and Todd, M. J. 2005. Thermodynamic stability of carbonic anhydrase: Measurements of binding affinity and stoichiometry using ThermoFluor. *Biochemistry* 44:5258–5266.

Matulis, D. and Lovrien, R. 1998. 1-Anilino-8-naphthalene sulfonate anion-protein binding depends primarily on ion pair formation. *Biophys. J.* 74:422–429.

Morkunaite, V., Gylyte, J., Zubriene, A., Baranauskiene, L., Kisonaite, M., Michailoviene, V., Juozapaitiene, V., Todd, M. J., and Matulis, D. 2014. Intrinsic thermodynamics of sulfonamide inhibitor binding to human carbonic anhydrases I and II. *J. Enzyme Inhib. Med. Chem.* 30:204–211.

Pantoliano, M. W., Petrella, E. C., Kwasnoski, J. D., Lobanov, V. S., Myslik, J., Graf, E., Carver, T., Asel, E., Springer, B. A., Lane, P., and Salemme, F. R. 2001. High-density miniaturized thermal shift assays as a general strategy for drug discovery. *J. Biomol. Screen.* 6:429–440.

Petrauskas, V., Gylyte, J., Toleikis, Z., Cimmperman, P., and Matulis, D. 2013. Volume of Hsp90 ligand binding and the unfolding phase diagram as a function of pressure and temperature. *Eur. Biophys. J.* 42:355–362.

Sudzius, J., Baranauskiene, L., Golovenko, D., Matuliene, J., Michailoviene, V., Torresan, J., Jachno, J., Sukackaite, R., Manakova, E., Grazulis, S., Tumkevicius, S., and Matulis, D. (2010). 4-[N-(Substituted 4-pyrimidinyl)amino]benzenesulfonamides as inhibitors of carbonic anhydrase isozymes I, II, VII, and XIII. *Bioorg. Med. Chem.* 18:7413–7421.

Todd, M. J. and Salemme, F. R. 2003. Direct binding assays for pharma screening. *Gen. Eng. News* 23:28–29.

Toleikis, Z., Cimmperman, P., Petrauskas, V., and Matulis, D. 2011. Determination of the volume changes induced by ligand binding to heat shock protein 90 using high-pressure denaturation. *Anal. Biochem.* 413:171–178.

Toleikis, Z., Cimmperman, P., Petrauskas, V., and Matulis, D. 2012. Serum albumin ligand binding volumes using high pressure denaturation. *J. Chem. Thermodyn.* 52:24–29.

Wiseman, T., Williston, S., Brandts, J. F., and Lin, L. N. 1989. Rapid measurement of binding constants and heats of binding using a new titration calorimeter. *Anal. Biochem.* 179:131–137.

Zubriene, A., Matuliene, J., Baranauskiene, L., Jachno, J., Torresan, J., Michailoviene, V., Cimmperman, P., and Matulis, D. 2009. Measurement of nanomolar dissociation constants by titration calorimetry and thermal shift assay–radicicol binding to Hsp90 and ethoxzolamide binding to CAII. *Int. J. Mol. Sci.* 10:2662–2680.

CHAPTER 15

Joining Thermodynamics and Kinetics by kinITC

Philippe Dumas

CONTENTS

15.1 INTRODUCTION

In chemistry, the use of isothermal titration calorimetry (ITC) as a kinetic technique has a very long history, and the link between the measured heat power and the kinetics of the reaction has long been the subject of theoretical investigations (Calvet and Prat 1963; Garcia-Fuentes et al. 1998; Lopez-Mayorga et al. 1987). However, due to rather long response times, the measurements were limited to slow and sometimes even very slow reactions followed experimentally for 20 days or more (Willson et al. 1995). In the latter publication, it was even mentioned that ITC is sensitive enough to follow first-order reactions with a rate constant of $10^{-11}\,\mathrm{s^{-1}}$ (half-life = 2200 years)! In biology, the usual time scale of interest is smaller by several orders of magnitude, and, in addition, the amount of available material is limited, not to mention the fact that a modest concentration (by usual chemical standards) of 1 mM usually implies more than 20 g $\mathrm{L^{-1}}$ for a macromolecule and would correspond to a physically absurd concentration of 2 kg $\mathrm{L^{-1}}$ for a ribosome. These obvious facts explain why the use of calorimetry long remained the domain of pioneers in biology (Buzzell and Sturtevant 1951; Watt and Sturtevant 1969; Wiseman et al. 1989; Johnson and Biltonen 1975; Langerman and Biltonen 1978; Morin and Freire 1991) and why its introduction as a daily technique in biological laboratories lagged so far behind its use in chemistry (for an interesting and comprehensive account, see Chaires 2008). Modern instruments are now available with cell volumes as small as 200 μL and faster response times, which fueled the (re)discovery of ITC as a kinetic technique in biology (Egawa et al. 2007; Burnouf et al. 2012; Vander Meulen and Butcher 2012). In a previous work (Burnouf et al. 2012), we coined the term kinITC and described how to take into account in a realistic way the different problems that cause the measured heat power to be different from the actual heat power evolved in the measurement cell (for an illustration, see http://www-ibmc.u-strasbg.fr:8080/webMathematica/kinITCdemo/). Our goal was to obtain a realistic simulation of the complete injection curves (which was the first original aspect of that work) and, from that, to obtain the best kinetic parameters. We showed that this can be achieved, not only for reactions involving a single kinetic step, such as $A + B \leftrightarrows C$, but also for more complex reactions involving a two-step kinetic mechanism, for example, the *induced-fit mechanism* (which was the second original aspect of that work). However, proving that a reaction proceeds via an induced-fit mechanism (binding first) or, alternatively, via a *conformational selection* mechanism (change of conformation first) is not trivial and not a part of routine experiments. This chapter, on the contrary, will focus on simple reactions and on a thorough analysis of a simplified kinITC method based on the evolution from injection to injection of the time of *return to baseline*. The method was recently introduced in the program AFFINImeter (S4SD, Santiago de Compostela, Spain; https://www.affinimeter.com/).

Such simple reactions correspond to by far to the most common and also the practically most useful situation. In addition, even in situations where a single kinetic step does not strictly apply, it may represent an excellent approximation. For example, the annealing of two DNA strands S_1 and S_2 leading to a duplex (as studied by P. L. Privalov in Chapter 11 in this book) proceeds through two consecutive kinetic steps according to

$$S_1 + S_2 \underset{k_{off}}{\overset{k_{on}}{\rightleftharpoons}} D^* \tag{15.1a}$$

$$D^* \underset{k_{unzip}}{\overset{k_{zip}}{\rightleftharpoons}} D \tag{15.1b}$$

where D^* is an unstable short duplex (corresponding to an initial nucleus of three to four base pairs), which is eventually transformed into a full stable duplex through a monomolecular zipping process (typically an induced-fit mechanism). This two-step mechanism was established long ago after two remarkable studies (Craig et al. 1971; Porschke and Eigen 1971). According to these studies, it can nevertheless be accurately described kinetically by a global single-step mechanism $S_1 + S_2 \underset{k_{off}(global)}{\overset{k_{on}(global)}{\rightleftharpoons}} D$ (with parameters $k_{on}(global) \approx k_{on}$ and $k_{off}(global) \neq k_{off}$), because the zipping process (Equation 15.1b) is extremely fast in comparison to the formation of the initial nucleus (Equation 15.1a).

15.2 MATHEMATICAL BASIS OF THE SIMPLIFIED kinITC METHOD

Here, we thus consider simple situations corresponding to reactions of the type

$$A + B \underset{k_{off}}{\overset{k_{on}}{\rightleftharpoons}} C \tag{15.2}$$

We consider that compound A, initially alone in the measurement cell, is the titrand, and compound B, in the syringe, is the titrant. The kinetics of such a reaction is represented by the differential equation

$$\frac{dC}{dt} = k_{on}\left[A\right]_{tot} AB - k_{off}C \tag{15.3a}$$

where:
  A   is the simplified notation for the reduced concentrations $[A]/[A]_{tot}$
  B   is the simplified notation for the reduced concentrations $[B]/[A]_{tot}$
  C   is the simplified notation for the reduced concentrations $[C]/[A]_{tot}$
  $[A]_{tot}$ is the total concentration of the titrand in the measurement cell

Equation 15.3a represents perfectly the kinetics of the reaction at any step of a titration if one considers the evolution of the concentrations after compound B has been injected (and correctly mixed). For this simplified kinITC method, we neglect the small concentration variations resulting from the course of the reaction during the injection period (in the full kinITC method, there is no such approximation), but the concentration variation due to dilution is not neglected. By using conservation equations $[A] + [C] = [A]_{tot}$ and $[B] + [C] = [B]_{tot}$, where $[A]_{tot}$ and $[B]_{tot}$ are the total concentrations of compounds A and B, respectively, after injection, one is led to

$$\frac{dC}{dt} = k_{on}\left[A\right]_{tot}(1-C)(B_{tot}-C) - k_{off}C \tag{15.3b}$$

where $B_{tot} = [B]_{tot}/[A]_{tot}$ is the stoichiometric ratio for the current injection. To avoid pointless complication, we keep the same notations $[A]_{tot}$, $[B]_{tot}$ for each injection, but $[B]_{tot}$ increases steadily and $[A]_{tot}$ decreases slightly at each injection of compound B. In a previous work on similar problems (Egawa et al. 2007), a simplification was introduced by considering a succession of first-order approximations (by neglecting in the kinetic equations for each injection the quadratic term $k_{on}\left[A\right]_{tot}(C-\tilde{C})^2$, where $\tilde{C} = C$ at equilibrium). Here, on the contrary, we consider the complete Equation 15.3b. By factorizing the polynomial, one obtains

$$\frac{dC}{dt} = k_{on}[A]_{tot}(C - C_1)(C - C_2) \tag{15.3c}$$

where C_1 and C_2 are the roots of $k_{on}[A]_{tot}(1-C)(B_{tot}-C)-k_{off}C=0$ (with $C_1>C_2$):

$$C_1 = \frac{1}{2}\left(p+\sqrt{p^2-4B_{tot}}\right), \quad C_2 = \frac{1}{2}\left(p-\sqrt{p^2-4B_{tot}}\right); \quad p=1+B_{tot}+\frac{K_d}{[A]_{tot}} \quad (15.3d)$$

Interestingly, the difference (C_1-C_2), which will appear often in this chapter, is equal to $\tilde{A}+\tilde{B}+c^{-1}$, where $\tilde{A}$, $\tilde{B}$ are the reduced equilibrium concentrations (i.e., divided by $[A]_{tot}$) of compounds A and B, respectively, and $c=[A]_{tot}/K_d$ is the Wiseman parameter (Wiseman et al. 1989). By integration of the differential Equation 15.3c following standard methods, we obtain

$$C(t)=C_1-\frac{C_1-C_2}{1-K^{-1}e^{-t/\tau}}; \quad \tau=\left[k_{on}[A]_{tot}(C_1-C_2)\right]^{-1}; \quad K^{-1}=\frac{(C_2-C_0)}{C_1-C_0} \quad (15.4)$$

with $C_0=C(t=0)$ for the current injection and τ a characteristic time of the evolution of concentrations. Note first that $C(t) \rightarrow C_2$ for $t \rightarrow \infty$, which means that C_2 is the value reached at equilibrium for the current injection (i.e., $C_2 = \tilde{C}$). Therefore, $C(t)$ evolves from $C(0)=C_0$ at $t=0$ to C_2 for $t \rightarrow \infty$, which implies that $K^{-1}<1$, since $C_1 > C_2$. Note also that the initial value $C(0)$ for each injection depends on the value $C(t \rightarrow \infty)$ reached at the end of the preceding injection and on the slight dilution resulting from the injected volume. Since the reaction is not first order (Equation 15.2), τ is not equally representative of the evolution of the concentrations from immediately after injection till the return to equilibrium. In fact, τ is representative of a single exponential only for large values of t/τ, since, from Equation 15.4, $C(t) \approx C_2 - (C_1-C_2)K^{-1}e^{-t/\tau}$ for $t/\tau \rightarrow \infty$. To make use of a quick and simplified kinITC method based on the determination of the time necessary to return to baseline, it is important to evaluate how the effective characteristic time evolves during the equilibration process, particularly close to the end of the equilibration process.

For this, we note that, with the notation $X=K^{-1}e^{-t/\tau}$, $C(t)$ from Equation 15.4 is of the form

$$C(t)=C_1-\frac{C_1-C_2}{1-X} \quad (15.5a)$$

Since $X<1$, the term $1/(1-X)$ is the sum of a geometric series, which yields

$$C(t) = C_1-(C_1-C_2)\sum_{n=0}^{n=\infty} X^n = C_1-(C_1-C_2)\sum_{n=0}^{n=\infty} K^{-n}e^{-nt/\tau} \quad (15.5b)$$

Mathematically, Equation 15.5b shows that the Laplace transform of $C(t)$, is discrete since $C(t)$ can be expressed as a sum of exponentials, the nth one being characterized by an amplitude $(C_1-C_2)K^{-n}$ and a characteristic time τ/n. I am unaware of whether or not this result has already been mentioned elsewhere. Equation 15.5b is quite informative, as it shows that the evolution of the concentrations is under the dependence of a set of discrete time values, that is, that there are no intermediates between the successive characteristic times $\tau, \tau/2, \tau/3, \tau/4...$. Since $K^{-1} < 1$, the successive terms in Equation 15.5b become less and less important, and, practically, a finite number of terms is sufficient. The interest in Equation 15.5b is of allowing us to determine the instantaneous heat power signal $P_s(t)$ evolved in the measurement cell:

$$P_s(t) = V_{cell} \, \Delta H \left[A\right]_{tot} \frac{dC}{dt} = V_{cell} \, \Delta H \left[A\right]_{tot} \left(C_1 - C_2\right) \tau^{-1} \sum_{n=1}^{n=\infty} n K^{-n} e^{-n\,t/\tau} \qquad (15.6a)$$

From the expression of τ (Equation 15.4), this can be transformed into

$$P_s(t) = V_{cell} \, \Delta H \, k_{on} \left[A\right]_{tot}^2 \left(C_1 - C_2\right)^2 \sum_{n=1}^{n=\infty} n K^{-n} e^{-n\,t/\tau} \qquad (15.6b)$$

Importantly, this yields the instantaneous heat power evolved in the measurement cell, and not the measured heat power due to the finite response time of the instrument. This effect is taken into account in the following section.

15.3 INFLUENCE OF THE INSTRUMENT RESPONSE TIME: PRACTICAL ASPECTS

It is well known that the response time is accounted for by the Tian equation (for accessible references, see Calvet and Prat 1963; Tachoire et al. 1986; Burnouf et al. 2012)

$$P_m(t) + \tau_{ITC} \frac{dP_m}{dt} = P_s(t) \qquad (15.7)$$

where $P_m(t)$ is the effectively measured heat power. Equation 15.7 shows that $P_m(t)$ is the convolution of $P_s(t)$ and of the response of the instrument to a unit impulse. In fact, this equation is in no way specific to calorimetry, as it accounts for the influence of finite response time for any instrument responding linearly to an excitation. It is important to emphasize that a single relaxation time may not be sufficient to describe fully the response of an ITC instrument (Garcia-Fuentes et al. 1998; Lopez-Mayorga et al. 1987; Tachoire et al. 1986). However, in practice, the real response is adequately described by one or two such relaxation times, which means that the *ideal* response to a very short heat burst of total energy E_{tot} (delivered at $t = 0$) is given either by

$$P_m(t) = \frac{E_{tot}}{\tau_{ITC}} e^{-t/\tau_{ITC}} \qquad (15.8a)$$

or by

$$P_m(t) = \left(\frac{\varepsilon_1}{\tau_1} e^{-t/\tau_1} + \frac{\varepsilon_2}{\tau_2} e^{-t/\tau_2} \right) E_{tot} \tag{15.8b}$$

where ε_1 and ε_2 are fractions of the total heat E_{tot} delivered in the measurement cell. Equations 15.8a and 15.8b are only ideal, as, in practice, the heat burst is not instantaneous, and Equations 15.8a and 15.8b are only applicable after a short transient period during which the absolute value of the heat power amplitude increases from zero and passes a maximum before the exponential return to the baseline. According to our experience with our instrument (an ITC200 from Microcal/Malvern), we observed that Equation 15.8a described reasonably well the response to a heat burst generated by a quick injection into water of a small volume (1–2 µL) of diluted methanol (1–2% v[MeOH]/v[H$_2$O]) with a mixing speed of 1500 rpm (this is to be tested for other instruments). Importantly, we observed that the value τ_{ITC} was extremely dependent on the cleanness of the instrument, as short values of 3.3–3.5 s were obtained with a freshly cleaned instrument, but values around 7 s were obtained with a dirtier instrument. ITC users must be warned that proteins may leave a thin film on the cell surface, which makes it "dirty." Of course, not all proteins are equal in this respect, and those that are more difficult to handle because of their tendency to aggregate are likely to be considered "dirtier." It is also well known that hydrophobic compounds, often used as protein ligands, have a strong wall-coating tendency. Needless to say, any "greasy" compounds used for micelle studies should be seen as potential "dirty" compounds. Nucleic acids seem to be "cleaner" compounds (but not all their ligands). Therefore, two rules of thumb should be kept in mind: (i) one should not consider the response time as an intrinsic property of the instrument in use, and (ii) an ITC instrument should be cleaned regularly and thoroughly (see manufacturer indications).

15.4 INFLUENCE OF INSTRUMENT RESPONSE TIME: THEORETICAL ASPECTS

Here, we will continue with the theoretical aspects linked to the influence of one single response time (τ_{ITC} in the following). Since Equation 15.7 is linear, its solution is the sum of all partial solutions obtained by considering in turn each term of the infinite sum in Equation 15.6b. It is thus obtained (by imposing $P_m(t=0)=0$) as

$$P_m(t) = V_{cell}\,\Delta H\,k_{on}\,[A]_{tot}^2\,(C_1 - C_2)^2 \sum_{n=1}^{n=\infty} n K^{-n} \frac{e^{-n\,t/\tau} - e^{-t/\tau_{ITC}}}{1 - n\,\tau_{ITC}/\tau} \tag{15.9a}$$

First, it should be noted that the comparison of Equations 15.9a and 15.6b provides us with an explicit expression of the influence of the instrument response time on the nth component of the infinite sum through the replacement of $e^{-nt/\tau}$ with $(e^{-n\,t/\tau} - e^{-t/\tau_{ITC}})/(1 - n\,\tau_{ITC}/\tau)$. In the particular situation where one term of the infinite sum in Equation 15.9a is such that the denominator $1 - n\tau_{ITC}/\tau = 0$ rigorously or approximately, the rational function

$(e^{-n\,t/\tau} - e^{-t/\tau_{TC}})/(1 - n\,\tau_{TC}/\tau)$ has the indefinite value 0/0, or may be poorly defined numerically. However, the ambiguity is resolved by the use of the *de l'Hôpital* rule, which yields the well-defined value $(t/\tau_{TC})e^{-t/\tau_{TC}}$.

Second, it can be shown, for example, with a software with symbolic capabilities such as *Mathematica* (Wolfram Research), that the infinite sum in Equation 15.9a can be expressed in closed form (albeit by referring to the rather unusual *hypergeometric* function $_2F_1[a,b,c,z]$ (Weisstein, http://mathworld.wolfram.com/HypergeometricFunction.html):

$$P_m(t) = P_m^0 \left(e^{-t/\tau}\ _2F_1\left[2,\,\theta,\,1+\theta,\,K^{-1}e^{-t/\tau}\right] - e^{-t/\tau_{TC}}\ _2F_1\left[2,\,\theta,\,1+\theta,\,K^{-1}\right]\right)$$

$$\text{with } \theta = \frac{\tau_{TC}-\tau}{\tau_{TC}} \quad \text{and } P_m^0 = V_{cell}\,\Delta H\,[A]_{tot}\,\frac{C_1-C_2}{K(\tau-\tau_{TC})} \tag{15.9b}$$

This result is essentially of theoretical interest, since, in practice, explicit summation of, at most, six to seven terms in Equation 15.9a yields a result indistinguishable from that from Equation 15.9b. Therefore, it does not seem worthwhile to give further explanation here. In addition, only the comparison of Equations 15.9a and 15.6b provides us with an explicit expression of the influence of the instrument response time on each component of the Laplace transform of $P_s(t)$. These considerations are illustrated by the simulation of a titration experiment and superimposition for each injection of the first seven components of Equation 15.9a and their sum (dashed curves), which is indistinguishable from the rigorous result of Equation 15.9b (Figure 15.1).

15.5 IMPORTANT PARAMETERS

To have a view of the evolution of the return to equilibrium for each injection, we have to see how the important parameters τ and K evolve from injection to injection. Since τ is proportional to $(C_1-C_2)^{-1}$ (Equation 15.4), the evolution of τ during a titration experiment is governed from Equation 15.3d by

$$(C_1-C_2)^{-1} = \left[\left(1+c^{-1}+s\right)^2 - 4s\right]^{-1/2} \tag{15.10a}$$

which implies

$$\tau^{-1} = k_{on}[A]_{tot}\left[\left(1+c^{-1}+s\right)^2 - 4s\right]^{1/2} \tag{15.10b}$$

where:
  $c = [A]_{tot}/K_d$ is the Wiseman parameter for the current injection, that is, by taking into account the dilution resulting from successive additions of compound B
  and s is the corresponding stoichiometric ratio $[B]_{tot}/[A]_{tot}$

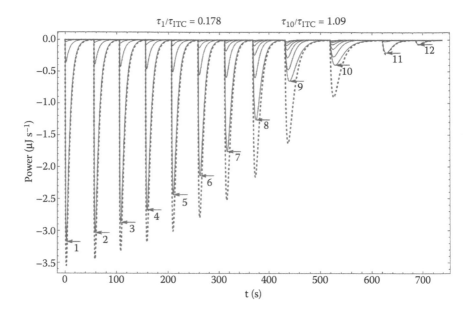

FIGURE 15.1 Illustration of the successive components in Equation 15.9a. The components from $n=1$ to $n=7$ in Equation 15.9a are superimposed (thin curves). Their sum (dashed curve) corresponds to the measured signal. The maximum amplitude of each first component is marked by an arrow. Note that at the scale of the figure, at most five components among seven are visible. The simulation was performed with the values: cell volume $= 203$ µL, injected volume at each injection $= 2$ µL, $[A]_0 = 30$ µM (cell), $[B]_0 = 300$ µM (syringe), $k_{on} = 3 \times 10^4$ M^{-1}s^{-1}, $k_{off} = 10^{-3}$ s^{-1}, $\Delta H = -60$ kJ mol^{-1}, $\tau_{ITC} = 7$ s. The terms τ_1 and τ_{10} correspond to the characteristic times defined in Equation 15.4 for injections #1 and #10, respectively (injection #10 is the closest to s$=1$).

Equation 15.10b corresponds to equation (4) in the Supplementary Information part of Burnouf et al. (2012). The function given by Equation 15.10a is shown in Figure 15.2 for c values ranging from 5 to 500 in equal logarithmic steps: it starts at $c/(1+c)$ for $s=0$ and reaches a maximum value equal to $1/2c^{1/2}$ for $s=1-1/c$ (vertical mark on each curve). This means that a high c value produces an important variation of the characteristic time of return to equilibrium with a marked maximum. This is quite understandable when one recalls that $C_1 - C_2 = \tilde{A} + \tilde{B} + c^{-1}$, which means that the sum of the concentrations of the free species A and B is minimum at $s=1-1/c$, and thus that their association rate is minimum too. Such a variation from injection to injection of the time needed to return to baseline is at the basis of the simplified kinITC technique.

Note that there is nothing new with this quantity $C_1 - C_2$, which has already been represented in Egawa et al. (2007) and Indyk and Fisher (1998), but it is important to show it here (in fact, its reciprocal) for good understanding.

Next, we examine the evolution of K governing the relative importance of the successive components (Equation 15.9a). From Equation 15.4, $K = (C_1 - C_0)/(C_2 - C_0) = 1 + (C_1 - C_2)/(C_2 - C_0)$. The term $(C_1 - C_2)$ is known from Equation 15.10a, and, in $(C_2 - C_0)$, C_0 is the value of C at t$=0$ for the current injection, and C_2 is the value of C after return to

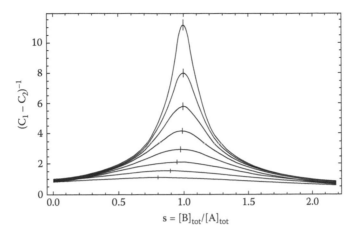

FIGURE 15.2 Evolution of $(C_1 - C_2)^{-1}$ for different values of c. The lower and upper curves correspond to $c = 5$ and $c = 500$, respectively. The maximum amplitude is equal to $1/2c^{1/2}$.

equilibrium. Essentially, C_0 is close to the value of C_2 reached at the end of the preceding injection, and therefore, $(C_2 - C_0)$ can be approximated by $(dC_2/ds)\delta s$, where the derivative can be calculated from the expression of C_2 in Equation 15.3d, and δs is the variation of the stoichiometric ratio between the two successive injections. Without giving the details of the calculation, it may be obtained (more easily with software such as *Mathematica* from Wolfram Research) by

$$K = 1 + \frac{2\left[1 + (s-1)^2 c^2 + 2(1+s)c\right]}{c(r+s)\left[-1 + c\left(1 - s + \sqrt{(1+1/c+s)^2 - 4s}\right)\right]} \frac{V_{cell}}{\delta V} \tag{15.11a}$$

with the same notations as for Equations 15.10a and 15.10b and δV the injected volume at each injection. Although not rigorous, this result describes well the evolution of this parameter for all injections of a titration experiment, as shown in Figure 15.3. Interestingly, it does not involve any kinetic parameters.

Comparison of these curves with those in Figure 15.3b and d of Egawa et al. (2007) shows a striking resemblance, which indicates that the parameter K^{-1} is related to the *practical* parameter r' defined in Egawa et al. (2007). With our notation, it may be shown that, to a very good approximation, $r' \approx (C_2 - C_0)/(C_1 - C_2 + c^{-1})$, whereas $K^{-1} = (C_2 - C_0)/(C_1 - C_0)$ may be written under the comparable form $(C_2 - C_0)/(C_1 - C_2 + C_2 - C_0)$, which differs from r' by $C_2 - C_0$ in the denominator (always a small term) replacing c^{-1}. It was obtained in Egawa et al. (2007) that r' has to be significantly smaller than 1 to justify their simplified method based on a succession of first-order approximations (see Section 15.2). This is in full agreement with the present study, showing that this criterion ($r' \approx K^{-1} \ll 1$) implies that, for sufficiently large values of t, $P_m(t)$ is well approximated by the first exponential term in Equation 15.9a. This approximation is illustrated in the following.

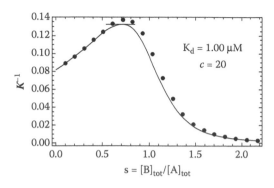
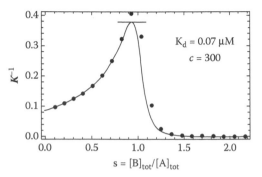

FIGURE 15.3 Evolution of K^{-1} during a titration experiment for different values of c. The dots correspond to the exact values for each injection from Equation 15.4, whereas the curves were obtained from the approximate Equation 15.11a. The short horizontal segments mark the maximum value of K^{-1} obtained with Equation 15.11b.

In addition, our study allows an approximate value of the maximum of K^{-1} to be obtained, which is important precisely to evaluate the time after which the first component ($n = 1$) of the infinite sum in Equation 15.9a is accurately representative of the whole sum. This maximum can be obtained from Equation 15.11a as

$$\left(K^{-1}\right)_{\text{max}} = \frac{3}{3 + \left(V_{\text{cell}}/\partial V\right)\dfrac{16\sqrt{3c}}{(1+r)3c - 2\sqrt{3c} - 3}} \qquad (15.11b)$$

where δV is the injected volume and $r = [B]_0 / [A]_0$ is the ratio (concentration of the titrant in the syringe)/(initial concentration of the titrand in the measurement cell). The estimate from Equation 15.11b is visualized in Figure 15.3 as short horizontal bars. To visualize the dependence on both c and δV, a two-dimensional plot is shown in Figure 15.4. It appears that $(K^{-1})_{\text{max}}$ is lower than 0.5 in most practical situations. Although Equation 15.11b does not show it immediately, numerical calculations reveal that the lines of equal values of $(K^{-1})_{\text{max}}$ are very close, such that $\sqrt{c}\,\delta V$ is constant (the approximation is less good for low c values).

Figure 15.4 shows that, for a given value of the Wiseman parameter c, a decrease of δV leads to a decrease of $(K^{-1})_{\text{max}}$ too. Since large values of K^{-1} are less favorable (see later in this section), large increments of stoichiometry between injections should be avoided as long as the measured heat power is high enough. Interestingly, since large values of $(K^{-1})_{\text{max}}$ correspond to large values of c and to a sharp maximum around the unit stoichiometry (see Figure 15.3), reducing the stoichiometry increments δs in this region of the titration would also yield a better sampling of the titration curve in the critical region of its inflexion point and a better determination of K_d. Clearly, there is room here for significant improvement of data collection software, since the necessary variation of the injected volume δV during the titration could be achieved automatically.

The effect of the evolution of K^{-1} during a titration experiment is illustrated in Figure 15.5, which shows that, as expected, the injection power curve is rapidly

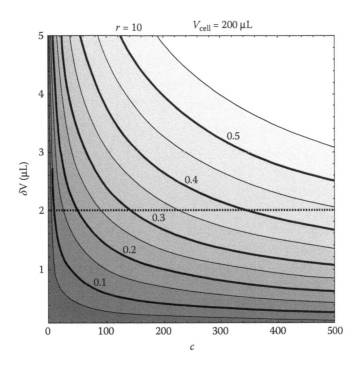

FIGURE 15.4 Two-dimensional plot of $(K^{-1})_{max}$ as a function of c and δV. The values of $(K^{-1})_{max}$ are indicated for curves of equal value of $(K^{-1})_{max}$. The thick curves marking the values from 0.1 to 0.5 correspond to the equation $\delta V = \alpha c^{-1/2}$ with, respectively, $\alpha = 6$, 14, 24, 37.5, and 56. With $V_{cell} = 200\ \mu L$ and the commonly used value $r = [B]_0 / [A]_0 = 10$, an injected volume $\delta V = 2\ \mu L$ (dashed line) is such that the unit stoichiometry $s = [B]_0 / [A]_0 \approx 1$ is reached after 10 injections.

dominated by the first term in Equation 15.9a for $K^{-1} \approx 0.1$, but that the following terms become significant for $K^{-1} \approx 0.3$.

15.6 VISUALIZATION OF INFLUENCE OF INSTRUMENT RESPONSE TIME

Here, we examine the practical consequences of the previous analysis. We first illustrate the effect of a variation of the response time from 1 to 10 s (Figure 15.6). The value of 1 s is hypothetical, as it is unattainable in practice for any present-day instrument; 3.5 s is attainable for a (clean!) ITC200, and 10 s is of the order of the response time for a VP-ITC, both from Microcal/Malvern. The dramatic influence of this variation appears clearly with an important decrease of the maximum amplitude of $P_m(t)$ and a concomitant increase of the time needed to return to flat baseline.

Next, we consider the combined influence of the ratio τ/τ_{ITC} and of K^{-1} on the relative importance of the different components $n K^{-n}(e^{-n\,t/\tau} - e^{-t/\tau_{ITC}})/(1 - n\,\tau_{ITC}/\tau)$ in the sum yielding $P_m(t)$ (Equation 15.9a). This is illustrated in Figure 15.7, where each component is normalized in such a way that the amplitude of the first one is equal to 1. In each case, a vertical arrow marks the effective end of the first ($t = t_1$) and second components ($t = t_2$) (somewhat arbitrarily determined by their normalized amplitude being equal to $e^{-4.5} = 0.011$):

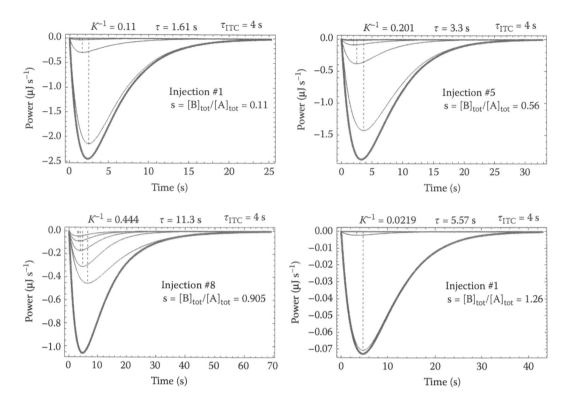

FIGURE 15.5 Illustration of the discrete power components. Four injection curves of a simulated titration experiment are shown. The lower thick curve is the power curve $P_m(t)$ for the corresponding injection. It is the sum of the successive thin curves corresponding to the discrete components characterized, from the larger to the smaller amplitude, by the characteristic times τ, $\tau/2$, $\tau/3$, $\tau/4$, ... (see Equation 15.9a). The figure highlights their amplitude decrease and concomitant shift toward lower time values. The curves were calculated with the values $k_{on} = 3.5 \times 10^4$ M^{-1} s^{-1}, $k_{off} = 1.5 \times 10^{-3}$ s^{-1}, $\Delta H = -40$ kJ mol^{-1}, cell volume $= 203$ μL, volume of successive injections $= 1.4$ μL, $[A]_0 = 20$ μM, $[B]_0 = 320$ μM.

beyond each time (t_1 or t_2), the contribution of each component is considered negligible. The time separation $t_1 - t_2$ between the two arrows is indicated and underlined when it is too small in comparison to t_1, that is, when the second component does not vanish rapidly enough in comparison to the first component. As expected, a low ratio $\tau/\tau_{ITC} \lesssim 0.2$ is not favorable unless K^{-1} is less than 0.2, and $\tau/\tau_{ITC} \approx 1$ is also borderline if $K^{-1} \geq 0.4 - 0.5$. However, most often, the first component is not significantly "polluted" by the second one (and even less by those of higher order) for $t \geq 2\tau$, which means that locating the effective end of a power curve $P_m(t)$ yields information on the effective end of its first component and, thus, on the time τ defined by Equation 15.4. In the following section, we examine in full the problem of determining accurately this "effective end" and also how to derive the time τ.

15.7 EQUILIBRATION TIME CURVE (ETC) AND kinITC-ETC METHOD

With the previous considerations, we have analyzed in depth the basis of the simplified kinITC method (*kinITC-ETC*) and the influence of different factors. In practice, each

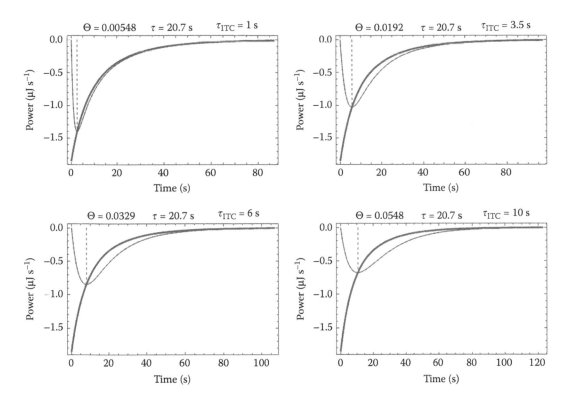

FIGURE 15.6 Influence of the instrument response time. Four simulations of the same injection (# 8) with identical kinetic parameters but different response times τ_{ITC} (1, 3.5, 6, 10 s) are shown. The exponential-like thick curve corresponds to $P_s(t)$ (Equation 15.6b), whereas the thin curves correspond to $P_m(t)$ (Equation 15.9a) obtained with the indicated response time. The vertical dashed line highlights the time when $P_m(t)$ reaches its extremum (here minimum) value, which corresponds exactly to the time when $P_s(t)$ and $P_m(t)$ intersect in agreement with Equation 15.7. The dimensionless parameter Θ is defined in Table 15.AI and discussed in Section 15.8. The characteristic time $\tau = 20.7$ s is given by Equation 15.4. The parameters for the simulations were $k_{on} = 1.2 \times 10^4$ M^{-1} s^{-1}, $k_{off} = 0.1 \times 10^{-3}$ s^{-1}, $\Delta H = -40 \times 10^3$ kJ mol^{-1}, cell volume = 203 µL, volume of successive injections = 2 µL, $[A]_0 = 25$ µM, $[B]_0 = 300$ µM.

titration experiment is processed classically, and in addition, an estimate of the "effective end" of each injection curve is obtained automatically, which yields the *equilibration time curve* (ETC). All necessary methods were first developed with *Mathematica* (Wolfram Research) and then implemented in AFFINImeter (S4SD, Santiago de Compostela, Spain; https://www.affinimeter.com/). Practically, each experimental injection curve is represented by a theoretical fit $P_m^{fit}(t)$ devoid of any physical meaning (which means that it is purely numerical and does not involve any kinetic considerations) (Figure 15.8). Here, it is not necessary to say more as the technical details are of no interest for the following. It suffices to know that, at each injection, the theoretical function $P_m^{fit}(t)$ is used to determine when its absolute value becomes less than some fraction of the local root mean square deviation (r.m.s.d) of the experimental points about $P_m^{fit}(t)$: the corresponding time is taken as the "effective end" of the injection. This allows this "effective end" to be determined

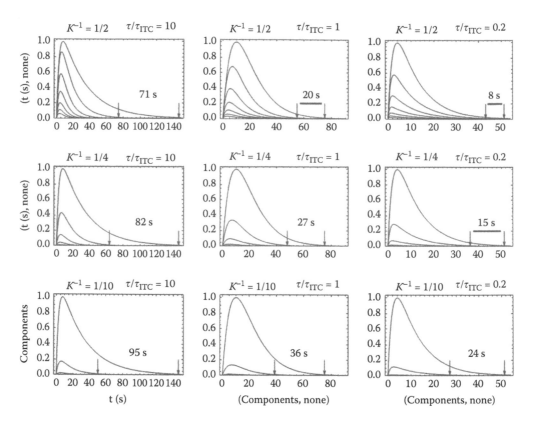

FIGURE 15.7 Influence of K^{-1} and of τ/τ_{ITC} on the successive components of $P_m(t)$. The seven first successive (normalized) components in Equation 15.9a are represented for various combinations of K^{-1} and τ/τ_{ITC}. Note that, at the scale of the figure, not all components are visible if K^{-1} is too low. See text for the meaning of the arrows and inserted time values.

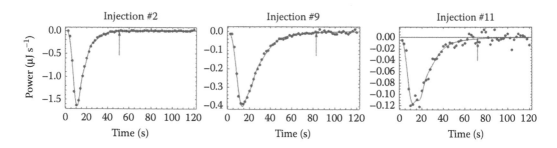

FIGURE 15.8 Determination of the "effective end" of each injection $P_m(t)$. Three injections from a titration at 6.1°C of carbonic anhydrase with the inhibitor 4CBS: $[A]_0 = 26\ \mu M$, $[B]_0 = 315\ \mu M$, injected volume $= 1.9\ \mu L$, cell volume $= 203.6\ \mu L$, sampling time $= 2$ s. Injection #9 corresponds to the injection with the longest equilibration time. The continuous curves are the fits of the experimental points, and the arrows mark the automatically determined "effective end" of each injection. Note that the cause of error in the determination of this "effective end" is the same for all these injections, as the increased noise at injection #11 is purely visual due to the "zooming."

automatically, and the result matches closely what would have been obtained "by eye" (arrows in Figure 15.8). It is obvious from inspection of Figure 15.8 that the uncertainty of the "effective end time" is strongly dependent on the signal/noise ratio (see Figure 15.9 for a real ETC).

One must realize that linking the "effective end time" $t_{end}(N_{inj})$ determined at injection N_{inj} to the time $\tau(N_{inj})$ defined by Equation 15.4 is not straightforward. If the return to baseline of the heat power signal were perfectly exponential, an arbitrary "effective end time" would be determined as the time corresponding to a decrease of the signal from its maximum absolute value to some fixed fraction of this maximum (e.g., 1 %). However, as we have seen (Figure 15.5), the heat power signal is the sum of components with different characteristic times, and such a simple criterion cannot be used. This is the reason why we consider as "effective end time" the minimum time after which the heat power signal may reasonably be considered as practically null. We have thus adopted the following pragmatic approximation:

$$t_{end}\left(N_{inj}\right) = \alpha\left[\tau\left(N_{inj}\right) + \tau_{ITC}\right] + \tau_{inj} + \tau_{mix} \tag{15.12}$$

where:

α was tuned to 4.5 after trial and error

τ_{inj} and τ_{mix} are, respectively, the injection time and the mixing time

The injection time is normally determined by a rule given by the manufacturer (e.g., 2 s per microliter injected with the ITC200 from Microcal/Malvern), and the mixing time is of the order of 1 s. Such an estimate should be seen merely as an order of magnitude, as the mixing time depends on the viscosity of the solution. This requires more detailed examination, but it is sufficient here as a correction for the simplified kinITC-ETC method.

Ideally, the response time τ_{ITC} should be experimentally measured. There are, however, two pitfalls: a practical one and a methodological one. The practical pitfall is that, in daily life, one does not measure the response time before all experiments potentially of interest for kinITC (see Section 15.3 for comments on the variability of τ_{ITC}). The methodological pitfall is that, even with the right value of τ_{ITC}, the best result is not obtained, since Equation 15.12 is only approximate. It has been observed empirically that considering τ_{ITC} as a fudge factor in the fit of the ETC with Equation 15.12 ($\tau(N_{inj})$ being given by Equation 15.10b) yields a better result (Figures 15.9a and 15.10). For this purpose, the kinITC-ETC results obtained with carbonic anhydrase (compound A) and its inhibitor 4CBS (compound B) were compared with the results from a surface plasmon resonance (SPR) benchmark (Navratilova et al. 2007). The conditions used for ITC were the same as for SPR, apart for the concentrations. The results obtained by kinITC-ETC are compared with those from SPR in the form of Arrhenius plots (Figure 15.10). One observes first that the straight lines are parallel within experimental error, which points to identical activation energies and, thus, identical temperature dependence of k_{on} and k_{off} by the two methods. Also, the maximum separation between the straight lines is 0.5, which corresponds to

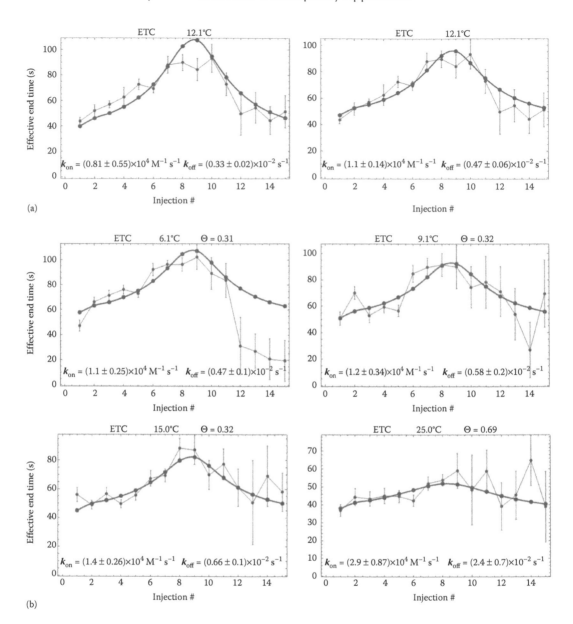

FIGURE 15.9 (a) Improvement of the fit of an ETC by considering τ_{ITC} as a fudge factor. The data are the same as in Figure 15.8, with $[A]_0 = 24.5\ \mu M$ and $[B]_0 = 315\ \mu M$. Either k_{on} or k_{off} has to be considered as a free parameter, since $K_d = k_{off}/k_{on}$, and K_d is known from the classical treatment of each titration curve (not shown). Left: τ_{ITC} was fixed at 3.5 s; right: τ_{ITC} was considered as an additional free parameter and adjusted to (6.6 ± 0.7) s. The kinetic parameters obtained from the fit are indicated and differ significantly depending on whether τ_{ITC} was fixed or not. (b) Fit of ETC for carbonic anhydrase at various temperatures. The data are from the titration of carbonic anhydrase with its inhibitor 4CBS with the same parameters as in Figure 15.8. The initial concentrations of carbonic anhydrase (compound A) were $[A]_0 = 24.4, 24.2, 24.5, 24.6$, and $17.6\ \mu M$ at 6.1°C, 9.1°C, 12.1°C, 15°C, and 25°C, respectively (see Figure 15.9a for 12.1°C). The concentration of 4CBS was $[B]_0 = 315\ \mu M$ at all temperatures. The parameter Θ indicated on top of each figure is defined in Table 15.AI and discussed in Section 15.8.

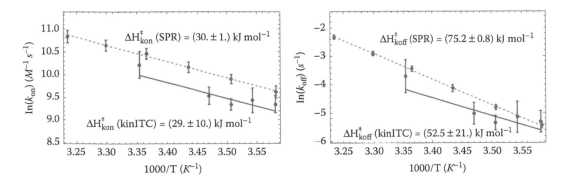

FIGURE 15.10 Arrhenius plots for the comparison of kinITC-ETC and SPR results. The kinITC-ETC results (continuous lines) are compared with the SPR results (dashed lines).

k_{on}(SPR)/k_{on}(kinITC-ETC) $\approx k_{off}$(SPR)/k_{off}(kinITC-ETC) ≤ 1.65 in the temperature range common to the two types of experiments (6.1–25°C). Notably, this ratio was significantly higher (ca. 2.5) when τ_{ITC} was not considered as a fudge factor (not shown). Whether or not the remaining small systematic difference by a factor of 1.65 is real, or results from imperfections of either kinITC-ETC or SPR, or both, is not known. In any case, such a discrepancy is small and usually not considered significant when comparing SPR results with other studies. It is true that the kinITC-ETC results are noisier than those from SPR. However, it is fair to recall that SPR data result from averaging a great number of independent experiments, whereas the kinITC-ETC results were obtained very rapidly from single experiments.

15.8 LIMITATIONS OF kinITC-ETC AND CONCLUSION

It is important to examine the limitations of the technique. For this purpose, several geometrical parameters of the bell-shaped ETC have been defined (see Table 15.AI and Figure 15.AI in Appendix). In addition, the parameter $\Theta = \sqrt{k_{off} k_{on} [A]_0} \, \tau_{ITC}$ is recalled. This parameter was obtained in Burnouf et al. (2012) as a way of quantifying the possibility of discerning the true kinetic signal from the kinetic response of the instrument. It was concluded, but not firmly, that Θ should lie below 1. With more experience with kinITC-ETC, it may be stated more firmly that Θ should lie below ca. 0.5. Of course, this is not a strict limitation, as the quality of the data (involving the ΔH) is, naturally, of great importance for our ability to recover the true kinetic signal. Interestingly, it is seen from Table 15.AI that the maximum equilibration time (i.e., the maximum value of the ETC) is equal to $\alpha \tau_{ITC}[1 + (2\Theta)^{-1}]$, which confirms that high Θ values lead to lower maximum equilibration time and thus to featureless ETC. This is well illustrated in Figure 15.9b, which shows that, indeed, the ETC at 25°C with $\Theta = 0.69$ is the flattest. Nevertheless, this high value did not prevent us from obtaining estimates of $k_{on} = (2.9 \pm 0.9) \times 10^4 \ M^{-1} \ s^{-1}$, $k_{off} = (2.4 \pm 0.7) \times 10^{-2} \ s^{-1}$, comparing decently with those from SPR (Figure 15.10), albeit with higher uncertainties, as expected. It should not be concluded from this particular example that k_{on} and k_{off} are more or less limited to such values, as other parameters influence Θ. Recent studies on the formation of DNA/DNA duplexes (not shown) correctly yielded k_{on} values of the order of 0.5×10^6 M^{-1}s^{-1}. In this respect, one should insist on the importance of having as many injections as possible (i.e., maintaining sufficient heat power) before the maximum of the ETC to tackle difficult situations corresponding to fast kinetics. There are two reasons for this. First, as explained in Section 16.5, small stoichiometry increments produce smaller values of K^{-1}, which is more favorable. Second, increasing the number of experimental points in the ETC increases the statistical significance.

The kinITC-ETC method has been analyzed in great detail on theoretical grounds. It has been shown that in many situations, the automatic determination of the effective end of each injection provides a bell-shaped ETC, which is sufficient to determine k_{on} and k_{off} immediately after determination of K_d by classical methods. Here, kinITC-ETC was illustrated with a real case that was used against an SPR benchmark involving several laboratories. This comparison with the best possible kinetic data showed that the results from kinITC-ETC differ only marginally from those from SPR. These methods are now readily available in the program AFFINImeter (S4SD, Santiago de Compostela, Spain; https://www.affinimeter.com/).

APPENDIX: GEOMETRICAL FEATURES OF ETC

The analysis performed using the kinITC-ETC method can be used to determine all geometrical characteristics of an ETC: minimum and maximum height, maximum variation of the equilibration time, and half-height width. These can be easily obtained from Equations 15.12 and 15.10b. To obtain the simplest results, we make the approximation

TABLE 15A.1   Geometrical Features of an ETC (Equations)

Min. equilibration time (at $s=0$)	$a\left[\dfrac{1}{k_{off}(c+1)}+\tau_{ITC}\right]$	Lower horizontal line ($\alpha \approx 4.5$)
Max. equilibration time (at $s=1-1/c$)	$a\left[\dfrac{1}{2k_{off}\,c^{1/2}}+\tau_{ITC}\right]=a\,\tau_{ITC}\left[1+\dfrac{1}{2\Theta}\right]$	Upper short line (Θ defined below)
Max. equilibration time variation:	$ak_{off}^{-1}\left[\dfrac{1}{2c^{1/2}}-\dfrac{1}{c+1}\right]$	Length of the vertical arrow (at $s=1-1/c$)
Half-height width	$\dfrac{4\left(1-c^{-1/2}\right)\sqrt{3+2c^{1/2}+3c}}{\left(1+c^{1/2}\right)^2}$	Length of the half-height dashed line
Parameter Θ	$\sqrt{k_{off}\,k_{on}[A]_0}\ \tau_{ITC}=c^{1/2}k_{off}\tau_{ITC}$	Defined in Burnouf et al. (2012)

Note: [A]$_0$: initial concentration in the cell; c: [A]$_0$/K$_d$ (Wiseman parameter); τ_{ITC}: response time of the instrument; s: current stoichiometric ratio [B]$_{tot}$/[A]$_{tot}$. It is remarkable that the maximum amplitude of equilibration time variation (arrow in Figure 15.11) does not depend on τ_{ITC} and that the half-height width only depends on c.

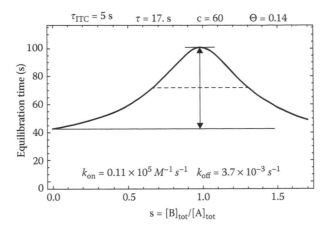

FIGURE 15A.1 Geometrical features of an ETC. The minimum equilibration time (ordinate of the lower horizontal line), the half-height width (length of the dashed line), the maximum equilibration time (ordinate of the upper short line), and the maximum amplitude (length of the vertical arrow) are highlighted. See Table 15A.1 for equations.

that the injection and mixing times are negligible. The results are given (without proofs) in Table 15A.1 with reference to Figure 15A.1.

ACKNOWLEDGMENTS

Developing methods would be useless, and even impossible, without being immersed in an environment of experiments on real biological problems. I am indebted to all present and past members of my laboratory who participated at one time or another in this work: G. Bec, D. Burnouf, C. Da-Veiga, E. Ennifar, S. Guedich, G. Hoffmann, B. Meyer, and B. Puffer-Enders. I am also indebted to F. Disdier for the development and M. Zerbib for the maintenance of the website (http://www-ibmc.u-strasbg.fr:8080/webMathematica/kinITCdemo/). I am also strongly indebted to the enthusiastic people (A. Piñeiro, E. Muñoz, J. Sabin, and J. Rial) from the company S4SD in Santiago de Compostela, Spain, for the incorporation and thorough testing of the presented methods in their software AFFINImeter.

REFERENCES

Burnouf, D., Ennifar, E., Guedich, S., Puffer, B., Hoffmann, G., Bec, G., Disdier, F., Baltzinger, M., and Dumas, P. 2012. kinITC: A new method for obtaining joint thermodynamic and kinetic data by isothermal titration calorimetry. *J. Am. Chem. Soc.* 134:559–565.

Buzzell, A., and Sturtevant, J. M. 1951. A new calorimetric method. *J. Am. Chem. Soc.* 73:2454–2458.

Calvet, E., and Prat, H. 1963. *Recent Progress in Microcalorimetry.* Pergamon: Oxford.

Chaires, J. B. 2008. Calorimetry and thermodynamics in drug design. *Annu. Rev. Biophys.* 37:135–151.

Craig, M. E., Crothers, D. M., and Doty, P. 1971. Relaxation kinetics of dimer formation by self-complementary oligonucleotides. *J. Mol. Biol.* 62:383–401.

Egawa, T., Tsuneshige, A., Suematsu, M., and Yonetani, T. 2007. Method for determination of association and dissociation rate constants of reversible bimolecular reactions by isothermal titration calorimeters. *Anal. Chem.* 79:2972–2978.

Garcia-Fuentes, L., Baron, C., and Mayorga, O. L. 1998. Influence of dynamic power compensation in an isothermal titration microcalorimeter. *Anal. Chem.* 70:4615–4623.

Indyk, L., and Fisher, H. F. 1998. Theoretical aspects of isothermal titration calorimetry. In Ackers, G. K. and Johnson, M. L. (Eds.) *Methods in Enzymology.* Academic: San Diego.

Johnson, R. E., and Biltonen, R. L. 1975. Determination of reaction rate parameters by flow microcalorimetry. *J. Am. Chem. Soc.* 97:2349–2355.

Langerman, N., and Biltonen, R. L. 1978. Microcalorimeters for biological chemistry: Applications, instrumentation and experimental design. *Methods Enzymol.* 61:287–311.

Lopez-Mayorga, O., Mateo, P. L., and Cortijo, M. 1987. The use of different input signals for dynamic characterisation in isothermal microcalorimetry. *J. Phys. E; Sci. Instrum.* 20:265–269.

Morin, P. E., and Freire, E. 1991. Direct calorimetric analysis of the enzymatic activity of yeast cytochrome c oxidase. *Biochemistry* 30:8494–8500.

Navratilova, I., Papalia, G. A., Rich, R. L., Bedinger, D., Brophy, S., Condon, B., Deng, T., et al. 2007. Thermodynamic benchmark study using Biacore technology. *Anal. Biochem.* 364:67–77.

Porschke, D., and Eigen, M. 1971. Co-operative non-enzymic base recognition. III. Kinetics of the helix-coil transition of the oligoribouridylic—oligoriboadenylic acid system and of oligoriboadenylic acid alone at acidic pH. *J. Mol. Biol.* 62:361–381.

Tachoire, H., Macqueron, J. L., and Torra, V. 1986. Traitement du signal en microcalorimétrie: applications en cinétique et thermodynamique. *Thermochim. Acta* 105:333–367.

Vander Meulen, K. A., and Butcher, S. E. 2012. Characterization of the kinetic and thermodynamic landscape of RNA folding using a novel application of isothermal titration calorimetry. *Nucleic Acids Res.* 40:2140–2151.

Watt, G. D., and Sturtevant, J. M. 1969. The enthalpy change accompanying the oxidation of ferrocytochrome c in the pH range 6-11 at 25 degrees. *Biochemistry* 8:4567–4571.

Weisstein, E. W. *Hypergeometric Function.* From MathWorld: A Wolfram Web Resource. http://mathworld.wolfram.com/HypergeometricFunction.html.

Willson, R. J., Beezer, A. E., Mitchell, J. C., and Loh, W. 1995. Determination of thermodynamic and kinetic parameters from isothermal heat conduction microcalorimetry: Application to long-term-reaction studies. *J. Phys. Chem.* 99:7108–7113.

Wiseman, T., Williston, S., Brandts, J. F., and Lin, L. N. 1989. Rapid measurement of binding constants and heats of binding using a new titration calorimeter. *Anal. Biochem.* 179:131–137.

CHAPTER 16

Calorimetric Approaches to Studying Complex Protein Structure–Function– Stability Relationships in Conformational Diseases

The Case of Cystathionine β-Synthase

Angel L. Pey, Tomas Majtan, and Jan P. Kraus

CONTENTS

16.1 ROLE OF PROTEIN STABILITY AND LIGAND BINDING IN HUMAN LOSS-OF-FUNCTION CONFORMATIONAL DISEASES

Protein folding *in vivo* is a complex process that depends on the interaction of the polypeptide with a wide range of proteins organized in macromolecular machineries acting as protein quality control systems (Balch et al. 2008; Powers and Balch 2013; Powers et al. 2009). This complex network comprises over 1000 proteins and is responsible for protein synthesis, folding, trafficking, and degradation (Calamini et al. 2011; Hartl et al. 2011; Powers and Balch 2013; Powers et al. 2009). The protein quality control system allows adaptation of the cell to various intrinsic or environmental challenges, such as genetic mutations and thermal or oxidative stress (Balch et al. 2008; Powers and Balch 2013; Powers et al. 2009). Nevertheless, a single genetic mutation may impact protein folding ability substantially and thus lead to a loss-of-function conformational disease (Balch et al. 2008; Martinez et al. 2008; Muntau et al. 2014; Pey 2013; Powers et al. 2009).

Protein folding efficiency *in vivo* is determined by a complex balance between protein thermodynamic stability and folding/unfolding/misfolding kinetics, and the action of protein quality control systems (Balch et al. 2008; Pey 2013; Powers et al. 2009). Therefore, this efficiency can be reduced by mutations at different levels. Consequently, the ability of a given protein to fold can be enhanced by boosting either the molecular chaperone activity (e.g., by protein homeostasis modulators) or the protein stability (e.g., by native-state ligands; Muntau et al. 2014; Pey 2013), some of the new approaches to pharmacologically targeting conformational diseases.

In this chapter, we present a combination of calorimetric methods and biochemical procedures to ascertain the role of S-adenosyl-methionine (SAM) binding in the activity and stability of human cystathionine β-synthase (human CBS or hCBS), an enzyme associated with classical homocystinuria (HCU) (Mudd et al. 2001; Pey et al. 2013). Moreover, application of these procedures to other eukaryotic CBS enzymes also shed light on evolutionary changes in the domain organization, regulation, and stability of CBS enzymes (Majtan et al. 2014). We also briefly discuss how calorimetric procedures may be useful for the future discovery and development of small ligands to treat homocystinuria and other diseases associated with hCBS dysfunction.

16.2 CYSTATHIONINE β-SYNTHASE: PHYSIOLOGICAL ROLES AND IMPLICATION IN HUMAN DISEASE

Methionine metabolism involves two main pathways that compete for the intermediate metabolite homocysteine (Hcy): the methionine (Met) cycle, which converts Hcy back to Met, and the transsulfuration pathway, by which Hcy is irreversibly converted to cysteine, the rate-limiting step in glutathione synthesis (Finkelstein 2006; Miles and Kraus 2004). hCBS (EC 4.2.1.22) is the enzyme that controls the commitment of Hcy to the transsulfuration pathway, and its deficiency causes a severe inherited metabolic disorder, HCU, clinically characterized by thromboembolism, mental retardation, and connective tissue defects (Mudd et al. 2001). Moreover, alterations in CBS activity and the flux of Hcy through the transsulfuration pathway may also be involved in liver disease and cancer (Prudova et al. 2006).

hCBS is a heme-containing, pyridoxal-5'-phosphate (PLP)-dependent enzyme that catalyzes the condensation of L-serine (Ser) and Hcy to yield cystathionine (Miles and Kraus 2004). Alternatively, hCBS also catalyzes different reactions involving sulfur amino acids (Hcy and Cys) that lead to either formation or consumption of H_2S (Majtan et al. 2014; Singh et al. 2009). H_2S has multiple roles in vasorelaxation, neuromodulation, apoptosis, and inflammation (Kabil et al. 2014). Binding of SAM to hCBS catalytically activates the enzyme up to five-fold, and thus stimulates the flux of Hcy through the transsulfuration pathway rather than through the competing methionine cycle (Finkelstein and Martin 1984; Majtan et al. 2014).

Structurally, hCBS forms native homotetramers, and each subunit of 551 amino acid residues displays three structural domains (Figure 16.1a): (1) a short N-terminal domain (residues 1–70) containing a B-type heme, likely involved in the regulation of hCBS activity on binding of low molecular ligands (O_2, CO, and NO·) and acting as a redox sensor (see Banerjee et al. 2003; Vicente et al. 2014) or affecting stability and folding of the enzyme

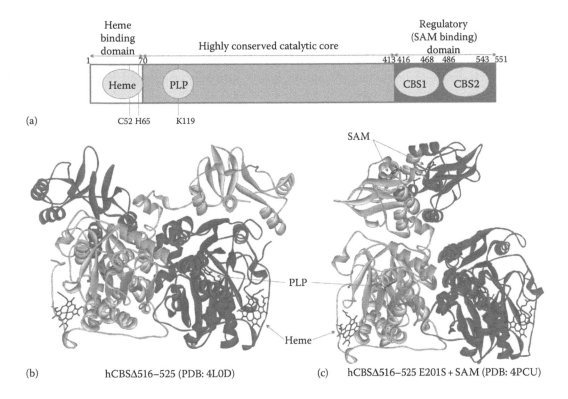

FIGURE 16.1 Domain organization and three-dimensional structure of hCBS. (a) Schematics of hCBS domain organization showing an N-terminal heme-binding domain, where residues C52 and H65 serve as heme axial ligands, a central catalytic core containing the PLP cofactor bound via Schiff bond to residue K119, and a C-terminal SAM-binding regulatory domain harboring a tandem of CBS domains designated as CBS1 and CBS2. (b–c) Crystal structures of an optimized full-length hCBS construct (hCBSΔ516–525) in basal (b) and SAM-bound activated (c) conformations. Light gray and dark gray colors designate the individual subunits in the dimers, while arrows point to cofactors (heme, PLP, SAM) shown as black sticks.

(Majtan et al. 2008, 2011); (2) a catalytic core (CD, residues 71–413) containing the substrate binding sites and a reactive PLP molecule bound to Lys119 (Miles and Kraus 2004); (3) a C-terminal regulatory domain (RD, residues 414–551) containing a tandem pair of CBS motifs that bind SAM, causing displacement of RD from the catalytic site and thus activating the enzyme (Ereno-Orbea et al. 2014). Removal of the C-terminal domain yields highly active CBS dimers no longer regulated by SAM (Kery et al. 1998). Very recently, several crystal structures have been reported for a "full-length" hCBS in the absence and presence of SAM, thus providing structural insight into the mechanism of allosteric activation of hCBS by SAM (Figure 16.1b and c) (Ereno-Orbea et al. 2013, 2014; McCorvie et al. 2014).

HCU is caused by mutations in hCBS, which are inherited in an autosomal recessive manner (Mudd et al. 2001). Over 160 HCU-causing mutations have been described so far (http://medschool.ucdenver.edu/krauslab), and 87% of them are missense variations. Missense mutations often perturb the folding and stability of hCBS *in vitro* and in cells (Hnizda et al. 2012; Kozich et al. 2010; Majtan et al. 2010; Singh et al. 2010). The levels of functional hCBS in missense mutations can be efficiently modulated by altering the activity of molecular chaperone and by the presence of small molecules (Kopecka et al. 2011; Majtan et al. 2010; Singh et al. 2010). Interestingly, mutations at the C-terminal regulatory domain often decrease the folding efficiency of hCBS and affect the activation mediated by SAM (Majtan et al. 2010; Mendes et al. 2014).

16.3 STABILITY OF hCBS: DOMAIN-SPECIFIC DESTABILIZATION BY DISEASE-CAUSING MUTATIONS IS OVERCOME UPON SAM BINDING

Wild-type hCBS displays two major and independent unfolding events as studied by differential scanning calorimetry (DSC): a low-temperature transition (with a T_m value of about 53°C), corresponding to denaturation of RDs, and a high-temperature transition (with a T_m value of about 71°C), reflecting denaturation of CDs (Figure 16.2a). These two unfolding events are kinetically controlled and behave as independent two-state irreversible processes, each described using Equation 16.1 (Pey et al. 2013; Rodriguez-Larrea et al. 2006):

$$C_{p(app)} = C_{p(pre)} + \left(C_{p(post)} - C_{p(pre)}\right) \cdot \left(1 - X_N\right) - \Delta H \cdot \left(\frac{dX_N}{dT}\right) \tag{16.1}$$

where:

$C_{p(app)}$ is the temperature-dependent apparent heat capacity

$C_{p(pre)}$ and $C_{p(post)}$ are the pre- and posttransition baselines (linearly depending on temperature)

ΔH is the denaturation enthalpy

X_N is the fraction of native protein (domain in our case)

The last term on the right-hand side of this expression corresponds to the excess heat capacity $C_{p(exc)}$ (i.e., the denaturation transition). This model assumes that only the

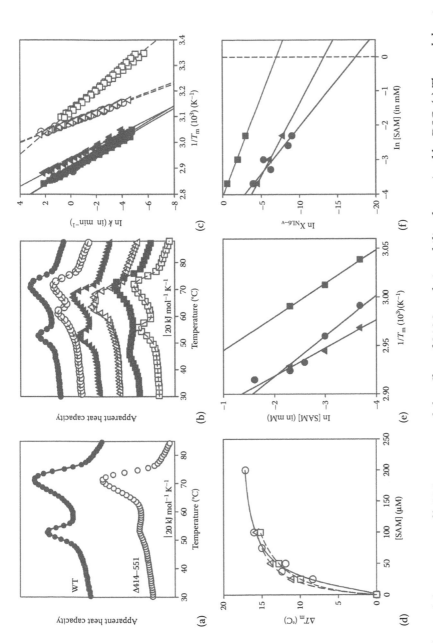

FIGURE 16.2 Thermal denaturation of hCBS variants and the effect of SAM on their stability determined by DSC. (a) Thermal denaturation of WT hCBS and the truncated hCBS Δ414–551. (b) Denaturation of WT (circles), P78R (triangles), and P422L (squares) variants in the absence (closed symbols) or the presence of 25 μM SAM (open symbols). (c) Arrhenius plots derived from DSC analyses for the RDs (open symbols) and CDs (closed symbols) of WT (circles), P78R (triangles), and P422L (squares). (d) SAM concentration dependence of the T_m for the RDs of WT (circles), P78R (triangles), and P422L (squares) based on a two-state kinetic model. (e) Analyses of SAM-mediated stabilization of RDs of WT (circles), R125Q (triangles), and P422L (squares) based on a three-state kinetic mechanism. (f) Analyses of SAM-mediated stabilization of WT (circles), P78R (triangles), and P422L (squares) based on a three-state kinetic model. Experiments were performed in 20 mM Na-Hepes pH 7.4 (Adapted from Pey, A. L., et al., *Biochem. J.*, 449:109–121, 2013.)

native (N) and the irreversibly denatured states (F) are significantly populated, following Model 16.1:

$$N \rightarrow F \qquad \text{(Model 16.1)}$$

and this conversion is characterized by a first-order rate constant k.

The expressions for X_N and dX_N/dT are given by Equations 16.2 and 16.3:

$$X_N = \exp\left[-\exp\left(\frac{E_a \cdot \Delta T}{R \cdot T_m^2}\right)\right] \qquad (16.2)$$

$$\frac{dX_N}{dT} = -\frac{E_a}{R \cdot T_m^2} \cdot \exp\left(\frac{E_a \cdot \Delta T}{R \cdot T_m^2}\right) \cdot \exp\left[-\exp\left(\frac{E_a \cdot \Delta T}{R \cdot T_m^2}\right)\right] \qquad (16.3)$$

where:

T_m holds for the temperature of maximum of the transition
E_a is the activation energy

Since these transitions are protein concentration independent (Pey et al. 2013), the first-order rate constants k can be derived from Equation 16.4 (Sanchez-Ruiz et al. 1988):

$$k = \frac{\tau \cdot C_{p(exc)}}{\Delta H - \langle H \rangle} \qquad (16.4)$$

where:

τ is the scan rate
$C_{p(exc)}$ and $\langle H \rangle$ are the excess heat capacity and denaturation enthalpies at a given temperature
ΔH is the total denaturation enthalpy

The temperature dependence of the first-order rate constants follows the Arrhenius equation.

This model describes very well the denaturation of full-length hCBS (Figure 16.2a and b). Moreover, the E_a values for the low- and high-temperature transitions yield consistent results using different procedures described by Sanchez-Ruiz and coworkers (423 ± 27 kJ mol^{-1} and 256 ± 32 kJ mol^{-1}, respectively; Pey et al. 2013; Sanchez-Ruiz et al. 1988). Our interpretation of the low- and high-temperature transitions as corresponding to the unfolding of regulatory and catalytic domains is further supported by the denaturation enthalpies determined for them (335 ± 23 kJ mol^{-1} and 934 ± 79 kJ mol^{-1}, respectively; Pey et al. 2013). These values agree very well with the theoretical denaturation enthalpies determined for a domain of 138 amino acids (regulatory domains; theoretical value of 347 kJ mol^{-1}) and 343 amino acids (catalytic domains; theoretical value of 1,221 kJ mol^{-1}) determined at their T_m

value using well-known structure–energetic correlations (Robertson and Murphy 1997). Aggregation of hCBS seems to occur on denaturation of the CDs (i.e., the high-T transition; Mendes et al. 2014).

This model allows the denaturation rate constants to be extrapolated to physiological temperature (Figure 16.2c). The corresponding half-lives for the denaturation of RDs and CDs are 30 ± 3 h and 515 ± 57 h, respectively. These results imply that the RDs are kinetically unstable at physiological temperature, which leads to the irreversible activation of full-length hCBS on a time scale of hours (Pey et al. 2013).

We have analyzed the thermal denaturation profiles of seven HCU-causing mutants. These mutants were selected because they represent different types of enzymatic and allosteric behavior: P49L, P78R, and A114V show wild type (WT)-like activation by SAM; R125Q and E176K cannot be activated by SAM; and P422L and S466L are constitutively activated in the absence of SAM (Majtan et al. 2010). For all HCU-causing mutants, the two-transition profile of WT hCBS is maintained, but the kinetic stability of their RDs and CDs is differentially affected (Figure 16.2b and c). Particularly, the kinetic stability of the RDs seemed to be reduced, from a mild 1.5–3-fold effect (P49L, P78R, and R125Q) to a large 200-fold decrease in P422L. Since some of the mutations occur at the CD (P78R, A114V, R125Q, and E176K), this suggests that mutational effects can propagate from CDs to RDs, leading to domain destabilization (Pey et al. 2013).

We therefore tested a possible stabilizing effect of SAM on a full-length hCBS. Addition of concentrations of SAM in the micromolar range have a strong and specific stabilizing effect on the RDs (Figure 16.2b and d) of both WT and HCU-causing mutants, thus suggesting that the ability of these mutants to respond to SAM in terms of activity might be uncoupled from its stabilizing effect. In addition, structural data suggest that these variants do not directly interfere with SAM binding to the RDs (Ereno-Orbea et al. 2014). These conclusions are further supported by the direct titration of CBS variants with SAM by isothermal titration calorimetry (ITC) (Figure 16.3).

16.4 EVIDENCE FOR TWO INDEPENDENT SAM-BINDING SITES RESPONSIBLE FOR ACTIVATION AND STABILIZATION OF hCBS

Direct calorimetric titrations of WT hCBS show specific binding to the RDs (Figure 16.3a). Interestingly, the binding isotherm systematically deviated from a model that considers the existence of a single type of independent binding sites (Figure 16.3a). This result contrasts with previous studies using radioactive filter binding assays, equilibrium dialysis, and fluorescent titrations (Frank et al. 2006; Janosik et al. 2001; Taoka et al. 1999), which might indicate the higher capability of ITC to detect small deviations from the simplest binding models (Freire et al. 2009). Moreover, fittings using a model with two different types of independent binding sites provide an excellent description (Figure 16.3a) with a set of high-affinity ($K_d \sim 10$ nM; two sites per CBS tetramer) and low-affinity sites ($K_d \sim 0.5$ μM; four sites per CBS tetramer) (Figure 16.3b). Under similar conditions, the concentration of SAM required for half-activation of WT hCBS is about 3 μM, suggesting that the low-affinity sites are responsible for SAM-mediated activation (Pey et al. 2013). Both types of sites also display different thermodynamic signatures, with binding enthalpies of -51 and -25 kJ mol^{-1}

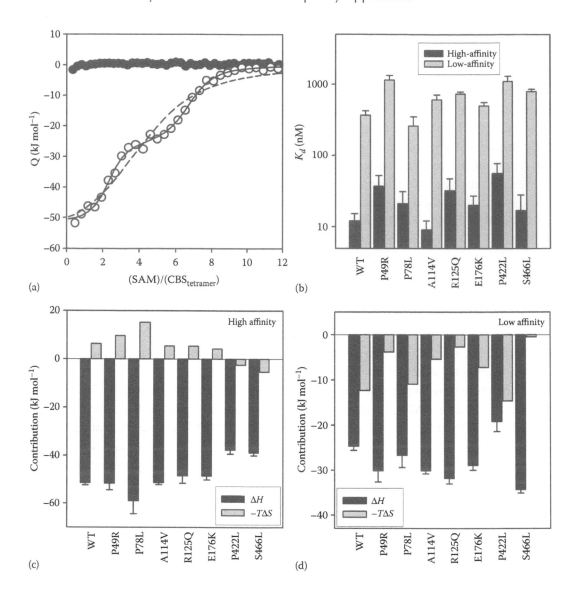

FIGURE 16.3   SAM binding to hCBS variants by ITC. (a) Binding isotherms for WT (open circles) and Δ414–D551 (closed circles) at 25°C. (b–d) Binding affinity (b), enthalpic (c), and entropic (d) contributions to binding to the high- and low-affinity sites (Adapted from Pey, A. L., et al., *Biochem. J.*, 449:109–121, 2013.)

at 25°C and binding heat capacities of −4.2 and −1.9 kJ mol⁻¹ K⁻¹, respectively (Pey et al. 2013). Since SAM-binding thermodynamics are buffer independent, binding enthalpies and heat capacity changes must be associated with structural differences in the binding sites and/or the conformational changes associated with SAM binding (Pey et al. 2013).

Titrations of HCU-causing variants with SAM show overall binding properties similar to those found for both types of sites in WT hCBS with only subtle differences in binding affinity and thermodynamics (Figures 16.3b–d). Therefore, the lack of activation in the presence of SAM for some disease-causing mutants (R125Q, E176K, P422L, and

S466L) cannot be explained by the lack of SAM binding, but rather by the perturbation of the transmission of the allosteric signal leading to hCBS activation by these mutations. Interestingly, a recent study has shown that other HCU-causing mutations occurring at the C-terminal domain may affect binding of SAM based on qualitative chromatographic experiments (Mendes et al. 2014). Therefore, there may exist different molecular mechanisms for the lack of SAM responsiveness in HCU mutants, some of them possibly involving changes in protein dynamics on mutation, as supported by changes in proteolytic susceptibility (Hnizda et al. 2012; Mendes et al. 2014).

To determine whether the high-affinity sites were responsible for SAM-mediated stabilization of RDs, we studied the stabilizing effect of SAM using a simple two-state irreversible kinetic model (Model 16.2):

$$NL_6 \rightarrow F \qquad \text{(Model 16.2)}$$

where the fully saturated hCBS tetramer (NL_6) undergoes the irreversible denaturation of the RDs to a final state F, and this process is characterized by a ligand concentration–dependent first-order rate constant (k_{app}). The rate law for Model 16.2 is provided by Equation 16.5:

$$\frac{d[NL_6]}{dt} = -k_{app}[NL_6] \qquad (16.5)$$

Previous theoretical work on the effect of ligands on irreversible and kinetically controlled protein denaturation have shown that in this context, the number of ligand molecules *released* prior to the denaturation rate-limiting step (v), and hence contributing to the stabilizing effect, can be determined from the ligand concentration dependence of the T_m values (Sanchez-Ruiz 1992; Equation 16.6):

$$\ln[\text{SAM}] = \text{constant} - \left(\frac{Ea}{\upsilon \cdot R \cdot T_m} \right) \qquad (16.6)$$

When applied to WT and HCU-causing mutants, calculation yields a value of $v = 1.4 \pm 0.3$ (average ± standard deviation for the eight CBS variants studied; see Figure 16.2e for some examples), suggesting that within experimental error, two (out of six) SAM-binding sites contribute to the kinetic stabilization of the RDs.

To determine whether the two binding sites involved in RD kinetic stabilization are those with high affinity, we analyzed the stabilizing effect exerted by SAM using the following kinetic mechanism (Model 16.3):

$$
\begin{array}{c}
K \\
NL_6 \leftrightarrow NL_{6-\upsilon} + \upsilon L \\
\downarrow k' \\
F
\end{array}
\qquad \text{(Model 16.3)}
$$

where:

NL_{6-v} is a CBS tetramer with $6-v$ SAM-bound molecules that is kinetically sensitive to denaturation (with a denaturation rate constant k')

K is the overall dissociation constant

The corresponding rate law for this mechanism is given by Equation 16.7:

$$\frac{d[NL_6]}{dt} = -k'[NL_{6-v}] \tag{16.7}$$

Considering that the binding equilibrium between NL_6 and NL_{6-v} is characterized by the dissociation constant K (Equation 16.8),

$$K = \frac{[NL_{6-v}] \cdot [L]^v}{[NL_6]} \tag{16.8}$$

Then, substituting $[NL_{6-v}]$ into Equation 16.7 gives

$$\frac{d[NL_6]}{dt} = -k' \cdot K \cdot [NL_6] \cdot [L]^{-v} \tag{16.9}$$

Combining Equations 16.5 and 16.9, we obtain

$$k_{app} = k' \cdot K \cdot [L]^{-v} \tag{16.10}$$

Considering those states with $6-v$ SAM molecules bound to display similar kinetic stability to unligated CBS, then $k \gg k'$, and by rearranging Equation 16.10 and taking logarithms, we obtain

$$\ln k_{app} - \ln k = \ln K - v \ln[L] \tag{16.11}$$

where k_{app} and k are available from DSC scans in the presence and absence of SAM. These plots, when applied to the eight CBS variants studied, yield a value of $v = 2.4 \pm 0.6$ molecules of SAM, which again supports the notion that about 2 mol of SAM kinetically stabilizes CBS tetramers (see Figure 16.2f for several examples). Moreover, these plots allow the estimation of $\ln K$, where K square root represents the averaged dissociation constants for the sites responsible for kinetic stabilization. Estimation of K_d using this procedure and extrapolating to 25°C gives an average value of 25 nM for the CBS variants studied, which is in very good agreement with the binding affinities determined for the high-affinity sites by ITC (Figure 16.3b).

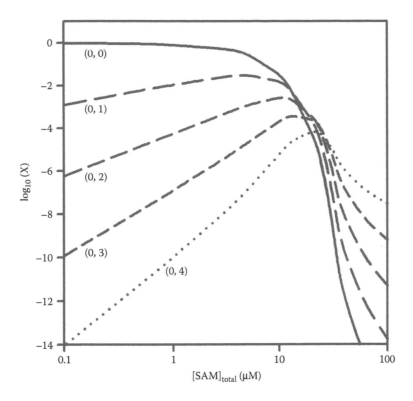

FIGURE 16.4 SAM concentration dependence of the population of kinetically sensitive species of the RDs toward denaturation based on a binding polynomial formalism and ITC data for WT hCBS (Adapted from Pey, A. L., et al., *Biochem. J.*, 449:109–121, 2013.)

Our analyses on SAM-mediated stabilization may be counterintuitive at first sight, because they may imply that those ligation species sensitive to denaturation must be formed on SAM *release* from the high-affinity sites. However, further analyses using a ligand-binding partition function formalism (Wyman and Gill 1990) strongly support that those species kinetically sensitive to denaturation already exist in the ligand-binding equilibrium (Figure 16.4). In these calculations, the fraction of different ligation species (notated as [*m*,*n*], indicating the number of high-affinity, *m*, and low-affinity, *n*, sites occupied by SAM) in the CBS tetramer at different free SAM concentrations ([SAM]) is provided by the following binding polynomial *P*:

$$P = \left(1 + K_{high} \cdot [\text{SAM}]\right)^2 \cdot \left(1 + K_{low} \cdot [\text{SAM}]\right)^4 \tag{16.12}$$

where K_{high} and K_{low} correspond to the association constants ($1/K_d$) for the high-affinity and low-affinity sites determined by ITC. The fractions of the different ligation species are calculated using Equation 16.13:

$$\chi_{[m,n]} = \frac{\omega_{[m,n]}}{P} \tag{16.13}$$

where $\omega_{[m,n]}$ is the statistical factor corresponding to a given $[m,n]$ state based on Equation 16.14:

$$\omega_{[m,n]} = \frac{2!}{(2-m)! \cdot m!} \left(K_{high} \cdot [\text{SAM}]\right)^m \cdot \frac{4!}{(4-n)! \cdot n!} \left(K_{low} \cdot [\text{SAM}]\right)^n \qquad (16.14)$$

As the total SAM concentration increases, the population of ligation species with the high-affinity sites occupied rapidly increases, while the population of those with these sites unoccupied (and assumed to be kinetically sensitive) rapidly decreases (Pey et al. 2013). However, at SAM concentrations such as those used in our DSC experiments, the most populated of the kinetically sensitive species is that with four SAM molecules bound to the low-affinity sites (Figure 16.4), in full agreement with our interpretation of the high-affinity sites as being responsible for the ligand-mediated kinetic stabilization.

16.5 SURFACE ELECTROSTATICS STRONGLY AFFECT hCBS STABILITY AND SAM-MEDIATED ACTIVATION AND STABILIZATION

The experiments and analyses on WT and HCU-causing variants described in Sections 16.3 and 16.4 show quantitative differences in the binding properties of SAM as well as the thermal stability of RDs and CDs described by other authors (Frank et al. 2006; Janosik et al. 2001; Mendes et al. 2014; Taoka et al. 1999). One of the main differences generally observed between our studies and those from other laboratories was that some of the studies have been carried out at vastly different ionic strengths. This discrepancy prompted us to investigate the effect of ionic strength on the properties of WT hCBS in calorimetric and biochemical experiments.

Addition of physiological salt concentrations did not cause significant changes in the native-state properties of hCBS as assessed by spectroscopic methods and dynamic light scattering, but decreased WT CBS activity by 30%–40% (Pey et al. 2014). At this ionic strength, the enzyme displayed a typical four-fold activation by SAM, even though the concentration of SAM required for hCBS half-activation increased four-fold. Overall, these results suggest that the affinity of the low-affinity sites must be reduced, which was confirmed by ITC experiments showing 2.5-fold higher K_d for the low-affinity sites (Figure 16.5a and Pey et al. 2014).

Regarding the stability of hCBS, we observed changes in kinetic stability for both RDs and CDs (Figure 16.5b and c). These changes are essentially independent of the nature of the ions used, as the salts used are found in different positions along the Hofmeister series (Pey et al. 2014), and they change exponentially with the salt concentration (Figure 16.5c), thus supporting the notion that surface electrostatic screening is the main source of the changes in kinetic stability. The effects of surface electrostatic screening on the stability of RDs and CDs are opposite to each other. The kinetic stability of RDs is greatly increased in the presence of physiological salt concentrations, while the kinetic stability of CDs is moderately decreased (Figure 16.5c). These results indicated that at physiological conditions (pH, ionic strength), both domains may show comparable kinetic stability.

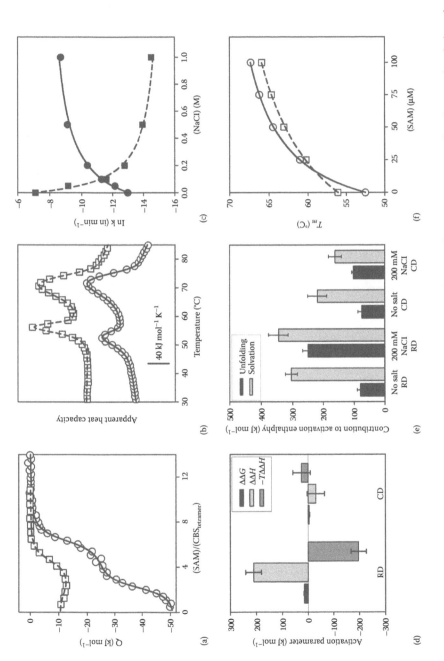

FIGURE 16.5 Salt effects on the SAM binding and stability of hCBS. (a) ITC binding isotherms at 25°C in the absence (circles) and presence of 200 mM NaCl (squares). (b) DSC scans of WT hCBS in the absence (circles) and presence of 200 mM NaCl (squares). (c) Salt concentration dependence of the kinetic stability of the RDs (squares) and CDs (circles) of hCBS. (d) Changes in the activation free energy, enthalpy, and entropy for the denaturation of RDs and CDs of hCBS in the presence of 200 mM NaCl. (e) Contributions from unfolding and solvation barriers to the activation enthalpy in the absence and the presence of NaCl for the RDs and CDs. (f) SAM concentration dependence of the T_m for RDs in the absence (circles) and presence of 200 mM NaCl (squares) (Adapted from Pey, A. L., et al., *Biochim. Biophys. Acta*, 1844:1453–1462, 2014.)

The change in the kinetic stability of RDs and CDs is the result of significant and compensating changes of enthalpic and entropic contributions to the denaturation free energy barriers (Figure 16.5d). The change in these contributions may arise from different structural effects on the denaturation transition state (TS). To provide an estimation of these structural differences, we determined the value of the kinetic m parameter, which is proportional to the difference in surface exposed to the solvent in the native structure and the denaturation TS, from DSC experiments performed in the presence of low nondenaturing urea concentrations, using the following expression (Rodriguez-Larrea et al. 2006; Equation 16.15):

$$m^{\ddagger} = -\frac{E_a}{T_m}\left(\frac{dT_m}{d[\text{urea}]}\right) - R \cdot T_m \cdot \left[\frac{d\ln\left(\frac{E_a}{R \cdot T_m^2}\right)}{d[\text{urea}]}\right] \tag{16.15}$$

These kinetic m values allow the determination of two contributions to the changes in activation energy (~enthalpy) found in the presence of salt (Figure 16.5d): a contribution from *unfolding barriers* ($\Delta H^{\ddagger}_{\text{UNF}}$), arising from the unfolding and solvation of native structure in the TS, and a contribution from *solvation barriers* (ΔH^*), arising from unfolded but not yet solvated regions in the TS. These two contributions are determined from Equations 16.16 and 16.17:

$$\Delta H^{\ddagger}_{\text{UNF}} = \Delta H \cdot \frac{m^{\ddagger}}{m} \tag{16.16}$$

$$\Delta H^* = E_a - \Delta H^{\ddagger}_{\text{UNF}} \tag{16.17}$$

where:

m is the equilibrium m value (simply determined from the size of the RDs and CDs using structure–energetic relationships from Myers et al. 1995)

ΔH is the denaturation enthalpy

These analyses reveal that the strong effect of salt on the kinetic stability of RDs mainly arises from changes in the structure of the denaturation TS (Figure 16.5e), which shows a 10% increase in solvent exposure compared to the absence of salt (Pey et al. 2014). Addition of physiological concentrations of salt also decreased the SAM concentration dependence of the kinetic stabilization of RDs (Figure 16.5f), which can be explained by the approximately eight-fold decreased binding affinity for SAM found in ITC titrations for the high-affinity sites (Figure 16.5a and Pey et al. 2014).

16.6 EUKARYOTIC CBS ENZYMES DIFFER IN OLIGOMERIZATION, STABILITY, AND SAM-MEDIATED ACTIVATION AND STABILIZATION

Studies on eukaryotic CBS such as *Drosophila melanogaster* CBS (dCBS) and yeast CBS (yCBS) have shown differences in their domain organization, oligomerization status, and basal activities compared with hCBS (Majtan et al. 2014). yCBS lacks the N-terminal heme-binding domain, while dCBS does not. yCBS and hCBS form functional tetramers, while dCBS forms homodimers. All of them contain a regulatory C-terminal domain with two potential SAM-binding sites per monomer, but only hCBS activity is enhanced in the presence of SAM, while the basal activities of dCBS and yCBS are high and comparable to that of SAM-activated hCBS in the canonical condensation of Hcy and Ser as well as in alternative H_2S-forming and H_2S-consuming reactions (Majtan et al. 2014). Removal of C-terminal regulatory domains of tetrameric hCBS and yCBS yields soluble dimers showing four- and two-fold, respectively, higher catalytic activity compared with the full-length WT enzymes, while the truncation of dCBS yields insoluble and inactive aggregates.

We have recently characterized the thermal unfolding behavior and SAM binding of hCBS, yCBS, and dCBS by calorimetric methods (Figure 16.6 and Majtan et al. 2014). In contrast to hCBS, both dCBS and yCBS show a main single transition, with widely different denaturation temperatures (about 60°C and 71°C for yCBS and dCBS, respectively; Figure 16.6a), and their denaturation is irreversible and kinetically controlled. The unfolding transitions of yCBS and dCBS are described well by a simple two-state irreversible model (see Equations 16.1 through 16.4). In the case of dCBS, the experimental denaturation enthalpy is 1623 ± 13 kJ mol^{-1}, in close agreement with the theoretical value of 1665 kJ mol^{-1} for the denaturation of both regulatory and catalytic domains (Majtan et al. 2014). Thus, the thermal denaturation of dCBS resembles that of hCBS under strongly stabilizing conditions in the presence of SAM (Majtan et al. 2014). Indeed, extrapolation of denaturation rates of dCBS to 37°C yields a half-life of over a thousand years. For yCBS, the experimental denaturation enthalpy is 661 ± 29 kJ mol^{-1}, which does not agree with the denaturation enthalpy of either the regulatory domain (381–536 kJ mol^{-1}) or the catalytic domain (950 kJ mol^{-1}). Even though thermal denaturation of yCBS occurs at temperatures close to the T_m determined for hCBS RD, it is likely that this thermal transition involves at

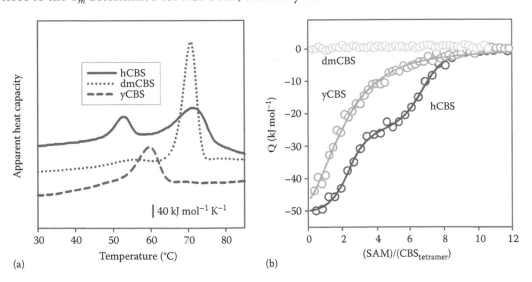

FIGURE 16.6 Stability and SAM binding to eukaryotic CBS proteins. (a) Thermal scans by DSC; (b) SAM-binding isotherms at 25°C (Adapted from Majtan et al., *PLoS One*, 9:e105290, 2014.)

least partial denaturation of the catalytic domains, causing enzyme inactivation (Majtan et al. 2014).

Since dCBS and yCBS show no activation in the presence of SAM, this prompted us to investigate whether SAM would bind to and stabilize these two CBS enzymes. Addition of SAM has no significant effect on the thermal denaturation profiles of either dCBS or yCBS, suggesting that the stabilizing binding sites for SAM are not present in these eukaryotic CBS enzymes (Majtan et al. 2014). Remarkably, ITC titrations revealed binding of SAM to yCBS with a stoichiometry of ~2 mol SAM/yCBS tetramer and a low binding affinity ($K_d = 5.0 \pm 0.5$ μM), while no binding to dCBS was detected (Figure 16.6b). We can speculate that these binding sites are analogous to those of high affinity in hCBS, and that the negligible thermal stabilization afforded to yCBS is a consequence of the low binding affinity (almost three orders of magnitude lower than to these sites in hCBS).

16.7 PHARMACOLOGICAL CHAPERONES TO TREAT CBS-RELATED DISEASES

The calorimetric and biochemical procedures presented in this chapter have provided unprecedented detail of the SAM-mediated regulation of the stability and activity of hCBS. Very recently, crystal structures of a full-length hCBS (lacking only residues 516–525 from the RDs) in the SAM-free basal and SAM-bound activated states have been solved (Figure 16.1b–c) (Ereno-Orbea et al. 2013; 2014, McCorvie et al. 2014). In these structures, only one of two potential SAM-binding sites per monomer is occupied (site S2), which contrasts with our calorimetric analyses of SAM binding to the full-length hCBS tetramer in solution. Several explanations have been proposed for this discrepancy, including the different experimental conditions, possible changes in oligomerization status on SAM binding, and the existence of cooperative effects between identical and/or different types of SAM-binding sites in the hCBS oligomers.

It is also known than SAM-binding sites in hCBS are capable of binding other ligands, such as S-adenosyl-homocysteine (SAH) and sinefungin (Frank et al. 2006; Kozich et al. 2010; McCorvie et al. 2014). Interestingly, these two ligands seem to bind to hCBS but display only marginal enzyme activation at supraphysiological concentrations, suggesting that their binding affinity is significantly lower than that of SAM (Frank et al. 2008). Limited activation of hCBS by SAH or sinefungin also suggests that the enzyme retains its basal conformation after the ligand binding (Figure 16.1b) rather than undergoing a conformational rearrangement with the formation of a disk-shaped CBS module, as in the case of SAM binding (Figure 16.1c). Study of 27 hCBS mutants suggests that manipulation of SAM or SAH levels could lead to rescue of the residual activity of these hCBS mutants (Kozich et al. 2010). Therefore, it is reasonable to propose that ligands might be found that specifically bind to the stabilizing sites at the RDs, enhancing the stability of human CBS. These ligands might selectively stabilize mutant forms of human CBS without interfering in its proper activation by SAM, thus allowing the development of new pharmacological therapies to treat homocystinuria and other disorders associated with hCBS dysfunction.

ACKNOWLEDGMENTS

This work was supported by Postdoctoral Fellowship 0920079G from the American Heart Association (to TM) and by a Ramón y Cajal research contract RYC2009-04147 from MINECO and University of Granada (to ALP). We acknowledge grant support by National Institutes of Health Grant HL065217, by American Heart Association Grant In-Aid 09GRNT2110159, by the Jerome Lejeune Foundation, by MINECO (grants CSD2009-00088 and BIO2012-34937), and by Junta de Andalucia (grant P11-CTS-07187).

REFERENCES

Balch, W. E., Morimoto, R. I., Dillin, A., and Kelly, J. W. (2008), Adapting proteostasis for disease intervention, *Science* 319, 916–919.

Banerjee, R., Evande, R., Kabil, O., Ojha, S., and Taoka, S. (2003), Reaction mechanism and regulation of cystathionine beta-synthase, *Biochim. Biophys. Acta* 1647, 30–35.

Calamini, B., Silva, M. C., Madoux, F., Hutt, D. M., Khanna, S., Chalfant, M. A., Saldanha, S. A., et al. (2011), Small-molecule proteostasis regulators for protein conformational diseases, *Nat. Chem. Biol.* 8, 185–196.

Ereno-Orbea, J., Majtan, T., Oyenarte, I., Kraus, J. P., and Martinez-Cruz, L. A. (2013), Structural basis of regulation and oligomerization of human cystathionine beta-synthase, the central enzyme of transsulfuration, *Proc. Natl. Acad. Sci. USA* 110, E3790–E3799.

Ereno-Orbea, J., Majtan, T., Oyenarte, I., Kraus, J. P., and Martinez-Cruz, L. A. (2014), Structural insight into the molecular mechanism of allosteric activation of human cystathionine beta-synthase by S-adenosylmethionine, *Proc. Natl. Acad. Sci. USA* 111, E3845–E3852.

Finkelstein, J. D. (2006), Inborn errors of sulfur-containing amino acid metabolism, *J. Nutr.* 136, 1750S–1754S.

Finkelstein, J. D., and Martin, J. J. (1984), Methionine metabolism in mammals. Distribution of homocysteine between competing pathways, *J. Biol. Chem.* 259, 9508–9513.

Frank, N., Kent, J. O., Meier, M., and Kraus, J. P. (2008), Purification and characterization of the wild type and truncated human cystathionine beta-synthase enzymes expressed in *E. coli*, *Arch. Biochem. Biophys.* 470, 64–72.

Frank, N., Kery, V., Maclean, K. N., and Kraus, J. P. (2006), Solvent-accessible cysteines in human cystathionine beta-synthase: Crucial role of cysteine 431 in S-adenosyl-L-methionine binding, *Biochemistry* 45, 11021–11029.

Freire, E., Schon, A., and Velazquez-Campoy, A. (2009), Isothermal titration calorimetry: General formalism using binding polynomials, *Methods Enzymol.* 455, 127–155.

Hartl, F. U., Bracher, A., and Hayer-Hartl, M. (2011), Molecular chaperones in protein folding and proteostasis, *Nature* 475, 324–332.

Hnizda, A., Majtan, T., Liu, L., Pey, A. L., Carpenter, J. F., Kodicek, M., Kozich, V., and Kraus, J. P. (2012), Conformational properties of nine purified cystathionine beta-synthase mutants, *Biochemistry* 51, 4755–4763.

Janosik, M., Kery, V., Gaustadnes, M., Maclean, K. N., and Kraus, J. P. (2001), Regulation of human cystathionine beta-synthase by S-adenosyl-L-methionine: Evidence for two catalytically active conformations involving an autoinhibitory domain in the C-terminal region, *Biochemistry* 40, 10625–10633.

Kabil, O., Motl, N., and Banerjee, R. (2014), H_2S and its role in redox signaling, *Biochim. Biophys. Acta* 1844, 1355–1366.

Kery, V., Poneleit, L., and Kraus, J. P. (1998), Trypsin cleavage of human cystathionine beta-synthase into an evolutionarily conserved active core: Structural and functional consequences, *Arch. Biochem. Biophys.* 355, 222–232.

Kopecka, J., Krijt, J., Rakova, K., and Kozich, V. (2011), Restoring assembly and activity of cystathionine beta-synthase mutants by ligands and chemical chaperones, *J. Inherit. Metab. Dis.* 34, 39–48.

Kozich, V., Sokolova, J., Klatovska, V., Krijt, J., Janosik, M., Jelinek, K., and Kraus, J. P. (2010), Cystathionine beta-synthase mutations: Effect of mutation topology on folding and activity, *Hum. Mutat.* 31, 809–819.

Majtan, T., Freeman, K. M., Smith, A. T., Burstyn, J. N., and Kraus, J. P. (2011), Purification and characterization of cystathionine beta-synthase bearing a cobalt protoporphyrin, *Arch. Biochem. Biophys.* 508, 25–30.

Majtan, T., Liu, L., Carpenter, J. F., and Kraus, J. P. (2010), Rescue of cystathionine beta-synthase (CBS) mutants with chemical chaperones: Purification and characterization of eight CBS mutant enzymes, *J. Biol. Chem.* 285, 15866–15873.

Majtan, T., Pey, A. L., Fernandez, R., Fernandez, J. A., Martinez-Cruz, L. A., and Kraus, J. P. (2014), Domain organization, catalysis and regulation of eukaryotic cystathionine beta-synthases, *PLoS One* 9, e105290.

Majtan, T., Singh, L. R., Wang, L., Kruger, W. D., and Kraus, J. P. (2008), Active cystathionine beta-synthase can be expressed in heme-free systems in the presence of metal-substituted porphyrins or a chemical chaperone, *J. Biol. Chem.* 283, 34588–34595.

Martinez, A., Calvo, A. C., Teigen, K., and Pey, A. L. (2008), Rescuing proteins of low kinetic stability by chaperones and natural ligands phenylketonuria, a case study, *Prog. Mol. Biol. Transl. Sci.* 83, 89–134.

McCorvie, T. J., Kopec, J., Hyung, S. J., Fitzpatrick, F., Feng, X., Termine, D., Strain-Damerell, C., et al. (2014), Inter-domain communication of human cystathionine beta synthase: Structural basis of S-adenosyl-L-methionine activation, *J. Biol. Chem.* 289, 36018–36030.

Mendes, M. I., Santos, A. S., Smith, D. E., Lino, P. R., Colaco, H. G., de Almeida, I. T., Vicente, J. B., Salomons, G. S., Rivera, I., Blom, H. J., and Leandro, P. (2014), Insights into the regulatory domain of cystathionine beta-synthase: Characterization of six variant proteins, *Hum. Mutat.* 35, 1195–1202.

Miles, E. W., and Kraus, J. P. (2004), Cystathionine beta-synthase: Structure, function, regulation, and location of homocystinuria-causing mutations, *J. Biol. Chem.* 279, 29871–29874.

Mudd, S. H., Levy, H. L., and Kraus, J. P. (2001), Disorders of transulfuration. In C. R. Scriver, A. Beaudet, W. Sly, and D. Valle (Eds.), *The Metabolic and Molecular Bases of Disease*, 2007–2056, McGraw-Hill: New York.

Muntau, A. C., Leandro, J., Staudigl, M., Mayer, F., and Gersting, S. W. (2014), Innovative strategies to treat protein misfolding in inborn errors of metabolism: Pharmacological chaperones and proteostasis regulators, *J. Inherit. Metab. Dis.* 37, 505–523.

Myers, J. K., Pace, C. N., and Scholtz, J. M. (1995), Denaturant m values and heat capacity changes: Relation to changes in accessible surface areas of protein unfolding, *Protein Sci.* 4, 2138–2148.

Pey, A. L. (2013), Protein homeostasis disorders of key enzymes of amino acids metabolism: Mutation-induced protein kinetic destabilization and new therapeutic strategies, *Amino Acids* 45, 1331–1341.

Pey, A. L., Majtan, T., and Kraus, J. P. (2014), The role of surface electrostatics on the stability, function and regulation of human cystathionine beta-synthase, a complex multidomain and oligomeric protein, *Biochim. Biophys. Acta* 1844, 1453–1462.

Pey, A. L., Majtan, T., Sanchez-Ruiz, J. M., and Kraus, J. P. (2013), Human cystathionine beta-synthase (CBS) contains two classes of binding sites for S-adenosylmethionine (SAM): Complex regulation of CBS activity and stability by SAM, *Biochem. J.* 449, 109–121.

Powers, E. T., and Balch, W. E. (2013), Diversity in the origins of proteostasis networks: A driver for protein function in evolution, *Nat. Rev. Mol. Cell Biol.* 14, 237–248.

Powers, E. T., Morimoto, R. I., Dillin, A., Kelly, J. W., and Balch, W. E. (2009), Biological and chemical approaches to diseases of proteostasis deficiency, *Annu. Rev. Biochem.* 78, 959–991.

Prudova, A., Bauman, Z., Braun, A., Vitvitsky, V., Lu, S. C., and Banerjee, R. (2006), S-adenosylmethionine stabilizes cystathionine beta-synthase and modulates redox capacity, *Proc. Natl. Acad. Sci. USA* 103, 6489–6494.

Robertson, A. D., and Murphy, K. P. (1997), Protein structure and the energetics of protein stability, *Chem. Rev.* 97, 1251–1268.

Rodriguez-Larrea, D., Minning, S., Borchert, T. V., and Sanchez-Ruiz, J. M. (2006), Role of solvation barriers in protein kinetic stability, *J. Mol. Biol.* 360, 715–724.

Sanchez-Ruiz, J. M. (1992), Theoretical analysis of Lumry-Eyring models in differential scanning calorimetry, *Biophys. J.* 61, 921–935.

Sanchez-Ruiz, J. M., Lopez-Lacomba, J. L., Cortijo, M., and Mateo, P. L. (1988), Differential scanning calorimetry of the irreversible thermal denaturation of thermolysin, *Biochemistry* 27, 1648–1652.

Singh, L. R., Gupta, S., Honig, N. H., Kraus, J. P., and Kruger, W. D. (2010), Activation of mutant enzyme function *in vivo* by proteasome inhibitors and treatments that induce Hsp70, *PLoS Genet.* 6, e1000807.

Singh, S., Padovani, D., Leslie, R. A., Chiku, T., and Banerjee, R. (2009), Relative contributions of cystathionine beta-synthase and gamma-cystathionase to H_2S biogenesis via alternative trans-sulfuration reactions, *J. Biol. Chem.* 284, 22457–22466.

Taoka, S., Widjaja, L., and Banerjee, R. (1999), Assignment of enzymatic functions to specific regions of the PLP-dependent heme protein cystathionine beta-synthase, *Biochemistry* 38, 13155–13161.

Vicente, J. B., Colaco, H. G., Mendes, M. I., Sarti, P., Leandro, P., and Giuffre, A. (2014), NO* binds human cystathionine beta-synthase quickly and tightly, *J. Biol. Chem.* 289, 8579–8587.

Wyman, J., and Gill, S. J. (1990), *Binding and Linkage. Functional Chemistry of Biological Macromolecules*, University Science Books: Mill Valley.

IV

Calorimetry as a Tool in Applied Fields

CHAPTER 17

An Account in Calorimetric Research

From Fundamentals to Pharmaceuticals

Patrick Connelly

CONTENTS

The steady blink of the white lightbulb immersed in the stirred water of the insulated orange and white plastic picnic cooler on the workbench commanded one's attention on first stepping foot into the cramped laboratory. A stainless steel contraption was suspended in the water from scaffolding that housed a stepping motor, the whole glorious thing decorated with a tangle of electrical wires. My eyes traced the leads to a Keithly voltmeter, a current source, and a Tronac temperature controller, then over to the incandescent bulb whose silent flashes beat a rhythm radiating outward from the calorimeter. It cast a glow on a bespectacled and rail-thin Stan Gill, his face adorned with graying mustache and brows, cheeks aflame and eyes alit with passion, and a smile that beamed of pure energy.

Professor Gill, a physical chemist, recounted the story that led to the design of his titration calorimeter—a high-precision instrument for measuring the binding of ligands to proteins, the precious quantities of which were often hard won by painstaking purification operations from natural sources. He pulled a mechanical pencil from his shirt pocket and produced a plain white piece of scrap paper. With the artistry and patience of a sculptor, he diagramed the calorimeter in clean lines, explaining each feature with such enthusiasm that I half expected the sketch to come alive with working parts—the glass cells fashioned from nuclear magnetic resonance (NMR) tubes blown out into spherical vessels to contain the titrand, twinned and sitting atop thermopiles deep within the copper block; the combination injector-stirrer for the titrant—a thin-walled glass pipette scarcely the size of a toothpick; the homebuilt electronic feedback current compensation controller; and for acquiring data, the analog strip chart recorder—a vestige of the recent past, and the HP-85 desktop computer for digital data recording with its tiny screen—a glimpse into the future. It was the spring of 1984 in the foothills of the Rocky Mountains in the basement of the Department of Chemistry and Biochemistry at the University of Colorado in Boulder. Ronald Regan was president, the Apple Macintosh computer and the Chrysler minivan were just introduced, there were approximately 1000 hosts on the Internet, and the virus responsible for causing AIDS had just been discovered. Our son Dylan took his first steps at the age of nine months, to the surprise of my wife Mary and me, the day we arrived in Colorado the previous summer. It was the first time that I really saw a calorimeter—or at least in the way that Stan Gill had shown me. Previously, I had fumbled with a freshman chemistry Styrofoam coffee cup and mercury thermometer device and took notice of a bomb calorimeter in the undergraduate physical chemistry laboratory, but this was altogether different. This was the view of things to come.

The Gill laboratory comprised two parts: a pure physical chemistry group, the focus of which was understanding the peculiar thermodynamics of dissolution of weakly soluble apolar substances in water—the so-called hydrophobic effect, and a biology group that concentrated on the thermodynamics of ligand binding, especially the phenomenon of allostery. It was a young Professor Tom Cech who had recruited me into the graduate program at CU Boulder at a time when I had only limited laboratory experience as an undergraduate working at Oxford University, and it was the language of thermodynamics, steeped in the mathematics of J. Williard Gibbs, described by Stan in a first course on the subject, that captured my imagination. Stan introduced me to the visiting Jefferies Wyman—the aging founder of the ligand-linked conformational change allosteric concept who had, together with John Edsall and Robert Oppenheimer, taken Percy Bridgeman's thermodynamics course at Harvard in the early 1920s. I was somewhat aware of Jeffries' theoretical work on the binding potential and the binding partition function, which offered a rigorous formalism that unified biochemistry and phenomenological thermodynamics. Both his writing and mathematical expression of ideas were pure poetry. Stan decided that it would be useful for me to have a desk next to Jeffries; however, it was the mathematician Bill Briggs with whom I would do my first work analyzing the zeros of binding polynomials of tetrameric hemoglobins. The ideas developed fast and, after some mentoring in the ways of scientific writing by Briggs and Gill, my first manuscript as a fledgling scientist was published (Connelly et al. 1986). It was a minor albeit novel result consisting of a theorem that allowed one to associate the patterns of zeros of binding polynomials for multisite macromolecules to degrees of interaction among the sites—applied to oxygen binding to hemoglobins from a variety of species; however, it opened my eyes to the notion that completely new insights could be extracted from the data collected by others, without ever donning a laboratory coat. Experimental work was still abstract to me at the time and really not my strength; however, I could see that the combination of experiment and theory was essential for science. I aspired to master both.*

---

* Frank Weinhold states that "science is concerned with observing the natural world by experimental means and expressing the underlying patterns and relationships by theoretical means." In the scientific realms where I felt most confident, theory is expressed by mathematics—as a model. I often see the model more clearly as expressed in mathematics. Further, when theory is expressed this way, the subsequent deductions become exact, so that the source of deviation between data and model is confined to the model's ability to capture reality in the first place. The burden is then placed on further experimentation to falsify the model so that it may be refined to become a closer approximation of reality. This perspective underscores the vital role that the imagination plays in science; as the faculties of imagination are called on to translate reality to model.

The underlying mechanism of positive cooperativity in O_2 binding to hemoglobin was still in fierce debate since the Monod–Wyman–Changeux (MWC) (Monod et al. 1965) and Koshland–Nemethy–Filmer (KNF) (Koshland et al. 1966) models were introduced in the mid-1960s. The MWC model applied to hemoglobin postulates that the tetrameric macromolecule exists in two overall structural states, in an equilibrium that preexists before oxygen binding, each state with a different affinity for oxygen, and each state binding oxygen noncooperatively. In this model, positive cooperativity develops as a result of ligand-linked global conformational change. The KNF model as applied to hemoglobin reflects a hypothesis whereby there is no preexisting equilibrium of quaternary states, but rather as each oxygen binds, it produces a conformational change in neighboring subunits. The Gill laboratory approached the exploration of the truth along experimental lines by seeking to measure thermodynamic response surfaces for hemoglobin through a plurality of intensive variable coordinates, for example, oxygen and carbon monoxide activities, and temperature, with its complement extensity, heat. And so the need for calorimetry. The approach was decidedly Popperian; that is to say, we sought to challenge the model by exploring binding behavior with multiple ligands to put hemoglobin through its molecular gymnastics. If the multidimensional data failed to be described by a particular allosteric model, the model could be falsified, and subsequently refined to identify one that could describe the totality of observations.

It was a rite of passage in the Gill laboratory to purify human hemoglobin, from our own blood, and measure a high-precision oxygen-binding curve with the famed Gill thin-layer spectroscopic apparatus, only to extract the four Adair constants under the carefully specified conditions of the experiment with home-fashioned software to perform the nonlinear regression analysis. Human hemoglobin was a touchstone of the laboratory. Numerous structures had been determined by x-ray diffraction, in their distinctly different oxy and deoxy states. It was a textbook example of a quaternary protein structure and the best studied example of cooperative binding in a biological system with the S-shaped oxygen-binding curve and steep Hill slope. Still, the structure–function mechanism responsible for its positive cooperativity was not fully elucidated. The answer bordered on an obsession for Stan and a few other devoted hemoglobinologists. The weight of facts was stacked in favor of the MWC allosteric model. Gill's unique mastery of theory and experiment coupled with a talent and penchant for the design and construction of instruments proved to be a powerful combination in the search for the deeper mysteries of macromolecular function in hemoglobin A (HbA). While I shared the laboratory's enthusiasm for quantitative approaches, it was a different class of respiratory proteins that was to draw my attention.

Hemocyanins are the copper-based oxygen transporters responsible for the blue blood found in molluscs and athropods. The oxygen-binding cooperativity in these systems was the highest known at the time (and still is) with Hill slopes approaching the steepness of phase transitions. The biophysical chemist Ken van Holde and biologist Karen Miller in Corvallis, Oregon, were working on the architecture of hemocyanin from a species of octopus procured from fishermen in the Puget Sound of the Pacific Northwest of the United States. The hemocyanin from *Octopus dofleini* was composed of an astonishing 70 oxygen-binding sites. Meanwhile, in the Netherlands, the group of Wim Hol was working on the atomic-level complete x-ray structure of the hexameric hemocyanin from the Mediterranean spiny lobster, *Panulirus interruptus*.*

Heinz Decker of Munich, Germany, was exploring the marvelous hierarchical structures of tarantulas and scorpions—the 12-site dodecamer and 24-subunit hemocyanins from these terrestrial arthropods. And so it was with a merry and experienced cadre of collaborators—theoreticians, marine biologists, structural chemists, physical chemists, mathematicians, and biochemists—from the United States and Europe that I cut my teeth as a PhD student from a base of operations in the shadow of Colorado's remarkable flatiron peaks stretching into famously sunny and cornflower blue skies.

Identical linkage is the term used to describe pure site-for-site competition in ligand binding—an eye for an eye, mutually exclusive ligand occupancy. Oxygen and carbon monoxide binding to the iron porphyrin sites on hemoglobin is one example. According to Haldane's laws—a principle put forth by J.B.S. Haldane in the late nineteenth century, the binding curves for oxygen and carbon monoxide are parallel; however, nobody had extracted the four binding constants of carbon monoxide with sufficient precision to seriously test the "law." It was a gap in our knowledge owing to the technical challenge of measuring the tight binding of carbon dioxide (CO). As part of the Gill laboratory's rite of passage, I stepped in to do it, and then together with the more highly skilled experimentalist Mike Doyle, a senior graduate student in the laboratory, and the illustrious Italian postdoc Enrico DiCera, we went on lay this academic question to rest (DiCera et al. 1987). Along the way, together with senior grad student Chuck Robert, we established some important though arcane numerical analyses procedures that proved to be crucial to the overall argument supporting the allosteric model for human hemoglobin (Gill et al. 1988). The test of the allosteric concept with HbA from an analysis of the identically linked ligands oxygen (O_2) and CO turned out not to be so satisfying since the ligands behaved so similarly. Haldane's law did not hold up; but since the cooperativity of the two ligands was remarkably similar, it was not an especially rigorous challenge of the MWC model. Exploring the linkage with heat as another test of the MWC model was fraught with technical difficulty since the irons of hemoglobin oxidized in the calorimeter during the experiment, giving rise to a baseline heat that confounded the desired signal, a battle well fought by Spanish visiting scientist Antonio Parody Morreale and American graduate student Gary Bishop. I can well recall the cacophony of lively discussions in the laboratory with Spanish (Antonio), Italian (Enrico), German (Heinz), and southern (Gary Bishop) accents. In contrast to the iron-based hemoglobins, copper-based hemocyanins are remarkably stable to oxidation and the difference in the binding of oxygen and carbon monoxide to hemocyanin was vastly different (CO binds noncooperatively; O_2 binds with exceptionally high

---

* A few of us in the hemocyanin field gathered at the Duke Marine Research Laboratory on the Atlantic coast to give talks and listen to Wim Hol describe the atomic-level features of what was then the largest protein structure ever to be solved—a staggering half million Daltons. After each day's session, Stan Gill, Chuck Robert, Wim Hol, Joe and Celia Boneventura, Jan Herman, and a host of others would gather by the picturesque harbor and drink beer before dinner. After the conference, Chuck Robert and I accompanied Jan Herman and his son on his sailboat for a trip to the Shakleford shoals off the Carolina coast. We anchored for the night and swam to the island to gather mussels for dinner while we spied wild goats and horses—progeny of survivors from hurricane-whipped farms that were ruined earlier in the twentieth. The experience underscored the value of social connectivity in science.

positive cooperativity). Therefore, I took up the work of fully exploring the identical linkage of oxygen and carbon monoxide in the octopus, tarantula, and spiny lobster species (Connelly et al. 1989a, 1989b; Decker et al. 1988). Linkage with heat as a coordinate would have to await my building a new titration calorimeter with design improvements that Stan had made in his endless search for increased precision.

And so it was on a snowy Colorado morning in the mid-1980s that Stan showed up to the laboratory with one of his famed calorimeter line sketches. With his new pencil drawing in hand, into the machine shop I strode to make plans with the affable metalsmith, Bill Ingino. It was a new world of spinning lathes and whirring milling machines heavy with the scent of lubricating oils heated by the friction of carbide tools that cut through bronze and brass and copper and aluminum and stainless steel like knives through butter. A cornucopia of materials, new to me, soon became familiar just by touch—Teflon, Delrin, and polyethylenes. The language of collets and chucks and taps and reamers formed a new branch of my vocabulary. From time to time, Stan would grace the shop with movements that expressed the inner glee of a taciturn child—wielding a calipers to check on a precision part before exclaiming "just about perfect" or "we might want to redo this piece." Though I was unaware of it at the time, the experience of design engineering and the ability to talk to machinists would come in handy for years to come.

Shortly after assembling and performing the initial calibrations of the new calorimeter, I was scheduled to give a talk at the North American Calorimetry Conference. At the time, the Calorimetry Conference was a meeting where you could meet a collection of academic, industrial, and government scientists from around the world representing chemistry, physics, and biology disciplines. We listened to talks that reflected scientific traditions that seemed to stretch nearly to the previous century. It was where I was first introduced to calorimetry noteworthies Lee Hansen from Utah, Ingemar Wadsö of Sweden, the American Ken Breslauer, and the Russian Peter Privalov. I presented the results of several of our manuscripts accepted for publication, or recently published, detailing our work on the thermodynamics of identical linkage in hemocyanin systems. The data were acquired using the thin-layer optical cell. In all systems, the MWC model, or an extension of it, termed a *nested model*, held up in describing the thermodynamics of oxygen and carbon monoxide binding. As for calorimetric results, I could only offer up to the audience the enviable traces of the baseline power output and "water into water" controls that foreshadowed the sensitivity of measurements to come. These baseline data caught the eye of an attendee at the conference—the well-known calorimetrist Julian Sturtevant from Yale University. Julian had been a professor in Yale's chemistry department since the late 1920s. He was the recipient of the longest running NIH grant—the "go-to guy" for differential scanning calorimetric measurements to determine the thermodynamics of protein folding transitions. Dr. Sturtevant was desirous of measuring binding constants. He would consider nothing less than the direct thermodynamic approach—a titration microcalorimeter. And so, he bought the new titration calorimeter I had just built for $15,000 complete with the orange and white picnic cooler, and offered me a postdoctoral post at Yale.

I wrote my thesis and defended it a few weeks later. Tragically, Stan had been diagnosed with stomach cancer and was undergoing treatment. I have memories tinged with mixed emotions of walking to his home to visit him to review portions of my thesis and pending manuscripts. I loved meeting with Stan to discuss the work; however, his battle with disease was all too evident, and a foreboding feeling would overtake me on the walks back to the laboratory. His kindness and attention despite his weakening condition revealed an inner strength that reflected his character. I realize in retrospect how truly remarkable was the education that I received from him and fellow members of the Gill laboratory. Add to that Tom Cech's mentorship, including his course in advanced biochemistry taught while performing the work on catalytic RNA that would earn him a Nobel Prize in a few years (1989), and it is easy to see how Boulder in the late 1980s was a perfect storm for learning the fundamentals of molecular biology and physical chemistry. With my thesis complete, it was time to take up new adventures in science and head to New Haven. My wife Mary, sons Dylan and Jude (born in Boulder) and I, in a car packed for travel, waved good-bye to a throng of friends who had come to send us off—the sun setting over the Flatirons and the snow-capped mountains of the continental divide vanishing from our sight as we made the long trek east to the familiar shores of southern New England.

Long Island Sound, a tidal estuary of the Atlantic Ocean, lies between the coasts of Westchester County and Connecticut on one side, and Long Island on the other, as it stretches from the East River of the Bronx in New York City to Block Island Sound, off the coast of Rhode Island. Long Island Sound was the place of my youth. Swimming in her waves and fishing her waters were activities of pure delight for me. Upon coming to Yale, we rented a small house on the beach in Branford, CT, very near to the Sturtevants—Julian and Elizabeth—and just north of New Haven, the site of the old, magnificent, darkened redstone, gothic Sterling Chemistry building on Science Hill. Dylan was entering kindergarten; Jude was too young yet for school but not too young for adventure. Together with Mary and me, both lads took to the exhilaration of living by the sea with great enthusiasm, skipping in the surf and paddling out to the Thimble Islands off Stony Creek in a canoe.

In the Sturtevant laboratory, I was immediately at home too. It was conveniently positioned across from the machine shop and down the hall from the glass-blowing shop. I spent the first few weeks tweaking the calorimeter to improve its performance. Each day at 4:00 pm, with ritual aplomb, Julian would call us to tea. It was self-serve with Lipton tea bags and hot water procured from an old aluminum teakettle heated atop a gas flame on the laboratory bench.* Tea with Julian became the one time of the day when you could be assured of science discussion and debate. At the age of 83, Julian was still remarkably active and competitive as an experimentalist. We often compared notes on productivity for the day. He was rarely topped, and it was not only in intellectual matters. He was fond of taking

---

* Oversight from any laboratory safety group was not apparent at the time. Preparing food in the laboratory, much less eating and drinking, would not be tolerated by today's standards. During WWII, Yale chemist Herbert Harned worked secretly on the electrolytic separation and extraction of Uranium-235 isotope as part of the Manhattan Project. If you lift up a certain heating grill at the end of the hall on the east side of Sterling Chemistry laboratory near the Sturtevant laboratory, a hidden staircase can be found. It leads to an abandoned space where it is believed that Harned conducted his secret work. Years after I was in the laboratory, hot spots were found in the floor of Sterling from radioactivity left over from Harned's days working in secret. The floors have since been replaced. The sense of a thriving science culture was palpable.

postdocs to the Yale gymnasium for an afternoon game of squash. Despite the fact that we were all nearly a third his age, it is perhaps embarrassing to admit that he was surprisingly competitive with a racquet.

Among the big problems of the day was that of protein folding. The DNA code had been deciphered, thereby specifying the primary sequence of amino acids in a protein from the corresponding sequence of DNA; however, the question remained: how did the primary sequence specify the unique three-dimensional fold of the protein? The year was 1989. The Berlin Wall was falling, protesting students and the troops of the People's Liberation Army clashed in Tiananmen Square, the first commercial Internet service providers emerged, the 14th Dalai Lama won the Nobel Peace Prize, and Pons and Fleishman shocked the world with their announcement of having achieved cold fusion.* The energetic and structural determinants of the folding of small globular proteins were not fully elucidated and were under intense experimental and theoretical interrogation. Among the model objects of focus were lysozyme from bacteriophage T4 (in the laboratory of Brian Matthews at the University of Oregon), staph nuclease (Dave Shortle's laboratory at Johns Hopkins), and barnase (Alan Fersht at the Medical Research Council [MRC] in Cambridge). Site-directed mutagenesis was employed to change single amino acids in the protein, one at a time, and the resulting structure and energetic consequences were determined. A great amount of data accumulated. Brian Matthews' laboratory led the way on T4 lysozyme, where dozens of mutant proteins were produced and structures were solved. Julian Sturtevant's laboratory produced the energetic measurements by differential scanning calorimetry (DSC). Anybody that passed through Sturtevant's laboratory in that era would be lured or coaxed into some aspect of the T4 project. I seized the opportunity to study some variants with replacements at threonine 157, learning the art of DSC with the instruments built by Valerian Plotnikov and Peter Privalov and the commercially available DSC from newly formed MicroCal Inc. Postdocs and visiting scientists in the laboratory provided me with their data on T157 substitutions and we worked together on the analysis and drafted a manuscript. These great colleagues included Lily Ghosaini, Cui-Quing Hu, Shinichi Kitamura, Akiyoshi Tanaka, and Julian (Connelly et al. 1991; Hu et al. 1992). In our paper, we introduced the concept of a calorimetric hill slope to express the cooperativity of protein folding in analogy to the ligand binding. It built on the framework of heat binding developed by Gill and Wyman as we defined a physical parameter for the molecular weight of the cooperative unit and showed how the binding capacity and excess heat capacity functions were mirrored mathematically in the underlying expressions for binding and thermal hill slopes. It emphasized that the binding and folding reactions of macromolecules were unified in more ways than one. The work concluded that we had a ways to go in understanding the effects of the chemical changes.

---

* One day, early in my tenure at the Sturtevant laboratory, a member of the Yale physics department came to visit me. Yale had joined the scientific community to reproduce the work of Pons and Fleishman on cold fusion. They heard that there was a new guy on campus in the form of me who could build calorimeters. A calorimeter was needed to reproduce the alleged cold fusion experiment. He pressed me to forget about the thermodynamics of macromolecules, and join the physics department to work on cold fusion. I listened intently, but turned him down. It proved to be a good professional decision. In months and years to come, the cold fusion results were never reproduced. It was a lesson that science is filled with blind alleys.

The trouble with all the work measuring the changes in enthalpy, entropy, heat capacity, and free energy due to amino acid substitutions was that it was difficult to extract measures of specific noncovalent interactions since one could not ascribe the relative energetic contributions of the folded and unfolded states to the net stability differences. We had no reference structures for the unfolded states. To improve on this situation, Rhagavan Varadarajan and Fred Richards in the molecular biology and biochemistry department at Yale turned their attention to the ribonuclease S system and sought out Julian and I for collaboration. It had been discovered some years prior, that when the end terminal 15 amino acids of ribonuclease—a sequence that formed an alpha helix in the folded structure—were cleaved to form two fragments, the S15-peptide comprised of residues 1–15 and the S-protein, the fragments could be combined to form a bimolecular complex called *Ribonuclease S* that had a structure isomorphic to ribonuclease A. By making amino acid changes to the S15-peptide via peptide synthesis, one could determine the energetic and structural consequences of the changes, with high precision, through isothermal titration calorimetry and x-ray crystallography. There would be no confounding effects of a denatured state with which to contend. It was a good strategy.

If Fred Richards was the king of x-ray crystallography and Julian was the thermodynamic titan, then Rhagavan and I were the appointed postdoctoral princes for the Ribonuclease S project. Rather than use the Gill titration calorimeter, I recommended we employ the new titration calorimeter that was soon to hit the market from Microcal—the Omega. The Omega was the creation of John Brandts and engineer Tom Wiseman (Wiseman et al. 1989). Brandts had placed a prototype of the Omega in the laboratory free of cost. The Omega drew heavily from Gill's design, but it was distinguished in many regards and it was far more user friendly. I had been testing the instruments head to head and forecasted we could move much faster with the Omega and be free of numerous visits to the machine shop needed to tweak the homebuilt instrument. The decision troubled me—I was torn between my allegiance to Stan and my obligation to execute the current work in the best way possible. Scientifically, the strategy was a smashing success. What followed was a marvelously productive time, as Rhags and I worked furiously to produce substantive data within a few weeks. We immediately used our data to challenge the standard approach for interpreting the thermodynamic properties of mutant proteins. Several investigators had been arguing that the changes should correlate with the transfer energetics of the amino acids from organic solvents to water. Our data did not fit this hypothesis—a notion that had its origin in the early work of Kauzmann and Charles Tanford. We argued that the energetics of noncovalent interactions involving folding and binding should not be approximated by organic solvent-to-water transfer energies, that there was something unique about the interactions in biological macromolecules, that they were special forms of matter (Connelly et al. 1990). Others aligned with our view or arrived at it independently—a conclusion that has held up to the challenge of lots of data in the years to come.

In the fall of 1989, after having been at Yale for just over a year, during a time that I was interviewing for professorship posts, a scientist from pharmaceutical giant Bristol-Myers visited the laboratory. At tea that afternoon, the discussion turned to the future and the visitor urged me to consider an alternative career path—one in the pharmaceutical industry. I had not previously considered such a direction; it simply was not in my view prior to that day. The recommendation led to an exploratory phone call with a scientist at Merck Research Laboratories that in turn led me to a newly formed company in Cambridge, MA, founded by an ex-Merck scientist with the support of a venture capitalist and a cadre of scientific advisors from Harvard University. The company was called Vertex Pharmaceuticals Incorporated. A phone call to Vertex led to an interview and soon I set foot for the first time in Cambridge and Boston. The cities separated by the Charles River were spectacular.* Vertex was nestled in the gritty biotech ghetto of Cambridgeport midway between Harvard and the Massachusetts Institute of Technology. If the exterior was not pretty, the interior made up for it with a phenomenal cast of scientists and new laboratory space. I gave a talk about our work on Ribonuclease S. Among the members of the interview team I faced included Vertex's founder Joshua Boger, the company's first employee John Thomson, and a member of the scientific advisory board, Harvard professor Jeremy Knowles. A few days later, I received an offer to be a founding scientist. It was a chance to start the first structure-based thermodynamics laboratory in the industry. I accepted the offer, and in the late summer of 1990, Mary and I and our sons moved to the Commonwealth of Massachusetts. It was the start of a new chapter in our lives. Cambridge was fast becoming the epicenter of medicine, biotechnology, and pharmaceuticals. It was phenomenally exciting.

Structure-based drug design (SBDD), a technology based on the idea that one could use the atomic-level structure of a protein to design drugs to be more potent and specific, had not been put to the test. Two start-up companies—Vertex on the East Coast and Agouron on the West Coast—were founded to do just that. The flagship program for Vertex focused on discovering a better, safer FK-506—a drug that was in the clinic at the time for ensuring that transplant recipients would not reject their donor organ. The tools of x-ray crystallography and computational chemistry had well evolved to pair with biology and medicinal chemistry in the hunt for new drugs. If anything, the theory on which molecular modeling was based, as applied to macromolecules, was far ahead of experimental validation. With my new colleagues, I sought to change that by ensuring we had a far better grounding in the fundamentals of SBDD.

Medicinal chemists optimize compounds for potency by coming up with hypotheses about how to enhance the overall binding free energy relative to a lead compound, then making new compounds that can be tested for potency. As scientists, the test would determine if the hypothesis is verified; as drug hunters, all you cared about was if the compound was more potent. It was not always necessary to know why. The pure drug-hunting view is shortsighted, though it often prevails in practice. Knowing if a hypothesis was verified and knowing why a compound was more potent could allow one to leverage present results to future programs. With the aid of a high-resolution structure of the protein complexed with

---

* I have always believed that the natural world in which one is embedded has a direct influence on one's creative abilities in science. Doing science in the picturesque city of Oxford along the Thames, in the shadow of the inspiring flatiron foothills of the Rocky Mountains in Boulder, or along the Atlantic shores in New Haven shaped my patterns of thought in ways that I may never understand. The skyline of Boston and the beauty of the old parts of Boston and Cambridge stimulated my imagination immediately.

a lead compound, the chemist can hypothesize that by adding a hydrogen-bonding acceptor substituent, for instance, it would form a hydrogen bond with a donor within reach on the protein and thereby enhance potency. The challenge is that one never simply just adds a hydrogen bond donor atom. It comes with other atoms on the whole substituent. Further, even if the structures of the lead compound and the improved compound bound to the protein target are determined, and it verified that a new protein–ligand hydrogen bond is formed, it is not clear from the atomic-level structures if the improved observed binding energy was actually due to the new hydrogen bond. Molecular modelers of the day were fond of providing computational verification of the hypothesis and observed result; however, validation of our ideas in the field were lacking. Clarity and rigor about what drives potency increments were sorely needed. It was core to Vertex's founding technology.

The truth is, nobody had ever satisfactorily answered the question, "what is the energy of formation of a protein–ligand hydrogen bond?" A few estimates existed from model compound systems, and we had the results of site-directed mutagenesis, but no results existed in real protein–ligand systems where full structures and the thermodynamics of reactants and products were known. We sought an answer in the system of FK-506 and its binding protein FKBP-12. We knew from the structure of the complex that an amide carbonyl oxygen on FK-506 formed a hydrogen bond with a tyrosine hydroxyl group on the protein. By changing the tyrosine residue to a phenylalanine via site-directed mutagenesis, we could effect a one atom change to the system of thousands of atoms (removal of an oxygen atom) and abolish a single protein–ligand hydrogen bond. It was an ideal system. With the support of great Vertex colleagues, we made the required materials, examined the structures of the mutant protein and its complex with FK-506, and measured the changes in free energy, enthalpy, entropy, and heat capacity on binding to the protein (Connelly et al. 1994). To make the measurements, we employed the Omega, which was being sold commercially for approximately $70,000.

The results of the experiment were fairly shocking to all at first glance. Abolishing a key hydrogen bond in the complex led to a large *stabilization* in the enthalpy of protein–ligand binding (a whopping ~16 kJ/mol)—not a destabilization as we had thought. The free energy change slightly dropped (~2 kJ/mol), which meant that the stabilization was due to entropy, and that the entropic effect had to be very large to compensate for the enthalpic effect. But why? The answer would reveal itself on comparing the difference density maps calculated from the crystal structures of the unliganded mutant and wild-type proteins. The tyrosine of the wild type had two well-ordered water molecules bound to it. In the mutant phenylalanine structure, there were no water molecules, but the structures of both bound and free, mutant and wild type, were otherwise no different at high resolution. It led us to the conclusion that it was the entropy from the waters that were displaced on binding that aided binding to the wild-type protein and that were absent in the case of the mutant. We wrote the paper and I sent it to Julian. He immediately sponsored it for the *Proceedings of the National Academy of Sciences* (Connelly et al. 1994) and I delivered the work in a talk at the summer Gordon Conference on Biopolymers in Newport, Rhode Island.

Shortly after the hydrogen bond result, computational chemist Dave Pearlman and I collaborated to use the data as a validation of high-precision, temperature-dependent, free-energy perturbation methods. The techniques originally developed and applied to nonpolar gases by Robert Zwanzig from Yale's Sterling Chemistry laboratory in the early 1950s were now being applied to proteins. David, or DAP as we called him, was able to use the structures of the mutant and wild-type, free, and FK-506-bound proteins to perform simulations to obtain results in excellent agreement with the experiment (Pearlman and Connelly 1995). Temperature-dependent simulations provided us with estimates of the enthalpy and entropy of the one atom perturbation. It seemed miraculous that we obtained even the proper sign of the enthalpy stabilization, let alone a value of the magnitude that was in reasonable agreement with the experiment.

As drug designers at Vertex, the hydrogen bond result woke us up to the subtleties of the energetics and to the realities of our limited understanding and capabilities of predicting changes made to compounds in the process of lead optimization—even in cases when structural information was available to guide us. It made us highly attentive to the role of water in binding. We recognized that in some cases, it would be beneficial to displace protein-bound water; in other cases, we would need to preserve the bound water molecules to capture interactions of a refined compound to the protein-bound water. We continued to produce fundamental work to provide further understanding, and forged ahead on applied work to take advantage of it in drug discovery. A few years later, John Ladbury—a friend and postdoctoral colleague from the Sturtevant laboratory days—and I edited a book on SBDD that brought together fundamental strategic principles procured from a diversity of investigators (Ladbury and Connelly 1997). It includes a theoretical framework, a structure-energetic guide, for considering the role of water binding in drug discovery. John and I developed a scientific correspondence around the effects of bound water on heat capacity changes while he was still at Yale. While our independent results were determined from very different protein systems, they reinforced our shared view. We often lectured in the same sections of conferences and symposia during the late 1990s.

During the time we were performing this work, it became clear that the approach we were taking to develop a better FK-506 had a fatal flaw. The recognition of this fact led us to abandon the discovery program. Before the decision was made to drop the program, a few of us were approached by Vertex leaders who wanted us to tackle the problem of HIV infection and AIDS. The death toll of AIDS patients was mounting. The world was on full alert. Most of the major pharmaceutical companies were in the hunt for an AIDS drug. Vertex and many others decided to go after inhibitors of HIV protease, a protein responsible for the life cycle of the virus. Without it, the virus is rendered uninfectious. A source of protein was needed and an assay to test compounds for inhibition. Before we had learned to clone, express, and purify the protein, we ordered chemically synthesized material costing thousands of dollars for a fraction of a milligram. It was enough to get started. I developed the initial high-performance liquid chromatography (HPLC) assay for the program, then turned my attention to designing and constructing an instrument that would enable us to do temperature-dependent studies from which we could extract the thermodynamics of binding via the van't Hoff relation (the limited amount of HIV protease available would preclude the use of titration calorimetry; we needed another solution to determine the thermodynamics parameters).

The instrument, which I called a *temperature gradient calorimeter* (TGC), or a van't Hoff calorimeter, was inspired by the 96-well microtiter plate format. I learned from my friends who were further developing the HIV protease assay, that they were using a high-throughput format. When I first set eyes on the white rectangular plate with wells that each held around a quarter of a milliliter of liquid, I nearly laughed and scoffed. I exclaimed to my colleagues, "you can't be serious about performing experiments in a plastic dish?" When I was assured that the data that came from spectroscopic scanning of the titer plate were of high quality, I was intrigued. Then, with a piece of white paper, I sketched a 12×8 titer plate fashioned with ports for the circulation of temperature-controlled water at the sides of the long dimension of the rectangle. The idea was to put a fixed temperature gradient across the plate in one dimension. In the other dimension, one could vary a chemical potential—pH or the concentration of a ligand. By putting protein in the wells, one could measure with fluorescence detection, the entire thermal unfolding transformation of a protein at several different conditions. If those conditions included a ligand that bound to the native state of the protein, one could extract the binding constant and the enthalpy of binding from its temperature dependence. I brought the sketch to the machinist at nearby MIT. The smell of the machine shop brought back memories of shops at CU and at Yale. The TGC was fashioned of copper, then plated gold for inertness and fitted with a quartz lid. We used it to measure protein folding transitions, binding phenomena, and temperature-dependent enzyme kinetics. At the suggestion of Joshua Boger, I prepared a patent application with the help of external patent counsel. Within a year, the widget received approval from the U.S. Patent Office. It would be my first patented invention and among Vertex's first issued patents.

Although we did not employ the TGC to support lead optimization in the HIV drug discovery program, we did leverage the results from the FKBP work. Our chemists and molecular modelers designed ligands that took advantage of water molecules bound in the active site of HIV protease. It proved to be a useful strategy, as it helped lead us to the discovery of amprenavir, which we put into human clinical trials with the support of Burroughs Welcome, soon to become Glaxo Welcome, then eventually GSK (GlaxoSmithKline). Amprenavir was approved by the Food and Drug Administration (FDA) in 1999. It boasted a convenient twice per day dosing; however, the dose was high and the pill burden enormous: eight very large 150 mg gel capsules or 24 small 50 mg gel capsules, twice a day! While amprenavir did indeed save lives and contribute to society's great work to reverse the accelerating death toll due to AIDS, it was a lesson that an efficacious molecule alone did not make a breakthrough medicine. Molecules need to be rendered as real solid materials that have the properties to be formulated into pills that can be easily self-administered by patients. The lesson would not be lost on Vertex as the future would bear out.

With the FKBP approach to an immunosuppressive compound thwarted, Vertex took notice of a drug that would gain approval from the FDA in 1995 for refractory kidney transplantation. The compound was a variant of mycophenolic acid, an inhibitor of inosine monophosphate dehydrogenase (IMPDH). Once again, we took a dual approach to drug discovery, hammering away at lead optimization to design a compound that intervened in the appropriate way with human biology, while at the same time delving deep into the molecular understanding of the IMPDH system (Bruzzese and Connelly 1997). Enter iso-thermal titration calorimetry again. Titrations of IMPDH with mycophenolic acid and inosine monophosphate aimed at examining the resulting ligand linkage were striking. They revealed rich functional behavior—it was an allosteric system.

During the time of working in the Sturtevant laboratory, Ruth Spolar and Tom Record and colleagues at the University of Wisconsin published a paper in *PNAS* demonstrating that the heat capacity change for protein folding could be well predicted from the changes in the amount of solvent-accessible nonpolar surface areas. The heat capacity change was the hallmark of hydrophobic interaction. When I first started working on FKBP, I showed that the relation also applied in a protein–ligand-binding reaction. It was a significant result in our biophysics subfield since there existed no binding system in which both heat capacity change and the high-resolution structures of the reactants and products were known. It allowed us to leverage a great body of work in the field of protein folding into the area of SBDD (Connelly and Thomson 1992). We extended the results to an examination of FK-506 binding in D_2O (Connelly et al. 1993). From the Gill laboratory days, I was always on the look out to falsify or strengthen support for a model by expanding the dimension-ality along the lines of the thermodynamic potentials of temperature and ligand activity. Light and heavy water were identically linked ligands, as were FK-506 and rapamycin.* We exploited these weapons in the battle fought for patients waiting for desperately needed new drugs discovered with the help of the principles of linkage thermodynamics and the technique of calorimetry.

When the first thermodynamic results of IMPDH were complete, one of the remark-able observations was that the heat capacity change on binding was temperature depen-dent—over a fairly narrow range from 5°C to 30°C. Why was that? We certainly did not see it with FK-506 binding to FKBP; however, the FKBP-FK506 system was lock and key with very few changes to the structure of the protein on binding. Rhags and I had indeed detected a temperature dependence to the heat capacity change in the RNaseS binding of the S15 peptide (Varadarajan et al. 1992). In that case, it made sense to us and it certainly fit the Spolar–Record accounting, since at high-temperature titrations, the S-protein and S-peptide were more unfolded than at lower temperatures. Consequently, more nonpolar surface area would be buried on binding at higher temperatures. It occurred to me that the temperature-dependent heat capacity of binding in the IMPDH system could be due to something similar. While the protein was fully folded at all temperatures of the binding study, perhaps the allosteric equilibria in the absence of ligand was temperature dependent. In this way, at each temperature, there would be a different proportion of allosteric states preexisting in equilibrium at each temperature. If the ligand bound preferentially to one state, as specified by the allosteric model, then one should see a temperature dependence of the heat capacity. A wonderful gentleman scientist from Boston's north end named Frank Bruzzese—the first scientist I ever hired—performed the titrations with great attention to detail. He was a great partner in the laboratory. When Frank and I started the work, we

---

* I presented the results at the 48th Annual Calorimetry Conference in 1993 at Duke University. Lee Hansen delivered the Christensen Memorial Award lecture that year entitled "Fundamentals of Calorimeter Design for Fun and Profit."

had never anticipated we would be making observations that would form the basis of a new argument that pointed to temperature-dependent heat capacity changes as a signature for ligand-coupled conformational equilibria—rigorous thermodynamic support for the concept of preexisting equilibrium among conformational states of a protein. As an added bonus to this work, John Ladbury, by now at Oxford University, invited me to give a lecture at a conference in Oxford. The venue happened to be the lecture theater at St. Catherine's College—right across the road from where I lived in the summer of 1982 as an under-graduate, the year England was at war in the Falkland Islands.*

Frank Bruzzese and I were keen to apply our fundamental results to the new IMPDH discovery effort; however, before we could get started in earnest, I had decided to leave Vertex. After having submitted proposals to expand Vertex's efforts to help patients in other ways and in other diseases, I was handed a business plan by Vertex's chief scien-tific officer, Vicki Sato. Vicki remarked that my recent proposals were outside the areas of health care where Vertex wanted to go, but that I might be able to secure venture funding. The business plan she handed me was an example of the form conventionally accepted by investors at the time. It opened up a new world for me. I decided to start a new pharmaceu-tical company with the knowledge acquired from five years of working for a start-up com-pany where we went from a handful of idealistic scientists to a publicly traded company on the move with a medicine bound for patients.

We called the new entity The Althexis Company. With the new enterprise, the apple did not fall too far from the tree, in many respects. I contacted my ex-colleague Manual Navia—head of Vertex's structural biology group and my cousin Mike Dailey—CEO of a private equity firm in Fairfield, CT—and his partner, Paul Mass, and invited them all to a meeting at the Boston Harbor hotel. We discussed a plan to apply structure-based drug design to another type of anti-infective need: antibacterials. Antibiotic resistance was on the rise; the world needed better antibiotics. In addition to applying our efforts to protein targets in the public domain, we endeavored to identify new targets emerging from the newly sequenced genomes of the major bacterial pathogens. We founded and grew the company through the late 1990s and into the next millennium. With seed capital from Dailey and Mass, we then secured a $15 million research deal with Pliva Pharmaceuticals headquartered in Zagreb, Croatia. They had invented the blockbuster antibiotic azithro-mycin and were flush with cash from licensing it to Pfizer.†

It was the era when Dolly—the first animal to be cloned from an adult somatic cell—first bleated, Harry Potter burst onto the scene, evidence of the expanding universe was

---

* Each day at 9:00 am and 3:00 pm while I was at Oxford, the faculty and students would gather for coffee and tea, and discuss the science of the day. At lunch, I would steal off to the biochemistry departmental library to read journals. Among the topics that distracted me, in an attractive way, was the phenomenon of ligand-receptor-induced patching and capping in the immune response—work described by Gerald Edelman and colleagues. I inquired about Edelman's research background and learned that he was a biologist and physical chemist. I always wondered if that fact influenced my direction in research when I entered graduate school a year later.

† Paul Mass was a lawyer turned entrepreneur who had never worked in the biotech field. He was a wonderful teacher when it came to business. After we pitched the antibiotic project in Zagreb to Pliva management, Paul literally drafted the term sheet and much of the research collaboration agreement in the Intercontinental Hotel in Zagreb. We served on the board together at Althexis and remained in touch for many years to come. I sought his business counsel often.

heralded, hip-hop became the top-selling music genre, and the gene responsible for cystic fibrosis was discovered. Our sons Dylan and Jude entered high school, the Bromfield School in Harvard, MA, and our daughter Hilah would enter kindergarten at Harvard Elementary School. The world was approaching the monumental announcement of a rough draft of the human genome. I was summoned to Bethesda for a special meeting of academics and industrialists at the National Institutes for Health (NIH). The purpose of the meeting was to identify the best paths forward for determining the function of genes that had been sequenced, but whose functions were unknown. In parallel with the human genome project, the NIH funded a number of structural genomics centers. The idea was that as genes were identified, the structures of their protein products could be determined, and from structure one could deduce function.

There was a problem: this idea was just not working. Structures were being solved, but their functions could not be determined from the information. It seemed an obvious eventuality. At Althexis, we had devised a calorimetric approach to determine the function of unknown proteins, particularly their enzymatic function. It was based on the concept that the calorimetric signal due to enzymatic turnover was given by the product of the rate and the heat of the enzymatic transformation. Unlike ligand binding, which is controlled by the number of moles of reactants mixed, enzymatic transformations greatly amplified the enzymatic turnover. Armed with the new, more sensitive Microcal VP-ITC (Microcal/Malvern), it was feasible to use a library of test substrates and extracts to hone in on the enzymatic function of an unknown protein. We were poised to apply it to partner with companies that had identified unknown essential genes in bacteria that could serve as targets of small molecule drugs. To do drug discovery, a functional assay would be needed. The search for function was on the critical path for such targets. With this calorimetric-genomics technology and our existing antibiotic drug discovery program with Pliva advancing, we went to raise $20 million in venture funding to expand our operations. In response to our solicitation for capital, three venture firms offered a total of $60 million if we would merge with Microcide Inc., a public company in Mountain View, CA, that was predicated on the identification of essential genes as targets for new antibiotics. We sold our company in a transaction that valued the tiny upstart, Althexis, at $22.2 million. It was a high value considering we had started it with approximately $250,000 a few years prior. The transition was bittersweet—we had produced some valuable scientific research, but we had not yet achieved the intended goal of introducing a new antibiotic, much less enter the clinic.

One of the sweet parts of the bittersweet reign of Althexis was the ability to interact with other scientists around the world. Among the people I met at the very start of Althexis was John Burns. Dr. Burns was the former head of research at Hoffman LaRoche where he reigned for 17 years during a very productive era for the company. John lived in Fairfield, CT, where I grew up. He was retired from Roche and was serving as the head of the scientific advisory board for HealthCare Ventures in Cambridge, MA. I called John and he invited me to his home in Connecticut. I brought a computer and projector, prepared to make a presentation on Althexis. His walls were filled with paintings so there was no place to project the slides. John was six foot five with a wide arm span. As I was searching for a space to project, Burns grabbed a large painting by the frame and removed it from the wall, exclaiming with a big smile "art gives way to science." After the presentation, I asked him to be the head of Althexis' scientific advisory board. He agreed. During my entire tenure at Althexis, I would make periodic visits to see Dr. Burns. He was a colorful character and a very generous spirit who taught me much about the management of science and scientists.

Another scientist with whom I interacted at the time was Margarida Bastos, a professor from the University of Porto on the coast of Portugal. She came to America to do a sabbatical, in part at Althexis and in part at the University of Iowa. I fondly recall her visit to our family home in Harvard, MA, and to the Althexis laboratories. We decided to work together on an old problem: the energetics of forming an alpha helix in water. The problem was that you could not study the reversible alpha helical formation and unfolding in isolation. Peptide sequences simply could not form a high enough helical content in the absence of stabilizing interactions within a protein. While I was at Yale, Susan Marqusee, Bob Baldwin, and colleagues at Stanford had discovered short alanine-rich peptides that had a surprisingly high helical content, but they did not appear sufficient to study the full energetics of helical formation. Around the same time period, David Wemmer at UC Berkeley had reported the synthesis and structure of a hybrid peptide containing the S15 and apamin, a component of snake venom. It allowed the S15 peptide to be fully stabilized into a helix as part of the overall hybrid scaffold. While at Yale, I wrote to David and explained that I thought it would be useful to titrate his peptide with the S-protein and compare the thermodynamics of binding with S15 as a means of deducing the energetics of alpha helix formation. He agreed. He sent the material and I made the measurements. I had prepared a manuscript and submitted it. The editors rejected it at first. I got busy writing patent applications and business plans and simply never followed up on the paper—that is, until Dr. Bastos came to visit. Margarida was able to lead an effort that included the addition of her theoretical flourishes to the work. Her efforts drove the results to publication as we provided an estimate of the full thermodynamics of the helix-to-coil transition of the S15 peptide determined calorimetrically (Bastos et al. 2001).

After the sale of Althexis, I formed another research-based company called Synchrony Biosciences with the support of a Small Business Innovative Research grant from the NIH and the help of a few trusted colleagues included Frank Bruzzese and my first cousin Greg Connelly—who had earned his PhD from Penn with Walter Englander and had completed postdoctoral posts in Vancouver and Toronto. We continued our calorimetric research and applied it to a drug discovery effort for neuroinflammatory conditions. It was at this time that I paid a visit to Peter Isakson, the scientist who had led the discovery and development effort of the blockbuster drug Celebrex—a selective cyclooxygenase-2 inhibitor—under Dr. Phil Needleman. Peter had become vice president of research strategy at Pfizer after the acquisition of Pharmacia, which included what remained of the Monsanto and Searle components. Peter liked the project we were working on at Synchrony and decided to join us. Peter and I then went to see Phil Needleman, who had returned to the University of Washington in St. Louis following his work as an executive at Searle. He had also just become a general partner at a venture investment firm. At St. Louis airport on a sunny morning in 2003, Peter and I described the drug discovery project that we were working on to Dr. Needleman. Phil liked it and recommended it as an investment to his venture capital firm. A few weeks after the meeting, I had breakfast with my old colleague and current head of Boston research at Vertex, John Thomson. Our meeting led to a presentation of the Synchrony project at Vertex to none other than my former colleagues Josh Boger and Vicki Sato (who had become president), and the new head of R&D, Peter Muller. Vertex also liked the project and offered an opportunity for me to rejoin Vertex. So I did.

The year was 2004. Dylan and Jude were in college; Hilah was making her way through grammar school; and Mary and I were well rooted with our family among great friends in the central Massachusetts community of Harvard. We were all enjoying the riches of outdoor life in New England, taking advantage of woodland adventures in central Massachusetts and the mountains of New England—hiking, camping, swimming, upland bird hunting, trout fishing, skiing—and pursuing sea adventures on the coast and the islands.

Vertex was an altogether different place from the start-up days of the early 1990s. Operations were spread out over numerous buildings in Cambridge. Research sites in San Diego, Iowa, and Oxfordshire were in full swing. The company had taken on a bold new strategy to develop and commercialize its own drugs, rather than work through partnerships with big pharma. The research efforts to identify a drug candidate to treat hepatitis C had been successful. The compound had recently entered a Phase I clinical trial in the Netherlands. In the study, Vertex's hepatitis C protease inhibitor VX-950 was administered at several different doses in an escalating fashion. At each dose, blood levels of the orally administered aqueous suspension of the white solid were measured. At each successive dose, blood levels increased, until the very highest doses, where the drug went undetected in the blood. How could that be? It was a major problem. We could not advance the program to efficacy trials to determine if it would help patients, if we could not be assured of delivering drug to the blood where it could be distributed to the cells of the body infected with the hepatitis C virus.

The head of the formulation development unit, a sharp south-African engineer—Trish Hurter—was in charge of the problem. She was urged to consult with me as it was believed the problem was of a physical chemical nature. The hypothesis was that the white powder, which was an amorphous material, crystallized at the highest dose levels. Crystalline VX-950 has a many-fold lower aqueous solubility than the amorphous form administered—so much so that it would not likely have been bioavailable. Trish and I discussed the problem on a return flight from Vertex's San Diego site. We decided to divide and conquer. She would lead a team to identify a formulation that was more physically stable so that we could run an efficacy trial; I would lead a team to investigate why the blood levels had dropped to zero and if the hypothesis surrounding the crystallization made any sense.

We needed a way to monitor a milky suspension of amorphous material suspended in water and measure its transformation to a crystalline state. We also required a way to characterize the solid forms of the starting and final states of the transformation. Amorphous materials are higher in energy, often formed from heating a pure crystalline solid past its melting point, then rapidly supercooling it below the melting point so that it precipitates. Another way to prepare amorphous materials was through the technique of spray drying them—rapidly cooling heated solutions of the drug in volatile organic solvents. Amorphous solids are devoid of long-range order and therefore they do not show sharp x-ray diffraction peaks like crystalline powders. Since amorphous solids are higher in energy than their corresponding crystalline forms, they are metastable. A characteristic of amorphous solids is a transition in their heat capacity with temperature—the so-called glass transition temperature marking the change from the solid amorphous state to its more rubbery supercooled liquid state. The glass transition temperature could be conveniently measured by DSC: on increasing the temperature of VX-950 in the DSC, it would undergo a glass transition just above 100°C into a supercooled liquid. Taking the temperature higher would lead to the crystallization of the supercooled liquid, and heating yet further would result in melting the crystalline solid. DSC would allow the characterization of the solid; but what of the material in an aqueous suspension on the bench and under more physiological conditions at 37°C?

When I returned to Vertex, I dusted off a MicroCal VP-ITC instrument that had been purchased at some point to replace the Omega. When the problem arose of monitoring the hypothesized amorphous to crystalline transition in water, I thought immediately of attempting it calorimetrically. Why not? Phase transitions take place with heat changes. We should be able to see it. The experiment was simple: place a suspension in the reaction cell of the ITC and stir. No injection needed. Trish Hurter came down at the end of the day with a sample from the formulation development laboratory. It was the same material that had been used in the clinical trial, albeit prepared as a fresh suspension. I introduced it into the calorimeter, waited for equilibration, hit the "Run" button, and went home.

The next morning when I arrived, I went to the calorimeter monitor to examine the power output versus time curve. There was a decided hump that started at approximately 7 h into the experiment and lasted approximately 4.5 h before coming back to baseline. Analysis of the initial and final samples revealed that we had indeed been monitoring the amorphous to crystalline transformation. Subsequent work showed that the amorphous material was dissolving and then crystals were growing from the dissolved state—a solution-mediated phase transition. We reproduced the experiment many times. The induction time for the transition varied, but the transition kinetics reproduced beautifully. We next examined the temperature dependence of the phase transition in suspension. The kinetics of the transition changed markedly with temperature.

The clinical investigation had been performed in the Netherlands during a heat wave. The temperature had reached 29°C. When the investigator had come into the clinic in the morning, he made up a fresh formulation suspension and let it stir on the bench top. It was only late in the day when the highest doses were administered from the stock formulation. By our calculations from the kinetics of the transition at the relevant temperature, determined from calorimetric results, we surmised that the VX-950 material had likely crystallized by the time it was administered; and if it was not completely crystallized by then, it would have finished crystallization in vivo prior to absorption. The calorimeter had struck again. Not only had it helped elucidate what happened, but the data led to the suggestion of what to do in the critical Phase Ib efficacy trial, a study designed to measure the effect of VX-950 on the amount of viral mRNA in the bloodstream—an excellent surrogate measure of its antiviral action. We changed the formulation protocol so that the suspension was prepared and administered immediately. It worked. The subsequent trial was an enormous success—a many-fold drop in viral mRNA in Hep-C patients was observed within a week. Jaws dropped. Vertex realized it had a drug and the race was on to bring it through the clinic. A few years later, final marketing authorization was received from the FDA. It was the spring of 2011. Approvals in Europe, Canada, and Australia followed. The path from the calorimetric and subsequent Phase Ib results to the clinic is the subject of another tale in and of itself. The most important part of the story is that Incivek helped to cure thousands of patients. In getting to that point, we had purchased three more VP-ITCs and employed them to monitor the suspension stabilities of amorphous materials. Colleagues Steve Johnson and Tapan Sanghvi ensured that the calorimeters were faithfully loaded each evening, with amorphous materials comprised of drug and polymers and surfactants of varying compositions and amounts. Data were available to examine nearly every day for many months until we identified what would become the composition of the active amorphous solid that would go into the final commercial drug product.

In the mid-2000s, the pharmaceutical industry was still fond of performing small molecule lead compound optimization by exhaustively synthesizing and testing compound variants. In any given drug discovery project, it was a common experience that as the potencies of compounds were optimized, their aqueous solubilities decreased. It was one of drug discovery's great catch-22s. We termed it *the potency-insolubility conundrum*. During these years, I called up numerous scientists and executives in the pharmaceutical industry and asked them how they approached the conundrum. Nearly all described how they invested heavily in medicinal chemistry resources to solve the problem through alteration of the structure and properties of the compound. The trouble was, they often were not able to identify a compound with the desired properties, regardless of the amount of resources they invested. It occurred to me that we could take more compounds from research into clinical testing if we rendered the active solids as specialized materials (e.g., amorphous dispersions and co-crystalline solids) that conferred higher aqueous solubility. In theory, a greater number of compounds could be declared drug candidates and be moved into clinical development without having to make additional investments in medicinal chemistry—that is, if we were successful at rendering the compounds as specialized solids with improved water solubility and that otherwise had appropriate solid-state stability and a path forward for manufacturing. It potentially offered an R&D operation with a far higher success rate than what our industry typically suffered. I proposed a new organizational model to execute the strategy. Vertex agreed with the approach and I went on to build a new department of materials discovery and characterization (MDC) comprising a staff of over 40 physical chemists, biophysicists, material scientists, engineers, and chemists. MDC was only a part of the overall Chemistry, Manufacturing & Controls group that had grown in expertise and stature under the leadership of Trish Hurter (Connelly et al. 2015a; Connelly et al. 2011). The creation of MDC was an innovation in the business model, the effect of which could be assessed by the end result after nearly 10 years of operation. By 2015, Vertex had launched three breakthrough medicines: Incivek, and two drugs for cystic fibrosis, Kalydeco and Orkambi. Incivek and Kalydeco were specialized amorphous materials; Orkambi was a combination of a specialized amorphous material and a high-energy crystalline material. No other pharmaceutical company had such a high fraction of its medicines embodied as specialized materials.

If the MDC "drugs as specialized materials" strategy had paid off, why did we need it in the first place? The fundamental science question remained: why was it, that as the intermolecular interactions between protein target and compounds were optimized, the crystalline forms of the compounds became less soluble? What was the origin of the potency-insolubility conundrum? Most chemists would say that there was no mystery, that hydrophobicity increased as optimization proceeded. There were no data to support this common contention, no rigorous account that hydrophobicity was driving the conundrum. We had an alternative hypothesis: as we optimized for protein–ligand interactions consciously, we were unwittingly optimizing for the interactions among the molecules in their neat crystalline state. It was testable. To explore this idea, we compared the structures of neat crystalline VX-950 with that of the hepatitis C protease–VX-950 complex. Further, we sought to compare the thermodynamics of VX-950–protease complex dissociation into ligand and protein with the dissolution of the neat crystalline VX-950 into water. I had enlisted a number of people from MDC to make the measurements of the thermodynamics of VX-950 into water. The solubility was very low, so it proved to be technically challenging. We needed someone with a thermodynamic light in his or her eyes—a research partner who had a desire for uncovering the conundrum.

That partner would come in the form of Phil Snyder, a staff member in the laboratory of George Whitesides at Harvard University, down the road from Vertex's original Cambridge site. Phil joined MDC and was able to make the final determinations of the temperature-dependent solubility. With a team of Vertexians, we assembled the complex story. The result showed that in the VX-950/Hep-C protease system, the same pattern of hydrogen bonds coincidentally drove binding and insolubility. It was not hydrophobicity. The argument leveraged studies in structural chemistry and biology, experimental thermodynamics, computational chemistry, solid-state chemistry, pharmacokinetics, and absorption modeling. We struggled to find a journal that could put together a set of referees spanning the disciplines. I contacted Enrico DiCera, my old colleague from the Gill laboratory. He had just become the chair of biochemistry at St. Louis Medical School and he was the editor of *Biophysical Chemistry*. He identified an appropriate review committee and with their comments, the paper was accepted (Connelly et al. 2015b).

Research lives, like lives more generally, take unexpected turns. Our children have grown and we now have grandchildren. Our beloved son Dylan died of cancer at the age of 30, just two years ago. Mary, Jude, Hilah, and I go on. The pursuit of answers to esoteric basic science questions can and often does lead to knowledge that is applied to real-world problems. I recently dusted off an old file cabinet in my garage. In it, I found a sketch from Stan Gill buried in a file that I had saved. It was a depiction of an injection system for a titration calorimeter. Underneath the sketch, the following words were written: "The calorimeter cell has fixed volume V. The injection of a volume v of titrant releases the same volume of the cell mixture. We assume that for a given step i the concentration of the different species in the ejected volume is the same as the concentrations in the preceding step i-1." Modern research in science demands curiosity, passion, and the desire to work with people in a transdisciplinary fashion. Science is dependent on a rich tradition of apprenticeship—the process of learning the craft and trade from lessons endured and paths well laid by those who have gone before, those who have a light in their eyes and the leadership skills to teach and mentor. The pursuit of a passion in research such as calorimetry can, and often does, lead to an evolution in one's field of inquiry. The process is incremental, though nonlinear, and the path to which the pursuit leads is unpredictable. In time, if applied to a broader human endeavor such as health care, the work could lend a hand in the battle against human disease, and even help save the lives of thousands of people. The research journey starts with step i. It is a necessary step. It is dependent on step i−1, drawing on the experience of the past, and proceeding to step i + 1, i + 2...i + j and into a future rich in possibility.

ACKNOWLEDGMENTS

Special thanks to the many mentors and colleagues for their passions and pedagogy, and to my family members, for their stellar support.

REFERENCES

Bastos, M., Pease, J. H. B., Wemmer, D. E., Murphy, K. P., and Connelly, P. R. (2001), Thermodynamics of the helix-coil transition: Binding of S15 and a hybrid sequence, disulfide stabilized peptide to the S-protein, *Proteins* 42, 523–530.

Bruzzese, F. J., and Connelly, P. R. (1997), Allosteric properties of inosine monophosphate dehydrogenase revealed through the thermodynamics of binding of inosine 5'-monophosphate and mycophenolic acid. Temperature dependent heat capacity of binding as a signature of ligand-coupled conformational equilibria, *Biochemistry* 36, 10428–10438.

Connelly, P. R., Aldape, R. A., Bruzzese, F. J., Chambers, S. P., Fitzgibbon, M. J., Fleming, M. A., Itoh, S., Livingston, D. J., Navia, M. A., Thomson, J. A., and Wilson, K. P. (1994), Enthalpy of hydrogen bond formation in a protein-ligand binding reaction, *Proc. Natl. Acad. Sci. USA* 91, 1964–1968.

Connelly, P., Ghosaini, L., Hu, C.-Q., Kitamura, S., Tanaka, A., and Sturtevant, J. M. (1991), A differential scanning calorimetric study of the thermal unfolding of seven mutant forms of phage T4 lysozyme, *Biochemistry* 30, 1887–1891.

Connelly, P. R., Gill, S. J., Miller, K. I., Zhou, G., and Van Holde, K. E. (1989a), Identical linkage and cooperativity of oxygen and carbon monoxide binding to *Octopus dofleini* hemocyanin, *Biochemistry* 28, 1835–1843.

Connelly, P. R., Johnson, C. R., Robert, C. H., Bak, H. J., and Gill, S. J. (1989b), Binding of oxygen and carbon monoxide to the hemocyanin from the spiny lobster, *J. Mol. Biol.* 207, 829–832.

Connelly, P. R., Quinn, B. P., Johnston, S., Bransford, P., Mudunuri, P., Peresypkin, A., Fawaz, M., Roday, S., Kuldipkumar, A., Wang, H.-R., Snyder, P., Katstra, J., Sanghvi, T., Rowe, B., and Hurter, P. (2015a), Translational development of amorphous dispersions. In A. Newman (Ed.), *Pharmaceutical Amorphous Solid Dispersions*, pp. 259–287. John Wiley: Hoboken, NJ.

Connelly, P. R., Robert, C. H., Briggs, W. E., and Gill, S. J. (1986), Analysis of zeros of binding polynomials for tetrameric hemoglobins, *Biophys. Chem.* 24, 295–309.

Connelly, P. R., Snyder, P. W., Zhang, Y., McClain, B., Quinn, B. P., Johnston, S., Medek, A., Tanoury, J., Griffith, J., Patrick Walters, W., Dokou, E., Knezic, D., and Bransford, P. (2015b), The potency-insolubility conundrum in pharmaceuticals: Mechanism and solution for hepatitis C protease inhibitors, *Biophys. Chem.* 196, 100–108.

Connelly, P. R., and Thomson, J. A. (1992), Heat capacity changes and hydrophobic interactions in the binding of FK506 and rapamycin to the FK506 binding protein, *Proc. Natl. Acad. Sci. USA* 89, 4781–4785.

Connelly, P. R., Thomson, J. A., Fitzgibbon, M. J., and Bruzzese, F. J. (1993), Probing hydration contributions to the thermodynamics of ligand binding by proteins. Enthalpy and heat capacity changes of tacrolimus and rapamycin binding to FK506 binding protein in D2O and H2O, *Biochemistry* 32, 5583–5590.

Connelly, P. R., Varadarajan, R., Sturtevant, J. M., and Richards, F. M. (1990), Thermodynamics of protein-peptide interactions in the Ribonuclease S system studied by titration calorimetry, *Biochemistry* 29, 6108–6114.

Connelly, P. R., Vuong, T. M., and Murcko, M. A. (2011), Getting physical to fix pharma, *Nat. Chem.* 3, 692–695.

Decker, H., Connelly, P. R., Robert, C. H., and Gill, S. J. (1988), Nested allosteric interaction in tarantula hemocyanin revealed through the binding of oxygen and carbon monoxide, *Biochemistry* 27, 6901–6908.

DiCera, E., Doyle, M. L., Connelly, P. R., and Gill, S. J. (1987), Carbon monoxide binding to human hemoglobin A0, *Biochemistry* 26, 6494–6502.

Gill, S. J., Connelly, P. R., Cera, E. D., and Robert, C. H. (1988), Analysis and parameter resolution in highly cooperative systems, *Biophys. Chem.* 30, 133–141.

Hu, C. Q., Kitamura, S., Tanaka, A., and Sturtevant, J. M. (1992), Differential scanning calorimetric study of the thermal unfolding of mutant forms of phage T4 lysozyme, *Biochemistry* 31, 1643–1647.

Koshland, D. E., Némethy, G., and Filmer, D. (1966), Comparison of experimental binding data and theoretical models in proteins containing subunits, *Biochemistry* 5, 365–385.

Ladbury, J. E., and Connelly, P. R. (1997), *Structure-Based Drug Design: Thermodynamics, Modeling, and Strategy*, Springer: Berlin.

Monod, J., Wyman, J., and Changeux, J.-P. (1965), On the nature of allosteric transitions: A plausible model, *J. Mol. Biol.* 12, 88–118.

Pearlman, D. A., and Connelly, P. R. (1995), Determination of the differential effects of hydrogen bonding and water release on the binding of FK506 to native and Tyr82 → Phe82 FKBP-12 proteins using free energy simulations, *J. Mol. Biol.* 248, 696–717.

Varadarajan, R., Connelly, P. R., Sturtevant, J. M., and Richards, F. M. (1992), Heat capacity changes for protein–peptide interactions in the ribonuclease S system, *Biochemistry* 31, 1421–1426.

Wiseman, T., Williston, S., Brandts, J. F., and Lin, L.-N. (1989), Rapid measurement of binding constants and heats of binding using a new titration calorimeter, *Anal. Biochem.* 179, 131–137.

CHAPTER 18

Calorimetry for Monitoring Mixed Bacterial Populations and Biofilms

Simon Gaisford

CONTENTS

18.1 INTRODUCTION

Isothermal microcalorimetry (IMC) is a technique that permits accurate measurement of extremely small powers (<1 μW) and which imposes no requirements on the physical nature of the sample under investigation (the sample can be solid, liquid, or inhomogeneous, for instance). These qualities make it ideally suited to the study of bacterial growth and metabolism, because measurements can be made in real time and in opaque or heterogeneous biological media. In this regard, IMC measurements can have the best *in vitro–in vivo* correlation of any analytical measurement. Further, the exquisite sensitivity of the technique means as few as 10^4 metabolically active cells can be detected (Braissant et al. 2010).

IMC has been used to follow microbiological activity in a wide range of applications, including for monitoring and quantifying soil microbial activity and contamination (Bravo et al. 2011; Guo et al. 2012); for the treatment of sewage (Dziejowski and Bialobrzewski 2011); in detecting infection in and contamination of clinical products and samples (Trampuz et al. 2007); in quantifying the efficacy of antimicrobial compounds (Li et al. 2000; O'Neill et al. 2003), including the mode of action of bacteriostatic or bactericidal compounds (von Ah et al. 2009) and antiviral compounds (Tan and Lu 1999; Heng et al.

2005); and in looking at the spoilage of food (Alklint et al. 2005). Most of these studies, however, look at the behavior of individual bacterial strains whereas *in vivo* the likelihood is that bacteria will coexist in complex, multispecies populations. This is especially relevant in understanding the growing importance of the role of gut microflora to general health and well-being. There have been far fewer studies looking at defined mixed cultures with the aim of investigating the relationship between two or more bacteria (i.e., Schaffer et al. 2004; Kong et al. 2012; Vazquez et al. 2014), yet IMC is ideally suited to such investigations. The aim of this chapter is to highlight the role that IMC can play in characterizing bacterial growth, especially in mixed cultures, and in the development of pharmaceutical products.

18.2 BACTERIAL GROWTH MEASURED WITH ISOTHERMAL MICROCALORIMETRY (IMC)

The basic principle in designing an IMC experiment to monitor bacterial populations is to inoculate a culture into a suitable medium that supports growth and then follow the rate of heat production with time. Experiments are typically conducted in hermetically sealed ampoules; this has the benefit of ensuring not only that there can be no bacterial contamination of the instrument or the laboratory (especially if the calorimeter is not housed within a dedicated microbiological facility) but also that there can be limited oxygen to support aerobic growth (of course, this is also an advantage if strict anaerobes are being studied). As noted previously, the minimum detectable number of organisms is around 10^4 cfu/mL although less sensitive instruments may require 10^6 cfu/mL metabolically active cells to record a quantifiable power.

The output from a calorimeter is a plot of power (µW or µJ/s) as a function of time (t) and it follows that the area under the curve is equal to the heat released (µJ). The calorimetric data associated with bacterial growth are often complex, reflecting the numerous metabolic pathways that bacteria may use to process carbohydrates, and so interpretation must be undertaken with care. For example, Figure 18.1d shows typical power–time data for *Staphylococcus aureus* inoculated into a nutrient broth (NB). The complexity of the data is apparent, the trace comprising a series of peaks and troughs that are assumed to correspond to different phases of microbial growth. One point to note here is that while experienced users of calorimetry are happy to interpret power–time traces, microbiologists often prefer to base their interpretation on growth curves. Since the area under the power–time trace gives the heat released, it is possible to plot the data as cumulative heat versus time (Figure 18.1c). This results in a plot very similar in profile to a classical growth curve determined by optical density (OD) or cell counts (Figure 18.1a,b). However, as is clear from a comparison of the plots in Figure 18.1, much of the complexity seen in the power–time trace is lost when plotted as cumulative heat; therefore, in this chapter, data will be plotted primarily as power–time.

Irrespective of how they are plotted, the data show that biphasic exponential growth occurs initially (0–6 h), followed by a stationary phase (6–14 h) before commencement of cell death. Because experiments are conducted in hermetically sealed ampoules, with a small headspace, it is assumed that the initial exponential phase (ca. 1 h) represents aerobic metabolism, after which the oxygen in the ampoule is exhausted and the bacteria switch to anaerobic metabolism, resulting in the second exponential phase (ca. 5 h). These

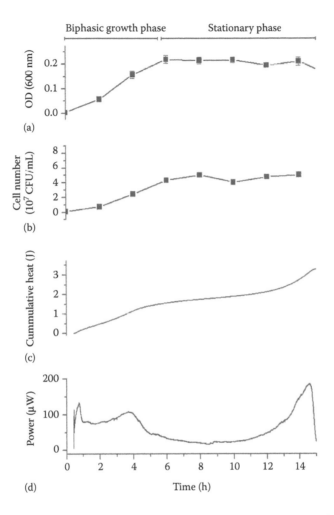

FIGURE 18.1 Comparison of power–time data (d) with plots of cumulative heat (c), cell number (b), and optical density (a) for *S. aureus* grown in NB. The growth and stationary phases are indicated.

growth phases are seen in the calorimeter as exothermic peaks over the same time period. Interestingly, while the OD and cell numbers remain constant during the stationary phase, the calorimeter records a broad, increasing exotherm over the same period. Since cell numbers do not increase during the stationary phase, this power must be associated with the bacteria utilizing an increasingly diverse range of nutrients. This effect is manifest in the cumulative heat versus time plot (Figure 18.1c), where biphasic exponential growth is clearly evident prior to the stationary phase (but the gradient of the stationary phase is not zero because of the effect of the exothermic heat measured during this period).

Two further points are of note. Firstly, because the calorimeter requires a minimum cell density before it can measure a detectable power, it is not possible to differentiate between minimum inhibitory concentration (MIC) and minimum bactericidal concentration (MBC) by calorimetric analysis alone, since experiments conducted with antibacterial agents at both inhibitory and bactericidal concentrations would result in a zero power signal. To verify a bactericidal effect, after each experiment, vials should be opened and viability counts performed. Secondly, there is a large effect from generational variability in using bacterial samples. For this reason, the results shown in this chapter are from aliquots of bacteria from the same batch. To preserve bacterial viability before use, samples are frozen and stored in liquid nitrogen. This approach means that the reproducibility of the growth curves obtained is typically around 6%.

Figure 18.2 shows growth curves for *S. aureus* in two media: NB (as previously mentioned) and simulated wound fluid (SWF). SWF (a mixture of maximum recovery diluent [MRD] and fetal bovine serum [FBS] in equal volumes) is a more complex medium than NB and it is apparent from the power–time traces that substrate availability (and thus utilization) is substantially different in the two media. In particular, anaerobic growth commences sooner, and lasts longer, in SWF. Plate counts on samples removed from the calorimeter once the power signal had returned to zero showed 1×10^7 cfu/mL viable cells in NB and 1.5×10^8 cfu/mL viable cells in SWF, confirming that the cells switch to a survival stasis mechanism once the nutrient supply is exhausted and suggesting that SWF is a better medium for supporting *S. aureus* growth. These data serve to show the power of IMC in highlighting subtle differences in growth behavior in different media.

Before discussing the growth of bacteria in mixed populations, it is worth highlighting an additional area in which IMC offers particular benefits: Testing products that claim antibacterial action. Unless they are clear solutions, testing antibacterial products with classical microbiological techniques is tricky, usually involving exposure of the bacteria to the product in question prior to incubation on agar to enable counting. IMC, conversely, does not require optical clarity of the sample and so it is often possible to arrange an experiment to test efficacy under conditions much closer to those expected in use. As an example, Figure 18.3 shows power–time traces for *S. aureus* growing in SWF in the presence of increasing masses of a silver-containing wound dressing (Aquacel Ag Hydrofiber, AAgH). It is apparent in all cases where the dressing is present that growth is significantly

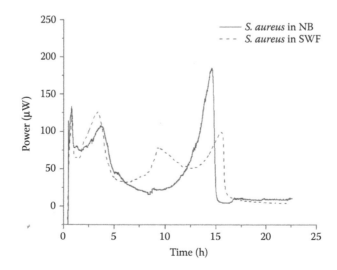

FIGURE 18.2 Calorimetric growth curves of *S. aureus* in NB and SWF.

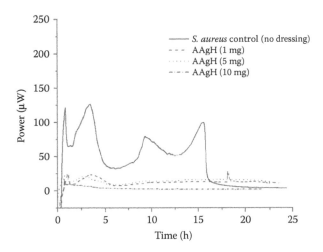

FIGURE 18.3 Calorimetric growth curves of *S. aureus* in SWF in the presence of increasing masses of AAgH.

reduced (by simple observation of a reduced power). These observations correlated with viable count data at the end of each experiment; 1×10^8, 1×10^8, and 1×10^6 cfu/mL with 1, 5, and 10 mg samples, respectively. It is important to note that this experiment was performed with the bacteria suspended in a biorelevant medium, with the wound dressing hydrated as it would be in use and that the data were recorded in real time.

18.3 BACTERIAL GROWTH IN MIXED POPULATIONS

In vivo it is highly unlikely that bacteria will exist in single colonies and so methods must be developed that permit the investigation of mixed bacterial populations. IMC has much potential in this area, since multiple species can be co-inoculated into the vial and the power–time trace can be recorded in real time. However, the technique is not without its drawbacks. Care must be taken to ensure that experimental factors do not accidentally inhibit the proliferation of one or more species (i.e., where bacteria are fastidious or strictly anaerobic). Further, as is often the case with IMC, deconvolution of the total power signal into its component parts (representing the power output of each species in this case) can be tricky. However, qualitative investigation of mixed bacterial populations is certainly possible.

The use of IMC for investigating mixed bacterial populations is highlighted here by consideration of three species: *S. aureus*, *Pseudomonas aeruginosa*, and *Escherichia coli* (*E. coli* and *S. aureus* are both facultative anaerobes while *P. aeruginosa* is an aerobe but is known to respire anaerobically). Growth curves were recorded for each species alone (providing a set of control experiments) and then samples were inoculated in binary pairs or a tertiary mixture. All experiments were performed in NB.

Figure 18.4 shows the power traces for *S. aureus*, *P. aeruginosa*, and their binary mixture (each inoculated initially to 10^6 cfu/mL). It is apparent that the power–time trace for the binary pair looks almost identical to the control trace for *P. aeruginosa*, suggesting that growth of this species predominates in mixed culture. In addition to a simple comparison

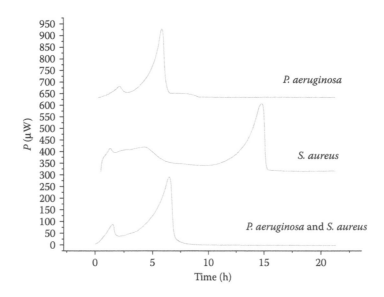

FIGURE 18.4 Comparison of the power–time traces of *P. aeruginosa* (top), *S. aureus* (middle), and their mixed culture (bottom). Data are displaced on the *y*-axis for clarity of presentation.

of data, the areas under the curves (AUC) can be calculated (which represent the total heat produced in each experiment). Considering the control experiments first, the AUC for *P. aeruginosa* was 2.04±0.21 J and the AUC for *S. aureus* was 4.40±0.32 J. The AUC for the binary pair was 2.30±0.25 J, again indicating that the growth of *P. aeruginosa* predominates but suggesting that the growth of *S. aureus* was not completely inhibited (confirmed with plate counts post-experimentation; *P. aeruginosa* 3×10^7 and *S. aureus* 8×10^5). *P. aeruginosa*, a Gram-negative organism, is known to produce bacteriocins, also known as virulence factors (Michel-Briand and Baysse 2002) that target and kill other organisms. Whatever metabolites *P. aeruginosa* produces, a simple reduction in the solution pH is not the mechanism of inhibition because the final pH of the medium after the calorimetric experiment was 6.8±0.2 for *P. aeruginosa* and 6.3±0.2 for *S. aureus* (the pH of the initial medium was 7.0).

To determine whether the *S. aureus* population could proliferate if the *P. aeruginosa* population was reduced, experiments were performed where the inoculum of *S. aureus* was kept constant (at 10^6 cfu/mL) while that of *P. aeruginosa* was decreased from 10^6 to 10^3 cfu/mL (Figure 18.5). The data show a gradual reversion of the growth curve from that characteristic of *P. aeruginosa* to that of *S. aureus* as the inoculum of *P. aeruginosa* is decreased. Plate counts showed that as the population of *P. aeruginosa* decreased relative to *S. aureus*, growth of *S. aureus* was enhanced (data not shown). When the inoculum of *P. aeruginosa* was kept constant (at 10^6 cfu/mL) and that of *S. aureus* decreased exponentially from 10^6 to 10^3 cfu/mL, the power–time trace was identical to that of *P. aeruginosa* in all cases. A similar trend is seen in co-mixed populations of *S. aureus* and *E. coli* (Figure 18.6), with *E. coli* growth predominating throughout.

The situation is more interesting in mixtures of *P. aeruginosa* and *E. coli*. Here, the power–time trace for the binary pair shows contributions from both species (Figure 18.7),

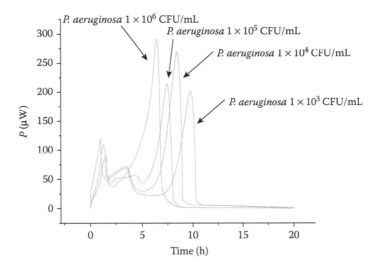

FIGURE 18.5 Comparison of the power–time traces of mixed cultures of *P. aeruginosa* and *S. aureus* when the inoculum of *S. aureus* is kept constant (10^6 cfu/mL) while that of *P. aeruginosa* is decreased (from 10^6 to 10^3 cfu/mL).

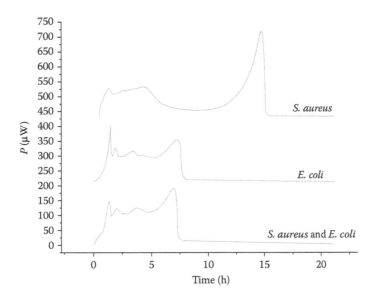

FIGURE 18.6 Comparison of the power–time traces of *S. aureus* (top), *E. coli* (middle), and their mixed culture (bottom). Data are displaced on the *y*-axis for clarity of presentation.

although cell counts post-experimentation found no viable *P. aeruginosa* cells. Both *P. aeruginosa* and *E. coli* are virulent organisms and both species produce bacteriocins (Gordon and O'Brien 2006) and so it is likely that both organisms are capable of inhibiting and thereby resisting each other, so they both grow in a mixed culture. However, *E. coli* is an extensive fermenter of carbohydrates, producing organic acids as products of fermentation. This effect produced a significant reduction in the pH of the medium (to 5.4 ± 0.1),

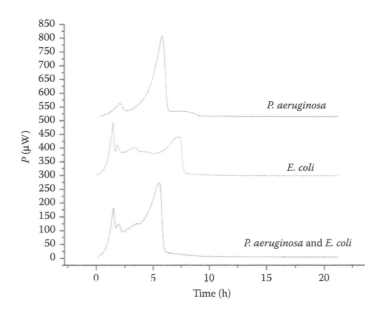

FIGURE 18.7 Comparison of the power–time traces of *P. aeruginosa* (top), *E. coli* (middle), and their mixed culture (bottom). Data are displaced on the *y*-axis for clarity of presentation.

which eventually poisoned the medium for *P. aeruginosa*, leading to the bactericidal effect. The AUC data confirmed that both species proliferated in combination (*P. aeruginosa* 2.04±0.21 J; *E. coli* 3.00±0.19 J; binary pair 3.01±0.20 J).

When the population of *P. aeruginosa* was kept constant (at 10^6 cfu/mL) and that of *E. coli* decreased exponentially from 10^6 to 10^3 cfu/mL, the resulting power–time traces showed a gradual reversion to the *P. aeruginosa* control as the population of *E. coli* decreased (Figure 18.8). Conversely, when the population of *E. coli* was kept constant and that of *P. aeruginosa* decreased exponentially, the power–time trace reverted to that of *E. coli* (Figure 18.9).

Commercial probiotic formulations provide another example in which mixed species are often found. Probiotics are defined as live microorganisms that when administered in adequate quantities confer a health benefit on the host and they are becoming increasingly popular as a food supplement to support a healthy gut microbiota. A typical example of a liquid probiotic formulation containing multiple species is Symprove, which contains *Lactobacillus rhamnosus*, *L. planatarum*, *L. acidophilus*, and *Enterococcus faecium*. When inoculated into growth medium, Symprove shows a strong growth curve, the profile of which varies depending on the medium used (Figure 18.10). The growth curve, which is already complex and presumably comprises powers from all four of the probiotic species, is significantly altered when the sample is spiked with small amounts of pathogenic species (Figure 18.11). This highlights the potential of IMC as a tool for routine batch-release testing of live probiotic products, since any contamination would alter the growth curve from that expected (although classical microbiological techniques would still be needed to identify and quantify the contaminants).

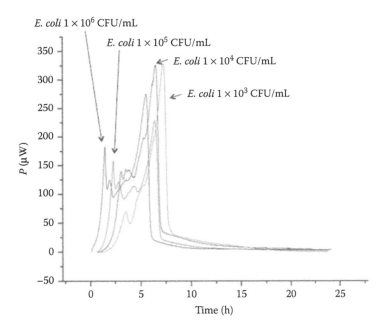

FIGURE 18.8 Comparison of the power–time traces of mixed cultures of *P. aeruginosa* and *E. coli* when the inoculum of *P. aeruginosa* is kept constant (10^6 cfu/mL) while that of *E. coli* is decreased (from 10^6 to 10^3 cfu/mL).

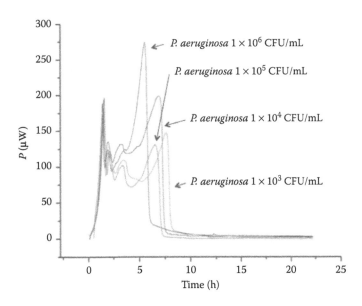

FIGURE 18.9 Comparison of the power–time traces of mixed cultures of *P. aeruginosa* and *E. coli* when the inoculum of *E. coli* is kept constant (10^6 cfu/mL) while that of *P. aeruginosa* is decreased (from 10^6 to 10^3 cfu/mL).

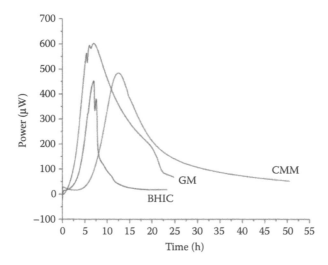

FIGURE 18.10 Growth curves for Symprove inoculated into growth medium (GM, MRS broth with L-cysteine), brain–heart infusion with L-cysteine (BHIC), and cooked-meat medium (CMM).

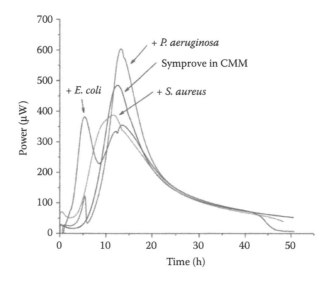

FIGURE 18.11 Growth curves for Symprove in cooked-meat medium (CMM) and spiked with trace numbers of *E. coli*, *P. aeruginosa*, and *S. aureus*.

Since Symprove contains a known number of species in defined ratios, it is interesting to attempt to deconvolute the overall power–time trace into its component parts. Figure 18.12 shows the result of curve fitting to four Gaussian curves. It is clear that the overall fit is extremely good, although some of the fine detail of the real data is lost. It may be the case that the individual growth curves of one or more of the species is not truly Gaussian (likely in fact, considering the data shown previously), but nonetheless the experiment illustrates how simple it may be to get (at least a qualitative) understanding of the contribution of each species to the total data. Of course, it would be necessary to determine the calorimetric

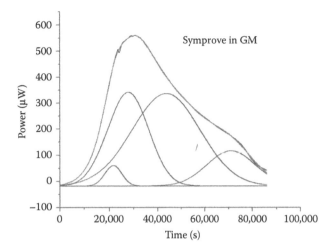

FIGURE 18.12 An attempt to deconvolute the power–time trace for Symprove into contributions from its four probiotic species by model-fitting to Gaussian curves. The overall fit is good, but some of the fine detail is lost.

growth curves for each of the individual probiotic species before the assignment of the Gaussian curves could be attempted.

There is much talk clinically of using probiotics to treat nosocomial (hospital-acquired) infections, such as *Clostridium difficile* and again IMC has some potential in investigating the nature of the species when inoculated together. Figure 18.13 shows the power–time traces for *L. acidophilus*, *C. difficile*, and their co-mixture. It is apparent that *C. difficile* is the slower growing species, and so it is outcompeted for nutrients in the limited environment of the calorimeter vial. Similar behavior is seen for the widely used probiotic species *Bifidobacterium bifidus* (Figure 18.14). While these experiments are obviously limited, they suggest that probiotic species can inhibit the growth of *C. difficile*, at least in a closed

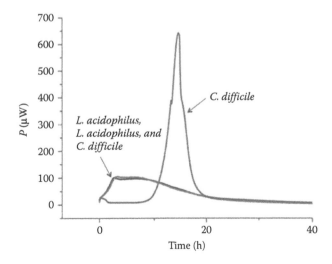

FIGURE 18.13 Power–time traces for *L. acidophilus*, *C. difficile*, and their co-mixture.

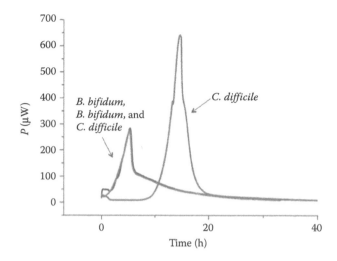

FIGURE 18.14   Power–time traces for *B. bifidum*, *C. difficile*, and their co-mixture.

environment and when inoculated to the same concentration. How probiotic species would compete in the gut, against an established *C. difficile* colony is yet to be determined.

18.4 MEASUREMENT OF BIOFILMS

Biofilms are sessile communities of microorganisms typically held together by an extracellular matrix. They are of increasing clinical interest because bacteria form biofilms on the surface of many medical implants and devices. The subsequent detachment of bacteria from the biofilm can lead to infections or, depending on the location of the implant, sepsis. Studies using *P. aeruginosa* have shown that biofilm development proceeds in stages including initial attachment, production of extrapolymeric substances (EPS), early development of biofilm architecture, maturation of biofilm architecture, and finally release of cells from the mature biofilm (Stoodley et al. 2002; Hall-Stoodley et al. 2004).

In vivo determination of biofilm formation and activity is difficult, so analytical methods are required to understand *in vitro* how biofilms form and to determine how effective antibiofilm technologies or eradication measures may be. Biofilms can be difficult to generate *in vitro*, because the drivers that direct bacteria to form biofilms are not always clear, but environmental stresses, including shear forces, are known to play a role. Biofilms are usually created in microbial bioreactors, typically one of three designs: closed batch systems with limited substrate, continuous flow stirred tank reactors (CFSTR), or plug flow reactors (PFR) (Coenye and Nelis 2010). With batch systems, the substrate concentration declines while there is a concomitant increase in biomass. With a CFSTR system, the quantities of substrate and biological products remain constant because a continuous flow of fresh substrate is introduced at the same rate as accumulating metabolic by-products are removed. A PFR system is "fed" initially and there is continuous flow to a waste vessel but conditions are otherwise similar to batch systems with accumulated biomass and reduced substrate over the course of experiments.

Irrespective of their basic design, flow models typically rely on shear forces to drive biofilm formation and utilize plastic inserts (or coupons) that act as a surface to initiate and support biofilm growth. The coupons are removable, which allows analysis of the biofilm with conventional microbiological methods such as microscopy (including scanning and transmission electron microscopy or confocal laser scanning microscopy), staining, or viable colony counting. One drawback of this approach is that they all require the removal of the biofilm from its natural environment prior to analysis, which may cause changes in structure or viability or both. It is also not possible to monitor the processes of biofilm growth and establishment in real time.

Direct, noninvasive methods of monitoring biofilms would, in principle, allow insight into the entire process of biofilm development in real time without interference. Methods have been developed based on differential turbidity, heat transfer, bioluminescence, computerized image analysis, and spectroscopy (Janknecht and Melo 2003). However, the challenge lies in correlating changes in measured parameters with actual processes in the biofilm. Flemming (2003) categorizes the techniques used to monitor biofilms as follows: (1) systems that detect biofilm by the deposition of material and changes in the thickness of the layer without differentiating between biotic and abiotic components, (2) systems that can distinguish between biotic and abiotic components, and (3) systems that provide detailed information about the chemical composition of the microorganisms involved. Janknecht and Melo (2003) add a fourth category, that of systems that monitor metabolic activity. IMC falls within the latter category (and may also be considered as part of the second category as abiotic components would not register any metabolic activity).

Several recent studies report the use of IMC for monitoring biofilm activity, suggesting that the technique has much potential (Buchholz et al. 2010). For instance, Clauss et al. (2013) monitored biofilms of S. aureus growing on human and bovine bone grafts, while Astasov-Frauenhoffer et al. (2012) investigated the variability and dynamics of a triple-species biofilm and determined the efficacy of amoxicillin and metronidazole combinations against biofilms of various species (Astasov-Frauenhoffer et al. 2014). Said et al. (2014a) determined the efficacy and mechanism of action of an antibiofilm wound dressing. Lerchner et al. (2008) note, however, that despite the widespread use of IMC for studying planktonic cells, its use for biofilm investigation is rarely considered. This may be because the data are complex to interpret, but may also be ascribed to experimental limitations. IMC experiments are typically conducted in closed vials, and this arrangement is not particularly suited to the study of biofilms as the limited nutrient supply may result in the bacteria entering a stationary phase and/or that metabolic by-products poison the medium, leading to cell death. It is also necessary to generate the biofilm on a coupon as described previously in this section prior to calorimetric measurement, so the processes of formation and growth are not observed as they occur prior to experimental measurement.

A more effective arrangement would allow fresh medium or bacterial culture or both to be circulated from an external reservoir through the calorimetric chamber, keeping nutrient levels high. Such systems are known as flow calorimeters, and fall under the category of PFR as defined previously in this section. They can be configured to circulate material back to the reservoir (continuous loop) or to flow material to waste. Flow calorimeters have many advantages that allow biofilms to form and to remain viable over extended time periods; nutrient levels or compositions or both can be optimized, planktonic cells can be continuously added, shear forces may exist in the circulating liquid, and it is possible to maintain or control pH, preventing conditions arising that limit cell growth. From the perspective of interrogating biofilms, it is possible to add agents that modify metabolic rates or pathways or which are designed to eradicate biofilms. Antimicrobial agents may also be titrated into the bioreactor, and so it is easy to get a direct measurement of antimicrobial efficacy, including the determination of MIC and MBC.

Despite these advantages, flow calorimetry has not been widely applied to biofilm investigation. Lerchner et al. (2008) discuss a flow calorimeter based on a solid-state (or chip) design and use it to study *Pseudomonas putida* biofilms, although in this case the biofilms were cultured prior to calorimetric measurement. Mariana et al. (2013) used the same chip calorimeter to monitor eradication of *P. putida* biofilms with antibiotics. Morais et al. (2014) have also used chip calorimeters to investigate the antimicrobial efficacy of biofilms. Peitzsch et al. (2008) used flow calorimetry to study biofilms of *E. coli*. Lerchner et al. (2008) discuss some of the potential limitations hindering widespread use of flow calorimetry; the cost of the calorimeter can be high, they are not suited to high-throughput studies, experiments may take days to weeks, and it can be difficult to manipulate and investigate substratum effects on the biofilm. Additionally, it can be difficult to maintain anaerobic conditions and there is a need to minimize the temperature difference between the external reservoir and the calorimeter.

Since many commercially available IMC instruments are designed to accept removable ampoules, it is possible to consider the development of a system that flows liquid through the ampoule from an external reservoir. The ampoule is housed in the calorimeter and so power data can be recorded from the medium flowing through it. Constructing a flow system in this way involves relatively low cost (assuming the calorimeter itself is already available). Such a system is shown schematically in Figure 18.15 (Said et al. 2014b). It comprises an external bioreactor, the contents of which are circulated through an ampoule housed within the calorimeter. The bioreactor is maintained at 37°C by a jacketed water bath and silicone tubing is used to attach the bioreactor to the ampoule. Two stainless steel tubes are set through the lid of the ampoule; in use, the tubing being tested (up to 15 cm in length) is connected to the inlet and outlet lines and coiled within the ampoule. The ampoule is filled with sterile distilled water (10 mL) to act as a heat transfer fluid enabling heat exchange with the calorimeter. This design maximizes the versatility of the system because tubing of different material can be connected or it is possible simply to fill the ampoule with pellets of a sample. A peristaltic pump is mounted in the outflow line and is used to circulate media at 4 mL/h. In this way, the system may be used to investigate biofilm formation on a wide range of materials used for medical tubing. It would be equally feasible to load the ampoule with pelletized samples to test biofilm formation on any material used for fabricating medical devices and implants.

Use of the flow system can be demonstrated with *S. aureus*. Flowing an *S. aureus* culture through the calorimeter results in an exothermic peak after a lag period of ca. 5 h, which increases to reach a plateau after ca. 12 h (Figure 18.16). Since, as shown in Section 18.2,

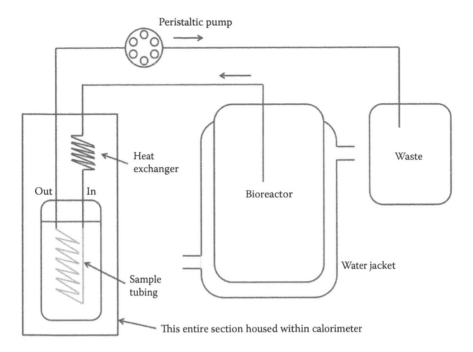

FIGURE 18.15 Schematic representation of a flow calorimeter system, suitable for investigating biofilm formation. (From Said J., et al., *Methods* 76, 35–40, 2014.)

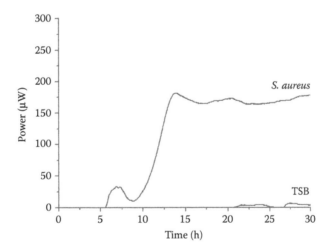

FIGURE 18.16 Power–time data showing the growth and biofilm formation of *S. aureus* in a flow calorimeter and a control experiment of TSB. (From Said J., et al., *Methods* 76, 35–40, 2014.)

it has been established that the growth of planktonic *S. aureus* cells in a closed ampoule results in an exothermic power that rapidly reduces to a zero power once nutrients in the medium are exhausted and/or the metabolites produced by the bacteria have poisoned the medium, the fact that the data recorded here reach a plateau, means there must be ongoing metabolic activity within the calorimetric ampoule from biofilm activity. Bacteria existing in a biofilm colony are not exponentially increasing in number and are shielded from toxic metabolites, but since fresh medium is constantly being circulated over the biofilm the bacteria have a nutrient source to sustain metabolic activity. IMC used in this way allows real-time observation of the formation and ongoing metabolic activity of a live biofilm, something that would be very challenging with another analytical method.

18.5 SUMMARY

The unique qualities of IMC (the ability to record powers from multiple species in opaque or heterogeneous media) mean that it is ideally suited to the investigation of mixed bacterial populations. The instrument can measure the powers from all species growing simultaneously and the growth curves produced are very sensitive to small changes in bacterial composition or media or both. In this regard, IMC offers great potential to qualitative investigations of complex bacterial systems. Quantitative interpretation is somewhat harder, simply because care is needed to ensure that experimental factors have not inhibited the proliferation of a particular species and because deconvolution of the total power–time signal into its component parts is challenging. The investigation of biofilm formation, maintenance, and eradication is feasible with flow calorimetry.

ACKNOWLEDGMENTS

The majority of the data contained within this chapter were recorded by Dr. Mansa Fredua-Agyeman and Dr. Jawal Said. Thanks also go to Mike Butler, Vivienne Clark, Dave Parsons, Barry Smith, Mike Walker, Dr. Paul Stapleton, and Professor Anthony Beezer for their contributions to data analysis and interpretation.

REFERENCES

Alklint, C., Wadsö, I., and Sjoholm, I. (2005), Accelerated storage and isothermal microcalorimetry as methods of predicting carrot juice shelf-life, *J. Sci. Food Agri.* 85, 281–285.

Astasov-Frauenhoffer, M., Braissant, O., Hauser-Gerspach, I., Daniels, A. U., Weiger, R., and Waltimo, T. (2012), Isothermal microcalorimetry provides new insights into biofilm variability and dynamics, *FEMS Microbiol. Lett.* 337, 31–37.

Astasov-Frauenhoffer, M., Braissant, O., Hauser-Gerspach, I., Weiger, R., Walter, C., Zitzmann, N. U., and Waltimo, T. (2014), Microcalorimetric determination of the effects of amoxicillin, metronidazole, and their combination on *in vitro* biofilm, *J. Periodontol.* 85, 349–357.

Braissant, O., Wirz, D., Gopfert, B., and Daniels, A. U. (2010), Use of isothermal microcalorimetry to monitor microbial activities, *FEMS Microbiol. Lett.* 303, 1–8.

Bravo, D., Braissant, O., Solokhina, A., Clerc, M., Daniels, A. U., Verrecchia, E., and Junier, P. (2011), Use of an isothermal microcalorimetry assay to characterize microbial oxalotrophic activity, *FEMS Microbiol. Ecol.* 78, 266–274.

Buchholz, F., Harms, H., and Maskow, T. (2010), Biofilm research using calorimetry: A marriage made in heaven? *Biotech. J.* 5, 1339–1350.

Clauss, M., Tafin, U. F., Bizzini, A. Trampuz, A., and Ilchmann, T. (2013), Biofilm formation by staphylococci on fresh, fresh-frozen and processed human and bovine bone grafts, *Eur. Cells Mater.* 25, 159–166.

Coenye, T., and Nelis, H. J. (2010), *In vitro* and *in vivo* model systems to study microbial biofilm formation, *J. Microbiol. Methods* 83, 89–105.

Dziejowski, J., and Bialobrzeski, I. (2011), Calorimetric studies of solid wastes, wastewaters and their effects on soil biodegradation processes, *J. Therm. Anal. Cal.* 104, 161–168.

Flemming, H.-C. (2003), Role and levels of real-time monitoring for successful anti-fouling strategies: An overview, *Water Sci. Tech.* 47, 1–8.

Gordon, D. M., and O'Brien, C. L. (2006), Bacteriocin diversity and the frequency of multiple bacteriocin production in *Escherichia coli*, *Microbiology* 152, 3239–3244.

Guo, H., Yao, J., Cai, M., Qian, Y., Guo, Y., Richnow, H. H., Blake, R. E., Doni, S., and Ceccanti, B. (2012), Effects of petroleum contamination on soil microbial numbers, metabolic activity and urease activity, *Chemosphere* 87, 1273–1280.

Hall-Stoodley, L., Costerton, J. W., and Stoodley, P. (2004), Bacterial biofilms: From the natural environment to infectious diseases, *Nat. Rev. Micro.* 2, 95–108.

Heng, Z., Congyi, Z., Cunxin, W., Jibin, W., Chaojiang, G., Jie, L., and Yuwen, L. (2005), Microcalorimetric study of virus infection: The effects of hyperthermia and a 1b recombinant homo interferon on the infection process of BHK-21 cells by foot and mouth disease virus, *J. Therm. Anal. Cal.* 79, 45–50.

Janknecht, P., and Melo, L. F. (2003), Online biofilm monitoring, *Rev. Environ. Sci. Biotech.* 2, 269–283.

Kong, W. J., Xing, X. Y., Xiao, X. H., Zhao, Y. L., Wei, J. H., Wang, J. B., Yang, R. C., and Yang, M. H. (2012), Effect of berberine on *Escherichia coli*, *Bacillus subtilis*, and their mixtures as determined by isothermal microcalorimetry, *Appl. Microbiol. Biotech.* 96, 503–510.

Lerchner, J., Wolf, A., Buchholz, F., Mertens, F., Neu, T. R., Harms, H., and Maskow, T. (2008), Miniaturized calorimetry: A new method for real-time biofilm activity analysis, *J. Microbiol. Methods*, 74, 74–81.

Li, X., Liu, Y., Zhao, R. M., Wu, J., Shen, X. S., and Qu, S. S. (2000), Microcalorimetric study of *Escherichia coli* growth inhibited by the selenomorpholine complexes, *Biol. Trace Element Res.* 75, 167–175.

Mariana, F., Buchholz, F., Lerchner, J., Neu, T. R., Harms, H., and Maskow, T. (2013), Chip-calorimetric monitoring of biofilm eradication with antibiotics provides mechanistic information, *Int. J. Med. Microbiol.* 303, 158–165.

Michel-Brian, Y., and Baysse, C. (2002), The pyocins of *Pseudomonas aeruginosa*, *Biochimie* 84, 499–510.

Morais, F., Buchholz, F., and Maskow, T. (2014), Chip calorimetry for evaluation of biofilm treatment with biocides, antibiotics, and biological agents, *Methods Mol. Microbiol.* 1147, 267–275.

O'Neill, M. A. A., Vine, G. J., Bishop, A. E., Hadgraft, J., Labetoulle, C., Walker, M., and Bowler, P. G. (2003), Antimicrobial properties of silver-containing wound dressings: A microcalorimetric study, *Int. J. Pharm.* 263, 61–68.

Peitzsch, M., Kiesel, B., Harms, H., and Maskow, T. (2008), Chip calorimetry and its use for biochemical and cell biological investigations, *Chem. Eng. Process.: Process Intensification* 47, 1000–1006.

Said, J., Walker, M., Parsons, D., Stapleton, P., Beezer, A. E., and Gaisford, S. (2014a), An *in vitro* test of the efficacy of an anti-biofilm wound dressing, *Int. J. Pharm.* 474, 177–181.

Said, J., Walker, M., Parsons, D., Stapleton, P., Beezer, A. E., and Gaisford, S. (2014b), Development of a flow system for studying biofilm formation on medical devices with microcalorimetry, *Methods* 76, 35–40.

Schaffer, B., Szakaly, S., and Lorinczy, D. (2004), Examination of the growth of probiotic culture combinations by the isoperibolic batch calorimetry, *Thermochim. Acta* 415, 123–126.

Stoodley, P., Sauer, K., Davies, D. G., and Costerton, J. W. (2002), Biofilms as complex differentiated communities, *Ann. Rev. Microbiol.* 56, 187–209.

Tan, A. M., and Lu, J. H. (1999), Microcalorimetric study of antiviral effect of drug, *J. Biochem. Biophys. Methods* 38, 225–228.

Trampuz, A., Salzmann, S., Antheaume, J., and Daniels, A. U. (2007), Microcalorimetry: A novel method for detection of microbial contamination in platelet products, *Transfusion* 47, 1643–1650.

Vazquez, C., Lago, N., Mato, M. M., Casas, L. M., Esarte, L., Legido, J. L., and Arias, I. (2014), Microcalorimetric performance of the growth in culture of *Escherichia coli, Proteus mirabilis* and their mixtures in different proportions, *J. Therm. Anal. Cal.* 116, 107–112.

Von Ah, U., Wirz, D., and Daniels, A. U. (2009), Isothermal micro calorimetry: A new method for MIC determinations: Results for 12 antibiotics and reference strains of *E. coli* and *S. aureus*, *BMC Microbiol* 9(106), 1–14.

CHAPTER 19

Isothermal Microcalorimetry for the Investigation of Clinical Samples

Past and Present

Olivier Braissant, Gernot Bonkat, and Alexander Bachmann

CONTENTS

19.1 INTRODUCTION

Modern clinical pathology and microbiology laboratories are expected to provide timely results at an affordable cost for the health-care system (Fournier et al. 2013; Graban 2011; Peterson et al. 2001). Such laboratories are dealing with samples of very different natures containing various types of pathogens or tumorous cells. Among those samples, urine and blood are the most commonly encountered (Evans and Fine 2013). In addition, other liquids such as pus and cerebrospinal fluid (CSF) are often sent to the microbiology laboratory (Sarathbabu et al. 2013). Solid samples such as biopsies and tissues are also commonly submitted for clinical investigations. Due to the variety of samples and pathogens encountered, automation is a very difficult task and new methods with potential for automation are of interest (Bourbeau and Ledeboer 2013). In this context, the recent development of isothermal microcalorimetry (IMC) could be of interest. In IMC, the detection of heat released by microbial or tumor cell metabolism, occurs passively in a label-free manner using a Pelletier element in contact but outside of the sample container (Braissant et al. 2010a,c; Wadsö 2002). Therefore, the physical nature of the sample is not a limiting factor and this technique can be readily applied to solid or opaque samples (Braissant et al. 2010b) such as stool or blood, for example. Moreover, because IMC is label-free, samples processed can be recovered and subjected to additional analyses (Braissant et al. 2010a,c) as they are almost undisturbed.

In addition to its versatility, IMC has been shown to be sensitive enough to detect very low numbers of metabolically active cells. For example, assuming a detection limit of 200 µW, only 100,000 bacteria or 1,000 rat hepatocytes are needed to reach the detection limit (Braissant et al. 2010a,c; Kemp and Lamprecht 2000). Although there can be strong variations in the heat production when different organisms or cell lines are considered, usually detection limits remain rather low compared with other techniques (Braissant et al. 2015a, Table 19.1). Considering these advantages, this chapter will discuss the potential applications of IMC in the clinical and biomedical fields. In addition, this chapter will provide examples of the practical use of IMC for biomedical applications. Finally, the drawbacks of current instruments and the potential solutions brought by new instruments will be discussed as well.

19.2 PURPOSE FOR WHICH ISOTHERMAL CALORIMETRY CAN BE USEFUL

IMC can provide a variety of different information this is useful for clinicians. The simplest application of microcalorimetry is to detect the growth of microorganisms (bacteria, fungi, or protozoan) in clinical samples (added with or without growth medium). In this context, the sensitivity of IMC allows it to detect very few growing cells and thus to be faster than conventional culture (Braissant et al. 2015a). In this approach, as soon as the heat production passes above a predetermined threshold (5–10 times the detection limit usually), and shows exponential growth, the sample is considered positive. Although such an approach is interesting for rapid screening purposes, combining it with the appropriate mathematical model allows gaining even more insight on the samples. Analyzing the data with a growth model allows the rapid determination of the sample's growth rate and

TABLE 19.1   Heat Production per Cell and Detection Limit of Different Types of Organisms and Cells Lines

Cell type	Heat Production Rate per Cell	Detection Limit (200 nW)
Human erythrocyte	0.01 pW/cell	20.0×10^6 cells
Human platelets	0.06 pW/cell	3.3×10^6 cells
Human neutrophils	2.5 pW/cell	8.0×10^4 cells
Human lymphocytes	5 pW/cell	4.0×10^4 cells
Human T-lymphoma	8 pW/cell	2.5×10^4 cells
Human adenocarcinoma (HeLa)	31 pW/cell	6.5×10^3 cells
Human white adipocytes	49 pW/cell	4.1×10^3 cells
Rat white adipocytes	40 pW/cell	5.0×10^3 cells
Rat hepatocytes	329 pW/cell	6.1×10^2 cells
Mouse fibroblasts	17 pW/cell	1.2×10^4 cells
Mouse lymphocyte hybridoma	30 pW/cell	6.7×10^3 cells
Mouse macrophage hybridoma	32 pW/cell	6.3×10^3 cells
Saccharomyces cerevisiae	4.1 pW/cell	4.9×10^4 cells
Fusarium roseum	40 pW/cell	5.0×10^3 cells
Escherichia coli	2.1 pW/cell	9.5×10^4 cells
Staphylococcus aureus	2 pW/cell	1.0×10^5 cells
Klebsiella aerogenes	2 pW/cell	1.0×10^5 cells
Bacillus megaterium	3 pW/cell	6.7×10^4 cells

Source: Data from Kemp, R. B., and Lamprecht, I., *Thermochim. Acta* 348:1–17, 2000; James 1987; Kimura and Takahashi 1985; Braissant et al. 2012.

the lag phase (Braissant et al. 2013). The growth rate might provide information on the aggressivity of the pathogen, especially in an undisturbed sample (urine for example). On the other hand, the lag phase provides an estimate of the bacterial load in the sample. Indeed, a longer time before growth leads to a detectable signal and is negatively correlated with the inoculum size (Braissant et al. 2015a; Trampuz et al. 2007c). Several studies have also demonstrated that the heat production pattern could be used to identify the pathogen. Although pathogen identification using IMC could be interesting, other rapid techniques such as mass spectrometry (MALDI-TOF) can provide a very accurate identification within minutes, thus clearly outcompeting IMC and many conventional techniques (Dierig et al. 2015; Dixon et al. 2015). Still, IMC can be very useful in screening positive samples prior to performing identification by mass spectrometry, which has a detection limit of ca 10^5 CFU/mL, thus ensuring good-quality identification.

A comparison between untreated samples and samples added with different compounds allows a rapid comparison of the efficacy of a molecule against a given pathogen or cancer type. Such a comparison can be useful to evaluate the activity of a new compound in vitro. In addition, this approach has been used to quickly discriminate between drug-susceptible and drug-resistant microbes (Baldoni et al. 2009; von Ah et al. 2008).

Although microbial infections represent a large portion of clinical studies involving IMC, one has to keep in mind that IMC is a technique that measures heat resulting from metabolic activity. Therefore, its application could be useful for a variety of other diseases. This could be of interest for studies on obesity and adipocyte metabolism (De Meis et al.

2012; Monti et al. 1980; Sorbris et al. 1982). In addition this could also be of particular interest for mitochondrial metabolism, which is involved in a variety of diseases ranging from heart failure to degenerative disease. Such an approach is often not as simple as it is to detect microbial growth and experimental conditions. Additionally, a threshold for diagnosing needs to be determined for each novel disease investigated.

19.3 APPLICATION TO CLINICAL SAMPLE

19.3.1 Liquid Samples

Liquid samples have received much more attention than their solid counterparts (see Table 19.2). This is mostly due to the easier handling of such samples. Among those, blood, urine, and CSF have received quite a lot of attention. The following subsections provide a review of the work done using such samples.

19.3.1.1 Blood, Blood Cells, and Blood Products

Blood, its components, and medical products derived from blood have received considerable attention over the last 40 years. A large part of the early work on blood has been reviewed by Monti (1999). Many of the different types of cells that can be found in blood can be easily separated from each other (Monti 1999); therefore, the following sections will deal not only with whole blood but also with its components.

19.3.1.2 Red Blood Cells

The heat production rate of erythrocytes has been shown to be affected in patient with renal insufficiency, liver disease, anemia, and obesity (Monti et al. 1987; Monti and Wadsö 1973, 1976a,b). In all these cases, the erythrocytes of sick patients had a higher heat production than the erythrocytes of healthy patients. Although the diagnostic value of an elevated heat production rate in erythrocytes is low because it does not point to a specific disease, several studies have pointed out that this could be used to follow the efficacy/progress of therapy.

19.3.1.3 White Blood Cells

Lymphocytes can be recovered not only from lymph nodes (Monti 1999) but also from blood (Murigande et al. 2009) and can be purified using density gradients (Histopaque 1077, for example). Elevated heat production in lymphocytes has been linked with non-Hodgkin's lymphoma (NHL) and different types of leukemia. For NHL, the heat production per lymphocytes has been clearly linked to the severity of the disease (i.e., high-grade NHL lymphocytes produced more heat than low-grade NHL lymphocytes) and to the survival rate of patients (i.e., patients who survived less than 2 years had lymphocytes with a higher heat production rate than patient who survived more than 2 years). Similarly, in cases of remission the heat production of patients' lymphocytes returned to values close to those of controls (Monti et al. 1981, 1986).

For patients with chronic lymphocytic leukemia (CLL), the heat production rate per lymphocyte was found to be lower than normal, indicating a potential responsible factor for the development of hypermetabolism (Brandt et al. 1979). As for erythrocytes, the

TABLE 19.2   Liquid Clinical Samples Investigated Using Calorimetry

Type of Sample	Disease/Application	Approach	Rational	Results
Blood	Detection of bacterial and fungal infections	Growth of sample "as is," or in medium	Bacterial growth will lead to sufficient heat production within a few hours	Rising heat flow signal that passes above a set threshold. Time to detection is calculated
Erythrocyte	Renal failure, liver disease, anemia, obesity	Erythrocytes activity measured directly	Low or high heat output of red blood cells indicates a disease	Heat flow signal outside of the normal range
Lymphocytes	Lymphoma or leukemia	Lymphocyte activity measured directly in plasma or M199 medium +10% serum	High heat output might indicate a lymphoma and low heat output might indicate a leukemia	Heat flow value outside of the normal range
Platelet	Detection of bacterial and fungal infections	Growth of sample "as is," or in medium	Bacterial growth will lead to sufficient heat production within a few hours	Rising heat flow signal that passes above a set threshold. Time to detection is calculated
Plasma	Different cancer types and microbial infection such as Lyme disease?	Dialyzed plasma is scanned through different temperature	Changes in protein concentration, or small peptide interaction with one of the major plasma proteins will lead to a different DSC pattern	Abnormal DSC pattern
Urine	Detection of bacterial and fungal infections (with detection of drug-resistant strains)	Growth of sample "as is," in artificial urine, or in urine added with a nutrient broth or an antimicrobial	Bacterial growth will lead to sufficient heat production within a few hours	Rising heat flow signal that passes above a set threshold. Time to detection is calculated
Synovial fluid	Detection of bacterial and fungal infections	Growth of sample diluted in TSB medium	Bacterial growth will lead to sufficient heat production within a few hours	Rising heat flow signal that passes above a set threshold. Time to detection is calculated
Sonication fluid	Detection of bacterial and fungal infections	Growth of sample diluted in TSB medium	Bacterial growth will lead to sufficient heat production within a few hours	Rising heat flow signal that passes above a set threshold. Time to detection is calculated
Semen	Viability of sperm	Metabolic activity of spermatozoid measured directly	Heat flow correlates with viability	Low heat flow values indicate a loss of viability

Note: The table briefly depicts the intended application, the methodological approach used, the rational for the measurement, and the results that can be expected from such measurement.

diagnostic value of an elevated heat production rate in lymphocytes is low because it does not point to a specific disease; however, again calorimetry could be used to follow the efficacy/progress of therapy.

19.3.1.4 Platelets

Platelets are often transfused to prevent bleeding especially in anticipation of a surgical procedure that causes some bleeding. Platelet products bear a special risk for bacterial contamination because they must be stored at room temperature (22°C±2°C) to preserve the functionality of the platelet. Such storage at ambient temperature in turn might favor the rapid replication of a wide range of microorganisms, especially bacteria (Salzmann et al. 2007; Trampuz et al. 2007b). Therefore, early detection of contamination is vital to avoid transfusion-associated sepsis that could be lethal (Salzmann et al. 2007; Trampuz et al. 2007b). In this context, platelets that were about to be discarded were artificially contaminated with different organisms at concentrations ranging from 1 to 10^5 CFU/mL. IMC of platelets (1.1 mL) diluted in tryptic soy broth (TSB) (1.9 mL) was shown to be sensitive enough to detect contamination within 2.8 h for fast-growing pathogens (*Streptococcus sanguinis*) to 24 h for the slowest one (*Propionibacterium acnes*) at the highest concentration. At an initial concentration of 10 CFU/mL, this time increased to 7.8 h for *S. sanguinis* and 73.5 h for *P. acnes*.

19.3.1.5 Blood Plasma

Although this section does not refer directly to the application of IMC (contrary to most of this chapter), it deals with important diagnostic aspects using differential scanning calorimetry (DSC) that could be performed with instruments that are usually used as isothermal microcalorimeters. Therefore, the recent work on blood plasma has been included in this chapter. Blood plasma can be obtained and easily purified. It contains a large amount of protein among which 16 proteins represent most of the protein content of the plasma. When temperature scans from 20°C to 110°C are performed on a healthy subject's plasma, the DSC profile is highly reproducible (Garbett et al. 2008). However, when changes in the plasma protein concentration occur or if small molecules or peptides bind to one of the major plasma proteins, the DSC profile is altered. Using such changes in the DSC profile, it was shown that rheumatoid arthritis, Lyme disease, and systemic lupus could be easily and accurately diagnosed. In a later study, several types of cancer such as cervical cancer and breast cancer were also successfully diagnosed and staged (Garbett et al. 2009, 2014; Zapf et al. 2011). In addition, a statistical analysis of the curves was proposed to aid the diagnosis (Fish et al. 2010).

19.3.1.6 Urine

As urinary tract infection (UTI) is one of the most common infections worldwide (Al-Hasan et al. 2010; Stamm 2002), it is not surprising that many studies have tried to use IMC in their diagnosis and treatment. Because urine is a potent growth medium (Brooks and Keevil 1997), it can be used directly. Alternatively, urine can be added with different substrates for bacterial growth to speed up the diagnosis. Finally, artificial urine (Brooks and Keevil 1997) is also used to dilute samples and thus standardize the approach. Thus, different approaches are used. As early calorimeters had a very limited number of channels (up to four usually), a first approach was to add nutrient broth to the samples and to incubate them separately in an oven at 37°C for 2 h. The samples were then processed using a flow-through system. After measuring, the sample was flushed and the next sample was introduced, thereby preserving a rather high throughput with a sample measurement being done in 15 min after its initial introduction into the measuring chamber (Beezer et al. 1978, 1974).

Flow-through systems are often quite difficult to clean or sterilize, therefore another approach was to use a static closed ampoule filled with urine, artificial urine, or a medium. Using closed glass ampoules prefilled with 3 mL of filtered sterilized urine or 3 mL of artificial urine and inoculated with different bacteria at concentrations ranging from 1 to 10^5 CFU/mL, detection time between 3 and 6 h in urine and 3 and 9 h in artificial urine (Bonkat et al. 2012, 2013a,b) was achieved at high concentration (i.e., 10^5 CFU is commonly admitted to be the threshold value for UTIs). Using a very similar approach, it was shown that by comparing the growth of bacteria in urine treated with an antibiotic and an untreated sample, it was possible to deliver an accurate antibiogram within 7 h (Braissant et al. 2014). The accuracy of the drug susceptibility pattern delivered by the calorimetric technique was compared with conventional methods (culture monitored by optical density [OD]) and the results were obtained using the automated VITEK® 2 system (Biomérieux) for 15 strains isolated from patients with a UTI. In 92% of the cases, the three techniques were in agreement. In an additional 6% of cases, calorimetry was in agreement with either the OD measurement or with the VITEK2 system. This approach could be very interesting in urosepsis cases where a high bacterial load is encountered with a low probability of mixed infection (i.e., only one pathogen [Braissant et al. 2014]).

19.3.1.7 Synovial Fluids

Synovial fluid samples from knee, hip, shoulder, ankle, and elbow were investigated for the presence of bacteria indicative of septic arthritis. Synovial fluid (1 mL) was collected and placed in a calorimetric vial prefilled with 2 mL of medium (TSB). A rising heat production rate of 10 μW above the lowest point observed after thermal equilibration was considered positive. Using this setup, detection was achieved between 2 and 8 h with an accuracy of 89% (Yusuf et al. 2015).

19.3.1.8 Cerebrospinal Fluid

When a fungal or bacterial meningitidis is suspected, a lumbar puncture can be performed and CSF collected. A study using infected infant rats as the animal model has shown that inoculating TSB medium with very small amount of CSF (between 1 and 10 uL) could lead to a pathogen detection time of between 1.5 and 13 h. The study investigated three pathogens, *Neisseria meningitidis*, *S. pneumoniae*, and *Listeria monocytogenes*, and concluded that IMC could be very useful in cases of meningitis. The study also emphasized that the technique was particularly interesting for newborns and infants whose CSF volumes are limited and that higher volumes of CSF collected in adults, for example, could lead to a faster detection time (Trampuz et al. 2007c).

19.3.1.9 Sonication Fluids

For solid samples that are too large to fit into a microcalorimetric vial, a combination of sonication with microcalorimetry can be helpful in detecting infection. Using this approach, the original sample is placed in an ultrasound bath with a sterile solution (Ringer solution is often used). The low-power and low-frequency ultrasonic waves generate microcurrents, shear forces, and oscillating cavitation bubbles that disperse the biofilm components. As a result, biofilm cells are detached from the surface of the sample and spread in the sonication fluid. It must be noted that the frequency and duration of the sonication must be tuned to avoid killing the bacterial cells. In most studies, the sonication is performed at 40 ± 2 kHz for 5 min (Trampuz et al. 2007a). The sonication fluid can then be used to inoculate a microcalorimetric vial prefilled with a medium such as TSB. This approach has been successfully applied to several implant types (Borens et al. 2013) and other medical devices (urinary stents—Bonkat et al. 2011). A study focusing on implant-related infection investigated different types of orthopedic implants (hip prosthesis, knee prosthesis, shoulder prosthesis, joint spacer, orthopedic screws, orthopedic nails, and orthopedic plates) and demonstrated that the detection of infection could be obtained within 1 and 21 h. The same study also showed that a combination of sonication and IMC provided better detection than sonication combined with a conventional culture of the sonication fluid (Borens et al. 2013). In a more recent case study performed on an infected breast implant, infection was diagnosed by microcalorimetry of sonication fluid in 1 h 21 min. The isolated organism from the sonication fluid and the purulent liquid collected was the same and was identified as S. pyogenes one day after clinical presentation (Yusuf et al. 2014).

19.3.1.10 Semen

Many studies in reproductive medicine have focused on sperm using microcalorimetry as a "nonsubjective" method (compared with microscopy) to evaluate its quality. The method was considered effective; however, in the 1970s the cost of the instrument was deemed too high (Linford et al. 1976). Still, several studies that have focused on bulls and horses have demonstrated that sperm viability could be investigated using IMC. Similarly, the effect of freezing or mechanical damage on spermatozoid could be related to a lower heat output (Fischer et al. 2007; Linford et al. 1976; Vasconcelos et al. 2009, 2010). Nevertheless, very few studies have considered using IMC for men, although it is considered a promising tool (Saymé et al. 2013). This is especially true considering that previous studies (Vasconcelos et al. 2009) have shown that parameters such as glucose concentration and spermatozoid concentration could easily be optimized to achieve good detection.

19.3.2 Solid Samples

Monitoring growth within solid samples often remains a challenge for cell biologist and microbiologist. Except for transparent materials for which optical methods can be applied, most samples require the use of destructive extraction and quantification methods. In this

TABLE 19.3   Solid Clinical Samples Investigated Using Calorimetry

Type of Sample	Disease	Approach	Rational	Results
Urogenital biopsies	Urogenital cancer of different types	Biopsies placed in Ringer solution or cell lines isolated and further investigated as cultures in RPMI medium	Heat production rate corresponds with the grading of the tumor	Elevated heat flow of biopsies and cell cultures compared with their healthy counterparts
Biofilm	Testing new drugs against biofilms, testing new materials	Biofilms are grown on a surface and placed in the calorimeter. Alternatively, surface/volume ratio is increase by adding glass beads in calorimetric vials. Finally, solid medium is used in conjunction with solid samples to avoid planktonic cells in the system.	Heat production rate on antimicrobial material will reflect the biofilm formation or activity on this material.	Heat flow signal will reflect the biofilm activity
Solid culture	Detection of bacterial and fungal infections (strictly aerobic organisms)	Solid medium inoculated with sample (sputum or other fluids)	Bacterial growth will lead to sufficient heat production within a few hours	Rising heat flow signal that passes above a set threshold. Time to detection is calculated

Note: The table briefly depicts the application intended, the methodological approach used, the rational for the measurement, and the results that can be expected from such measurement.

context, microcalorimetry offers a relatively simple way to assess the growth of cells or microbes within a solid and often opaque material (Braissant et al. 2015a). Similarly, it offers an easy way to test for the antimicrobial properties of clinical materials (Braissant et al. 2015b). Table 19.3 summarizes the type of solid samples that have been processed by this technique over the last years.

19.3.2.1 Solid Tumors

Histological examination remains the gold standard in determining the malignancy of tumors. Histological evaluation is slow and usually does not provide much information on the dynamic behavior of the tumor cells (Kallerhoff et al. 1996). In this context, several studies have investigated biopsies. The direct measurement of biopsy heat production in Ringer solution demonstrated that tumor tissue could be discriminated from nontumor tissue (bladder, prostate, kidney, and testicle) based on their heat production rate. Indeed, tumorous tissues consistently produced more heat than their nontumorous counterparts (Kallerhoff et al. 1996). Similarly, the drug susceptibility of isolated renal cancer cell lines could also be investigated by IMC. Cells incubated with 5-fluoruracil, interleukin-2, and alpha-interferon-2a had a lower heat production rate than untreated cell lines. The results indicated the cytostatic effect of the different agents used on the sensible cell lines (Bluthnerhassler et al. 1995).

The investigation of microtissues opens a new area of cancer research. Using isolated cell lines, it is relatively easy to generate microtissues simulating a tumor using "hanging drop technology" (Kelm and Fussenegger 2004; Walker and Rapley 2000). The microtissues offer cell-to-cell contact and diffusion properties that are closer to a real tumor than a conventional monolayer culture. However, monitoring the growth and effect of specific compounds on the microtissues requires either a destructive assay (protein or DNA assay) or measurement of the microtissue diameter. The latter provides very little information on the metabolism of the microtissue. In this context, IMC has been shown to accurately monitor the growth of microtissue in a specific medium. As microtissues have a relatively low metabolic activity compared with monolayer tissue, the closed environment of the microcalorimeter vial was not found to be an issue. Indeed, without medium change, the hepatocarcinoma microtissue was shown to remain active for 15–30 days (Braissant et al. 2015c).

19.3.2.2 Biofilms

Biofilms are a major issue in the medical routine. They can form on many surfaces and threaten patients' lives (Costerton et al. 1987, 1995; Flemming and Wingender 2010; Hall-Stoodley et al. 2004; LaFleur et al. 2006). This is of special importance because biofilms are known to be more resistant to antimicrobial than planktonic cells. Many clinical laboratories are involved in testing new treatments against biofilm and materials that prevent or delay biofilm formation. While microscopy provides a nice way to study biofilm on a transparent or flat surface, it becomes much more difficult to investigate porous and opaque materials (Stewart and Franklin 2008). In addition, many staining procedures for fluorescence microscopy are destructive and can only be used as endpoint assays (Stewart and Franklin 2008). Using IMC, several studies have investigated biofilm activity in real time and have collected valuable results. Different methods have been used to form a biofilm in a calorimetric vial. One way to investigate biofilms in a calorimeter is to artificially increase the surface-to-volume ratio in the calorimetric vial, favoring thus the number of attached cells as compared with the planktonic cells. A simple way to increase the surface-to-volume ratio is to add glass beads or various scaffolds in a calorimetric vial prefilled with medium (TSB or Luria broth [LB], for example). Although this technique was used in several studies, it is difficult to assess the proportion of signal that can be attributed to planktonic cells growing in the added liquid medium (Corvec et al. 2013; Furustrand Tafin et al. 2015; Maiolo et al. 2014; Mihailescu et al. 2014; Said et al. 2014a,b). In experiments where large volumes of liquid are added to the beads, scaffolds, or any material, the presence of planktonic cells should not be neglected, in particular when the antimicrobial agent is not soluble in the medium. In another approach, flow-through isothermal microcalorimetric measurements were used to assess the antimicrobial properties of catheters. The biofilm was first grown at a low flow rate. Such a low flow rate allowed planktonic cells to be flushed out and only biofilm activity was monitored (Said 2014). Using this approach, it was possible to also circulate antimicrobials (in this case AgNO$_3$ solution) in the system and estimate the biofilm response. Indeed, established biofilms kept their metabolic activity at a silver concentration that fully inhibited planktonic cells (Said 2014).

Finally, inoculating a porous surface and placing that surface on a solid medium prepared in a microcalorimetric vial allowed monitoring the development of a biofilm. Avoiding the use of liquid medium and using such a "solid–solid" interface provides a nice way to prevent having planktonic cells. This approach shares some similarities with the technique of the colony biofilm where a biofilm is grown on a sterile filter placed on a solid medium (Merritt et al. 2005; Peterson et al. 2011). This approach was used to test the antimicrobial effect of new implant coatings (Braissant et al. 2015b). In this case, silver-based antimicrobial materials were shown to have a prolonged lag phase due to the initial killing of the inoculum (i.e., its bactericidal effect).

19.4 LIMITATIONS

Although microcalorimetry appears to be widely applicable, there are some limitations that need to be discussed. A physical limitation that one often encounters is density issues. For example, fat cells might float above the medium, whereas blood cells, yeast, and even bacteria might settle at the bottom of the tube (Maskow et al. 2014). To avoid the sample floating, a porous cap might be added at the top of the sample to keep it fully immersed in medium. Polypropylene foam seems to be quite appropriate for this purpose (authors' personal observations). Avoiding sedimentation can be easily achieved using a commercial product such as Percoll or Ficoll, which are used for creating density gradients during centrifugation (Maskow et al. 2014). Another limitation of the instrument is the fixed temperature. Indeed, testing a biological process (such as growth) at several different temperatures requires quite a long time because most instruments need to be set, equilibrated, and calibrated each time the temperature is changed. However, some prototype instruments are available that can maintain a stable temperature gradient through the different measuring positions (Wadsö et al. 2011). Some researchers also used the DSC mode to scan across the temperature range of interest (Mukhanov et al. 2012), but DSC instruments usually have a limited throughput (Braissant et al. 2015a; Mukhanov et al. 2012).

As microcalorimeter vials are closed systems, gas exchange does not happen as in a conventional culture that can be aerated by shacking or by bubbling air. In particular, oxygen depletion is an issue in some cases, especially because it dissolves poorly in aqueous media and because its diffusion is slow in those media. For most bacteria, such oxygen depletion is not an issue because they can switch to anaerobic metabolism (fermentation or nitrate reduction, for example), which is usually visible by a different peak in the heat flow pattern (Braissant et al. 2010c, 2015a; Johansson and Wadsö 1999). On the other hand, for strict aerobes, oxygen depletion is a severe issue. In this case, the use of a solid medium is of interest because the bacteria (and bacterial colonies) are directly at the medium–air interface. This approach was successfully used for fungi (Oriol et al. 1987) and for mycobacteria (Braissant et al. 2010b; Howell et al. 2012; Rodriguez et al. 2011). In addition to such a simple approach, it was shown that using silicon to seal microcalorimetric vials allowed gas exchange, but not water evaporation, thus preventing the appearance of an endothermic signal due to water loss from the system (Sparling 1983).

In addition to physicochemical factors, some biological factors can also make the microcalorimetry measurements more difficult to interpret. Among these, biofilm formation and mixed microbial cultures might be very difficult to interpret. Biofilms were shown to have high heterogeneity (Stewart and Franklin 2008), which, in turn, is reflected by a high variability in their metabolic activity (Astasov-Frauenhoffer et al. 2012) as measured by IMC, even when coming from the same flow cell (i.e., prepared in the same batch of medium, with the same inoculum at the same time).

19.5 CURRENT INSTRUMENTS AND THEIR POTENTIAL FOR CLINICAL APPLICATIONS

A limited number of isothermal calorimeters are available on the market (Table 19.4). Among them, the sensitivity and the number of channels vary quite a lot. Therefore, it is interesting to evaluate their potential use in a clinical environment. The calorimeters offering the highest sensitivity usually have a rather small sample volume and a higher number of measuring channels. This makes such calorimeters quite interesting for processing a large number of liquid samples that will fit into the small calorimetric vials of these instruments. On the other hand, instruments with lower sensitivity and larger volumes could be of interest for more specific application such as testing the contamination of small implants (screws or small plates, for example) where high throughput is less important. In any case, the choice of calorimetry instrument should be dictated by the final use of the instrument and the sample that will be processed.

TABLE 19.4 Commercially Available Isothermal Microcalorimeters and Their Major Characteristics

Instrument	Sensitivity (μW)	Sample Size (mL)	Channels	DSC Capability	Use in Clinic
TA instruments TAM air	5	20	8	No	+ (low throughput)
TA instruments TAM III	0.02–3[b]	1–125[b]	4–24[b]	Yes	++
TA instruments TAM 48	0.2	4	48	Yes	+++
TA instrument MC DSC	0.2	3	3	Yes	+ (low throughput)
Symcel Calscreener	0.05	0.4	32–47[a]	No	+++
THT[a] uMC	0.1	4	14	No	++ (low throughput)
C3 MC CAL	5–20	40	12	No	–
Calmetrix Biocal 4000	5	125	4	No	+/– (for large samples)
Calmetrix Biocal 2000	5	125	2	No	+/– (for large samples)
Setaram MS80	0.1	12.5–100	2–4	(No)	+/– (for large samples)
Setaram C80	0.1	12.5	1	Yes	– (too low throughput)
TTP Labtech ChipCal	0.03–0.085	0.015	1 (flow-through)	No	++

[a] Thermal hazard technology.
[b] Depending on the type of calorimeters installed in the thermostat.

19.6 SOME RECOMMENDATIONS FOR THE USE OF CALORIMETRY IN THE CLINIC

As yet, calorimetric measurements are not used in clinical routine although some clinical trials have recently been approved. As a result, when developing a new assay or a new type of measurement, it is worth investing some time to translate conventional assays into microcalorimetric assays. In our experience, it is worth spending time on finding the right conditions for the cells' metabolic activity to be easily detected in the calorimetric assays. It is especially useful to look at different medium volumes and compositions. Lower volumes will allow better oxygen diffusion (Maskow et al. 2014; Zaharia et al. 2013). Further, since microcalorimetric vials are closed systems, some extra buffer might be useful to compensate for quick pH changes (usually acidification). Data management is also a crucial issue with microcalorimetric assays. Usually, they are not endpoint assays in this area, but rather a dynamic measurement of a process, and the amount of data generated can be quite high. Therefore, careful planning of the data analysis is as crucial as the experimental setup planning if one wants to get the most out of the data.

19.7 CONCLUSIONS

This chapter has illustrated that IMC and calorimetry in general can be applied to a wide variety of assays. Many clinical uses have already been described, but many researchers are still exploring the potential of calorimetry for other clinical uses. It seems that in many cases the data analysis remains a bottleneck as biologists and microbiologists are usually not exposed to microcalorimery during their academic studies. In addition, very few studies have investigated thermograms in detail or associated a pattern (or a section of the heat flow pattern) to a specific process. As a result, it is sometimes difficult to translate calorimetric data into meaningful biological events. With the increasing number of peer-reviewed articles in which calorimetry is used, a strong knowledge base might soon be available, helping researchers to interpret their results. This will undoubtedly provide a bright future for calorimetry use in the medical area.

REFERENCES

Al-Hasan, M. N., Eckel-Passow, J. E., and Baddour, L. M. 2010. Bacteremia complicating gram-negative urinary tract infections: A population-based study. *J. Infect.* 60:278–285.

Astasov-Frauenhoffer, M., Braissant, O., Hauser-Gerspach, I., Daniels, A. U., Weiger, R., and Waltimo, T. 2012. Isothermal microcalorimetry provides new insights into biofilm variability and dynamics. *FEMS Microbiol. Lett.* 337:31–37.

Baldoni, D., Hermann, H., Frei, R., Trampuz, A., and Steinhuber, A. 2009. Performance of microcalorimetry for early detection of methicillin resistance in clinical isolates of *Staphylococcus aureus. J. Clin. Microbiol.* 47:774–776.

Beezer, A. E., Bettelheim, K. A., Al-Salihi, S., and Shae, E. J. 1978. The enumeration of bacteria in culture media and clinical specimens of urine by microcalorimetry. *Science Tools* 25:2510–2512.

Beezer, A. E., Bettelheim, K. A., Newell, R. D., and Stevens, J. 1974. The diagnosis of bacteriuria by flow microcalorimetry. *Science Tools* 21:13–15.

Bluthnerhassler, C., Karnebogen, M., Schendel, W., Singer, D., Kallerhoff, M., Zoller, G., and Ringert, R. H. 1995. Influence of malignancy and cyctostatic treatment on microcalorimetric behavior of urological tissue samples and cell-cultures. *Thermochim. Acta* 251:145–154.

Bonkat, G., Braissant, O., Rieken, M., Solokhina, A., Widmer, A. F., Frei, R., van der Merwe, A., Wyler, S., Gasser, T. C., and Bachmann, A. 2013a. Standardization of isothermal microcalorimetry in urinary tract infection detection by using artificial urine. *World J. Urol.* 31:553–557.

Bonkat, G., Braissant, O., Widmer, A. F., Frei, R., Rieken, M., Wyler, S., Gasser, T. C., Wirz, D., Daniels, A. U., and Bachmann, A. 2012. Rapid detection of urinary tract pathogens using microcalorimetry: Principle, technique and first results. *BJU Int.* 110:892–897.

Bonkat, G., Rieken, M., Rentsch, C. A., Wyler, S., Feike, A., Schafer, J., Gasser, T., Trampuz, A., Bachmann, A., and Widmer, A. F. 2011. Improved detection of microbial ureteral stent colonisation by sonication. *World J. Urol.* 29:133–138.

Bonkat, G., Wirz, D., Rieken, M., Gasser, T. C., Bachmann, A., and Braissant, O. 2013b. Areas of application of isothermal microcalorimetry in urology. An overview. *Urologe* 52:1092–1096.

Borens, O., Yusuf, E., Steinrucken, J., and Trampuz, A. 2013. Accurate and early diagnosis of orthopedic device-related infection by microbial heat production and sonication. *J. Orthop. Res.* 31:1700–1703.

Bourbeau, P. P., and Ledeboer, N. A. 2013. Automation in clinical microbiology. *J. Clin. Microbiol.* 51:1658–1665.

Braissant, O., Bachmann, A., and Bonkat, G. 2015a. Microcalorimetric assays for measuring cell growth and metabolic activity: Methodology and applications. *Methods.* 76:27–34.

Braissant, O., Bindschedler, S., Daniels, A. U., Verrecchia, E. P., and Cailleau, G. 2012. Microbiological activities in moonmilk monitored using isothermal microcalorimetry (Cave of Vers chez leBrandt, Neuchatel, Switzerland). *J. Cave Karst Studies* 74: 116–126.

Braissant, O., Bonkat, G., Wirz, D., and Bachmann, A. 2013. Microbial growth and isothermal microcalorimetry: Growth models and their application to microcalorimetric data. *Thermochim. Acta* 555:64–71.

Braissant, O., Chavanne, P., Wild, M., Pieles, U., Stevanovic, S., Schumacher, R., Straumann, L., Wirz, D., Gruner, P., and Bachmann, A. 2015b. Novel microcalorimetric assay for antibacterial activity of implant coatings: The cases of silver-doped hydroxyapatite and calcium hydroxide. *J. Biomed. Mater. Res. B. Appl. Biomat.* 103:1161–1167.

Braissant, O., Keiser, J., Meister, I., Bachmann, A., Wirz, D., Gopfert, B., Bonkat, G., and Wadsö, I. 2015c. Isothermal microcalorimetry accurately detects bacteria, tumorous microtissues, and parasitic worms in a label-free well-plate assay. *Biotechnol. J.* 10:460–468.

Braissant, O., Muller, G., Egli, A., Widmer, A., Frei, R., Halla, A., Wirz, D., Gasser, T. C., Bachmann, A., Wagenlehner, F., and Bonkat, G. 2014. Seven hours to adequate antimicrobial therapy in urosepsis using isothermal microcalorimetry. *J. Clin. Microbiol.* 52:624–626.

Braissant, O., Wirz, D., Gopfert, B., and Daniels, A. U. 2010a. Biomedical use of isothermal microcalorimeters. *Sensors* 10:9369–9383.

Braissant, O., Wirz, D., Gopfert, B., and Daniels, A. U. 2010b. The heat is on: Rapid microcalorimetric detection of mycobacteria in culture. *Tuberculosis* 90:57–59.

Braissant, O., Wirz, D., Gopfert, B., and Daniels, A. U. 2010c. Use of isothermal microcalorimetry to monitor microbial activities. *FEMS Microbiol. Lett.* 303:1–8.

Brandt, L., Ikomikumm, J., Monti, M., and Wadsö, I. 1979. Heat-production by lymphocytes in chronic lymphocytic-leukemia. *Scand. J. Haematol.* 22:141–144.

Brooks, T., and Keevil, C. W. 1997. A simple artificial urine for the growth of urinary pathogens. *Lett. Appl. Microbiol.* 24:203–206.

Corvec, S., Furustrand Tafin, U., Betrisey, B., Borens, O., and Trampuz, A. 2013. Activities of fosfomycin, tigecycline, colistin, and gentamicin against extended-spectrum- -lactamase-producing *Escherichia coli* in a foreign-body infection model. *Antimicrob. Agents Chemother.* 57:1421–1427.

Costerton, J. W., Cheng, K. J., Geesey, G. G., Ladd, T. I., Nickel, J. C., Dasgupta, M., and Marrie, T. J. 1987. Bacterial biofilms in nature and disease. *Ann. Rev. Microbiol.* 41:435–464.

Costerton, J. W., Lewandowski, Z., Caldwell, D. E., Korber, D. R., and Lappin-Scott, H. M. 1995. Microbial biofilms. *Ann. Rev. Microbiol.* 49:711–745.

De Meis, L., Ketzer, L. A., Camacho-Pereira, J., and Galina, A. 2012. Brown adipose tissue mitochondria: Modulation by GDP and fatty acids depends on the respiratory substrates. *Biosci. Rep.* 32:53–59.

Dierig, A., Frei, R., and Egli, A. 2015. The fast route to microbe identification matrix assisted laser desorption/ionization-time of flight mass spectrometry (MALDI-TOF MS). *Pediatr. Infect. Dis. J.* 34:97–99.

Dixon, P., Davies, P., Hollingworth, W., Stoddart, M., and MacGowan, A. 2015. A systematic review of matrix-assisted laser desorption/ionisation time-of-flight mass spectrometry compared to routine microbiological methods for the time taken to identify microbial organisms from positive blood cultures. *Eur. J. Clin. Microbiol. Infect. Dis.* 34:863–876.

Evans, R. C., and Fine, B. R. 2013. Time to detection of bacterial cultures in infants aged 0 to 90 days. *Hosp. Pediatr.* 3:97–102.

Fischer, C., Scherfer-Brahler, V., Muller-Schlosser, F., Schroder-Printzen, I., and Weidner, W. 2007. A thermodynamic study on bovine spermatozoa by microcalorimetry after Percoll density-gradient centrifugation: Experimental probe of its utility in andrology. *Aktuelle Urol.* 38:237–242.

Fish, D. J., Brewood, G. P., Kim, J. S., Garbett, N. C., Chaires, J. B., and Benight, A. S. 2010. Statistical analysis of plasma thermograms measured by differential scanning calorimetry. *Biophys. Chem.* 152:184–190.

Flemming, H.-C., and Wingender, J. 2010. The biofilm matrix. *Nat. Rev. Microbiol.* 8: 623–633.

Fournier, P. E., Drancourt, M., Colson, P., Rolain, J. M., La Scola, B., and Raoult, D. 2013. Modern clinical microbiology: New challenges and solutions. *Nat. Rev. Microbiol.* 11:574–585.

Furustrand Tafin, U., Betrisey, B., Bohner, M., Ilchmann, T., Trampuz, A., and Clauss, M. 2015. Staphylococcal biofilm formation on the surface of three different calcium phosphate bone grafts: A qualitative and quantitative *in vivo* analysis. *J. Mater. Sci. Mater. Med.* 26:130.

Garbett, N. C., Mekmaysy, C. S., Helm, C. W., Jenson, A. B., and Chaires, J. B. 2009. Differential scanning calorimetry of blood plasma for clinical diagnosis and monitoring. *Exp. Mol. Pathol.* 86:186–191.

Garbett, N. C., Merchant, M. L., Helm, C. W., Jenson, A. B., Klein, J. B., and Chaires, J. B. 2014. Detection of cervical cancer biomarker patterns in blood plasma and urine by differential scanning calorimetry and mass spectrometry. *Plos One* 9.

Garbett, N. C., Miller, J. J., Jenson, A. B., and Chaires, J. B. 2008. Calorimetry outside the box: A new window into the plasma proteome. *Biophys. J.* 94:1377–1383.

Graban, M. 2011. *Lean Hospitals: Improving Quality, Patient Safety, and Employee Satisfaction*. Boca Raton, FL: CRC Press.

Hall-Stoodley, L., Costerton, J. W., and Stoodley, P. 2004. Bacterial biofilms: From the natural environment to infectious diseases. *Nat. Rev. Microbiol.* 2:95–108.

Howell, M., Wirz, D., Daniels, A. U., and Braissant, O. 2012. Application of a microcalorimetric method for determining drug susceptibility in *Mycobacterium* species. *J. Clin. Microbiol.* 50:16–20.

James, M. A. 1987. Growth and metabolism of bacteria. In M. A. James (Ed.), *Thermal and Energetic Studies of Cellular Biological Systems*, 69–105, Butterworth-Heinemann: Bristol.

Johansson, P., and Wadsö, I. 1999. An isothermal microcalorimetric titration/perfusion vessel equipped with electrodes and spectrophotometer. *Thermochim. Acta* 342:19–29.

Kallerhoff, M., Karnebogen, M., Singer, D., Dettenbach, A., Gralher, U., and Ringert, R. H. 1996. Microcalorimetric measurements carried out on isolated tumorous and nontumorous tissue samples from organs in the urogenital tract in comparison to histological and impulse-cyto-photometric investigations. *Urol. Res.* 24:83–91.

Kelm, J. M., and Fussenegger, M. 2004. Microscale tissue engineering using gravity-enforced cell assembly. *Trends Biotechnol.* 22:195–202.

Kemp, R. B., and Lamprecht, I. 2000. La vie est donc un feu pour la calorimetrie: Half a century of calorimetry–Ingemar Wadsö at 70. *Thermochim. Acta* 348:1–17.

Kimura, T. and Takahashi, K. 1985. Calorimetric studies of soil microbes: Quantitative relation between heat evolution during microbial degradation of glucose and changes in microbial activity in soil. *J. Gen. Micr.* 131:3083–3089.

LaFleur, M. D., Kumamoto, C. A., and Lewis, K. 2006. *Candida albicans* biofilms produce antifungal-tolerant persister cells. *Antimicrob. Agents Chemother.* 50:3839–3846.

Linford, E., Glover, F. A., Bishop, C., and Stewart, D. L. 1976. The relationship between semen evaluation methods and fertility in the bull. *J. Reprod. Fert.* 47:283–291.

Maiolo, E. M., Furustrand Tafin, U., Borens, O., and Trampuz, A. 2014. Activities of fluconazole, caspofungin, anidulafungin, and amphotericin b on planktonic and biofilm candida species determined by microcalorimetry. *Antimicrob. Agents Chemother.* 58:2709–2717.

Maskow, T., Morais, F. M., Rosa, L. F. M., Qian, Y. G., and Harnisch, F. 2014. Insufficient oxygen diffusion leads to distortions of microbial growth parameters assessed by isothermal microcalorimetry. *RSC Adv.* 4:32730–32737.

Merritt, J. H., Kadouri, D. E., and O'Toole, G. A. 2005. Growing and analyzing static biofilms. In R. Coico, T. Kowalik, J. Quarles, B. Stevenson, and R. Taylor (Eds), *Current Protocols in Microbiology*, Hoboken, NJ: John Wiley.

Mihailescu, R., Furustrand Tafin, U., Corvec, S., Oliva, A., Betrisey, B., Borens, O., and Trampuz, A. 2014. High activity of fosfomycin and rifampin against methicillin-resistant *Staphylococcus aureus* biofilm *in vitro* and in an experimental foreign-body infection model. *Antimicrob. Agents Chemother.* 58:2547–2553.

Monti, M. 1999. Calorimetric studies in medicine. In R. B. Kemp (Ed.), *Handbook of Calorimetry and Thermal Analysis: From Macromolecules to Man*, 657–710, Amsterdam: Elsevier.

Monti, M., Brandt, L., Ikomikumm, J., and Olsson, H. 1986. Microcalorimetric investigation of cell-metabolism in tumor-cells from patients with non-Hodgkin lymphoma (Nhl). *Scand. J. Haematol.* 36:353–357.

Monti, M., Brandt, L., Ikomikumm, J., Olsson, H., and Wadsö, I. 1981. Metabolic-activity of lymphoma-cells and clinical course in non-Hodgkin lymphoma (Nhl). *Scand. J. Haematol.* 27:305–310.

Monti, M., Hedner, P., Ikomi-Kumm, J., and Valdemarsson, S. 1987. Erythrocyte metabolism in hyperthyroidism: A microcalorimetric study on changes in the Embden-Meyerhof and the hexose monophosphate pathways. *Acta Endocrinol. (Copenh)* 115:87–90.

Monti, M., Nilsson-Ehle, P., Sorbris, R., and Wadsö, I. 1980. Microcalorimetric measurement of production heat in isolated human adipocytes. *Scand. J. Clin. Lab. Invest.* 40:581–587.

Monti, M., and Wadsö, I. 1973. Microcalorimetric measurements of heat production in human erythrocytes. I. Normal subjects and anemic patients. *Scand. J. Clin. Lab. Invest.* 32:47–54.

Monti, M., and Wadsö, I. 1976a. Microcalorimetric measurements of heat production in human erythrocytes. II. Hyperthyroid patients before, during and after treatment. *Acta Med. Scand.* 200:301–308.

Monti, M., and Wadsö, I. 1976b. Microcalorimetric measurements of heat production in human erythrocytes. III. Influence of pH, temperature, glucose concentration, and storage conditions. *Scand. J. Clin. Lab. Invest.* 36:565–572.

Mukhanov, V. S., Hansen, L. D., and Kemp, R. B. 2012. Nanocalorimetry of respiration in microorganisms in natural waters. *Thermochim. Acta* 531:66–69.

Murigande, C., Regenass, S., Wirz, D., Daniels, A. U., and Tyndall, A. 2009. A comparison between (3H)-thymidine incorporation and isothermal microcalorimetry for the assessment of antigen-induced lymphocyte proliferation. *Immunol. Invest.* 38:67–75.

Oriol, E., Conteras, R., and Raimbault, M. 1987. Use of microcalorimetry for monitoring the solid state culture of *Aspergillus niger*. *Biotechnol. Techn.* 1:79–84.

Peterson, L. R., Hamilton, J. D., Baron, E. J., Tompkins, L. S., Miller, J. M., Wilfert, C. M., Tenover, F. C., and Thomson, R. B. 2001. Role of clinical microbiology laboratories in the management and control of infectious diseases and the delivery of health care. *Clin. Infect. Dis.* 32:605–610.

Peterson, S. B., Irie, Y., Borlee, B., Murakami, K., Harrison, J., Colvin, K., and Parsek, M. 2011. Different methods for culturing biofilms *in vitro*. In Th. Bjarnsholt, P. Østrup Jensen, C. Moser and N. Høiby (Eds), *Biofilm Infections*, 251–266, New York: Springer.

Rodriguez, D., Daniels, A. U., Urrusti, J. L., Wirz, D., and Braissant, O. 2011. Evaluation of a low-cost calorimetric approach for rapid detection of tuberculosis and other mycobacteria in culture. *J. Appl. Microbiol.* 111:1016–1024.

Said, J. 2014. *The Application of Static and Flow-Through Isothermal Microcalorimetry to the Antimicrobial Analysis of Medical Materials and Devices*, London: School of Pharmacy, University College London.

Said, J., Dodoo, C. C., Walker, M., Parsons, D., Stapleton, P., Beezer, A. E., and Gaisford, S. 2014a. An *in vitro* test of the efficacy of silver-containing wound dressings against *Staphylococcus aureus* and *Pseudomonas aeruginosa* in simulated wound fluid. *Int. J. Pharm.* 462:123–128.

Said, J., Walker, M., Parsons, D., Stapleton, P., Beezer, A. E., and Gaisford, S. 2014b. An *in vitro* test of the efficacy of an anti-biofilm wound dressing. *Int. J. Pharm.* 474:177–181.

Salzmann, S., Antheaume, J., Steinhuber, A., Frei, R., Daniels, A. U., and Trampuz, A. 2007. Microcalorimetry: A novel screening method for microbial contamination of platelet concentrates. *Swiss Medical Weekly* 137:51S–51S.

Sarathbabu, R., Rajkumari, N., and Ramani, V. T. 2013. Characterization of coagulase negative staphylococci isolated from urine, pus, sputum and blood samples. *Int. J. Pharm. Sci. Inv.* 2:37–46.

Saymé, N., Krebs, T., and Maas, D. H. A. 2013. Anwendungsperspektiven für die Mikrokalorimetrie in der Reproduktionsmedizin zur Charakterisierung komplexer Systeme. Paper read at 20. Ulm-Freiberger Kalorimetrietage, at Freiberg/Sa. Germany.

Sorbris, R., Monti, M., Nilsson-Ehle, P., and Wadsö, I. 1982. Heat production by adipocytes from obese subjects before and after weight reduction. *Metabolism* 31:973–978.

Sparling, G. P. 1983. Estimation of microbial biomass and activity in soil using microcalorimetry. *J. Soil Sci.* 34:381–390.

Stamm, W. E. 2002. Scientific and clinical challenges in the management of urinary tract infections. *Am. J. Med.* 113:1s–4s.

Stewart, P. S., and Franklin, M. J. 2008. Physiological heterogeneity in biofilms. *Nat. Rev. Microbiol.* 6:199–210.

Trampuz, A., Piper, K. E., Jacobson, M. J., Hanssen, A. D., Unni, K. K., Osmon, D. R., Mandrekar, J. N., Cockerill, F. R., Steckelberg, J. M., Greenleaf, J. F., and Patel, R. 2007a. Sonication of removed hip and knee prostheses for diagnosis of infection. *N. Engl. J. Med.* 357:654–663.

Trampuz, A., Salzmann, S., Antheaume, J., and Daniels, A. U. 2007b. Microcalorimetry: A novel method for detection of microbial contamination in platelet products. *Transfusion* 47:1643–1650.

Trampuz, A., Steinhuber, A., Wittwer, M., and Leib, S. L. 2007c. Rapid diagnosis of experimental meningitis by bacterial heat production in cerebrospinal fluid. *BMC Infect. Dis.* 7:116.

Vasconcelos, A. B., Souza, P. C., Varago, F. C., Lagares, M. A., and Santoro, M. M. 2009. Determination of optimal glucose concentration for microcalorimetric metabolic evaluation of equine spermatozoa. *Braz. Arch. Biol. Technol.* 52:1129–1136.

Vasconcelos, A. B., Santana, M. A., Santos, A. M., Santoro, M. M., and Lagares, M. A. 2010. Metabolic evaluation of cooled equine spermatozoa. *Andrologia* 42:106–11.

von Ah, U., Wirz, D., and Daniels, A. U. 2008. Rapid differentiation of methicillin-susceptible *Staphylococcus aureus* from methicillin-resistant *S. aureus* and MIC determinations by isothermal microcalorimetry. *J. Clin. Microbiol.* 46:2083–2087.

Wadsö, I. 2002. Isothermal microcalorimetry in applied biology. *Thermochim. Acta* 394:305–311.

Wadsö, L., Salamanca, Y., and Johansson, S. 2011. Biological applications of a new isothermal calorimeter that simultaneously measures at four temperatures. *J. Therm. Anal. Calorim.* 104:119–126.

Walker, J. M., and Rapley, R. 2000. *Molecular Biology and Biotechnology.* Cambridge: Royal Society of Chemistry.

Yusuf, E., Hügle, T., Daikeler, T., Voide, C., Borens, O., and Trampuz, A. 2015. The potential use of microcalorimetry in rapid differentiation between septic arthritis and other causes of arthritis. *Eur. J. Clin. Microbiol. Infect. Dis. Sep.* 34:461–465.

Yusuf, E., Steinrucken, J., Nordback, S., and Trampuz, A. 2014. Necrotizing fasciitis after breast augmentation rapid microbiologic detection by using sonication of removed implants and microcalorimetry. *Am. J. Clin. Pathol.* 142:269–272.

Zaharia, D. C., Muntean, A. A., Popa, M. G., Steriade, A. T., Balint, O., Micut, R., Iftene, C., Tofolean, I., Popa, V. T., Baicus, C., Bogdan, M. A., and Popa, M. I. 2013. Comparative analysis of *Staphylococcus aureus* and *Escherichia coli* microcalorimetric growth. *BMC Microbiol.* 13:171.

Zapf, I., Fekecs, T., Ferencz, A., Tizedes, G., Pavlovics, G., Kalman, E., and Lorinczy, D. 2011. DSC analysis of human plasma in breast cancer patients. *Thermochim. Acta* 524:88–91.

CHAPTER 20

Calorimetric Assays for Heterogeneous Enzyme Catalysis

Hydrolysis of Cellulose and Biomass

Peter Westh and Kim Borch

CONTENTS

20.1 INTRODUCTION

Finding reliable methods for activity measurements remains a major challenge within enzymology, and in some cases, isothermal calorimetry has proved to be an important supplement to the arsenal of other methods used in this field. Compared with a range of so-called high-throughput methods (Goddard and Reymond 2004; Reymond 2004), which are mainly based on detecting changes in spectral properties, calorimetry is quite a slow and labor-intensive approach. However, as the principle of quantifying the rate of a reaction through the heat flow it generates is fundamentally different from other assay principles, many cases can be envisioned where calorimetry could be an effective tool for enzymologists. This potential has been documented by isothermal flow calorimetry (Beezer et al. 1974; Eftink et al. 1981) and later shown to be more conveniently exploited in titration instruments (Todd and Gomez 2001; Williams and Toone 1993).

Any kinetic study using isothermal calorimetry utilizes that the reaction rate, v, may be expressed as

$$v = \frac{dQ/dt}{\Delta_r H V_{sam}} \tag{20.1}$$

where:

$\Delta_r H$ is the enthalpy change of the reaction at the conditions in the calorimetric cell
V_{sam} is the volume of the sample
dQ/dt is the heat flow (in J/s) required to keep the sample isothermal (detected by the instrument)

One important consequence of Equation 20.1 is that calorimetry can in principle be applied to any reaction with $\Delta_r H \neq 0$, and hence does not rely on the selection of substrate or conditions where spectral or other properties change as the reaction progresses. Moreover, the method does not need any labeling or post-experiment procedures and it provides real-time data, which readily elucidate the time–course of the reaction. Another implication of Equation 20.1 is that the rate of reaction is proportional to the primary observable of the method (dQ/dt), and this is unique to calorimetry. Other approaches detect a concentration (typically of a product), and the rate must be calculated from changes occurring over two or more measurements (or the slope of a continuous curve). The concept of a direct rate measurement offers special advantages, particularly when kinetic measurements must be made against a background of high product concentration. Examples of this include analyses of product inhibition or kinetic studies late in the course of the reaction, where significant conversion of the substrate has occurred. Under such conditions, it is hard if not impossible to derive reaction rates from concentration measurements because the changes over limited time intervals are small compared with the accumulated (or added) background of the product. The calorimetric signal, on the other hand, is almost independent of the product concentration and calorimetry may therefore be particularly beneficial for this type of work.

The sentence should be instead: "Key aspects of calorimetric methods in enzymology are discussed in Chapter 5; thus, the current chapter will rather focus on one particular advantage of isothermal titration calorimetry (ITC), namely, its ability to measure the heat flow from complex samples including suspensions and other multiphase systems, opaque material, and samples with high concentrations of products or nonreacting components." The development of calorimetric approaches within this area is motivated by two factors. Firstly, most methods suffer from physical or chemical interference under these conditions, and the demand for reliable quantitative assays is particularly high. Secondly, enzymology under conditions of high solid contents and multiple phases is becoming increasingly important from both biological and technological points of view. Regarding the former, the protein concentration in the cytoplasm of living cells may be as high as 20%–40% (w/w), and excluded volume effect and intermolecular interactions in this "crowded" environment may result in enzyme kinetics that are quite different from those in dilute buffer

(Ellis 2001; Minton 1998; Zimmerman and Trach 1991). Such systems are readily assayed by calorimetry (Olsen 2006; Olsen et al. 2007) and the method may hence provide an avenue for a better understanding of *in vivo* enzymology. Industrial application of enzymes also regularly involves complex systems with high solid loadings, and indeed in many cases heterogeneous conversion of insoluble substrates such as complex carbohydrates, lipids, or precipitated proteins (Kirk et al. 2002; van Beilen and Li 2002). This application of calorimetry appears to hold a significant potential, but it has only been sporadically exploited. One early example includes the work by Beran and Paulicek (1992), who outlined how isothermal calorimetry could be used to assay enzyme kinetics on macromolecular substrates. More recently, Lonhienne et al. (2001) made a more thorough calorimetric study of chitinases acting on insoluble chitin. In this case, the enthalpy change, $\Delta_r H$, was measured under different experimental conditions, and used to convert the heat flow into absolute reaction rates (c.f. Equation 20.1). This, in turn, allowed the calculation of both catalytic rate constants and activation parameters for the reaction, and it was emphasized that the universality of the method made it generally promising for studies of macromolecular substrates. The two former examples were both concerned with enzymes from the glycoside hydrolase family, but some calorimetric studies of heterogeneous catalysis using other enzymes including lipases and proteases have also been published (Sotoft et al. 2010; Cenciani et al. 2011). In the following, we will discuss this application of calorimetry with special focus on cellulolytic enzymes.

20.2 CELLULOSE AND CELLULASES

Cellulose is an unbranched homopolymer of glucopyranose moieties connected by β-1,4-glycosidic bonds. It is one of the major components in the plant cell wall, and this makes it the most prevalent polymer in the biosphere. The total annual biosynthesis has been estimated to be on the order of 10^{11} t (Pauly and Keegstra 2008), and this makes cellulose—in so-called lignocellulosic biomass—highly attractive as feedstock for the sustainable production of, for example, liquid fuels and alternatives to petrochemical products (Dutta and Pal 2014; Himmel et al. 2007). Specifically, it is planned to use readily available lignocellulosic residue from agriculture and forestry in novel industries, and a central step in the process is the conversion of cellulose into soluble sugars (Wyman 2007; Yang et al. 2011). This may be done via the so-called biochemical platform, where the biomass is treated with a cocktail of enzymes that converts cellulose to its constituent glucose molecules (Wilson 2009). Subsequent fermentation converts glucose into ethanol or other chemicals. One of the key challenges for the implementation of this industry is the physical and chemical stability of cellulose (Wolfenden et al. 1998). This recalcitrance toward the breakdown of cellulose arises primarily from a network of inter- and intramolecular hydrogen bonds, which promotes the formation of a highly stable crystal structure (Nishiyama et al. 2002). The stability and limited accessibility of crystalline cellulose make its deconstruction (so-called saccharification) a rather slow process, and current protocols require many days of hydrolysis to achieve high (>80%) degrees of conversion (Hodge et al. 2009; Kristensen et al. 2009; Lu et al. 2010). It follows that the discovery and engineering of better enzymes for saccharification is of primary importance for the implementation of sustainable industries based on lignocellulosic feedstock (Wilson 2009; Zhang and Lynd 2004).

Cellulolytic enzymes are divided into three main classes (Horn et al. 2012). Exoglucanases (or cellobiohydrolases, CBHs) attack the end of a cellulose strand and often utilize a processive mechanism, in which the enzyme makes several consecutive hydrolytic cycles as it slides along the cellulose chain without dissociating. In contrast to this, endoglucanases (EGs) attack the chains randomly and cleave internally located glycosidic bonds in the polysaccharide. Both of these so-called depolymerizing enzymes produce soluble cellooligosaccharides (the EGs also produce longer insoluble oligosaccharides) by heterogeneous catalysis at the cellulose–water interface. Subsequently, the soluble oligomers are hydrolyzed to glucose in the aqueous phase by a group of enzymes called β-glucosidases. Most recently, a novel group of cellulolytic oxidoreductases (so-called LPMOs) has been discovered (Hemsworth et al. 2014; Horn et al. 2012).

Many mechanistic and regulatory aspects of cellulolytic enzymes remain incompletely resolved, and one of the major obstacles for a better understanding is a shortage of precise quantitative assays (Zhang et al. 2006). In the following, we discuss the potentials and limitations for the use of isothermal calorimetry in this area.

20.3 SIMPLE CALORIMETRIC ASSAYS

The most convenient way to detect the activity of cellulolytic enzymes is to use soluble chromogenic or fluorogenic substrate analogs, which on hydrolysis become directly detectable in simple spectroscopic methods. However, this approach appears to be of little value both for the selection of industrial enzymes and the elucidation of heterogeneous enzyme catalysis as the correlation between activities measured on such substrate analogs and on the real insoluble substrate is quite poor (Zhang et al. 2006). For investigations using insoluble substrates, the most common approach is to measure the concentration of the product in quenched samples. The concentration measurement is typically either a colometric detection of reducing ends of oligosaccharides (Ghose 1987) or a chromatographic analysis of the supernatant. These methods have been utilized extensively and have formed the basis for much improvement in cellulase enzymology, but they are still associated with a number of limitations including their noncontinuous principle and interference from other components in biomass.

Calorimetry may offer an attractive alternative for some types of cellulase activity measurements. This was noted in the aforementioned study by Beran and Paulicek (1992), who studied the activity of cellulases against macromolecular substrates. In this work, the calorimetric experiments only included soluble substrates such as carboxymethyl cellulose, but the broader potential of the method was discussed. Actual insoluble substrates including both pure cellulose and biomass were studied by Murphy et al. (2010b). The basic observation in this work was that a commercial cellulase cocktail added to 1 mL suspensions of either pure cellulose or biomass at room temperature generated a heat flow of some 5–10 μW. This was enough to follow the course of the hydrolytic reaction in different types of commercial calorimeters. After several hours, however, the heat flow had fallen to well below 1 μW and baseline drift and other sources of error made further quantification unfeasible. This low signal occurred despite the fact that only about 20% of the cellulose substrate had been hydrolyzed, and these observations capture both the main advantage and limitation of the method. Considering the advantages, it was possible to obtain a real-time picture of the cellulolytic reaction over several hours and it was documented that this information could be used to address some of the mechanistic questions pertaining to cellulases. For example, repeated or slow dosage of an enzyme was used to study the origin of the so-called nonlinear kinetics, which is ubiquitous to cellulolytic enzymes (Bansal et al. 2009). Nonlinear kinetics implies that the rate of reaction continuously decreases as the hydrolysis progresses even under conditions where the effects from, for example, product inhibition and substrate depletion can be ruled out. The origins of this slowdown remain controversial (Bansal et al. 2009), but most analyses have inferred that it relies either on the heterogeneity of the substrate or a gradual accumulation of unproductive enzyme on the cellulase surface (Zhang et al. 1999; Yang et al. 2006). Regarding the former interpretation, heterogeneity may couple to nonlinear kinetics if the most reactive parts of the insoluble substrate are hydrolyzed first and activity subsequently drops as less reactive material accumulates in the sample. Calorimetric monitoring during repeated enzyme dosage turned out to be an effective tool to clarify this question as fresh enzyme would increase the activity proportionally if inactivation was the cause of the slowdown, but not if it resulted from a depletion of the reactive substrate. Both mechanisms were found to be relevant. Specifically, the slowdown in the early phase (first ~5 h) was ascribed to enzyme inactivation, while the importance of substrate heterogeneity became increasingly important at later stages and dominant after about 24 h (Murphy et al. 2010b). The principle of elucidating mechanistic aspects of heterogeneous catalysis through repeated enzyme injection into the same sample was used earlier (Lonhienne et al. 2001). In this study, a strategy of repeated injection of chitinase into suspensions of insoluble chitin was used to elucidate kinetic saturation behavior and the availability of reactive sites on the particle surface. Another attractive application of simple calorimetric assays for heterogeneous enzyme catalysis may be studies of product inhibition. This important phenomenon is inherently challenging to address experimentally and often requires labeling of the substrate with either fluorophores or radioactive isotopes (Teugjas and Valjamae 2013). As calorimetry directly detects the reaction rate, no labeling is required, and this has been utilized in the characterization of product inhibition for a number of cellulases (Murphy et al. 2013).

Returning to the limitations of the simple calorimetric assay, the major challenge was the sensitivity, which was too low to monitor the process at high degrees of substrate conversion. The sensitivity is primarily dictated by the product of the catalytic rate constant, k_{cat}, and the reaction enthalpy, $\Delta_r H$. For most cellulases, k_{cat} values are on the order of 1 s^{-1} (Ryu et al. 1984; Baker et al. 1998; Nidetzky et al. 1994; Cruys-Bagger et al. 2012; Igarashi et al. 2009), and this is at the low end of what is typically observed for hydrolytic enzymes including other glycoside hydrolases. Moreover, cellulases show the aforementioned nonlinear kinetics and this implies that the specific rate of the reaction gradually falls to levels far below k_{cat} even if the substrate is in excess. The enthalpy of hydrolysis for a β-1,4 glycosidic bond is about −2.5 kJ/mol (Bohlin et al. 2010; Jeoh et al. 2005; Tewari and Goldberg 1989; Krokeide et al. 2007; Lonhienne et al. 2001). While this is a typical value for a hydrolytic reaction (Goldberg et al. 2004), it is low compared with most other reactions, and it follows that the product $k_{cat}\Delta_r H$ for cellulases typically falls in the range 1–10 kJ/(mol·s). This is quite small compared with other enzyme reactions and makes up the main limitation for calorimetric cellulase assays, particularly in experiments where the reaction needs to be monitored for long periods of time.

Stirring makes up a lesser, but still relevant limitation, particularly for substrates that form hydrogels such as amorphous or bacterial cellulose at loads above 10–20 g/L. In this case, stirring will generate a significant frictional heat, which may interfere with the heat flow from the hydrolytic reaction. While frictional heats are generally manageable as long as they are constant, problems arise when the hydrolytic reaction changes the viscosity of the gel and hence, in turn, the frictional heat flow and the location of the calorimetric baseline. Such problems can be circumvented by using substrates with larger particles such as microcrystalline cellulose (Avicel) and most types of industrially pretreated biomass. These suspensions are not gel-like and only show negligible changes in the heat of stirring, over typical hydrolysis trials. If gel-like substrates have to be used, the course of the baseline can be measured in separate experiments, where samples of known degrees of conversion are transferred to the calorimeter and tested for the heat flow associated with stirring (without adding enzyme) (Murphy et al. 2010b). Alternatively, the hydrolysis of problematic substrates can be monitored without stirring after the reaction has been initiated by short mixing either in the calorimetric cell (Murphy et al. 2013) or outside the calorimeter (Olsen et al. 2011).

20.4 HIGH SOLID ASSAYS

Calorimetric assays for cellulolytic enzymes may be particularly promising for experiments with high solid loadings (typically 10%–35% dry matter). This is partly because the sensitivity issues discussed in Section 20.3 are solved or at least postponed to later stages of the reaction, when the concentration of reactants is high. More importantly, solid loads as high as 30% are planned for the saccharification process in upcoming industries, and no other assay technology offers real-time measurements for such samples, which have the appearance of moist soil and an abundance of different compounds that tend to interfere with more specific analytical methods. The stirring of high solid samples in a calorimeter is not feasible as it would generate uncontrollable frictional heats, but since the relevant timescale for saccharification is tens of hours or even days, the course of the hydrolytic reaction can be monitored in samples that are started by thorough mixing outside the calorimeter and subsequent transfer to the instrument (Olsen et al. 2011). This approach was used in a series of dose–response experiments to identify rate-limiting steps for the hydrolysis of samples with up to 29% dry matter. The results suggested that at low degrees of cellulose conversion (<10% hydrolyzed), the hydrolytic process was limited by the scarcity of attack sites for the enzyme on the cellulose surface (substrate limitation). At later stages, this changed and the rate was limited by the amount of enzyme. This suggests that (partial) hydrolysis makes more attack sites accessible to the enzymes at least over the studied degree of cellulose conversion (0%–30%). These high solid assays used a crude extract of fungal cellulases, and as such the experiments are closely related to earlier calorimetric studies of the decomposing activity of living fungi on wood (Bjurman and Wadso 2000; Verma et al. 2008). In one study, it was first demonstrated that calorimetry could readily quantify the decay rates at different temperatures and subsequently illustrated how this information could be applied in discussions of the degradation of wood constructions, and the removal of recalcitrant toxic chemicals by white rot fungi (Bjurman and Wadso 2000).

20.5 REACTION ENTHALPY, $\Delta_r H$

In many cases, calorimetric data can be analyzed in a relative way, but if absolute reaction rates are required, the enthalpy change, $\Delta_r H$, defined in Equation 20.1, must be known. In cases where complete substrate conversion can be achieved in a manageable time, $\Delta_r H$ can be readily determined by mixing a small amount of substrate and a high enzyme concentration in a titration calorimeter (Jeoh et al. 2005; Krokeide et al. 2007). Then, we may simply assume $\Delta_r H = Q/n_s$, where Q is the integral of the ITC peak and n_s is the number of moles of substrate added in the experiment. For cellulases, this strategy generally cannot be used because full conversion of even small samples is only approached very slowly. Instead, calorimetric measurements of Q have to be combined with an independent analytic method such as chromatography, and samples either retrieved directly from the calorimeter or run in parallel must be analyzed to obtain n_s. This is quite cumbersome and generally requires dozens of samples with variable degrees of conversion to get good estimates of $\Delta_r H$. One shortcut may be to use literature values for the relevant reaction (i.e., $\Delta_r H = -2.5$ kJ/mol in the hydrolysis of the β-1,4 glycosidic bond [Tewari and Goldberg 1989]), but if high precision of the kinetic parameters is required, $\Delta_r H$ must be measured separately. This is particularly true for an insoluble substrate such as cellulose, which may occur in different physical states including different crystalline allomorphs (Nishiyama et al. 2010). Thus, in addition to the actual hydrolytic reaction catalyzed by the enzyme, other changes such as the dissolution of the crystal and the associated collapse in the matrices of other components in the biomass may add significantly to the apparent enthalpy change; especially so as the intrinsic $\Delta_r H$ for the hydrolysis of the β-1,4-glucosidic bond is quite small. Previous works have shown pronounced variations in $\Delta_r H$ for the hydrolysis of cellulose. Hydrolysis of pure so-called microcrystalline cellulose (Avicel) by an enzyme cocktail showed $\Delta_r H = -4.3$ kJ/mol, and was thus much (~70%) more exothermic than the hydrolysis of soluble oligosaccharides. For pure cellulose in an amorphous state and biomass, the analogous values were, respectively, –2.7 and –6.7 kJ/mol, and this further emphasizes the need to calibrate the method against each new substrate, even when studying the same chemical reaction. The aforementioned variation may reflect the sizable lattice energy of crystalline cellulose (Calvet and Hermans 1951; Dale and Tsao 1982), which contributes differently to $\Delta_r H$ depending on the physical state of the substrate. Another source of significant variation in $\Delta_r H$ was documented by Karim et al. (2005; Karim and Kidokoro 2004), who studied mutarotation in the product of the enzymatic reaction. This effect arises because a cellulase exclusively makes one of the anomeric forms of the product (i.e., one of the two stereoisomers of the anomeric [C-1] carbon in the pyranose ring). Some cellulases are so-called retaining enzymes that keep the anomeric form of the substrate (β-glycopyranose) in their product, while others are inverting and produce the α-anomer.

On release, these anomerically pure products relax toward α/β equilibrium at a rate that depends on the experimental conditions, and as the enthalpy of this mutarotation is comparable to the heat of hydrolysis, the relaxation may interfere with the hydrolytic heat flow depending on the relative rates of these two processes. Karim et al. (2005) studied both a retaining and an inverting cellulase and found that mutarotation was negligibly slow at pH 4, whereas it strongly influenced the measured heat flow at pH 7. As a result, the apparent enthalpy change associated with hydrolysis at the latter pH was only about −0.8 kJ/mol. Another complication of determining $\Delta_r H$ was described in a study of three EGs from the filamentous fungus *Trichoderma reesei* (Murphy et al. 2012). Although all measurements were made on the same amorphous substrate and all enzymes were of the retaining type, the measured $\Delta_r H$ values ranged from −1.7 to −5.4 kJ/mol, depending on which enzyme was used. It was suggested (but not documented) that this variation could rely on the profile of products (Murphy et al. 2012). Thus, these enzymes produced both soluble and insoluble cellooligosaccharides of different average length, and this influenced the heat of hydration (Dale and Tsao 1982; Taylor 1957), and therefore possibly the apparent molar enthalpy change for the different enzyme reactions.

20.6 COUPLED CALORIMETRIC ASSAYS

As discussed in Section 20.4, one of the limitations of calorimetric assays for cellulases and other enzymes hydrolyzing insoluble substrates is a low $\Delta_r H$, and the associated sensitivity problems. In some cases, including studies of enzyme cocktails and EGs, the signal is high enough to be monitored directly (Murphy et al. 2012, 2010b), but for many mono-component cellulases, including the important group of CBHs, limited sensitivity may prevent even initial rate measurement. One way to compensate for this is to use coupled assays, where additional enzymes are added to the sample so that the products of the cellulolytic reaction are modified further after their release from the insoluble substrate. CBHs primarily release the disaccharide cellobiose as a product, and this opens a possibility for coupled assays. Thus, if the reaction occurs in the presence of a β-glycosidase (cellobiase), which converts the disaccharide into two glucose molecules, the associated breakage of a second β-1,4 glycosidic bond will approximately double $\Delta_r H$, and hence also the calorimetric sensitivity. Moreover, much larger signal amplification can be achieved if the reaction medium also contains the enzymes glucose oxidase and catalase. The former uses oxygen to convert glucose to the corresponding glucono-lactone and also produces hydrogen peroxide. In the last coupled step, hydrogen peroxide is converted to water and molecular oxygen by catalase, and the two coupled redox reactions dramatically increase $\Delta_r H$ to a value of around −360 kJ/mol of cellobiose originally released by the cellulase. This means that the coupled reactions increase $\Delta_r H$ by more than two orders of magnitude, and this entirely changes the situation and makes calorimetry an excellent and highly sensitive assay for mono component CBHs even at low (10–100 nM) enzyme concentrations (Murphy et al. 2010a). The results for two of the most studied CBHs, Cel7A and Cel6A from *T. reesei*, showed that both the so-called initial burst and the steady-state rate could be characterized on different types of cellulose using this approach (Murphy et al. 2010a; Praestgaard et al. 2011). An analogous strategy has been used to study the kinetics of xylanases on their natural polymeric substrate (Baumann et al. 2011). Again, the amplified calorimetric signal provided high sensitivity and hence the ability to quantify reaction rates at lower substrate loads than other methods. For the xylanases, this was important because their (low) Michaelis constant could be determined with reasonable precision (Baumann et al. 2011). The basic idea of this type of thermal signal amplification has previously been used in a related method called *calorimetric* (or *thermal*) *biosensors* (Blum and Coulet 1991; Ramanathan and Danielsson 2001; Zhang and Tadigadapa 2004). In this latter technique, the purpose is not to characterize enzyme kinetics, but rather to quantify a soluble analyte such as glucose or urea through the temperature change that occurs when a solution is passed through a column with immobilized enzymes that modify the analyte.

The substantial sensitivity improvement in this type of coupled assay does not come without extra limitations. As in any coupled assay, it is important that the secondary enzyme reactions are faster than the reaction that is under study. If this is not the case, one may observe the kinetics of the coupled reactions rather than the activity of the cellulase. Systematic criteria for assessments of this effect have been stipulated (Storer and Cornish-Bowden 1974) and must be thoroughly tested for any amplification system. In the cases discussed earlier in this section, these criteria turned out to be met, but it requires comprehensive experimental work to test a new signal amplification system (or changes in an existing one) before it can be used. Other limitations include the accumulation of inhibitors and the depletion of oxygen in the reaction mixture. The most important inhibitor in the β-glucosidase/glucose oxidase/catalase system discussed earlier in this section was the glucono-lactone, which is produced by glucose oxidase. This compound shares some structural similarity with the transition state in the β-glycosidase reaction and is hence an inhibitor of this enzyme (de Melo et al. 2006). In practical terms, the β-glucosidase activity became too low for the amplification system to work properly when the lactone concentration exceeded ~0.1 mM. Another potential problem in the reaction sequence was the depletion of oxygen (only half of the oxygen consumed in the glucose oxidase reaction is returned in the catalase reaction). When the conventional method of vacuum degassing samples for the calorimeter was used, this problem essentially precluded the use of the signal amplification system, but if the samples were instead degassed by ultrasound, oxygen depletion occurred later in the run than glucono-lactone inhibition (Murphy et al. 2010a). All this considered, experiments could easily be designed so that the coupled reaction could be monitored for one or a few hours without interference from inhibition or oxygen depletion.

20.7 CONCLUSIONS

Molecular mechanisms underpinning heterogeneous enzyme catalysis remain poorly understood in comparison with its homogeneous counterpart. Judging from its prevalence both in biology and biotechnology, deeper insight into biocatalysis at interfaces seems important, but a shortage of quantitative, real-time assay technologies has (among other challenges) slowed progress in this area. Isothermal calorimetry appears to offer some advantages including the ability to handle high solid suspensions, and multicomponent and opaque samples. Moreover, calorimetric methods are applicable at high concentrations of the product and they readily detect the overall activity of enzymes, such as the endoglycanases discussed here, which produce a mixture of soluble and insoluble products. These potential advantages of calorimetry were discussed with special focus on cellulolytic enzymes, and it was found that under some conditions calorimetry did indeed offer important benefits. The main limitation of the method for cellulases was the sensitivity. Thus, the product of the catalytic constant and apparent enthalpy change, $k_{cat}\Delta_r H$, was about 1–10 kJ/(mol s), and this meant that several hours of reaction could typically be monitored, while longer experiments that approached full conversion of the substrate were not amenable to calorimetric measurements. In some cases, the apparent enthalpy change could be noticeably increased by coupled reactions and this allowed kinetic investigations at very low enzyme concentration, but not prolonged monitoring. It appears that isothermal calorimetry may become an important tool in studies of heterogeneous enzyme catalysis, particularly in cases where the product $k_{cat}\Delta_r H$ is higher than for cellulases.

ACKNOWLEDGMENTS

This work was supported by the Danish Council for Strategic Research, Program Commission on Sustainable Energy and Environment Grant 2104-07-0028 and 11-116772 (to PW) and by the Carlsberg Foundation, grant 2013-01-0208 (to PW). Valuable input from Drs. Leigh Murphy, Martin Bauman, Nina Lei, Jens E. Olsen, Nicolaj Cruys-Bagger, and Søren Olsen is gratefully acknowledged.

REFERENCES

Baker, J. O., Ehrman, C. I., Adney, W. S., Thomas, S. R., and Himmel, M. E. 1998. Hydrolysis of cellulose using ternary mixtures of purified celluloses. *Appl. Biochem. Biotechnol.* 70–2:395–403.

Bansal, P., Hall, M., Realff, M. J., Lee, J. H., and Bommarius, A. S. 2009. Modeling cellulase kinetics on lignocellulosic substrates. *Biotechnol. Adv.* 27:833–848.

Baumann, M. J., Murphy, L., Lei, N., Krogh, K. B. R. M., Borch, K., and Westh, P. 2011. Advantages of isothermal titration calorimetry for xylanase kinetics in comparison to chemical-reducing-end assays. *Anal. Biochem.* 410:19–26.

Beezer, A. E., Steenson, T. I., and Tyrrell, H. J. V. 1974. Application of flow-microcalorimetry to analytical problems. 2. Urea-urease system. *Talanta* 21:467–474.

Beran, M., and Paulicek, V. 1992. Flow-microcalorimetric determination of enzymatic-activities of the *Trichoderma viridae*–cellulase complex. *J. Therm. Anal.* 38:1979–1988.

Bjurman, J., and Wadso, L. 2000. Microcalorimetric measurements of metabolic activity of six decay fungi on spruce wood as a function of temperature. *Mycologia* 92:23–28.

Blum, L. J., and Coulet, P. R. 1991. *Biosensor Principles and Applications.* Marcel Dekker: New York.

Bohlin, C., Olsen, S. N., Morant, M. D., Patkar, S., Borch, K., and Westh, P. 2010. A comparative study of activity and apparent inhibition of fungal beta-glucosidases. *Biotechnol. Bioeng.* 107:943–952.

Calvet, E., and Hermans, P. H. 1951. Heat of crystallization of cellulose. *J. Polym. Sci.* 6:33–38.

Cenciani, K., Freitas, S. D., Critter, S. A. M., and Airodi, C. 2011. Enzymatic activity measured by microcalorimetry in soil amended with organic residues. *Rev. Bras. Cienc. Solo* 35:1167–1175.

Cruys-Bagger, N., Ren, G., Tatsumi, H., Baumann, M. J., Spodsberg, N., Andersen, H. D., Gorton, L., Borch, K., and Westh, P. 2012. An amperometric enzyme biosensor for real-time measurements of cellobiohydrolase activity on insoluble cellulose. *Biotechnol. Bioeng.* 109:3199–3204.

Dale, B. E., and Tsao, G. T. 1982. Crystallinity and heats of crystallization of cellulose: A microcalorimetric investigation. *J. Appl. Polym. Sci.* 27:1233–1241.

de Melo, E. B., Gomes, A. D., and Carvalho, I. 2006. Alpha- and beta-glucosidase inhibitors: Chemical structure and biological activity. *Tetrahedron* 62:10277–10302.

Dutta, S., and Pal, S. 2014. Promises in direct conversion of cellulose and lignocellulosic biomass to chemicals and fuels: Combined solvent-nanocatalysis approach for biorefinary. *Biomass Bioenerg.* 62:182–197.

Eftink, M. R., Johnson, R. E., and Biltonen, R. L. 1981. The application of flow micro-calorimetry to the study of enzyme-kinetics. *Anal. Biochem.* 111:305–320.

Ellis, R. J. 2001. Macromolecular crowding: Obvious but underappreciated. *Trends Biochem. Sci.* 26:597–604.

Ghose, T. K. 1987. Measurement of cellulase activities. *Pure Appl. Chem.* 59:257–268.

Goddard, J. P., and Reymond, J. L. 2004. Enzyme assays for high-throughput screening. *Curr. Opin. Biotechnol.* 15:314–322.

Goldberg, R. N., Tewari, Y. B., and Bhat, T. N. 2004. Thermodynamics of enzyme-catalyzed reactions: A database for quantitative biochemistry. *Bioinformatics* 20:2874–2877.

Hemsworth, G. R., Henrissat, B., Davies, G. J., and Walton, P. H. 2014. Discovery and characterization of a new family of lytic polysaccharide monooxygenases. *Nat. Chem. Biol.* 10:122–126.

Himmel, M. E., Ding, S. Y., Johnson, D. K., Adney, W. S., Nimlos, M. R., Brady, J. W., and Foust, T. D. 2007. Biomass recalcitrance: Engineering plants and enzymes for biofuels production. *Science* 315:804–807.

Hodge, D. B., Karim, M. N., Schell, D. J., and McMillan, J. D. 2009. Model-based fed-batch for high-solids enzymatic cellulose hydrolysis. *Appl. Biochem. Biotechnol.* 152:88–107.

Horn, S. J., Vaaje-Kolstad, G., Westereng, B., and Eijsink, V. G. H. 2012. Novel enzymes for the degradation of cellulose. *Biotechnol. Biofuels* 5:45.

Igarashi, K., Koivula, A., Wada, M., Kimura, S., Penttila, M., and Samejima, M. 2009. High speed atomic force microscopy visualizes processive movement of *Trichoderma reesei* cellobiohydrolase i on crystalline cellulose. *J. Biol. Chem.* 284:36186–36190.

Jeoh, T., Baker, J. O., Ali, M. K., Himmel, M. E., and Adney, W. S. 2005. Beta-d-glucosidase reaction kinetics from isothermal titration microcalorimetry. *Anal. Biochem.* 347:244–253.

Karim, N., and Kidokoro, S. 2004. Precise and continuous observation of cellulase-catalyzed hydrolysis of cello-oligosaccharides using isothermal titration calorimetry. *Thermochim. Acta* 412:91–96.

Karim, N., Okada, H., and Kidokoro, S. 2005. Calorimetric evaluation of the activity and the mechanism of cellulases for the hydrolysis of cello-oligosaccharides accompanied by the mutarotation reaction of the hydrolyzed products. *Thermochim. Acta* 431:9–20.

Kirk, O., Borchert, T. V., and Fuglsang, C. C. 2002. Industrial enzyme applications. *Curr. Opin. Biotechnol.* 13:345–351.

Kristensen, J. B., Felby, C., and Jorgensen, H. 2009. Yield-determining factors in high-solids enzymatic hydrolysis of lignocellulose. *Biotechnol. Biofuels* 2:11.

Krokeide, I. M., Eijsink, V. G. H., and Sorlie, M. 2007. Enzyme assay for chitinase catalyzed hydrolysis of tetra-n-acetylchitotetraose by isothermal titration calorimetry. *Thermochim. Acta* 454:144–146.

Lonhienne, T., Baise, E., Feller, G., Bouriotis, V., and Gerday, C. 2001. Enzyme activity determination on macromolecular substrates by isothermal titration calorimetry: Application to mesophilic and psychrophilic chitinases. *Biochim. Biophys. Acta-Protein Struct. Molec. Enzym.* 1545:349–356.

Lu, Y. F., Wang, Y. H., Xu, G. Q., Chu, J., Zhuang, Y. P., and Zhang, S. L. 2010. Influence of high solid concentration on enzymatic hydrolysis and fermentation of steam-exploded corn stover biomass. *Appl. Biochem. Biotechnol.* 160:360–369.

Minton, A. P. 1998. Molecular crowding: Analysis of effects of high concentrations of inert cosolutes on biochemical equilibria and rates in terms of volume exclusion. *Methods Enzymol.* 295:127–149.

Murphy, L., Baumann, M. J., Borch, K., Sweeney, M., and Westh, P. 2010a. An enzymatic signal amplification system for calorimetric studies of cellobiohydrolases. *Anal. Biochem.* 404:140–148.

Murphy, L., Bohlin, C., Baumann, M. J., Olsen, S. N., Sorensen, T. H., Anderson, L., Borch, K., and Westh, P. 2013. Product inhibition of five hypocrea jecorina cellulases. *Enzyme. Microb. Technol.* 52:163–169.

Murphy, L., Borch, K., McFarland, K. C., Bohlin, C., and Westh, P. 2010b. A calorimetric assay for enzymatic saccharification of biomass. *Enzyme. Microb. Technol.* 46:141–146.

Murphy, L., Cruys-Bagger, N., Damgaard, H. D., Baumann, M. J., Olsen, S. N., Borch, K., Lassen, S. F., Sweeney, M., Tatsumi, H., and Westh, P. 2012. Origin of initial burst in activity for *Trichoderma reesei* endo-glucanases hydrolyzing insoluble cellulose *J. Biol. Chem.* 287:1252–1260.

Nidetzky, B., Steiner, W., and Claeyssens, M. 1994. Cellulose hydrolysis by the cellulases from *Trichoderma reesei*: Adsorptions of 2 cellobiohydrolases, 2 endocellulases and their core proteins on filter-paper and their relation to hydrolysis. *Biochem. J.* 303:817–823.

Nishiyama, Y., Langan, P., and Chanzy, H. 2002. Crystal structure and hydrogen-bonding system in cellulose 1 beta from synchrotron x-ray and neutron fiber diffraction. *J. Am. Chem. Soc.* 124:9074–9082.

Nishiyama, Y., Langan, P., Wada, M., and Forsyth, V. T. 2010. Looking at hydrogen bonds in cellulose. *Acta Crystallogr. D* 66:1172–1177.

Olsen, S. N. 2006. Applications of isothermal titration calorimetry to measure enzyme kinetics and activity in complex solutions. *Thermochim. Acta* 448:12–18.

Olsen, S. N., Lumby, E., McFarland, K., Borch, K., and Westh, P. 2011. Kinetics of enzymatic high-solid hydrolysis of lignocellulosic biomass studied by calorimetry. *Appl. Biochem. Biotechnol.* 163:626–635.

Olsen, S. N., Ramlov, H., and Westh, P. 2007. Effects of osmolytes on hexokinase kinetics combined with macromolecular crowding test of the osmolyte compatibility hypothesis towards crowded systems. *Comp. Biochem. Phys.* 148:339–345.

Pauly, M., and Keegstra, K. 2008. Cell-wall carbohydrates and their modification as a resource for biofuels. *Plant J.* 54:559–568.

Praestgaard, E., Elmerdahl, J., Murphy, L., Nymand, S., McFarland, K. C., Borch, K., and Westh, P. 2011. A kinetic model for the burst phase of processive cellulases. *FEBS J.* 278:1547–1560.

Ramanathan, K., and Danielsson, B. 2001. Principles and applications of thermal biosensors. *Biosens. Bioelectron.* 16:417–423.

Reymond, J. L. 2004. Spectrophotometric enzyme assays for high-throughput screening. *Food Technol. Biotechnol.* 42:265–269.

Ryu, D. D. Y., Kim, C., and Mandels, M. 1984. Competitive adsorption of cellulase components and its significance in a synergistic mechanism. *Biotechnol. Bioeng.* 26:488–496.

Sotoft, L. F., Westh, P., Christensen, K. V., and Norddahl, B. 2010. Novel investigation of enzymatic biodiesel reaction by isothermal calorimetry. *Thermochim. Acta* 501:84–90.

Storer, A. C., and Cornish-Bowden, A. 1974. Kinetics of coupled enzyme reactions: Applications to assay of glucokinase, with glucose-6-phosphate dehydrogenase as coupling enzyme. *Biochem. J.* 141:205–209.

Taylor, J. B. 1957. The water solubilities and heats of solution of short chain cellulosic oligosaccharides. *T. Faraday Soc.* 53:1198–1203.

Teugjas, H., and Valjamae, P. 2013. Product inhibition of cellulases studied with 14c-labeled cellulose substrates. *Biotechnol. Biofuels* 6:104.

Tewari, Y. B., and Goldberg, R. N. 1989. Thermodynamics of hydrolysis of disaccharides: Cellobiose, gentiobiose, isomaltose, and maltose. *J. Biol. Chem.* 264:3966–3971.

Todd, M. J., and Gomez, J. 2001. Enzyme kinetics determined using calorimetry: A general assay for enzyme activity? *Anal. Biochem.* 296:179–187.

van Beilen, J. B., and Li, Z. 2002. Enzyme technology: An overview. *Curr. Opin. Biotechnol.* 13:338–344.

Verma, P., Dyckmans, J., Militz, H., and Mai, C. 2008. Determination of fungal activity in modified wood by means of micro-calorimetry and determination of total esterase activity. *Appl. Microbiol. Biotechnol.* 80:125–133.

Williams, B. A., and Toone, E. J. 1993. Calorimetric evaluation of enzyme-kinetic parameters. *J. Org. Chem.* 58:3507–3510.

Wilson, D. B. 2009. Cellulases and biofuels. *Curr. Opin. Biotechnol.* 20:295–299.

Wolfenden, R., Lu, X. D., and Young, G. 1998. Spontaneous hydrolysis of glycosides. *J. Am. Chem. Soc.* 120:6814–6815.

Wyman, C. E. 2007. What is (and is not) vital to advancing cellulosic ethanol. *Trends Biotechnol.* 25:153–157.

Yang, B., Dai, Z., Ding, S.-Y., and Wyman, C. E. 2011. Enzymatic hydrolysis of cellulosic biomass. *Biofuels* 2:421–450.

Yang, B., Willies, D. M., and Wyman, C. E. 2006. Changes in the enzymatic hydrolysis rate of avicel cellulose with conversion. *Biotechnol. Bioeng.* 94:1122–1128.

Zhang, S., Wolfgang, D. E., and Wilson, D. B. 1999. Substrate heterogeneity causes the nonlinear kinetics of insoluble cellulose hydrolysis. *Biotechnol. Bioeng.* 66:35–41.

Zhang, Y. H. P., Himmel, M. E., and Mielenz, J. R. 2006. Outlook for cellulase improvement: Screening and selection strategies. *Biotechnol. Adv.* 24:452–481.

Zhang, Y. H. P., and Lynd, L. R. 2004. Toward an aggregated understanding of enzymatic hydrolysis of cellulose: Noncomplexed cellulase systems. *Biotechnol. Bioeng.* 88:797–824.

Zhang, Y. Y., and Tadigadapa, S. 2004. Calorimetric biosensors with integrated microfluidic channels. *Biosens. Bioelectron.* 19:1733–1743.

Zimmerman, S. B., and Trach, S. O. 1991. Estimation of macromolecule concentrations and excluded volume effects for the cytoplasm of *Escherichia coli. J. Mol. Biol.* 222:599–620.

Index

9 780367 870287